Neurogenic Inflammation in Health and Disease

Neuroimmune Biology, Volume 8

Neuroimmune Biology

Series Editor

I. Berczi

Neurogenic Inflammation in Health and Disease

Volume Editor

Gábor Jancsó
Department of Physiology, University of Szeged, Hungary

ELSEVIER

AMSTERDAM • BOSTON • HEIDELBERG • LONDON • NEW YORK • OXFORD
PARIS • SAN DIEGO • SAN FRANCISCO • SINGAPORE • SYDNEY • TOKYO

Elsevier
Radarweg 29, PO Box 211, 1000 AE Amsterdam, The Netherlands
Linacre House, Jordan Hill, Oxford OX2 8DP, UK
32 Jamestown Road, London, NWI 7BY, UK
The Boulevard, Langford Lane, Kidlington, Oxford OX5 1GB, UK

First edition 2009

British Library Cataloguing in Publication Data
A catalogue record for this book is available from the British Library

Library of Congress Cataloging-in-Publication Data
A catalog record for this book is available from the Library of Congress

ISBN: 978-0-444-53229-9
ISSN: 1567-7443

For information on all Elsevier Science publications
visit our website at www.elsevierdirect.com

Printed and bound in the United Kingdom

Transferred to Digital Print 2011

Contents

Foreword: Neurogenic Inflammation Coming of Age

Observations with regards to the neural regulation of vasodilatation and vascular permeability have been already made during the 19th century. The role of nerves in initiating an inflammatory response has been clearly delineated for the first time by Miklos Jancso. It was observed that the stimulation of capsaicin-sensitive sensory nerve fibers could induce an inflammatory response, whereas denervation inhibited the inflammatory effect of irritants, such as capsaicin. The inflammation elicited by sensory nerves has been coined by Jancso as "neurogenic inflammation" (NIF) (reviewed in Ref. [1]).

Numerous denervation strategies over many years have shown that in most animal models arthritis either does not develop or is significantly attenuated in denervated joints [2–10]. It is also firmly established that patients who develop inflammatory joint disease after nerve injury, hemiplegia, or poliomyelitis show sparing of the denervated joints [11–14]. These observations indicate that nerve supply is necessary for the development of arthritis at a site distant to the initial lesion. In one study where bilateral cartilage degradation was found in monoarthritis, spinal cord compression injury inhibited contralateral cartilage degradation [15]. This indicates the role of direct segmental spinal connection between homologous primary afferents. Denervation experiments and analogous observations in patients clearly indicate the significant role of neural regulation of inflammatory reactions [2].

Today it is clear that sensory nerves bearing nociceptors have the capacity to detect various insults and respond by inducing inflammation, which is coupled with pain sensation and hypersensitivity to heat and cold. Such sensory nerves innervate all tissues and organs of the body. These nerves are heterogenous with various nociceptor capability. There are also major species-specific differences. For instance in pig and human skin, the key neurons are mechanically insensitive (heat nociceptors and/or silent nociceptors) and polymodal nociceptors are not involved. In the rat and rabbit, a proportion of polymodal nociceptors produce antidromic vasodilatation and, in the rat, also plasma extravasation. In the pig, the heat nociceptors that are the only group causing antidromic vasodilatation have relatively long duration axonal action potentials [16].

The mediators involved may stimulate or inhibit the inflammatory process. Substance P, neurokinin A and B, collectively known as "tachykinins," and calcitonin gene-related peptide (CGRP) are proinflammatory mediators capable of eliciting an inflammatory response. These mediators also enhance various immune responses. On the other hand, somatostatin and galanin are anti-inflammatory mediators [1,17,18].

Blood vessels receive peptidergic perivascular innervation. This perivascular nervous system acts both under the control of the central nervous system and independently by responding to local conditions through the release of peptides. Peptides released by perivascular nerves not only act locally by altering blood flow but also exert a systemic effect by interacting with the components of other body systems, which results in an integrated response. Conversely, dysregulation of peptidergic innervation may be considered as an important contributor to the pathology of a variety of common medical conditions [19].

Somatostatin reaches distant parts of the body by the circulation and exerts systemic anti-inflammatory effects via G-protein-coupled membrane receptors. Five receptor subtypes are known ($sstr_1$–$sstr_5$). $sstr_1$ and $sstr_4$ are presumed to mediate anti-inflammatory actions. Somatostatin influences NIF both by presynaptic and postsynaptic actions. It inhibits proinflammatory neuropeptide release from sensory nerve endings and also affects vascular endothelial, inflammatory, and immune cells [1,17].

Recently, members of the galanin peptide family, e.g., galanin, galanin-like peptide (GALP), and alarin, have emerged as vasoconstrictors and inhibitors of cutaneous extravasation during inflammation in several species. Galanin also inhibited cell proliferation [20,21]. Chronic cutaneous denervation resulted in a significant elevation of the permeability-enhancing effect of histamine. In the denervated skin, this response was significantly inhibited by galanin. This indicates that in the normally innervated skin, endogenous galanin exerts a neurogenic tonic inhibitory effect on histamine-induced plasma leakage. Therefore, sensory nerves possess not only proinflammatory but also anti-inflammatory (inhibitory) sensory-efferent functions [22].

It is apparent that there is a multitude of receptors on target cells for these mediators [18,19,23–25]. Thus, the variability of sensory nerves involved and the multitude of mediators and of receptors present in the system indicates that we are dealing here with a redundant system, which is designed to resort to alternative pathways in case of failure of a particular effector mechanism. The immune system is characterized by such redundancy, which is extremely useful to assure the maintenance of adequate host defense, even in the face of severe damage to the host. However, redundancy poses a problem when it comes to the treatment of certain diseases, such as rheumatoid arthritis. Here, the inhibition of a single receptor or mediator would not be expected to have the desired effect.

Mast cells (MCs) are inflammatory cells that are present in various tissues and organs. It is well established that MCs are innervated and that MCs do release their proinflammatory mediators after nerve stimulation or after exposure to proinfammatory neuropeptides. Conversely, somatostatin inhibits MC discharge [26–29]. Therefore, MCs may be considered as a neuroeffector cell for inflammation. It has been established that this unit has physiological functions as well, such as the maintenance of normal function of the gastrointestinal tract, for instance [18,30].

MCs are also important effector cells for the adaptive immune system. They bind specifically IgE antibodies and degranulate after exposure to the specific antigen, which leads to a local inflammatory response at the site of antigen exposure. An exaggerated response to antigen (allergen) leads to allergic reactions. So far, immunologists blame IgE antibodies for the development of allergy [31]. However, allergy cannot be transferred with IgE antibodies [32], which illustrates that such antibodies are not involved in the pathogenesis of this condition.

Capsaicin treatment depletes the peptide mediators stored in sensory nerves, which has a similar effect on denervation. Neutrophil recruitment by ovalbumin after pulmonary challenge was exacerbated following capsaicin treatment [21,30,33]. Capsaicin-treated mice showed enhanced contact sensitivity reactions to various haptens [34,35] that could not be transferred to naïve mice by effector T cells [34]. Interleukin-1 release was significantly higher (by over 20-fold) in capsaicin-treated animals compared to controls [35]. UV light exposure inhibited contact hypersensitivity reactions in control but not in capsaicin-treated mice [36]. These results indicate that C-type nerve fibers regulate cutaneous delayed-type hypersensitivity reactions in the mouse. CGRP, but not tachykinin receptor antagonists, had the effects of capsaicin denervation in this experimental model [26,36].

NK1 receptor knockout mice show no plasma extravasation response in an immune-complex airway inflammation model. This indicates that SP is a significant mediator of airway inflammatory response [37]. Twenty eight percent decrease was found in intestinal mucosal MCs in the small bowel of capsaicin-treated rats compared with littermate controls [38]. A similar decrease can be induced by truncal vagotomy (a 25% decrease in jejunal MC density) [39]. Therefore, the capsaicin-sensitive innervation of the gut, which is associated with MCs, is vagal in origin [23]. Capsaicin evokes the degranulation of MCs in the human skin via release of sensory neuropeptides [23,39].

It has been known for some time that pulmonary allergy is associated with "bronchial hypersensitivity," i.e., the bronchus of allergic animals or of patients would react to an inflammatory mediator or antigen in an exaggerated fashion. It was also observed that histamine production in the bone marrow of rats and mice is regulated by the sympathetic nervous system [40]. These observations indicate that the histamine content of tissues is neurally regulated.

Bronchial hypersensitivity was elicited in experimental animals by tachykinins and could be prevented by treatment with the tachykinin-specific NK2 receptor selective receptor antagonist, nepudutant [18,41]. The exposure of sensitized animals to antigens, such as ovalbumin, produced tachykinin-mediated bronchoconstriction, plasma protein extravasation, and bronchial hypersensitivity [18,42–44]. Therefore, allergy as well as asthma and atopic disease in general *cannot be transferred from one animal to the other*, because neurogenic mechanisms are involved in the pathogenesis of these diseases, which are unique to the allergic/asthmatic host.

Inflammation also serves as an important effector mechanism for the innate immune system. Here, cytokines produced by macrophages and by other inflammatory cells play a role in the initiation of the inflammatory response. Neuroendocrine mechanisms play an important role in the regulation of innate immunity. As for neurogenic mechanisms, the role of catecholamines is well recognized in the acute phase response. The elevation of C-reactive protein and of other acute phase proteins during inflammatory disease, such as arthritis, indicates the involvement of the innate immune system [44]. There is good evidence for the involvement of NIF in arthritis as discussed above. Taken together, these facts indicate that neurogenic mechanisms also regulate inflammatory responses initiated by the innate immune system.

Currently, there is much information with regards to the involvement of NIF in the skin [1] in the gastrointestinal tract and in other internal organs [18], in the heart [45], and in the central nervous system. Apparently, migraine headache is due to the malfunction of the neural-MC unit [46], and NIF is definitely involved in the pathogenesis of arthritis [2]. Neurogenic components may also be involved in asthma, allergic rhinitis and other allergic upper airway diseases, cystitis, and skin diseases, such as contact dermatitis or psoriasis [1,47].

One should caution against a simplistic view of the control of the inflammatory response. There is much evidence for the anti-inflammatory effect of the beta-adrenergic system, whereas cholinergic mechanisms support inflammation. Also growth and lactogenic hormones support inflammation, whereas the various hormones of the hypothalamus–pituitary–adrenal axis are anti-inflammatory [47–51]. In addition, immune-derived cytokines may have a proinflammatory (e.g., IL-1, IL-6, and TNF-alpha) or anti-inflammatory (IL-10 and TGF-beta) effect [52].

Endogenous opioids, released into the systemic circulation by the pituitary gland or produced locally by immunocompetent cells, can suppress NIF. Opioid receptors are expressed by sensory neurons. Endogenous opioid peptides released from macrophages or lymphocytes in chronically inflamed tissue can act on sensory nerves to inhibit pain and the release of excitatory proinflammatory neuropeptides and thus limit the perpetuation of NIF. Opioids present within airway-innervating nerve fibers influence airway functions via modification of neural transmission [25,53,54].

Bradykinin-induced synovial inflammation was shown to be dependent on sympathetic postganglionic neurons. Sympathetic C-fiber afferent neuron integration plays a role in acute and chronic inflammation in the synovium and in other tissues. Similar experiments demonstrated sympathetic C-fiber coupling in skin inflammatory response [52,55,56]. Apparently, sympathetic efferent activity acts on peripheral adrenergic receptors to enhance C-fiber sensitization, thereby augmenting the generation of dorsal root reflexes that produce vasodilatation [57].

Current evidence indicates that sensory nerves are not directly stimulated after antigen challenge, but indirectly by endogenous kinins, to initiate NIF. This response to bradykinin is mediated by B_2 receptors. Nerve growth factor (NGF), released from inflammatory cells, not only enhances synthesis, axonal transport, and release of SP and CGRP in arthritis, but it also sensitizes the nerve terminals. Kallikreins also play a role in the regulation of inflammation and immunity [29,58,59].

The activation of protein kinase C mediates the neuronal excitatory and heat-sensitizing actions of bradykinin, while Ca^{2+} accumulation induces the formation of nitric oxide (NO) within sensory neurons. NO is involved – together with receptor downregulation – in the development of tachyphylaxis, a B_2 receptor-mediated effect of bradykinin. Nitric oxide, however, may also contribute to the excitatory and sensitizing actions of bradykinin. Cyclooxygenase metabolites of arachidonic acid (prostanoids) may also contribute to both the excitatory and the sensitizing effects of bradykinin [58].

The gastrointestinal tract has several communication channels transmitting to the spinal cord and brain, which include vagal afferent neurons, spinal afferent neurons, intestinofugal enteric neurons, and endocrine mechanisms. The vagus nerves have long been thought to play a role in the regulation of digestion by efferent reflex. Now it is clear that the majority of fibers in the vagus are afferent and serve an unprecedented variety of physiological and pathophysiological roles. Vagal afferents have a function in chemonociception and interoception (the sense of well-being), modify affective-emotional processes in the brain, can cause nausea and vomiting, and carry messages relevant to appetite and nutrition. Vagal afferents serve as a sensory interface between the peripheral immune system and the central as well as autonomic nervous system. Vagal afferents can respond to proinflammatory cytokines, such as interleukin-1beta, and contribute to the sickness response to infection. The vagovagal anti-inflammatory reflex results in the stimulation of cholinergic vagal efferents. In turn, acetylcholine activates alpha-7 subunit-containing nicotinic receptors on immune cells, which inhibit further release of proinflammatory cytokines and suppress inflammation. These advances in the understanding of vagal nerve function offer novel therapeutic opportunities for the management of nociception and inflammation [60].

Clearly, there is ample redundancy in the regulation of immune function and of inflammation at various levels, which interact, fine tune, compensate, and correct for each other during the inflammatory process. Ultimately, the hypothalamus controls this fascinating interacting and inter-dependent Neuroimmune Supersystem of Integrative regulation in higher animals and humans, which is fundamental to the entire life cycle of these organisms [29,61,62]. NIF is yet another example for the neural control of immune and inflammatory responses, which is presented in detail in this volume.

István Berczi

REFERENCES

1. Jancsó G, Katona M, Horváth V, Sántha P, Nagy J. Sensory nerves as modulators of cutaneous inflammatory reactions in health and disease. In: Jancsó G, ed., Berczi I, Szentiványi A, series ed. Neurogenic Inflammation in Health and Disease. Amsterdam: Elsevier, 2009; pp. 3–36.
2. Donaldson LF. Neurogenic mechanisms in arthritis. In: Jancsó G, ed., Berczi I, Szentiványi A, series ed. Neurogenic Inflammation in Health and Disease. Amsterdam: Elsevier, 2009; pp. 211–242.
3. Cruwys SC, Garrett NE, Kidd BL. Sensory denervation with capsaicin attenuates inflammation and nociception in arthritic rats. Neurosci Lett 1995;193(3):205–207.
4. Lam FY, Ferrell WR. Capsaicin suppresses substance P-induced joint inflammation in the rat. Neurosci Lett 1989;105(1–2):155–158.
5. Lam FY, Ferrell WR. Neurogenic component of different models of acute inflammation in the rat knee joint. Ann Rheum Dis 1991;50(11):747–751.
6. Levine JD, Dardick SJ, Basbaum AI, Scipio E. Reflex neurogenic inflammation. I. Contribution of the peripheral nervous system to spatially remote inflammatory responses that follow injury. J Neurosci 1985;5(5):1380–1386.
7. Mapp PI, Kerslake S, Brain SD, Blake DR, Cambridge H. The effect of intra-articular capsaicin on nerve fibres within the synovium of the rat knee joint. J Chem Neuroanat 1996;10(1):11–18.
8. Donaldson LF, McQueen DS, Seckl JR. Neuropeptide gene expression and capsaicin-sensitive primary afferents in the maintenance and spread of adjuvant arthritis in the rat. J Physiol 1995;486(2):473–482.
9. Carleson J, Kogner P, Bileviciute I, Theodorsson E, Appelgren A, Appelgren B et al. Effects of capsaicin in temporomandibular joint arthritis in rats. Arch Oral Biol 1997;42(12):869–876.
10. Ahmed M, Srinivasan GR, Theodorsson E, Schultzberg M, Kreicbergs A. Effects of surgical denervation on substance P and calcitonin gene-related peptide in adjuvant arthritis. Peptides 1995;16(4):569–579.
11. Glick E. Asymmetrical rheumatoid arthritis after poliomyelitis. Br Med J 1967;3:26–29.
12. Glynn J, Clayton M. Sparing effect of hemiplegia on tophaceous gout. Ann Rheum Dis 1976;35:534–535.
13. Thompson M, Bywaters E. Unilateral rheumatoid arthritis following hemiplegia. Ann Rheum Dis 1962;21:370–377.
14. Veale D, Farrell M, Fitzgerald O. Mechanism of joint sparing in a patient with unilateral psoriatic-arthritis and a longstanding hemiplegia. Br J Rheumatol 1993;32(5):413–416.
15. Decaris E, Guingamp C, Chat M, Philippe L, Grillasca JP, Abid A et al. Evidence for neurogenic transmission inducing degenerative cartilage damage distant from local inflammation. Arthritis Rheum 1999;42(9):1951–1960.
16. Lynn B. The nature and electrophysiological properties of the afferent neurones involved in neurogenic inflammation in the skin. In: Jancsó G, ed., Berczi I, Szentiványi A, series ed. Neurogenic Inflammation in Health and Disease. Amsterdam: Elsevier, 2009; pp. 39–48.
17. Pintér E, Helyes Z, Németh J, Szolcsányi J. Somatostatin as an anti-inflammatory neuropeprtide: from physiological basis to drug development. In: Jancsó G. ed., Berczi I, Szentiványi A, series ed. Neurogenic Inflammation in Health and Disease. Amsterdam: Elsevier, 2009; pp. 121–134.

18. Pattachini R, Maggi CA. Tachykinins and neurogenic inflammation at visceral level. In: Jancsó G, ed., Berczi I, Szentiványi A, series ed. Neurogenic Inflammation in Health and Disease. Amsterdam: Elsevier, 2009; pp. 289–320.
19. Scott TM, Scott MM. Peptidergic innervation of blood vessels. In: Jancsó G, ed., Berczi I, Szentiványi A, series ed. Neurogenic Inflammation in Health and Disease. Amsterdam: Elsevier, 2009; pp. 49–74.
20. Sántha P, Pierau FK, Jancsó G. Evidence for an inhibition by endogenous galanin of neurogenic cutaneous vasodilatation in the pigeon. Neurosci Lett 1998;243(1–3):101–104.
21. Bauer JW, Lang R, Jakab M, Kofler B. Galanin – 25 years with a multitalented neuropeptide : galanin family of peptides in skin function. Cell Mol Life Sci 2008;65(12):1820–1825.
22. Jancsó G, Sántha P, Horváth V, Pierau F. Inhibitory neurogenic modulation of histamine-induced cutaneous plasma extravasation in the pigeon. Regul Pept 2000;95(1–3):75–80.
23. Cortright DN, Szallasi A. The role of the vanilloid and related receptors in nociceptor function and neuroimmune regulation. In: Jancsó G, ed., Berczi I, Szentiványi A, Series ed. Neurogenic Inflammation in Health and Disease. Amsterdam: Elsevier, 2009; pp. 101–118.
24. Donnerer J. The inhibition of neurogenic inflammation. In: Jancsó G, ed., Berczi I, Szentiványi A, series ed. Neurogenic Inflammation in Health and Disease. Amsterdam: Elsevier, 2009; pp. 169–190.
25. Holzer P. Neurogenic vasodilatation and plasma leakage in the skin. Gen Pharmacol 1998;30:5–11.
26. Nagy I. Molecular mechanisms of capsaicin receptor mediated nociception. In: Jancsó G, ed., Berczi I, Szentiványi A, series ed. Neurogenic Inflammation in Health and Disease. Amsterdam: Elsevier, 2009; pp. 75–100.
27. Domeij S, Carlsöö B, Dahlqvist A, Forsgren S. Occurrence of mast cells in relation to the distribution of nerve fibers in the rat larynx. Acta Otolaryngol 1991;111:981–989.
28. McKay DM, Bienenstock J. The interaction between mast cells and nerves in the gastrointestinal tract. Immunol Today 1994;15:533–538.
29. Berczi I, Szentiványi A. Nerve growth factor, leptin and neuropeptides. In: Berczi I, Szentiványi A, eds. Neuroimmmune Biology, Volume 3: The Immune-Neuroendocrine Circuitry. History and Progress. Amsterdam: Elsevier 2003; pp. 181–189.
30. Franco-Penteado CF, Desouza IA, Camargo EA, Teixeira SA, Muscara MN, De Nucci G et al. Mechanisms involved in the enhancement of allergic airways neutrophil influx by permanent C-fibers degeneration in rats. J Pharmacol Exp Ther 2005;313:440–448.
31. Poole JA, Matangkasombut P, Rosenwasser LJ. Targeting the IgE molecule in allergic and asthmatic diseases: review of the IgE molecule and clinical efficacy. J Allergy Clin Immunol 2005;115:S376–S385.
32. Slavin RG, Fischer VW, Levine EA, Tsai CC, Winzenburger P. A primate model of allergic bronchopulmonary aspergillosis. Int Arch Allergy Appl Immunol 1978;56(4):325–333.
33. Medeiros MV, Macedo-Soares MF, De Luca IM, Hyslop S, De Nucci G, Antunes E. Contribution of C-fibers to leukocyte recruitment in bronchoalveolar lavage fluid and pleural cavity in the rat. Eur J Pharmacol 2001;421:133–140.
34. Girolomoni G, Tigelaar RE. Capsaicin-sensitive primary sensory neurons are potent modulators of murine delayed-type hypersensitivity reactions. J Immunol 1990;145:1105–1112.
35. Veronesi B, Williams WC, Smialowicz RJ, Sailstad DM, Doerfler D, Selgrade MJ. Neuropeptide denervation alters both the elicitation and induction phases of contact hypersensitivity in mice. Toxicol Appl Pharmacol 1998;153:243–249.

36. Garssen J, Buckley TL, Van Loveren H. A role for neuropeptides in UVB-induced systemic immunosuppression. Photochem Photobiol 1998;68:205–210.
37. Boric CR, Lu B, Hopken UE, Gerard C, Gerard NP. Neurogenic amplification of immune complex inflammation. Science 1996;273:1722–1725.
38. Gottwald T, Lhotak S, Stead RH. Effect of truncal vagotomy and capsaicin on mast cells and IgA-positive plasma cells in rat jejunal mucosa. Neurogastroenterol Motil 1997;9:25–32.
39. Bunker CB, Cerio R, Bull HA, Evans J, Dowd PM, Foreman JC. The effect of capsaicin application on mast cells in normal human skin. Agents Actions 1991;33:195–196.
40. Szentiványi A, Berczi I, Pitak D, Goldman A. Studies of the hypothalamic regulation of histamine synthesis. In: Berczi I, Gorczynski R, eds. Neuroimmune Biology Volume 1: New foundation of Biology. Amsterdam, Netherlands: Elsevier, 2001; pp. 47–55.
41. Saria A, Lundbert JM, Skofitsch G, Lembeck F. Vascular protein leakage in various tissues induced by SP, Capsaicin, bradykinin, serotonin, histamine and by antigen challenge. Naunyn Schmiedeberg's Arch Pharmacol 1983;324:212–218.
42. Tramontana M, Santicioli P, Giuliani S, Catalioto R-M, Lecci A, Carini F et al. Role of tachykinins in sephadex-induced airway hyperractivity and inflammation in guinea-pigs. Eur J Pharmacol 2002;439:149–158.
43. Lundberg JM, Alving K, Karlsson JA, Matran R, Nilsson G. Sensory neuropeptide involvement in animal models of airway irritation and of allergen-evoked asthma. Am Rev Resp Dis 1991;143:1429–1431.
44. Maghni K, Taha R, Afif W, Hamid Q, Martin JG. Dichotomy between neurokinin receptor actions in modulating allergic airway responses in an animal model of helper T cell type 2 cytokine-associated inflammation. Am J Respir Crit Care Med 2000;162:1068–1074.
45. Ferdinandy P, Jancsó G. Capsaicin-sensitive sensory nerves in myocardial ischemia-reperfusion injury and ischemic stress adaptation: role of nitric oxide and calcitonin-gene related peptide. In: Jancsó G, ed., Berczi I, Szentiványi A, series ed. Neurogenic Inflammation in Health and Disease. Amsterdam: Elsevier, 2009; pp. 267–288.
46. Dux M, Messlinger K. Neurogenic vascular responses in the dura mater and their relevance for the pathophysiology of headaches. In: Jancsó G, ed., Berczi I, Szentiványi A, series ed. Neuroimmune Biology, Volume 8: Neurogenic Inflammation in Health and Disease. Amsterdam: Elsevier, 2009; pp. 193–210.
47. Meggs WJ. Neurogenic inflammation and sensitivity to environmental chemicals. Environ Health Perspect 1993;101:234–238.
48. Besedovsky HO, del Rey A. Processing of cytokine signals at cns levels: relevance for immune-HPA axis interactions. In: Del Ray A, Chrousos G, Besedovsky HO, eds, Berczi I, Szentiványi A, series eds. The Hypothalamus-Pituitary-Adrenal Axis. Neuroimmune Biology, Volume 7. Amsterdam: Elsevier, 2008; pp. 227–240.
49. Gabry KE, Chrousos G, Gold PW. The hypothalamus-pituitary-adrenal (HPA) axis. A major mediator to the adaptive response to stress. In: Berczi I, Szentiványi A, eds. Neuroimmmune Biology, Volume 3: The Immune-Neuroendocrine Circuitry. History and Progress. Amsterdam: Elsevier, 2003; pp. 379–414.
50. Szentiványi A, Berczi I, Nyanteh H, Goldman A. Altered effector responses. In: Berczi I, Szentiványi A, eds. Neuroimmmune Biology, Volume 3: The Immune-Neuroendocrine Circuitry. History and Progress. Amsterdam: Elsevier, 2003; pp. 21–29.

51. Berczi I, Szentiványi A. Growth and lactogenic hormones, insulin-like growth factor and insulin. In: Berczi I, Szentiványi A, eds. Neuroimmmune Biology, Volume 3: The Immune-Neuroendocrine Circuitry. History and Progress. Amsterdam: Elsevier, 2003; pp. 129–153.
52. Berczi I, Szentiványi A. Cytokines and chemokines. In: Berczi I, Szentiványi A, eds. Neuroimmmune Biology, Volume 3: The Immune-Neuroendocrine Circuitry. History and Progress. Amsterdam: Elsevier, 2003; pp. 191–220.
53. Stein C. The control of pain in peripheral tissue by opioids. New Engl J Med 1995;332:1685–1690.
54. Groneberg DA, Fischer A. Endogenous opioids as mediators of asthma. Pulm Pharmacol Ther 2001;14:383–389.
55. Wang J, Ren Y, Zou X, Fang L, Willis WD, Lin Q. Sympathetic influence on capsaicin-evoked enhancement of dorsal root reflexes in rats. J Neurophysiol 2004;92:2017–2026.
56. Lin Q, Zou X, Fang L, Willis WD. Sympathetic modulation of acute cutaneous flare induced by intradermal injection of capsaicin in anesthetized rats. J Neurophysiol 2003;89:853–861.
57. Green PG. Neurogenic regulation of bradykinin induced synovitis. In: Jancsó G, ed., Berczi I, Szentiványi A, series eds. Neurogenic Inflammation in Health and Disease. Amsterdam: Elsevier, 2009; pp. 243–266.
58. Pethő G, Reeh PW. Effects of Bradykinin on Nociceptors. In: Jancsó G, ed., Berczi I, Szentiványi A, series eds. Neuroimmune Biology, Volume 8: Neurogenic Inflammation in Health and Disease. Amsterdam: Elsevier, 2009; pp. 135–168.
59. Sabbadini ER, Berczi I. Regulatory enzymes. In: Berczi I, Szentiványi A, eds. Neuroimmmune Biology, Volume 3: The Immune-Neuroendocrine Circuitry. History and Progress. Amsterdam: Elsevier, 2003; pp. 271–276.
60. Holzer P. The role of the vagus nerve in efferent signaling and homeostasis during visceral inflammation. In: Jancsó G, ed., Berczi I, Szentiványi A, series eds. Neurogenic Inflammation in Health and Disease. Amsterdam: Elsevier, 2009; pp. 321–338.
61. Berczi I, Quintanar-Stephano A, Kovacs K. Immunoconversion in the Acute phase response. In: Plotnikoff N, ed. Cytokines, Stress and Immunity. Boca Raton: CRC Press, 2006; pp. 215–254.
62. Berczi I. Integration and regulation of higher organisms by the neuroimmune supersystem. Int J Integrative Biol 2007;1(3):216–231.
63. Lin Q, Zou X, Ren Y, Wang J, Fang L, Willis WD. Involvement of peripheral neuropeptide Y receptors in sympathetic modulation of acute cutaneous flare induced by intradermal capsaicin. Neuroscience 2004;123:337–347.

Preface

The participation of the nervous system in inflammatory reactions was suggested over a century ago, yet nervous elements as potentially important contributors to inflammatory mechanisms have been largely ignored for a long time. However, the discovery of the highly specific pharmacological effects of capsaicin in the 1950s and the selective neurotoxic action of vanilloid compounds in the 1970s on nociceptive sensory neurons have revived interest in studying neurogenic inflammatory processes. Early observations furnished firm evidence for the existence of an inflammatory response induced by a purely neurogenic route, resulting from the activation of nociceptive afferents. Morphological studies utilizing the neurotoxic/neurodegenerative actions of capsaicin demonstrated a widespread system of peptidergic and nonpeptidergic capsaicin-sensitive afferent nerves which innervate the skin, mucous membranes, and most visceral organs and tissues. From studies on different organs, tissues, and cells, a complex system of primary sensory neurons has emerged, which parallels the autonomic nervous system not only in its extent, but probably also in its significance. Afferent nerves, once believed to serve merely as sensory receptors conveying impulses generated by noxious stimuli, have evolved into key players in a complicated local regulatory system that participates in the contractile, glandular, vascular, inflammatory, immune, protective, restorative and trophic functions of somatic, and visceral tissues. Neuropeptides released from nociceptive afferents in response to tissue injury or to a wide variety of chemical stimuli involving inflammatory mediators and tissue and mast cell-derived agents play a pivotal role in these processes and are potent modulators of inflammatory reactions. New facets of capsaicin-sensitive neuronal and cellular responses were revealed following the cloning of the capsaicin/vanilloid receptor, now known as the transient receptor potential vanilloid type 1 receptor (TRPV1) which is primarily expressed in nociceptive primary sensory neurons but also in some other neurons and cells. A number of contributions in this volume focus on the characterization and functional traits of nociceptor neurons and on the mechanisms that activate them. Further chapters deal with the parts played by primary sensory neurons in inflammatory reactions and in the regulation/modulation of the functions of various organs and tissues under physiological and pathological conditions. Certain chapters touch upon the therapeutic implications offered by the use of vanilloids, novel nonpeptide antagonists of peptide and vanilloid receptors. It is the editor's hope that this volume will contribute to and initiate new interest in an understanding of the diverse roles fulfilled by primary afferent neurons in the functioning of the body in health and disease.

G. Jancsó
Szeged, May 26, 2008

List of Contributors

Number in parentheses indicate the pages on which the authors' contributions begin.

István Berczi (vii)
Department of Immunology, University of Manitoba, Winnipeg, Canada

Daniel N. Cortright (101)
Department of Biochemistry and Molecular Biology, Neurogen Corporation, Branford, CT, USA

Lucy F. Donaldson (211)
Department of Physiology and Pharmacology, School of Medical Sciences, University of Bristol, Bristol, UK

Josef Donnerer (169)
Institute of Experimental and Clinical Pharmacology, Medical University of Graz, Graz, Austria

Mária Dux (193)
Department of Physiology, University of Szeged, Szeged, Hungary

Péter Ferdinandy (267)
Cardiovascular Research Group and Pharmahungary™ Group, Department of Biochemistry, University of Szeged, Szeged, Hungary

Paul G. Green (243)
Department of Oral and Maxillofacial Surgery, University of California San Francisco, San Francisco, CA, USA

Zsuzsanna Helyes (121)
Department of Pharmacology and Pharmacotherapy, Faculty of Medicine, University of Pécs, Pécs, Hungary

Peter Holzer (321)
Research Unit of Translational Neurogastroenterology, Institute of Experimental and Clinical Pharmacology, Medical University of Graz, Graz, Austria

Ulrike Holzer-Petsche (169)
Institute of Experimental and Clinical Pharmacology, Medical University of Graz, Graz, Austria

Viktor Horváth (3)
First Department of Internal Medicine, University of Szeged, Szeged, Hungary

Gábor Jancsó (xv, 3, 267)
Department of Physiology, University of Szeged, Szeged, Hungary

Márta Katona (3)
Department of Pediatrics, University of Szeged, Szeged, Hungary

Bruce Lynn (39)
Department of Physiology, University College London, London, UK

Carlo A. Maggi (289)
Pharmacology Department, Menarini Ricerche, Florence, Italy

Karl Messlinger (193)
Institute for Physiology and Pathophysiology, University of Erlangen-Nürnberg, Erlangen, Germany

József Nagy (3)
Department of Traumatology, Semmelweis Hospital, Kiskunhalas, Hungary

Istvan Nagy (75)
Department of Anaesthetics and Intensive Care, Imperial College, Faculty of Medicine, Chelsea and Westminster Hospital, London, UK

József Németh (121)
Department of Pharmacology and Pharmacotherapy, Faculty of Medicine, University of Pécs, Pécs, Hungary

Riccardo Patacchini (289)
Department of Pharmacology, Chiesi Pharmaceuticals SpA, Parma, Italy

Cleoper C. Paule (75)
Department of Anaesthetics and Intensive Care, Imperial College, Faculty of Medicine, Chelsea and Westminster Hospital, London, UK

Gábor Pethő (135)
Department of Pharmacology and Pharmacotherapy, Faculty of Medicine, University of Pécs, Pécs, Hungary

Erika Pintér (121)
Department of Pharmacology and Pharmacotherapy, Faculty of Medicine, University of Pécs, Pécs, Hungary

Peter W. Reeh (135)
Institute for Physiology and Pathophysiology, University of Erlangen-Nürnberg, Erlangen, Germany

Péter Sántha (3)
Department of Physiology, University of Szeged, Szeged, Hungary

Thomas M. Scott (49)
Schulich School of Medicine and Dentistry, Windsor Program, University of Windsor, Windsor, Ontario, Canada

Michael M. Scott (49)
Department of Internal Medicine, Center for Hypothalamic Research, The University of Texas Southwestern Medical Center, Dallas, TX, USA

Arpad Szallasi (101)
Department of Pathology, Monmouth Medical Center, NJ, USA

János Szolcsányi (121)
Department of Pharmacology and Pharmacotherapy, Faculty of Medicine, University of Pécs, Pécs, Hungary

John P. M. White (75)
Department of Anaesthetics and Intensive Care, Imperial College, Faculty of Medicine, Chelsea and Westminster Hospital, London, UK

I. INTRODUCTION

Edited by Gábor Jancsó

Sensory Nerves as Modulators of Cutaneous Inflammatory Reactions in Health and Disease

GÁBOR JANCSÓ[1], MÁRTA KATONA[2], VIKTOR HORVÁTH[3], PÉTER SÁNTHA[1], and JÓZSEF NAGY[4]

[1] *Department of Physiology, University of Szeged, Szeged, Hungary*
[2] *Department of Pediatrics, University of Szeged, Szeged, Hungary*
[3] *First Department of Internal Medicine, University of Szeged, Szeged, Hungary*
[4] *Department of Traumatology, Semmelweis Hospital, Kiskunhalas, Hungary*

ABSTRACT

Chemosensitive afferent nerves expressing the capsaicin/TRPV1 (transient receptor potential vanilloid receptor-1) receptor are not only involved in the transmission of nociceptive impulses toward the central nervous system, but also play pivotal roles in the initiation and modulation of vascular, inflammatory, and immune reactions in a variety of organs, including the skin. These sensory nerves exert their efferent/local regulatory functions primarily via the release of vasoactive neuropeptides such as calcitonin gene-related peptide and substance P from their terminals upon antidromic or orthodromic stimulation. The chemical changes induced in the neural microenvironment by tissue injury or inflammation may promote the release of proinflammatory sensory neuropeptides and lead to augmentation of the inflammatory response. In turn, the release of anti-inflammatory peptides from sensory nerves or from inflammatory cells may result in an inhibition of cutaneous vascular reactions. The available experimental evidence implicates capsaicin-sensitive sensory nerves in the pathogenesis of certain skin diseases and maladies with cutaneous manifestations. The development of potent nonpeptide antagonists of sensory neuropeptides, the capsaicin/TRPV1 receptor, and the proteinase-activated receptors may offer new possibilities for the management of certain skin diseases and the itch and pain associated with altered sensory nerve functions.

1. INFLAMMATION AND SENSORY NERVES

The possible contributions of neural elements to the mechanisms of inflammatory reactions were largely ignored in early studies on inflammation, despite the fact that sensory nerve stimulation and irritation had been shown to produce changes characteristic of inflammation. As early as 1874, Goltz [1] observed that stimulation of the sciatic nerve induced vasodilatation.

Shortly thereafter, in 1876, Stricker [2] demonstrated that antidromic electrical stimulation of the peripheral end of transected sacral dorsal roots induced peripheral vasodilatation in tissues served by those roots in the dog. Later studies by Bayliss [3,4] confirmed this finding (which was in apparent contradiction with the Bell–Magendie law), by showing that the vasodilatatory nerves responsible for antidromic vasodilatation are identical with sensory fibers originating from dorsal root ganglia. It was suggested that vasodilatation develops in consequence of the activation of collaterals of the stimulated sensory nerve fibers which terminate in the vicinity of nearby arterioles and cause them to dilate (axon reflex theory). In the human skin, the development of the flare component of Lewis' triple response, which can be elicited by chemical irritants or other noxious stimuli, was explained by a similar axon reflex mechanism [5]. The studies by Spiess [6] showed that the application of irritant chemicals such as mustard oil produced a cutaneous inflammatory reaction, which he attributed to a spinal reflex mechanism. The experiments by Bruce [7,8] revealed that the inflammatory response elicited by mustard oil was markedly reduced or abolished after chronic denervation. Accordingly, he suggested that this phenomenon may be explained by an axon reflex mechanism of sensory nerves. However, subsequent investigations into the mechanisms of inflammation largely ignored the possible significance of sensory nerves in the development of inflammation. It was not until the 1950s that the role of sensory nerves in inflammatory processes was beginning to be clarified. The investigations by Jancsó and Jancsó [9] and N. Jancsó [10,11] demonstrated that capsaicin, the pungent principle of red pepper (paprika), causes pain and inflammation, which can eventually be completely prevented by prior repeated local or systemic administration of this agent, a phenomenon termed "capsaicin desensitization." Capsaicin desensitization or chronic sensory denervation prevented the inflammatory effect of a variety of chemical irritants, including capsaicin and mustard oil. It was concluded that many chemical irritants exert their phlogogenic and pain-producing effects by the stimulation of capsaicin-sensitive sensory nerve endings. These, in turn, release a vasoactive "neurohumor" which induces vasodilatation and an increase in vascular permeability, collectively termed the neurogenic inflammatory response [9–13].

2. THE NATURE OF SENSORY NERVES INVOLVED IN VASCULAR RESPONSES

Although the chemical sensitivity of capsaicin-sensitive afferent nerves was characterized pharmacologically [10], the nature of the postulated neurohumor released by these nerves remained unknown [14]. Identification of this mediator seemed especially important in light of the early suggestion by Dale [15] that the substance released from the peripheral sensory nerve endings, which mediates the axon reflex flare, may be identical with the transmitter released from the central terminals of these afferent fibers mediating pain.

The discovery of the rather selective neurotoxic action of capsaicin and related sensory neurotoxins [16–18] permitted the direct morphological identification of capsaicin-sensitive primary sensory neurons which mediate the neurogenic inflammatory response. Importantly, these observations pointed to the possibility that they "may also promote investigations to elucidate, for example, immunohistochemically, which of the putative transmitters may be involved in mediation of chemogenic pain" and, consequently, neurogenic inflammatory responses [16]. Hence, subsequent studies revealed populations of peptidergic capsaicin-sensitive sensory neurons [19–21] and provided evidence for the functional role of some of the peptides contained in these neurons in the mediation of nociceptive and neurogenic inflammatory responses (for reviews, see Refs [22–31]). In addition to earlier observations, experiments on rats treated (neonatally) with capsaicin led to the finding that capsaicin-sensitive afferent nerves are involved in the transmission of nociceptive impulses evoked not only by chemical irritants, but also by heat [32–34]. Recent studies have clearly confirmed the significance of this observation: experiments on the functional characteristics of the cloned capsaicin receptor disclosed that this receptor may be activated by and involved in the transmission of impulses generated by noxious heat under physiological conditions [35–37].

Making use of the neurotoxic action of capsaicin on a morphologically well-defined population of primary sensory neurons permitted the identification and characterization of a distinct division of the somatosensory system involved in the transmission of nociceptive information and the mediation of local regulatory functions. Histological studies revealed that capsaicin-sensitive sensory ganglion cells comprise about 50% of all primary sensory neurons, are mostly to be found in the small-sized cell population, and give rise almost exclusively to unmyelinated axons [16,24,38–40]. These capsaicin-sensitive primary afferents terminate predominantly in the superficial (marginal layer and substantia gelatinosa) layers of the spinal and medullary dorsal horn, in some well-defined deeper regions of the spinal cord (lamina V and intermedio-medial region), and in visceral sensory nuclei of the solitary tract and the area postrema [16,41,42]. These morphological studies supported the existence of both somatic and visceral capsaicin-sensitive sensory nerves [38,41] and strongly suggested [38,41,43] that capsaicin-sensitive afferents may have similar functions in somatic and visceral tissues. Immunohistochemical investigations revealed the presence of capsaicin-sensitive afferent nerves in most somatic and visceral organs and tissues (for reviews, see Refs [22–27,29,30,38,44–46]), including the skin [28,47–53].

In the past few decades, a vast body of experimental evidence has accumulated in support of and extending the original concept of N. Jancsó [10,11,14] and N. Jancsó et al. [54] on the dual function of capsaicin-sensitive afferent nerves. First, via the transmission of nociceptive information toward the central nervous system, these particular afferent nerves fulfill an afferent function. Second, through the secretion of (vasoactive) peptides, they exert a local regulatory or sensory-efferent function by the modulation of vascular and inflammatory reactions [32,55,56], smooth muscle contractile activity [46,57,58], glandular secretion [59–61], tissue repair processes [62], immune responses [63], and the cardiac function [43,46,64–66]. Experiments showing that selective chemical lesion of the central branches of (trigeminal) nociceptive sensory ganglion neurons, leaving the soma and the peripheral branch intact, resulted in a complete abolition of the transmission of nociceptive impulses but left the neurogenic inflammatory response intact, clearly illustrated the special dual functional character of these sensory neurons [67–69], and it was proposed that they be termed "secreto-sensory nociceptive neurons" [67,68]. These findings also suggest that enhanced dorsal root reflexes may not play a decisive role in the development of neurogenic plasma extravasation induced by (chemical) tissue injury (cf. Refs [70,71]).

As concerns the type(s) of afferent fibers involved in sensory neurogenic cutaneous vascular responses, it is generally accepted that only capsaicin-sensitive C-fiber polymodal nociceptors mediate plasma extravasation in the rat [72,73]. Neurogenic sensory vasodilatation is primarily mediated by capsaicin-sensitive C-fibers, but the activation of Aδ-afferent fibers may also produce vasodilatation [73]. In the pig, cutaneous sensory neurogenic vasodilatation, i.e., the axon reflex flare, is mediated by a specific population of C heat nociceptor afferents and not by polymodal nociceptors [74]. In the human skin, the sensory fibers mediating the flare response are known to be capsaicin-sensitive [10,11,54,55] and have been identified as mechano-insensitive C-fiber afferents [75]. Neurogenic sensory vasodilatation, at least in the rat dura mater, may also be produced by the stimulation of C-fiber afferents which are not sensitive to capsaicin [76]. However, the direct chemical stimulation of C-fiber afferents with capsaicin at very low (nanomolar) concentrations has also been found to induce a significant increase in dural blood flow in rats [77].

3. MEDIATORS OF NEUROGENIC INFLAMMATORY RESPONSES

3.1 Sensory neuropeptides

Rats treated neonatally with capsaicin provided an excellent model for studies on the function of peptidergic afferent neurons: the lack or impairment of sensory functions in these animals may be attributed to the irreversible destruction of capsaicin-sensitive sensory ganglion neurons [16]. Experiments in this line indicated that substance P (SP) is a likely candidate as a mediator of antidromic vasodilatation [78] and neurogenic plasma extravasation evoked by antidromic electrical stimulation of the sensory nerves [20]. Although these reactions were practically completely abolished in rats treated with capsaicin [16,20], this does not necessarily imply that they are mediated by SP, for several other peptides are localized or colocalized in subpopulations of capsaicin-sensitive sensory neurons [19,23,26,44]. The discovery and localization of the potent vasodilatator calcitonin gene-related peptide (CGRP) in sensory neurons [79] led to a reexamination of this issue. Several lines of evidence indicated that CGRP plays a decisive role in antidromic and neurogenic sensory vasodilatation [28,30,31,77,80]. CGRP elicits marked vasodilatation in various organs, and electrical or chemical stimulation of sensory nerves results in release of the peptide [31,77]. SP and the related tachykinin neurokinin A (also localized in sensory nerves) are now generally believed to mediate neurogenic plasma extravasation through the activation of NK1 receptors [30,31,81–86] situated on the endothelial cells of postcapillary venules [87,88].

Sensory neuropeptides released from sensory nerves may interact with each other or with other mediators of inflammation, resulting in a complex modulation of cutaneous inflammatory reactions. CGRP has been shown to increase plasma extravasation elicited by SP and histamine [82,89], leaving the effects of serotonin (5-HT) and bradykinin unaffected [82].

An intriguing aspect relating to the role of sensory neuropeptides in cutaneous inflammation is the indirect modulating role exerted by SP upon the vascular actions of CGRP. Hence, SP released from sensory nerves triggers the release of mast cell mediators, including mast cell proteases, which in turn, participate in the degradation of CGRP, leading to a shortening/reduction of its vasodilatatory effect [90].

3.2 Histamine

Although histamine derived from mast cells has also been implicated in the mediation of this response [50,91,92], it appears that histamine does not play a significant role in the initiation of the neurogenic inflammatory reaction. Histamine is released from mast cells by SP in the human skin [93], but it does not act as a final mediator of the axon reflex flare [94]. It has been suggested, however, that histamine may contribute to the flare by initiation of the axon reflex [93]. The possible role of histamine in the neurogenic inflammatory response has recently been investigated in microdialysis experiments in the human skin. These studies clearly showed that the development of the flare response is independent of histamine action irrespective of the chemical nature of the stimulants applied [95,96]. Hence, the cascade theory of neurogenic inflammation cannot be confirmed in the human skin [97].

Studies on the mechanisms of neurogenic cutaneous plasma extravasation revealed vascular responses of similar magnitude in wild-type and mast cell-deficient mice [98]. Similarly, neurogenic plasma extravasation induced by mustard oil was not affected in mast cell-deficient mice [86]. Pharmacological analysis of capsaicin-induced mouse ear edema disclosed that it is primarily mediated by SP via NK1 receptors [99], although involvement of 5-HT through action on 5-HT2 receptors cannot be excluded [100]. A possible contribution of histamine to the neurogenic inflammatory responses of some tissues cannot be ruled out because there are marked differences between different species and organs as regards the involvement of mast cells [47].

3.3 Nitric oxide

The local control of the cutaneous circulation includes a vasodilator response to local, non-painful pressure application [101]. This neurally mediated phenomenon was observed in humans [101], rats [102], and mice [103]. This mechanism, termed pressure-induced vasodilatation, depends on the activation of capsaicin-sensitive nerve fibers leading to the release of CGRP, vasoactive intestinal peptide, and/or pituitary adenylate cyclase-activating polypeptide [101,102,104]. Neonatal treatment with capsaicin or inhibition of CGRP or nitric oxide synthase (NOS) abolished pressure-induced vasodilatation [102]. These results empasize the role of nitric oxide in the process but further investigations are necessary to evaluate the cellular mechanisms underlying this phenomenon [105].

Nitric oxide is also implicated in the release of CGRP from the sensory nerves: in the rabbit skin, administration of an NOS inhibitor significantly reduced the neurogenic vasodilator response to capsaicin [106]. Studies utilizing nitric oxide donors indicate that nitric oxide may act through a pathway which involves prostaglandins [107].

3.4 Proteinase-activated receptors

The proteinase-activated receptors (PARs) are new types of cellular signaling molecules which are targeted by proteinases such as thrombin, trypsin, or human mast cell tryptase. PARs have been localized on sensory ganglion neurons [108,109]. In particular, out of the four known subtypes of PARs [110–112], PAR1 and PAR2 have been shown to be localized on sensory neurons expressing the capsaicin/TRPV1 (transient receptor potential vanilloid receptor-1) receptor [113,114], which also contain SP and/or CGRP [108]. PAR2 is highly expressed in the skin [115] and its agonists induce many signs of inflammation [108,116]. PAR agonists have been shown to release sensory neuropeptides in different peripheral tissues, presumably from capsaicin-sensitive afferent nerve endings [108]. PAR2 agonists have been shown to release these sensory neuropeptides in different peripheral tissues, presumably from capsaicin-sensitive afferent nerve endings [108]. Further studies have shown that PAR2 may play a role in neurogenic inflammatory responses; the inflammatory effect of PAR2 activation was inhibited by capsaicin pretreatment and by NK1 receptor antagonists [109]. These findings indicate that the activation of PARs may be involved in the initiation and/or modulation of neurogenically mediated vascular reactions. Also, in an experimentally induced contact dermatitis model of mice swelling responses, plasma extravasation and leukocyte adherence were significantly attenuated in PAR2 null mutant ($PAR2^{-/-}$), and PAR2 was capable of the regulation of inflammatory reactions of skin via neurogenic mechanisms [117]. PAR2 also participates in the pathomechanism of pruritus in inflammatory diseases of the skin, so its modulation may contribute to the treatment of pruritus [118,119]. However, other possible contributions of PARs to the development of skin pathologies (like autoimmune diseases) and the significance of these receptors as putative therapeutic targets await further investigations.

4. NEURONAL CAPSAICIN SENSITIVITY AND NEUROGENIC INFLAMMATION

The identification [27,120] and subsequent molecular cloning of the capsaicin/vanilloid receptor type 1 (VR1 [35]; now generally referred to as the TRPV1) permitted the further functional and pharmacological characterization of capsaicin-sensitive primary afferent neurons. It is interesting to address the issue of whether the local regulatory, efferent functions of sensory nerves may be linked with capsaicin sensitivity, i.e., the expression of the TRPV1 receptor. Studies on avian species revealed that these are largely insensitive to the irritant action of capsaicin [121] and lack the capsaicin/TRPV1 receptor [122]. However, the sensory ganglion neurons of these species contain vasoactive peptides implicated in the mediation of sensory nerve-mediated vascular reactions [123,124]. Furthermore, exogenous application of these peptides has been shown to produce cutaneous vasodilatation, albeit of moderate extent in these species [123]. Antidromic electrical stimulation of the afferent nerves, or the direct application of mustard oil onto the skin, induced moderate, but significant neurogenic vasodilatation in the pigeon [125–127]. Moreover, histamine and bradykinin applied intracutaneously through plasmapheresis capillaries produced neurogenic plasma extravasation [127]. The magnitude of these responses is modulated by endogenous inhibitory peptide(s) including galanin [125–128]. Taken together, these findings indicate that sensitivity to capsaicin, i.e., the expression of the TRPV1 receptor and the molecular mechanisms underlying the local regulatory (efferent) function of sensory nerves are differentially regulated in sensory ganglion neurons. The finding that mustard oil induces neurogenic plasma extravasation in birds [126,127], i.e., in species lacking TRPV1 receptors [122], is especially interesting in light of the recent observation that mustard oil elicits neurogenic plasma extravasation in TRPV1 receptor knockout rats [129,130]. The demonstration of the TRP channel TRPA1/ANKTM1 in sensory neurons responding specifically to mustard oil provides the molecular basis of the irritant action of this compound [131]. It has been demonstrated however that in the mouse and the rat the TRPA1 protein is expressed in neurons which are also sensitve to capsaicin [132–134]. These neurons may form a specific subpopulation of the capsaicin-sensitive primary nociceptors responding not only to heat and capsaicin, but also to various exogenous (i.e., mustard oil, acrolein, and unsaturated aldehydes) and endogenous (bradykinin) chemical irritants [132]. This fact also explains previous findings showing the complete abolition of mustard oil-induced cutaneous neurogenic inflammatory reaction following capsaicin-induced deletion of chemosensitive sensory nerve fibers in rodents [16,34]. Although morphological data on the coexpression of TRPV1 and TRPA1 are lacking, functional observations indicate a similar overlap of capsaicin- and mustard oil-sensitive populations of sensory neurons also in humans [10,54,135,136].

Although the expression of the capsaicin receptor may not be a prerequisite for the development of neurogenic sensory vascular reactions, these may be significantly modulated through the activation of capsaicin receptors. Acidic conditions [137] and/or some mediators of inflammation, e.g., bradykinin and prostaglandin E_2 (PGE_2) [138], interleukin-1 [139], neuropeptide Y

[140], some growth factors like nerve growth factor (NGF), glial cell line-derived neurotrophic factor (GDNF), and brain-derived neurotrophic factor (BDNF) [141], and activation of PARs [108] have been shown to increase neuropeptide release from sensory ganglion neurons and this, in turn, may result in an increased inflammatory response. The (patho)physiological significance of these interactions as regards cutaneous vascular responses awaits further experimentation, but it seems likely that metabolites produced by ischemic tissue are involved in the mediation of reactive cutaneous hyperemia through the activation of capsaicin-sensitive afferent nerves [142,143].

In chronic inflammatory conditions, increased local production of neurotrophic factors such as NGF might produce long-term changes in the morphology (innervation density), chemical phenotype, and excitability of capsaicin-sensitive cutaneous afferents, which finally results in an augmentation of the neurogenic component of the inflammatory process. [144–150].

5. SPECIES AND ORGAN DIFFERENCES IN NEUROGENIC INFLAMMATORY RESPONSES

The similarity of neurogenic sensory cutaneous vascular responses in different species may bear of some relevance as regards the biological significance of these reactions. As yet, however, the possible species and organ differences in neurogenic inflammatory responses have not been dealt with in detail.

Interestingly, a clear-cut, visible axon reflex flare response can be evoked only in primates, except for the pig. In this species, as in humans, local electrical, chemical (capsaicin), or heat stimulation elicits a characteristic flare reaction [151–153]. Hence, the pig skin is an excellent model of human neurogenic inflammatory responses. Similarly, as in humans, antidromic stimulation of the sensory nerves or direct stimulation of the nerve endings with capsaicin failed to produce plasma extravasation in the pig skin [151,152]. In contrast, the cutaneous application of mustard oil, which is a specific chemical stimulus in rodents and which elicits neurogenic plasma extravasation by the activation of afferent fibers, produced a marked flare response and plasma extravasation in the pig. This latter response was confined to the area of direct contact with the irritant; in particular, signs of increased vascular permeability, as measured with the Evans blue or the vascular labeling technique utilizing colloidal silver, were not observed in the zone of the flare [153]. These findings suggested that mustard oil may produce a plasma extravasation response in the pig skin without the activation of capsaicin receptors of sensory nerves. More recently, Caterina et al. [37] demonstrated that TRPV1 receptor knockout mice retain their sensitivity to mustard oil, which produces similar inflammatory responses in wild-type and TRPV1 knockout mice. In agreement with this finding, using TRPV1 knockout animals, the existence of mustard oil-induced TRPV1-independent neurogenic inflammation in the mouse skin was demonstrated [129].

The most numerous and detailed studies on neurogenic inflammatory responses have been performed on rodents, especially rats. In the rat skin, both direct chemical stimulation of the sensory nerve endings and antidromic electrical stimulation of the sensory nerves result in vasodilatation and plasma extravasation. These responses can also be elicited in the mucous membranes and most visceral organs including, for example, the urogenital, respiratory, and gastrointestinal systems (cf. Refs [10,11,14,22,23,26,43,46,54,57,154]).

Neurogenic sensory vascular responses have additionally been studied in some other mammalian species, including humans. In the cat, no extravasation could be observed in the skin in response to antidromic nerve stimulation or topical application of chemical irritants (mustard oil). However, extravasation was noted in the trachea upon antidromic stimulation of the vagus nerve (Jancsó, G. and Such, G., unpublished observations). In the rabbit, antidromic sensory vasodilatation can be readily demonstrated; the response is mediated by peptidergic afferent nerves sensitive to capsaicin [82,155,156].

In the guinea pig, neurogenic inflammatory responses have been detected in various visceral organs [154], but not in the skin; the application of mustard oil, a chemical irritant which produces neurogenic inflammation in the rat skin, failed to evoke a similar response in the guinea pig (Jancsó G., unpublished observations).

In the pig, antidromic electrical stimulation of the afferent nerves and the topical application of chemical irritants led to vasodilatation and flare. It is worth mentioning that the pig is one of the few mammalian species which gives a visible cutaneous flare response resembling that seen in humans [74,151,157]. The topical application of mustard oil in the pig also produced a marked increase in cutaneous vascular permeability, as assessed quantitatively by measuring the amount of extravasated Evans blue-labeled plasma albumin, or histologically by using the vascular labeling technique [153]. However, neurogenic cutaneous plasma extravasation in the pig may not be mediated by afferent nerves sensitive to capsaicin, since capsaicin failed to produce plasma extravasation [151], and local desensitization of the skin with capsaicin (Jancsó, G. and Pierau, Fr.-K., unpublished observations) did not inhibit mustard oil-induced extravasation.

Earlier studies on the human skin were focused mainly on the mechanisms of the flare response, which is abolished after local capsaicin treatment [10,11,158]. Less attention has been paid to the other component of the response, i.e., neurogenic plasma extravasation, which is so predominant in the rodent skin. Although cutaneous vasodilatation and flare responses may additionally be elicited by relatively mild chemical, physical, or electrical stimuli, the production of neurogenic skin edema apparently requires excessive stimulation [10,55,159,160].

The available experimental data clearly indicate that orthodromic or antidromic stimulation of the cutaneous sensory nerve endings readily induce neurogenic sensory vasodilatation in mammalian species. In contrast, neurogenic plasma extravasation may not necessarily be associated with the vasodilatatory reaction and may vary in the different organs, even in the same species. Indeed, neurogenic cutaneous plasma extravasation seems to be a less widespread phenomenon as compared with neurogenic cutaneous vasodilatation. It is unlikely that the apparent lack of neurogenic sensory plasma extravasation in some species and organs reflects methodological difficulties in the detection of modest increases in vascular permeability; the vascular labeling method used in our experiments with colloidal silver is superior in sensitivity to the commonly used Evans blue technique and permits an exact morphological demonstration of small increases in vascular permeability [10,38,59,161,162]. These differences in the neurogenic sensory plasma extravasation response may rather be related to variations in skin type, the density and pattern of cutaneous innervation, the amount(s) of the peptide(s) released, the expression, localization, and density of vascular peptide receptors, and the presynaptic mechanisms modulating the release of vasoactive peptides from the sensory nerve endings.

6. MODULATION OF THE INFLAMMATORY PROCESS BY CHEMOSENSITIVE AFFERENT NERVES

The discovery of the neurotoxic/neurodegenerative action of capsaicin [16] provided direct evidence that the neurogenic inflammatory response is initiated by capsaicin-sensitive afferent nerves. Indeed, selective degeneration and consequent loss of small-diameter afferent nerve fibers following neonatal capsaicin treatment resulted in a permanent, lifelong elimination of neurogenic inflammation in the rat [16,38]. The utilization of this experimental paradigm offered a unique approach to study *inter alia* the possible significance of these particular afferents in inflammatory reactions.

6.2 Augmentation of inflammatory responses by the activation of sensory nerves

Early observations indicated that a variety of chemically different irritant compounds (e.g., capsaicin, mustard oil, xylene, or chloracetophenone) are capable of eliciting a neurogenic inflammatory response, which can be inhibited or even completely abolished by pretreatment with capsaicin or related pungent agents producing "capsaicin desensitization" [10]. It has been suggested that certain chemical agents act exclusively through a neurogenic route involving sensory nerves to evoke an inflammatory reaction, whereas other phlogogens bring about the inflammatory reaction without the contribution of afferent nerves. Subsequent studies, however, revealed that sensory nerves are intimately involved in the mechanism of inflammatory reactions evoked by a variety of substances involving mediators of acute inflammatory responses. Studies on adult rats treated with capsaicin as neonates disclosed that cutaneous inflammatory reactions elicited by exogenously injected or endogenously released histamine were strongly reduced. Similarly, the plasma extravasation evoked by intracutaneously injected bradykinin, 5-HT, or PGE_2 was markedly inhibited by capsaicin pretreatment [32,38]. In the human skin, wheal and flare responses evoked by cold in patients with cold urticaria were strongly reduced in skin areas pretreated with capsaicin [135,159,160]. The flare, but not the wheal response of a delayed hypersensitivity response was abolished in the capsaicin-treated human skin [135,159,163].

These findings indicated that the functional significance of capsaicin-sensitive afferent nerves is far beyond the initiation of the neurogenic inflammatory response. Indeed, this particular class of afferent nerves possess a significant proinflammatory effect and contribute to a variety of inflammatory reactions. Accordingly, it has been suggested that capsaicin-sensitive sensory nerves are potent modulators of inflammatory responses and provide an augmentation of the primary inflammatory reaction by secreting proinflammatory peptides, such as SP, Neurokinin A (NKA), and/or CGRP [32,38,55,82,135]. These vasoactive peptides may enhance the inflammatory reaction by (i) direct vasodilatatory and permeability-increasing action on the blood vessels, (ii) an indirect action on mast cells, resulting in an additional release of histamine and possibly other vasoactive mast cell mediators, (iii) by interfering with the metabolism of the neurogenically released peptide(s), and (iv) local recruitment of circulating leukocytes through the activation of endothelial cell adhesion molecules [164,165]. Hence, sensory nerves, blood-borne leukocytes and tissue macrophages, and mast cells may form a self-sustained inflammatory cycle. An important aspect of this amplification process concerns the factors which trigger the release of vasoactive peptides from sensory nerve endings. There is ample evidence that, in both animal and human skin, nociceptive nerve endings can be stimulated by inflammatory mediators including vasoactive biogenic amines (e.g., histamine, 5-HT, [166,167]), peptides (e.g., bradykinin [166–169]), eicosanoids (e.g., PGE_2 [167]), and by acidic pH [170,171]. In turn, bradykinin, PGE_2 [138], and acidic conditions [137] have been shown to facilitate the release of neuropeptides from sensory neurons. Since a variety of agents, including cytokines, prostaglandins, and leukotrienes, capable of stimulating capsaicin-sensitive nociceptive nerve endings may be present in inflammed tissues (cf. Ref. [28]), this amplification mechanism may operate under a variety of pathophysiological conditions (cf. Refs [32,55,82,135]).

Further studies have revealed that capsaicin-sensitive afferent nerves may also influence the development and course of immune reactions [63,172] and chronic inflammatory reactions involving arthritis [173,174], airway inflammation [47,58,61,175,176], and inflammatory reactions of the gastrointestinal system [29,30,45,57,177].

6.3 Inhibitory modulation of inflammatory responses by sensory nerves

The efferent, local regulatory functions of cutaneous sensory nerves are generally considered to be stimulatory, i.e., proinflammatory in nature. However, the presence of peptides which exert inhibitory effects on inflammatory or other physiological/pathological processes in a variety of tissues/organs suggested that cutaneous sensory nerves may exert efferent inhibitory actions in addition to, or in parallel with their activatory effect. The most important candidates for the mediation of such inhibitory regulatory functions of sensory neurons are the neuropeptides galanin and somatostatin. Although both peptides are present in a relatively small population of the sensory neurons in mammals, amounting to 5–10% of dorsal root ganglion cells in the rat [178–182], the existence of galanin- and somatostatin-containing nerves has been demonstrated in many organs, including the skin [179,183–185]. In the rat, a majority of the sensory ganglion neurons which express galanin [178,186] or somatostatin [19,174,187–189] are sensitive to capsaicin, indicating that functionally these cells correspond to chemosensitive nociceptors which express the vanilloid/TRPV1 receptor [35].

Investigations into the possible functional significance of galanin in the mechanism of inflammatory responses in the rat skin have demonstrated that the intracutaneous injection of galanin strongly reduces the plasma extravasation induced by antidromic electrical stimulation of the sensory nerves [190]. Furthermore, when galanin was applied in high concentrations, the permeability-increasing effect of i.v. injected SP was reduced, suggesting that galanin might act directly on cutaneous microvessels in the skin. Coperfusion of galanin with capsaicin, but not with a combination of SP and CGRP, prevented the plasma protein leakage induced by selective chemical stimulation of joint nociceptors [191]. It has been concluded that both exogenous galanin and somatostatin (at 10 times higher concentration!) can effectively inhibit the capsaicin-induced release of vasoactive neuropeptides from synovial afferents, acting prejunctionally on their terminals [191]. Although these studies clearly proved that galanin may inhibit sensory nerve-mediated responses, the involvement of endogenous galanin was not demonstrated.

The proportion of sensory ganglion cells exhibiting galanin immunoreactivity varies considerably from species to species. It has been found immunohistochemically that in avian species the numbers of small- and medium-sized galanin-containing dorsal root ganglion neurons are considerably higher than in the rat, accounting for about 20% of the total cell population in the sensory ganglia [124]. In avian species, neither direct chemical nor antidromic electrical stimulation of the cutaneous sensory nerve endings produces significant plasma protein extravasation or vasodilatation [121,123,192,193]. This cannot be attributed to a lack of vasoactive sensory peptides in these species, since, similarly as in mammals, the existence of SP and CGRP has been demonstrated in avian sensory ganglion cells [194,195] and cutaneous nerves [192,196]. Furthermore, the responsiveness of cutaneous blood vessels to the vasodilatatory effects of SP and CGRP has been demonstrated in the avian skin [123].

Investigations using a high-affinity galanin receptor antagonist, M35 [197], revealed that the close arterial injection of M35 significantly and dose-dependently enhanced the cutaneous vasodilatatory responses elicited by either the epicutaneous application of mustard oil or antidromic electrical stimulation of the cutaneous nerves [125–127]. Similarly, pretreatment with M35 enhanced the vasodilatation induced by the intracutaneous perfusion of histamine, and to a lesser extent that of bradykinin [127]. The M35-induced augmentation of the vasodilatatory responses elicited by either bradykinin or histamine was completely abolished by prior chronic denervation of the skin. Additionally, in the intact, but not in the chronically denervated skin, the permeability-increasing effects of histamine and bradykinin were strongly augmented when these inflammatory mediators were coadministered with the galanin antagonist M35 [127,128]. These findings indicated that endogenous galanin may exert a significant (tonic) inhibitory local efferent action on sensory nerve-mediated vascular reactions, and that this inhibition is of neural origin. These observations were corroborated by recent data which showed a similar reduction in the mustard oil-induced cutaneous plasma extravasation in a galanin-overexpressing mouse model [198]. In further studies, it has been demonstrated that chronic denervation *per se* in the pigeon skin enhances the permeability-increasing effect of histamine, as revealed by either using the microcapillary perfusion technique or the direct histological visualization of leaky blood vessels with the vascular labeling technique. In the denervated skin, the coadministration of galanin significantly reduced the histamine-induced plasma protein extravasation, pointing to a possible postjunctional action of galanin in this process [128]. Direct vascular actions (vasoconstriction and inhibition of the edema formation) of galanin and galanin-like peptide have also been demonstrated recently in murine skin [199].

It should be mentioned that galanin expression was observed also in nonneuronal elements of the skin, in particular in the keratinocytes of the rat [200], humans [201], pigeon (Jancsó and Sántha, unpublished observations), and the epithelial layer of the rat gingiva [202]. The biological significance of epithelial galanin, especially its role in the regulation of the sensory efferent functions of afferent nerves remains to be elucidated.

These findings provided direct evidence for an endogenous galanin-mediated modulation of cutaneous vascular reactions induced by either the antidromic or the orthodromic activation of sensory nerves or by the application of inflammatory mediators. Since a large proportion of the sensory neurons in the pigeon, including those innervating the skin, contain galanin, it has been proposed that, at least in this species, cutaneous sensory nerves exert anti-inflammatory (inhibitory) local regulatory (sensory-efferent) functions which may contribute to the maintenance of the functional and structural integrity of the skin. Although studies in galanin-overexpressing mice suggested the existence of a similar mechanism in mammals [198], the involvement of endogenous galanin in the modulation of sensory efferent functions in mammals still remains to be elucidated. Immunohistochemical findings showing that galanin-immunoreactive unmyelinated axons constitute the largest population of peptidergic nerve fibers in the rat dorsal roots suggest a significant role for galanin in the function of capsaicin-sensitive sensory nerves [203]. Recent observations resulting from the administration of high-affinity galanin receptor antagonists, indicated that endogenous galanin may inhibit the activation of joint receptors [204]. However, intracutaneous application of M35 did not influence the capsaicin-induced pain reaction (flinching) in the rat hind paw [185]. Furthermore, it should be noted that galanin might exert not only inhibitory, but also excitatory effects on a certain population of capsaicin-sensitve cutaneous afferents, most probably on those expressing the type 2 but not of the type 1 and 3 galanin receptors [185,205,206].

Another candidate for the transmission of sensory nerve-mediated inhibitory efferent functions is somatostatin. This peptide has an inhibitory influence on a large variety of biological processes and it is present in primary sensory neurons [181]. In the rat, somatostatin-containing neurons amount to 8–15% [179,181,207–209] of the lumbar dorsal root ganglion cells and are capsaicin-sensitive [19,187,188,210]. This was confirmed by demonstration of the coexpression of somatostatin and VR1 receptor RNAs in dorsal root ganglion cells [189]. The first observation of an inhibitory effect of somatostatin on sensory nerve-mediated vascular reactions (in the dental pulp of the cat) was reported by Gazelius et al. [211]; the intra-arterial injection of somatostatin strongly reduced the vasodilatation induced by antidromic electrical stimulation of the alveolar nerve. Since somatostatin also prevented the depletion of the SP-like immunoreactivity from the dental pulp following long-lasting nerve stimulation, the authors concluded that somatostatin might act prejunctionally on the release of SP from the stimulated sensory nerve terminals. Lembeck et al. [212] reported that somatostatin can also inhibit the cutaneous vasodilatation and plasma protein extravasation induced in the rat hind paw skin induced by antidromic electrical stimulation of the saphenous nerve. Similarly, Green et al. [191] detected an inhibitory effect of somatostatin on the capsaicin-induced plasma protein leakage in the perfused knee joint of the rat. In both cases, the somatostatin-mediated inhibition appeared to be prejunctional, since the coadministration of somatostatin with SP and a combination of SP and CGRP did not produce any change in the vascular effects of these vasoactive sensory neuropeptides.

The phenomenon of "counterirritation," i.e., the anti-inflammatory effect of a local inflammatory reaction (irritation) on an ongoing inflammation process in other parts of the body, has been known for centuries [213]. The inhibition of the inflammatory reactions caused by the primary local irritation was earlier accounted for by the activation of nociceptive reflexes eliciting an increased sympathetic activity or the release of glucocorticoids from the adrenal cortex [214]. The possibility of cross-activation of the contralateral nociceptors through an unidentified intrasegmental (spinal) neural pathway was also postulated [174]. An alternative explanation emerged from the observation that a neurogenic inflammatory reaction in the rat hind paw strongly reduced the permeability-increasing effect of subsequent stimulation on the contralateral side. This inhibition can be elicited even after dorsal root transection [215] or an acute bilateral denervation of both hind paws [216]. It was therefore concluded that the observed inhibition is not mediated through a central reflex pathway, but is transmitted by a humoral factor released upon stimulation of the nociceptive nerve endings. It was suggested that the systemic anti-inflammatory effect was mediated by somatostatin, released from a subset of chemosensitive sensory nerve endings into the circulation. Indeed, the authors reported an up to fourfold increase in the plasma somatostatin-like immunoreactivity 2 min after bilateral antidromic electrical stimulation of both sciatic nerves [217] and a 40% increase 10 min after the epicutaneous application of mustard oil to the hind paw skin [216].

However, there is some degree of controversy as regards the somatostatin-mediated systemic anti-inflammatory functions of primary afferents. If somatostatin is released because of the activation of peptidergic afferents, the first site of action (where it also reaches the highest concentration) should be the site of the release, i.e., in the inflammed skin region served by the stimulated nerve. Nevertheless, a reduction of the primary response is not observed after pretreatment of the animals with somatostatin antagonists/chelators [217]. The antinociceptive and anti-inflammatory effects evoked by the first stimulation of sciatic afferents lasted for about 1 h and 20 min, respectively. These time intervals seem too long if it is considered that the plasma half-life of the circulating somatostatin is only a few minutes [218,219]. Furthermore, antidromic vasodilatation evoked by sciatic nerve stimulation was not affected either by somatostatin or by prior contralateral sciatic nerve stimulation [217], which is at variance with the earlier findings that antidromic vasodilatation was inhibited by somatostatin in both the cat dental pulp and rat skin [211,212].

7. CUTANEOUS NEUROGENIC SENSORY VASCULAR REACTIONS IN PATHOLOGIES AFFECTING THE SKIN

Cutaneous sensory nerves may be implicated in pathologies involving the skin in essentially two ways: diseases of the skin may affect cutaneous sensory nerves, resulting in an impaired sensory function, while perturbations of sensory neuron functions, e.g., by systemic diseases, may interfere with the mechanisms involved in the maintenance of the structural and functional integrity of the skin.

7.1 Diseases affecting cutaneous sensory nerves

Diabetes mellitus is a common metabolic disorder which is often associated with neuropathic alterations. In a rat model of experimental diabetes, a marked reduction of cutaneous neurogenic plasma extravasation has been revealed [82,220]. Subsequent studies demonstrated that this may be accounted for by decreases in the levels of sensory neuropeptides such as CGRP and tachykinins, which mediate neurogenic vascular responses [221–223]. Assessment of the flare response has been shown to be a reliable measure of the structural and functional integrity of cutaneous nociceptive afferent nerves in humans [224]. Hence, analysis of the neurogenic vascular reactions elicited by chemical agents, heat, or electrical stimulation has become a technique that is widely used to evaluate the functional condition of nociceptive afferent nerves [55,135,160,225–228]. Studies in diabetic patients disclosed a significant diminution of the axon reflex flare in diabetic patients [228–230]. This observation indicated that capsaicin-sensitive afferent nerves are impaired in diabetes, and this might contribute to the development of the trophic disturbances, neuropathic changes, and smooth muscle dysfunction characteristic of diabetes.

Reductions of axon reflex flare responses elicited by histamine have also been observed in uremic patients with or without pruritus. Interestingly, the itch responses to histamine were greater in patients with pruritus [231], suggesting that itch and flare responses are mediated by different populations of afferent nerves.

7.2 Involvement of capsaicin-sensitive sensory nerves in the pathogenesis of some skin diseases and maladies with cutaneous manifestations

Congenital insensitivity to pain has been found to be associated with a loss of the cutaneous flare responses elicited by mustard oil or histamine [135,232]. Electron microscopy revealed a substantial loss of unmyelinated axons from cutaneous nerves [135]. These findings provided further evidence that capsaicin-sensitive C-fiber afferent nerves are responsible for mediation of the axon reflex flare of the human skin. The results also imply that a lack of the capsaicin receptor, which is normally expressed in human sensory ganglion neurons [120,233,234], may explain some of the symptoms of congenital analgesia.

Sensory denervation of the skin in humans or in experimental animals results in the development of cutaneous lesions [38,235], impaired hair growth [235,236], and affected wound healing [62], indicating an important trophic role of capsaicin-sensitive sensory nerves in the skin.

Herpes zoster has long been known to be associated with the degeneration of cutaneous nerves and impaired cutaneous sensation [236]. The demonstration of reduced cutaneous flare responses to mustard oil, histamine [225,237], and capsaicin [238] indicated a severe dysfunction of the afferent nerves after herpes zoster infection.

A diminution of the neurogenic inflammatory response has also been demonstrated in patients with psoriasis [226,239]. This is in line with observations of a reduction in the number of epidermal nerve fibers in both involved and uninvolved psoriatic skin. Moreover, the extent of fiber loss correlated with the severity of the disease. It has been suggested that epidermal nerve fiber degeneration and the consequent loss of contact between the afferent nerves and the immunocompetent cells of the skin may play important roles in the self-maintenance of the disease [240]. However, reports on the density of epidermal nerve fibers in psoriatic skin are controversial. A significant increase in nerve fiber density was found in psoriatic skin, with a particular increase in the proportion of SP- and CGRP-containing nerves [241]. In further elegant studies, in which the severe combined immunodeficient mouse–human skin model was used, an increased production of NGF by keratinocytes was revealed, and it was suggested that this induced the hyperinnervation of the psoriatic skin. Additionally, a marked upregulation of SP and CGRP has been revealed in cutaneous nerves. It has been postulated that the NGF present in keratinocytes and peptidergic cutaneous nerves contributes significantly to the development of psoriatic lesions by promoting chemokine production, the recruitment of inflammatory cells, and the stimulation of keratinocyte proliferation [242,243]. The role of (sensory) nerves in the pathomechanism of psoriasis is also supported by the observation that cutaneous denervation results in a diminution or even the complete disappearance of the psoriatic lesions [244]. Furthermore, the topical application of capsaicin-containing preparations proved beneficial for the treatment of pruritic psoriasis and resulted in a significant reduction in combined psoriasis severity scores presumably by depleting sensory neuropeptides from cutaneous nerves [158,245,246]. Obviously, neurogenic inflammatory mechanisms are implicated in the pathogenesis of psoriasis, but further studies are needed to clarify the apparently contrasting findings and the exact roles played by sensory nerves in the pathogenesis of this disease.

The proliferation of dermal peptidergic afferent nerves and an elevated number of mast cells apparently innervated by these nerves have been demonstrated in prurigo nodularis [247,248]. An increase in the number of NGF-immunoreactive, possibly inflammatory cells has also been seen in the dermis, and this may contribute to the neurohyperplasia observed in prurigo nodularis [248,249].

Studies on patients with acquired cold and heat urticaria revealed that capsaicin-sensitive afferent nerves play a fundamental role in the initiation of whealing; neither flare nor urtication developed in response to a thermal challenge in skin areas pretreated with capsaicin [55,135,160,250]. Since histamine-induced whealing is unaffected in capsaicin-treated skin [159,160], the failure of thermal stimulation to produce whealing in capsaicin-treated skin may possibly be attributed to the failure of SP and/or CGRP to release mast cell histamine due to the depletion of these potent mast cell activators from the sensory nerves. This is supported by the finding that capsaicin-induced mast cell degranulation is effected through SP and/or CGRP released from sensory nerves [93]. These results suggest that an increased chemosensitivity of sensory nerves and/or mast cells in acquired cold and heat urticaria may contribute significantly to the pathogenesis of the disease.

Peptidergic sensory nerves and neurogenic inflammation have also been implicated in the pathomechanism of a variety of other diseases affecting the skin, including atopic dermatitis [226,227,251], alopecia areata [252], cutaneous herpesvirus infection [253], and notalgia paresthetica [254,255].

8. THERAPEUTIC POTENTIAL OF VANILLOIDS

The potential therapeutic use of capsaicin-type compounds has not been considered for a long time after the discovery of its selective activating and subsequent desensitizing effect on nociceptive nerve endings. The demonstration of a selective and long-lasting chemical and thermal analgesia resulting from a single application of capsaicin onto peripheral nerves has led to the suggestion that capsaicin may be considered as a promising tool in relieving certain pains of peripheral origins [34,136,256]. Studies on cutaneous sensation in humans [54,55,159,160,228,257,258] and animals [259–262] showed that repeated application of capsaicin onto the skin resulted in a significant and selective increase in heat pain threshold and a decrease in neurogenic vascular reactions. These studies indicated that the effect of topical application of capsaicin in inhibiting nociceptive functions is transient; recovery of the nociceptive afferent and the local regulatory, efferent functions of sensory nerve endings occurs within a few days or weeks after cessation of the treatment [55,159,258,260,262,263]. Importantly, trophic lesions of the skin, which may occur after neonatal capsaicin treatment in rats [38,235,264,265], were not observed after topical capsaicin neither in animal [259,260] nor in the human skin [55,159,160,258,263]. Based on these findings, capsaicin-containing creams and ointments have become widely used in a number of pathological conditions associated with pain and inflammation. Preparations containing 0.025–0.075% capsaicin have been used in the treatment of herpes zoster [266–269], neck pain [270,271], postthoracotomy and postmastectomy pain [272,273], amputation stump pain [274,275], diabetic neuropathy [276–282], notalgia paresthetica [283], and pain caused by skin tumor [284]. Capsaicin and its derivates may be used also in the treatment of pruritus [159,285,286]. The rationale of the therapeutic use of capsaicin in painful neuropathies is further supported by findings showing the presence of TRPV1 receptors on dermal and intraepidermal nerve fibers in pathologies associated with pain [49].

Although permanent elimination of C-fiber afferent nerves by various methods utilizing capsaicin has become a routine approach in the study of sensory mechanisms, for a long time, except for neonatal treatment with capsaicin, degeneration of sensory ganglion neurons and/or their axons has not been seriously considered as a significant mechanism of the pharmacological effects of capsaicin. Systemic capsaicin treatment of newborn animals, including small laboratory rodents and dogs results in a permanent elimination of capsaicin-sensitive spinal and cranial sensory ganglion neurons [16,24,43,287]. The exact mechanism of capsaicin-induced cell death and permanent sensory neuron loss is still unclear. Systemic administration of capsaicin [16,43,288–291] or resiniferatoxin [17,292–295] to rodents or to cultures of dorsal root ganglion cells results in a clear-cut degeneration and/or apoptosis of many sensory ganglion neurons. This is evidenced not only by the electron microscopic findings but also by the radiochemically and histochemically detectable massive intracellular accumulation of ionic calcium indicative of neuronal degeneration and/or apoptosis [287,289,296]. However, a substantial population of primary sensory neurons may be lost as a consequence of a delayed degeneration process most probably due to an impaired retrograde intraneural transport of trophic factors, in particular, NGF [24,297–299]. *In vivo* and *in vitro* studies revealed that direct application of capsaicin to the skin and some visceral tissues resulted in an extensive degeneration of peptidergic and nonpeptidergic cutaneous dermal and epidermal afferent nerves [50,51,259,300], as well as sensory axons of the ureter [301], the uterus [302], and the dura mater [77]. The loss of epidermal nerve fibers after topical [50,51,259,300] or perineural [53] application of capsaicin has been suggested to be causally related to the development of cutaneous hypoalgesia to noxious mechanical, heat, and chemical stimuli [50,51,53,259,300]. This is supported by the findings demonstrating a parallel loss and recovery of cutaneous nerve fibers and pain sensation after topical treatment of the skin with capsaicin [50]. The reduction of epidermal axonal numbers is transient after topical [50], but apparently permanent after perineural [53] treatment with capsaicin. Interestingly, nerve fiber degeneration was not observed 24 h after intravesical administration of vanilloid compounds [303]. Therefore, these authors suggest that vanilloid compounds may exert their effects also by alternative mechanisms involving inhibition of the retrograde axonal transport of trophic factors, e.g., NGF and consequent depletion of sensory peptides such as SP and CGRP, which participate in the transmission of noxious stimuli. In addition, downregulation of the TRPV1 receptor and consequent loss of neuronal capsaicin sensitivity which have been observed after peripheral nerve injury [189,304,305] may also contribute to the effects of intravesical application of vanilloids. It should be noted, however, that capsaicin's neurotoxic action may develop within minutes after the application and degenerated structures may disappear within 24 h [16,18,41,287,289,296,306].

An apparent common difficulty in controlled trials of the analgesic actions of capsaicin and related pungent agents is the lack of appropriate placebo with similar irritant properties. Obviously, the introduction of preparations containing capsaicin and a local anesthetic with a controlled release sequence to reduce irritation and/or the development of stable vanilloid analogs with reduced irritancy would greatly promote the topical use of vanilloid-like compounds in the treatment of various painful conditions. A recent report on the critical evaluation of the therapeutic effectiveness of topically applied capsaicin suggests that it may be a useful adjunct or even sole therapy for patients with musculoskeletal or neuropathic pain who are unresponsive to or intolerant of other treatments [307].

Modulation and/or inhibition of vanilloid receptor function through an interaction with the molecular mechanisms of TRPV1 channel activation by the development of vanilloid analogs of reduced irritancy, improved metabolic profile, and increased potency may represent a possible future direction in achieving antinociception involving vanilloid action. In addition, investigations into the mechanisms of nociceptive signaling via vanilloid receptors and endovanilloids [308] point to novel perspectives in pain management involving the development of competitive [309] and noncompetitive [310] antagonists of TRPV1 or drugs affecting mechanisms which may modulate the activity of TRPV1 by, for example, phosphorylation of the receptor [311,312]. A common feature of these new types of TRPV1 antagonists is their ability to attenuate both neuropathic and inflammatory pain. The data demonstrating the high potency and oral bioavailability of some of these compounds indicate that these antagonists are not only promising experimental tools for the study of TRPV1 in nociceptive mechanisms but may bear significant therapeutic potential [309].

9. CONCLUSIONS AND PERSPECTIVES

Investigations into the morphology, biochemistry, and function of cutaneous sensory nerves have revealed new facets of the biology of the skin. Cutaneous sensory nerves, once believed to serve merely as sensory receptors conveying impulses generated by innocuous and noxious environmental stimuli, have emerged as key players in a complicated local regulatory system involved in the vascular, inflammatory, immune, protective, restorative, and trophic functions of the skin. Neuropeptides released from a subpopulation of nociceptive afferents in response to tissue injury or to a wide variety of chemical stimuli involving inflammatory mediators and tissue and mast cell-derived agents fulfill pivotal roles in these processes and are potent modulators of inflammatory reactions. They may augment the primary inflammatory reaction by the release of proinflammatory peptides such as SP and CGRP, which act either directly on blood vessels or indirectly by the activation of mast cells and immune cells. Additionally, by the release of anti-inflammatory peptides such as galanin or somatostatin, they may reduce inflammation by prejunctional and/or postjunctional mechanisms. Neurally released peptides also play important roles in the leukocyte, keratinocyte, fibrocyte, and immune cell functions. Rapidly accumulating evidence from clinical data indicates that sensory nerves are intimately involved in the pathomechanisms of a variety of skin diseases. The development of potent nonpeptide antagonists of sensory neuropeptides, the capsaicin/vanilloid, and the PARs, may offer new possibilities in the management of certain skin diseases and the itch and pain associated with perturbations of the sensory nerve functions.

ACKNOWLEDGMENTS

This work was supported in part by grants from OTKA (T 046469, T 034964), ETT (569/2003), and RET. P. Sántha was supported by the Bólyai János Postdoctoral Fellowship.

REFERENCES

1. Goltz F. Über gefässerweiternde Nerven. Pflügers Arch Ges Physiol 1874;9:174–190.
2. Stricker S. Untersuchungen über die Gefässwurzeln des Ischiadicus. Sitz Kaiserl Akad Wiss 1876;3:173–185.
3. Bayliss WM. On the origin from the spinal cord of the vasodilator fibres of the hind limb and on the nature of these fibres. J Physiol (Lond) 1901;26:173–209.
4. Bayliss WM. The Vasomotor System. London: Longmans, Green and Co Inc., 1923.
5. Lewis T, Marvin HM. Observations relating to vasodilatation arising from antidromic impulses to herpes zoster and trophic effects. Heart 1927;14:27–42.
6. Spiess G. Die Bedeutungder Anasthäsie in der Entzündungstherapie. Münch med Wschr 1906;8:345–351.
7. Bruce AN. Über die Beziehung der sensiblen Nervenendigungen zum Entzündungsvorgang. Arch Exptl Pathol Pharmakol 1910;63:424–433.
8. Bruce AN. Vaso-dilator axon reflexes. Q J Exp Physiol 1913;6:339–354.
9. Jancsó N, Jancsó A. Érzőidegvégződések desensibilizálása (Desensitization of sensory receptors). Kísérletes Orvostudomány 1949;2:15.
10. Jancsó N. Role of the nerve terminals in the mechanism of inflammatory reactions. Bull Millard Fillmore Hosp 1960;7:53–77.
11. Jancsó N. Desenzitisation with capsaicin as a tool for studying the function of pain receptors. Proceedings of the 3rd International Pharmacology Meeting 1966. Oxford: Pergamon Press, 1968; pp. 33–55.
12. Jancsó N. Speicherung. Stoffanreichung im Retikuloendothel und in der Niere. Budapest: Akadémiai Kiadó, 1955.
13. Jancsó N, Jancsó-Gábor A. Dauerausschaltung der chemischen Schmerzempfindlichkeit durch Capsaicin. Arch Exp Path Pharmak 1959;236:142.

14. Jancsó N. Neurogenic inflammatory responses. Acta Physiol Acad Sci Hung 1964;(Suppl 23):3–4.
15. Dale HH. Pharmacology and nerve-endings. Proc R Soc Med 1935;28:319–332.
16. Jancsó G, Kiraly E, Jancsó-Gábor A. Pharmacologically induced selective degeneration of chemosensitive primary sensory neurones. Nature 1977;270:741–743.
17. Jancsó G. Degenerative changes induced by capsaicin and related vanilloids in primary sensory neurones. J Physiol (Lond) 2000;526:9S–10S.
18. Jancsó G, Király E. Sensory neurotoxins: chemically induced selective destruction of primary sensory neurons. Brain Res 1981;210:83–89.
19. Jancsó G, Hökfelt T, Lundberg JM, Király E, Halász N, Nilsson G et al. Immunohistochemical studies on the effect of capsaicin on spinal and medullary peptide and monoamine neurons using antisera to substance P, gastrin/CCK, somatostatin, VIP, enkephalin, neurotensin and 5-hydroxytryptamine. J Neurocytol 1981;10:963–980.
20. Gamse R, Holzer P, Lembeck F. Decrease of substance P in primary afferent neurones and impairment of neurogenic plasma extravasation by capsaicin. Br J Pharmacol 1980;68:207–213.
21. Jessell TM, Iversen LL, Cuello AC. Capsaicin-induced depletion of substance P from primary sensory neurones. Brain Res 1978;152:183–188.
22. Holzer P. Local effector functions of capsaicin-sensitive sensory nerve endings: involvement of tachykinins, calcitonin gene-related peptide and other neuropeptides. Neuroscience 1988;24:739–768.
23. Buck SH, Burks TF. The neuropharmacology of capsaicin: review of some recent observations. Pharmacol Rev 1986;38:179–226.
24. Jancsó G. Pathobiological reactions of C-fibre primary sensory neurones to peripheral nerve injury. Exp Physiol 1992;77:405–431.
25. Nagy I, Sántha P, Jancsó G, Urbán L. The role of the vanilloid (capsaicin) receptor (TRPV1) in physiology and pathology. Eur J Pharmacol 2004;500:351–369.
26. Nagy JI. Capsaicin: a chemical probe for sensory neuron mechanisms. In: Iversen LL, Iversen SD, eds. Handbook of Psychopharmacology. New York: Plenum Press, 1982; pp. 185–235.
27. Szállási Á, Blumberg PM. Vanilloid (Capsaicin) receptors and mechanisms. Pharmacol Rev 1999;51:159–212.
28. Holzer P. Neurogenic vasodilatation and plasma leakage in the skin. Gen Pharmacol 1998;30:5–11.
29. Holzer P. Gastrointestinal afferents as targets of novel drugs for the treatment of functional bowel disorders and visceral pain. Eur J Pharmacol 2001;429:177–193.
30. Holzer P. Peptidergic sensory neurons in the control of vascular functions: mechanisms and significance in the cutaneous and splanchnic vascular beds. Rev Physiol Biochem Pharmacol 1992;121:49–146.
31. Brain SD. Sensory neuropeptides in the skin. In: Geppetti P, Holzer P, eds. Neurogenic Inflammation. Boca Raton, New York, London, Tokyo: CRC Press, 1996; pp. 229–244.
32. Jancsó G. Sensory nerves as modulators of inflammatory reactions. In: Chahl LA, Szolcsányi J, Lembeck F, eds. Antidromic Vasodilatation and Neurogenic Inflammation. Budapest: Akadémiai Kiadó, 1984; pp. 207–222.
33. Jancsó G, Jancsó-Gábor A. Effect of capsaicin on morphine analgesia – possible involvement of hypothalamic structures. Naunyn Schmiedebergs Arch Pharmacol 1980;311:285–288.

34. Jancsó G, Király E, Jancsó-Gábor A. Direct evidence for an axonal site of action of capsaicin. Naunyn Schmiedebergs Arch Pharmacol 1980;313:91–94.
35. Caterina MJ, Schumacher MA, Tominaga M, Rosen TA, Levine JD, Julius D. The capsaicin receptor: a heat-activated ion channel in the pain pathway. Nature 1997;389:816–824.
36. Tominaga M, Caterina MJ, Malmberg AB, Rosen TA, Gilbert H, Skinner K et al. The cloned capsaicin receptor integrates multiple pain-producing stimuli. Neuron 1998;21:531–543.
37. Caterina MJ, Leffler A, Malmberg AB, Martin WJ, Trafton J, Petersen-Zeitz KR et al. Impaired nociception and pain sensation in mice lacking the capsaicin receptor. Science 2000;288:306–313.
38. Jancsó G. Chemosensitive primary sensory neurones: some morphological and functional characteristics. In: Broen AG, Réthelyi M, eds. Spinal Cord Sensation. Edinburgh: Scottish Academic Press, 1981; pp. 55–56.
39. Nagy JI, Iversen LL, Goedert M, Chapman D, Hunt SP. Dose-dependent effects of capsaicin on primary sensory neurons in the neonatal rat. J Neurosci 1983;3:399–406.
40. Scadding JW. The permanent anatomical effects of neonatal capsaicin on somatosensory nerves. J Anat 1980;131:471–482.
41. Jancsó G, Király E. Distribution of chemosensitive primary sensory afferents in the central nervous system of the rat. J Comp Neurol 1980;190:781–792.
42. Jancsó G, Maggi CA. Distribution of capsaicin-sensitive urinary bladder afferents in the rat spinal cord. Brain Res 1987;418:371–376.
43. Jancsó G, Király E, Such G, Joó F, Nagy A. Neurotoxic effect of capsaicin in mammals. Acta Physiol Hung 1987;69:295–313.
44. Holzer P. Capsaicin: cellular targets, mechanisms of action, and selectivity for thin sensory neurons. Pharmacol Rev 1991;43:143–201.
45. Holzer P, Schluet W, Lippe IT, Sametz W. Involvement of capsaicin-sensitive sensory neurons in gastrointestinal function. Acta Physiol Hung 1987;69:403–411.
46. Maggi CA, Meli A. The sensory-efferent function of capsaicin-sensitive sensory neurons. Gen Pharmacol 1988;19:1–43.
47. Baluk P. Neurogenic inflammation in skin and airways. J Investig Dermatol Symp Proc 1997;2:76–81.
48. Guo A, Vulchanova L, Wang J, Li X, Elde R. Immunocytochemical localization of the vanilloid receptor 1 (VR1): relationship to neuropeptides, the P2X3 purinoceptor and IB4 binding sites. Eur J Neurosci 1999;11:946–958.
49. Stander S, Moormann C, Schumacher M, Buddenkotte J, Artuc M, Shpacovitch V et al. Expression of vanilloid receptor subtype 1 in cutaneous sensory nerve fibers, mast cells, and epithelial cells of appendage structures. Exp Dermatol 2004;13:129–139.
50. Nolano M, Simone DA, Wendelschafer-Crabb G, Johnson T, Hazen E, Kennedy WR. Topical capsaicin in humans: parallel loss of epidermal nerve fibers and pain sensation. Pain 1999;83:135–145.
51. Simone DA, Nolano M, Johnson T, Wendelschafer-Crabb G, Kennedy WR. Intradermal injection of capsaicin in humans produces degeneration and subsequent reinnervation of epidermal nerve fibers: correlation with sensory function. J Neurosci 1998;18:8947–8959.
52. Sann H, McCarthy PW, Jancsó G, Pierau FK. RT97: a marker for capsaicin-insensitive sensory endings in the rat skin. Cell Tissue Res 1995;282:155–161.

53. Dux M, Sann H, Schemann M, Jancsó G. Changes in fibre populations of the rat hairy skin following selective chemodenervation by capsaicin. Cell Tissue Res 1999;296:471–477.
54. Jancsó N, Jancsó-Gábor A, Szolcsányi J. The role of sensory nerve endings in neurogenic inflammation induced in human skin and in the eye and paw of the rat. Br J Pharmacol 1968;33:32–41.
55. Jancsó G, Obál F, Jr., Tóth-Kása I, Katona M, Husz S. The modulation of cutaneous inflammatory reactions by peptide-containing sensory nerves. Int J Tissue React 1985;7:449–457.
56. Jancsó G, Király E, Jancsó-Gábor A. Chemosensitive pain fibres and inflammation. Int J Tissue React 1980;2:57–66.
57. Barthó L, Benkő R, Patacchini R, Pethő G, Holzer-Petsche U, Holzer P et al. Effects of capsaicin on visceral smooth muscle: a valuable tool for sensory neurotransmitter identification. Eur J Pharmacol 2004;500:143–157.
58. Barnes PJ, Belvisi MG, Rogers DF. Modulation of neurogenic inflammation: novel approaches to inflammatory disease. Trends Pharmacol Sci 1990;11:185–189.
59. Bari F, Jancsó G. Ruthenium red antagonism of capsaicin-induced vascular changes in the rat nasal mucosa. Eur Arch Otorhinolaryngol 1994;251:287–292.
60. Schultz HD, Roberts AM, Bratcher C, Coleridge HM, Coleridge JC, Davis B. Pulmonary C-fibers reflexly increase secretion by tracheal submucosal glands in dogs. J Appl Physiol 1985;58:907–910.
61. Baraniuk JN. Sensory, parasympathetic, and sympathetic neural influences in the nasal mucosa. J Allergy Clin Immunol 1992;90:1045–1050.
62. Smith PG, Liu M. Impaired cutaneous wound healing after sensory denervation in developing rats: effects on cell proliferation and apoptosis. Cell Tissue Res 2002;307:281–291.
63. Nilsson G, Alving K, Ahlstedt S. Effects on immune responses in rats after neuromanipulation with capsaicin. Int J Immunopharmacol 1991;13:21–26.
64. Csont T, Csonka C, Kovács P, Jancsó G, Ferdinándy P. Capsaicin-sensitive sensory neurons regulate myocardial nitric oxide and cGMP signaling. Eur J Pharmacol 2003;476:107–113.
65. Ferdinándy P, Csont T, Csonka C, Török M, Dux M, Németh J et al. Capsaicin-sensitive local sensory innervation is involved in pacing-induced preconditioning in rat hearts: role of nitric oxide and CGRP? Naunyn Schmiedebergs Arch Pharmacol 1997;356:356–363.
66. Katona M, Boros K, Sántha P, Ferdinándy P, Dux M, Jancsó G. Selective sensory denervation by capsaicin aggravates adriamycin-induced cardiomyopathy in rats. Naunyn Schmiedebergs Arch Pharmacol 2004;370:436–443.
67. Jancsó G. Intracisternal capsaicin: selective degeneration of chemosensitive primary sensory afferents in the adult rat. Neurosci Lett 1981;27:41–45.
68. Gamse R, Jancsó G, Király E. Intracisternal capsaicin: a novel approach for studying nociceptive sensory neurons. In: Chahl LA, Szolcsányi J, Lembeck F, eds. Neurogenic Inflammation and Antidromic Vasodilatation. Budapest: Akadémiai Kiadó, 1984; pp. 93–110.
69. Lembeck F. Sir Thomas Lewis's nocifensor system, histamine and substance-P-containing primary afferent nerves. Trends Neurosci 1983;6:106–108.
70. Lin Q, Zou X, Willis WD. Adelta and C primary afferents convey dorsal root reflexes after intradermal injection of capsaicin in rats. J Neurophysiol 2000;84:2695–2698.
71. Lin Q, Wu J, Willis WD. Dorsal root reflexes and cutaneous neurogenic inflammation after intradermal injection of capsaicin in rats. J Neurophysiol 1999;82:2602–2611.

72. Kenins P. Identification of the unmyelinated sensory nerves which evoke plasma extravasation in response to antidromic stimulation. Neurosci Lett 1981;25:137–141.
73. Janig W, Lisney SJ. Small diameter myelinated afferents produce vasodilatation but not plasma extravasation in rat skin. J Physiol 1989;415:477–486.
74. Lynn B, Schutterle S, Pierau FK. The vasodilator component of neurogenic inflammation is caused by a special subclass of heat-sensitive nociceptors in the skin of the pig. J Physiol (Lond) 1996;494:587–593.
75. Schmelz M, Michael K, Weidner C, Schmidt R, Torebjork HE, Handwerker HO. Which nerve fibers mediate the axon reflex flare in human skin? NeuroReport 2000;11:645–648.
76. Peitl B, Pethő G, Pórszász R, Németh J, Szolcsányi J. Capsaicin-insensitive sensory-efferent meningeal vasodilatation evoked by electrical stimulation of trigeminal nerve fibres in the rat. Br J Pharmacol 1999;127:457–467.
77. Dux M, Sántha P, Jancsó G. Capsaicin-sensitive neurogenic sensory vasodilatation in the dura mater of the rat. J Physiol 2003;552:859–867.
78. Lembeck F, Holzer P. Substance P as neurogenic mediator of antidromic vasodilation and neurogenic plasma extravasation. Naunyn Schmiedebergs Arch Pharmacol 1979;310:175–183.
79. Brain SD, Williams TJ, Tippins JR, Morris HR, MacIntyre I. Calcitonin gene-related peptide is a potent vasodilator. Nature 1985;313:54–56.
80. Louis SM, Jamieson A, Russell NJ, Dockray GJ. The role of substance P and calcitonin gene-related peptide in neurogenic plasma extravasation and vasodilatation in the rat. Neuroscience 1989;32:581–586.
81. Lembeck F, Donnerer J, Tsuchiya M, Nagahisa A. The non-peptide tachykinin antagonist, CP-96,345, is a potent inhibitor of neurogenic inflammation. Br J Pharmacol 1992;105:527–530.
82. Gamse R, Posch M, Saria A, Jancsó G. Several mediators appear to interact in neurogenic inflammation. Acta Physiol Hung 1987;69:343–354.
83. Stjarne P, Rinder J, Delay-Goyet P. Effects of NK1 receptor antagonists on vasodilation induced by chemical and electrical activation of sensory C-fibre afferents in different organs. Acta Physiol Scand 1992;152:153–161.
84. Andrews PV, Helme RD. Tachykinin-induced vasodilatation in rat skin measured with a laser-Doppler flowmeter: evidence for receptor-mediated effects. Regul Pept 1989;25:267–275.
85. Inoue H, Nagata N, Koshihara Y. Effect of the tachykinin receptor antagonists, SR 140333, FK 888, and SR 142801, on capsaicin-induced mouse ear oedema. Inflamm Res 1996;45:303–307.
86. Inoue H, Asaka T, Nagata N, Koshihara Y. Mechanism of mustard oil-induced skin inflammation in mice. Eur J Pharmacol 1997;333:231–240.
87. Bowden JJ, Garland AM, Baluk P, Lefevre P, Grady EF, Vigna SR, Bunnett NW, McDonald DM. Direct observation of substance P-induced internalization of neurokinin 1 (NK1) receptors at sites of inflammation. Proc Natl Acad Sci U S A 1994;91:8964–8968.
88. Ruocco I, Cuello AC, Shigemoto R, Ribeiro-da-Silva A. Light and electron microscopic study of the distribution of substance P-immunoreactive fibers and neurokinin-1 receptors in the skin of the rat lower lip. J Comp Neurol 2001;432:466–480.
89. Brain SD, Williams TJ. Inflammatory oedema induced by synergism between calcitonin gene related peptide (CGRP) and mediators of increased vascular permeability. Br J Pharmacol 1985;86:855–860.

90. Brain SD, Williams TJ. Substance P regulates the vasodilator activity of calcitonin gene related peptide. Nature 1988;335:73–75.
91. Erjavec F, Lembeck F, Florjanc-Irman T, Skofitsch G, Donnerer J, Saria A et al. Release of histamine by substance P. Naunyn Schmiedebergs Arch Pharmacol 1981;317:67–70.
92. Foreman JC. Substance P and calcitonin gene-related peptide: effects on mast cells and in human skin. Int Arch Allergy Appl Immunol 1987;82:366–371.
93. Bunker CB, Cerio R, Bull HA, Evans J, Dowd PM, Foreman JC. The effect of capsaicin application on mast cells in normal human skin. Agents Actions 1991;33:195–196.
94. Barnes PJ, Brown MJ, Dollery CT, Fuller RW, Heavey DJ, Ind PW. Histamine is released from skin by substance P but does not act as the final vasodilator in the axon reflex. Br J Pharmacol 1986;88:741–745.
95. Petersen LJ, Church MK, Skov PS. Histamine is released in the wheal but not the flare following challenge of human skin in vivo: a microdialysis study. Clin Exp Allergy 1997;27:284–295.
96. Petersen LJ, Winge K, Brodin E, Skov PS. No release of histamine and substance P in capsaicin-induced neurogenic inflammation in intact human skin in vivo: a microdialysis study. Clin Exp Allergy 1997;27:957–965.
97. Schmelz M, Petersen LJ. Neurogenic inflammation in human and rodent skin. News Physiol Sci 2001;16:33–37.
98. Kowalski ML, Sliwinska-Kowalska M, Kaliner MA. Neurogenic inflammation, vascular permeability, and mast cells. II. Additional evidence indicating that mast cells are not involved in neurogenic inflammation. J Immunol 1990;145:1214–1221.
99. Inoue H, Nagata N, Koshihara Y. Involvement of substance P as a mediator in capsaicin-induced mouse ear oedema. Inflamm Res 1995;44:470–474.
100. Inoue H, Nagata N, Koshihara Y. Participation of serotonin in capsaicin-induced mouse ear edema. Jpn J Pharmacol 1995;69:61–68.
101. Fromy B, Abraham P, Saumet JL. Non-nociceptive capsaicin-sensitive nerve terminal stimulation allows for an original vasodilatory reflex in the human skin. Brain Res 1998;811:166–168.
102. Fromy B, Merzeau S, Abraham P, Saumet JL. Mechanisms of the cutaneous vasodilator response to local external pressure application in rats: involvement of CGRP, neurokinins, prostaglandins and NO. Br J Pharmacol 2000;131:1161–1171.
103. Siguado-Roussel D, Demiot C, Fromy B, Koitka A, Leftheriotis G, Abraham P et al. Early endothelial dysfunction severely impairs skin blood flow response to local pressure application in streptozotocin-diabetic mice. Diabetes 2004;53:1564–1569.
104. Fizanne L, Siguado-Roussel D, Saumet JL, Fromy B. Evidence for the involvement of VPAC1 and VPAC2 receptors in pressure-induced vasodilatation in rodents. J Physiol 2004;554:519–528.
105. Garry A, Siguado-Roussel D, Merzeau S, Dumont O, Saumet JL, Fromy B. Cellular mechanisms underlying cutaneous pressure-induced vasodilation: in vivo involvement of potassium channels. Am J Physiol Heart Circ Physiol 2005;289:H174–H180.
106. Hughes SR, Brain SD. Nitric oxide-dependent release of vasodilatator quantities of calcitonin gene-related peptide from capsaicin-sensitive nerves in rabbit skin. Br J Pharmacol 1994;111:425–430.
107. Holzer P, Jocic M, Peskar BA. Mediation by prostaglandins of the nitric oxide-induced neurogenic vasodilatation in rat skin. Br J Pharmacol 1995;116:2365–2370.

108. Steinhoff M, Vergnolle N, Young SH, Tognetto M, Amadesi S, Ennes HS et al. Agonists of proteinase-activated receptor 2 induce inflammation by a neurogenic mechanism. Nat Med 2000;6:151–158.
109. de Garavilla L, Vergnolle N, Young SH, Ennes H, Steinhoff M, Ossovskaya VS et al. Agonists of proteinase-activated receptor 1 induce plasma extravasation by a neurogenic mechanism. Br J Pharmacol 2001;133:975–987.
110. Macfarlane SR, Seatter M, Kanke T, Hunter GD, Plevin R. Proteinase-activated receptors. Pharmacol Rev 2001;53:245–282.
111. Coughlin SR. Thrombin signalling and protease-activated receptors. Nature 2000;407:258–264.
112. Dery O, Corvera CU, Steinhoff M, Bunnett NW. Proteinase-activated receptors: novel mechanisms of signaling by serine proteases. Am J Physiol 1998;274: C1429–C1452.
113. Dai Y, Moriyama T, Higashi T, Togashi K, Kobayashi K, Yamanaka H et al. Proteinase-activated receptor 2-mediated potentiation of transient receptor potential vanilloid subfamily 1 activity reveals a mechanism for proteinase-induced inflammatory pain. J Neurosci 2004;24:4293–4299.
114. Amadesi S, Nie J, Vergnolle N, Cottrell GS, Grady EF, Trevisani M et al. Protease-activated receptor 2 sensitizes the capsaicin receptor transient receptor potential vanilloid receptor 1 to induce hyperalgesia. J Neurosci 2004;24:4300–4312.
115. Steinhoff M, Corvera CU, Thomas MS, Kong W, McAlpine BE, Caughey GH et al. Proteinase-activated receptor-2 in human skin: tissue distribution and activation of kertinocytes by mast cell tryptase. Exp Dermatol 1999;8:282–294.
116. Vergnolle N. Proneinase-activated receptor-2-activating peptides induced leukocyte rolling, adhesion, and extravasation in vivo. J Immunol 1999;163:5064–5069.
117. Seeliger S, Derian CK, Vergnolle N, Bunnett NW, Naworth R, Schmelz M et al. Proinflammatory role of proteinase-activated receptor-2 in humans and mice during cutaneous inflammation in vivo. FASEB J 2003;17:1871–1885.
118. Ständer S, Steinhoff M, Schmelz M, Weisshaar E, Metze D, Luger T. Neurophysiology of puritus: cutaneous elicitation of itch. Arch Dermatol 2003;39:1463–1470.
119. Steinhoff M, Neisius U, Ikoma A, Fartasch M, Heyer G, Skov PS et al. Proteinase-activated receptor-2 mediates itch: a novel pathway for pruritus in human skin. J Neurosci 2003;23:6176–6180.
120. Szállási Á, Blumberg PM, Nilsson S, Hökfelt T, Lundberg JM. Visualization by [3H]resiniferatoxin autoradiography of capsaicin-sensitive neurons in the rat, pig and man. Eur J Pharmacol 1994;264:217–221.
121. Sann H, Harti G, Pierau FK, Simon E. Effect of capsaicin upon afferent and efferent mechanisms of nociception and temperature regulation in birds. Can J Physiol Pharmacol 1987;65:1347–1354.
122. Jordt SE, Julius D. Molecular basis for species-specific sensitivity to "hot" chili peppers. Cell 2002;108:421–430.
123. Sann H, Friedrich R, Pierau FK. Substance P and calcitonin gene-related peptide in the chicken skin: distribution and cardiovascular effects. Neuropeptides 1986;30:273–281.
124. Hildesheim I, Jancsó G, Pierau F-K. Changes of chemical coding of pigeon DRG neurons after axotomy. Neurobiology 1996;4:147.
125. Sántha P, Pierau FK, Jancsó G. Inhibitory modulation of cutaneous vascular responses by endogenous galanin in the pigeon. Neurosci Lett 1999;273:64–66.

126. Sántha P, Pierau FK, Jancsó G. Evidence for an inhibition by endogenous galanin of neurogenic cutaneous vasodilatation in the pigeon. Neurosci Lett 1998;243:101–104.
127. Sántha P, Pierau FK, Jancsó G. Galanin mediated inhibitory nervous modulation of cutaneous vascular reactions. Acta Physiol Hung 1999;86:279–285.
128. Jancsó G, Sántha P, Horváth V, Pierau F-K. Inhibitory neurogenic modulation of histamine-induced cutaneous plasma extravasation in the pigeon. Regul Pept 2000;95:75–80.
129. Bánvölgyi Á, Pozsgai G, Brain SD, Helyes ZS, Szolcsányi J, Ghosh M et al. Mustard oil induces a transient receptor potential vaniloid 1 receptor-independent neurogenic inflammation and a non-neurogenic cellular inflammatory component in mice. Neuroscience 2004;125:449–459.
130. Grant AD, Pintér E, Salmon A-ML, Brain SD. An examination of neurogenic mechanisms involved in mustard oil-induced inflammation in the mouse. Eur J Pharmacol 2005;507:273–280.
131. Jordt SE, Bautista DM, Chuang HH, McKemy DD, Zygmunt PM, Hogestatt ED et al. Mustard oils and cannabinoids excite sensory nerve fibres through the TRP channel ANKTM1. Nature 2004;427:260–265.
132. Bautista DM, Jordt S-E, Nikai T, Tsuruda PR, Read AJ, Poblete J et al. TRPA1 mediates the inflammatory actions of environmental irritants and proalgesic agents. Cell 2006;124:1269–1282.
133. Kobayashi K, Fukuoka T, Obata K, Yamanaka H, Dai Y, Tokunaga A et al. Distinct expression of TRPM8, TRPA1, and TRPV1 mRNAs in rat primary afferent neurons with a-delta/c-fibers and colocalization with trk receptors. J Comp Neurol 2005;493:596–606.
134. Nagata K, Duggan A, Kumar G, Garcia-Anoveros J. Nociceptor and hair cell transducer properties of TRPA1, a channel for pain and hearing. J Neurosci 2005;25:4052–4061.
135. Tóth-Kása I, Katona M, Obál F, Jr.., Husz S, Jancsó G. Pathological reactions of human skin: involvement of sensory nerves. In: Chahl LA, Szolcsányi J, Lembeck F, eds. Antidromic Vasodilatation and Neurogenic Inflammation. Budapest: Akadémiai Kiadó, 1984; pp. 317–328.
136. Jancsó G, Ferencsik M, Such G, Király E, Nagy A, Bujdosó M. Morphological effects of capsaicin and its analogues in newborn and adult mammals. In: Hankanson R, Sundler F, eds. Tachykinin Antagonists. London: Elsevier, 1985; pp. 35–44.
137. Kress M, Guthmann C, Averbeck B, Reeh PW. Calcitonin gene-related peptide and prostaglandin E2 but not substance P release induced by antidromic nerve stimulation from rat skin in vitro. Neuroscience 1999;89:303–310.
138. Vasko MR, Campbell WB, Waite KJ. Prostaglandin E2 enhances bradykinin-stimulated release of neuropeptides from rat sensory neurons in culture. J Neurosci 1994;14:4987–4997.
139. Inoue A, Ikoma K, Morioka N, Kumagai K, Hashimoto T, Hide I et al. Interleukin-1beta induces substance P release from primary afferent neurons through the cyclooxygenase-2 system. J Neurochem 1999;73:2206–2213.
140. Gibbs JL, Flores CM, Hargreaves KM. Attenuation of capsaicin-evoked mechanical allodynia by peripheral neuropeptide Y Y1 receptors. Pain 2006;124:167–174.
141. Price TJ, Louria MD, Candelario-Soto D, Dussor GO, Jeske NA, Patwardhan AM et al. Treatment of trigeminal ganglion neurons in vitro with NGF, GDNF or BDNF: effects on neuronal survival, neurochemical properties and TRPV1-mediated neuropeptide secretion. BMC Neurosci 2005;24:4.

142. Domoki F, Sántha P, Bari F, Jancsó G. Perineural capsaicin treatment attenuates reactive hyperaemia in the rat skin. Neurosci Lett 2003;341:127–130.
143. Lembeck F, Donnerer J. Postocclusive cutaneous vasodilatation mediated by substance P. Naunyn Schmiedebergs Arch Pharmacol 1981;316:165–171.
144. Lindsay RM, Harmar AJ. Nerve growth factor regulates expression of neuropeptide genes in adult sensory neurons. Nature 1989;337:362–364.
145. Donnerer J, Schuligoi R, Stein C, Amann R. Upregulation, release and axonal transport of substance P and calcitonin gene-related peptide in adjuvant inflammation and regulatory function of nerve growth factor. Regul Pept 1993;46:150–154.
146. Donnerer J, Schuligoi R, Stein C. Increased content and transport of substance P and calcitonin gene-related peptide in sensory nerves innervating inflamed tissue: evidence for a regulatory function of nerve growth factor in vivo. Neuroscience 1992;49:693–698.
147. Stucky CL, Koltzenburg M, Schneider M, Engle MG, Albers KM, Davis BM. Overexpression of nerve growth factor in skin selectively affects the survival and functional properties of nociceptors. J Neurosci 1999;19:8509–8516.
148. Koltzenburg M, Bennett DL, Shelton DL, McMahon SB. Neutralization of endogenous NGF prevents the sensitization of nociceptors supplying skin. Eur J Neurosci 1999;11:1698–1704.
149. Bowles WR, Sabino M, Harding-Rose C, Hargreaves KM. Chronic nerve growth factor administration increases the peripheral exocytotic activity of capsaicin-sensitive cutaneous neurons. Neurosci Lett 2006;403:305–508.
150. Dallos A, Kiss M, Polyanka H, Dobozy A, Kemény L, Husz S. Effects of the neuropeptides substance P, calcitonin gene-related peptide, vasoactive intestinal polypeptide and galanin on the production of nerve growth factor and inflammatory cytokines in cultured human keratinocytes. Neuropeptides 2006;40:251–263.
151. Alving K, Matran R, Lundberg JM. Capsaicin-induced local effector responses, autonomic reflexes and sensory neuropeptide depletion in the pig. Naunyn Schmiedebergs Arch Pharmacol 1991;343:37–45.
152. Pierau FK, Szolcsányi J. Neurogenic inflammation: axon reflex in pigs. Agents Actions 1989;26:231–232.
153. Jancsó G, Pierau FK, Sann H. Mustard oil-induced cutaneous inflammation in the pig. Agents Actions 1993;39:31–34.
154. Lundberg JM, Brodin E, Hua X, Saria A. Vascular permeability changes and smooth muscle contraction in relation to capsaicin-sensitive substance P afferents in the guinea-pig. Acta Physiol Scand 1984;120:217–227.
155. Lynn B, Shakhanbeh J. Neurogenic inflammation in the skin of the rabbit. Agents Actions 1988;25:228–230.
156. Lynn B, Shakhanbeh J. Substance P content of the skin, neurogenic inflammation and numbers of C-fibres following capsaicin application to a cutaneous nerve in the rabbit. Neuroscience 1988;24:769–775.
157. Pierau FK, Ernst R, Sann H, Barthó L. Endogenous opioids and ruthenium red inhibit the flare reaction in the pig skin by different mechanisms. Ann N Y Acad Sci 1991;632:460–463.
158. Carter RB. Topical capsaicin in the treatment of cutaneous disorders. Drug Dev Res 1991;22:109–123.
159. Tóth-Kása I, Jancsó G, Bognár A, Husz S, Obál F, Jr. Capsaicin prevents histamine-induced itching. Int J Clin Pharmacol Res 1986;6:163–169.

160. Tóth-Kása I, Jancsó G, Obál F, Jr., Husz S, Simon N. Involvement of sensory nerve endings in cold and heat urticaria. J Invest Dermatol 1983;80:34–36.
161. Dux M, Jancsó G. A new technique for the direct demonstration of overlapping cutaneous innervation territories of peptidergic C-fibre afferents of rat hindlimb nerves. J Neurosci Methods 1994;55:47–52.
162. Vass Z, Steyger PS, Hordichok AJ, Trune DR, Jancsó G, Nuttall AL. Capsaicin stimulation of the cochlea and electric stimulation of the trigeminal ganglion mediate vascular permeability in chochlear and vertebro-basilar arteries: a potential cause of inner ear dysfunction in headache. Neuroscience 2001;103:189–201.
163. Wallengren J, Moller H. The effect of capsaicin on some experimental inflammations in human skin. Acta Derm Venereol 1986;66:375–380.
164. Smith CH, Barker JN, Morris RW, MacDonald DM, Lee TH. Neuropeptides induce rapid expression of endothelial cell adhesion molecules and elicit granulocytic infiltration in human skin. J Immunol 1993;151:3274–3282.
165. Quinlan KL, Song IS, Naik SM, Letran EL, Olerud JE, Bunnett NW et al. VCAM-1 expression on human dermal microvascular endothelial cells is directly and specifically up-regulated by substance P. J Immunol 1999;162:1656–1661.
166. Lang E, Novak A, Reeh PW, Handwerker HO. Chemosensitivity of fine afferents from rat skin in vitro. J Neurophysiol 1990;63:887–901.
167. Lischetzki G, Rukwied R, Handwerker HO, Schmelz M. Nociceptor activation and protein extravasation induced by inflammatory mediators in human skin. Eur J Pain 2001;5:49–57.
168. Juan H, Lembeck F. Action of peptides and other algesic agents on paravascular pain receptors of the isolated perfused rabbit ear. Naunyn Schmiedebergs Arch Pharmacol 1974;283:151–164.
169. Beck PW, Handwerker HO. Bradykinin and serotonin effects on various types of cutaneous nerve fibers. Pflugers Arch 1974;347:209–222.
170. Steen KH, Issberner U, Reeh PW. Pain due to experimental acidosis in human skin: evidence for non-adapting nociceptor excitation. Neurosci Lett 1995;199:29–32.
171. Ugawa S, Ueda T, Ishida Y, Nishigaki M, Shibata Y, Shimada S. Amiloride-blockable acid-sensing ion channels are leading acid sensors expressed in human nociceptors. J Clin Invest 2002;110:1185–1190.
172. Chancellor-Freeland C, Zhu GF, Kage R, Beller DI, Leeman SE, Black PH. Substance P and stress-induced changes in macrophages. Ann N Y Acad Sci 1995;771:472–484.
173. Colpaert FC, Donnerer J, Lembeck F. Effects of capsaicin on inflammation and on the substance P content of nervous tissues in rats with adjuvant arthritis. Life Sci 1983;32:1827–1834.
174. Donaldson LF, McQueen DS, Seckl JR. Neuropeptide gene expression and capsaicin-sensitive primary afferents: maintenance and spread of adjuvant arthritis in the rat. J Physiol 1995;486(Pt 2):473–482.
175. Germonpre PR, Joos GF, Pauwels RA. Characterization of the neurogenic plasma extravasation in the airways. Arch Int Pharmacodyn Ther 1995;329:185–203.
176. Lammers JW, Barnes PJ, Chung KF. Nonadrenergic, noncholinergic airway inhibitory nerves. Eur Respir J 1992;5:239–246.
177. Holzer P, Lippe II, Raybould HE, Pabst MA, Livingston EH, Amann R et al. Role of peptidergic sensory neurons in gastric mucosal blood flow and protection. Ann N Y Acad Sci 1991;632:272–282.

178. Skofitsch G, Jacobowitz DM. Galanin-like immunoreactivity in capsaicin sensitive sensory neurons and ganglia. Brain Res Bull 1985;15:191–195.
179. Ju G, Hökfelt T, Brodin E, Fahrenkrug J, Fischer JA, Frey P et al. Primary sensory neurons of the rat showing calcitonin gene-related peptide immunoreactivity and their relation to substance P-, somatostatin-, galanin-, vasoactive intestinal poly peptide- and cholecystokinin-immunoreactive ganglion cells. Cell Tissue Res 1987;247:417–431.
180. Hökfelt T, Skirboll L, Dalsgaard C-J, Johansson O, Lundberg JM, Norell G et al. Peptide neurons in the spinal cord with special reference to descending systems. In: Sjülund B, Bjürklund A, eds. Brainstem Control of Spinal Mechanisms. Amsterdam: Elsevier, 1982; pp. 89–117.
181. Hökfelt T, Elde R, Johansson O, Luft R, Nilsson G, Arimura A. Immunohistochemical evidence for separate populations of somatostatin-containing and substance P-containing primary afferent neurons in the rat. Neuroscience 1976;1:131–136.
182. Hökfelt T, Vincent S, Dalsgaard CJ, Skirboll L, Johansson O, Schultzberg M et al. Distribution of substance P in brain and periphery and its possible role as a co-transmitter. Ciba Found Symp 1982;84–106.
183. Johansson O, Vaalasti A. Immunohistochemical evidence for the presence of somatostatin-containing sensory nerve fibres in the human skin. Neurosci Lett 1987;73:225–230.
184. Jimenez-Andrade JM, Zhou S, Yamani A, Valencia de Ita S, Castaneda-Hernandez G, Carlton SM. Mechanism by which peripheral galanin increases acute inflammatory pain. Brain Res 2005;1056:113–117.
185. Jimenez-Andrade JM, Lundstrom L, Sollenberg UE, Langel U, Castaneda-Hernandez G, Carlton SM. Activation of peripheral galanin receptors: differential effects on nociception. Pharmacol Biochem Behav 2006;85:273–280.
186. Ch'ng JL, Christofides ND, Anand P, Gibson SJ, Allen YS, Su HC et al. Distribution of galanin immunoreactivity in the central nervous system and the responses of galanin-containing neuronal pathways to injury. Neuroscience 1985;16:343–354.
187. Gamse R, Petsche U, Lembeck F, Jancsó G. Capsaicin applied to peripheral nerve inhibits axoplasmic transport of substance P and somatostatin. Brain Res 1982;239: 447–462.
188. Noguchi K, De Leon M, Nahin RL, Senba E, Ruda MA. Quantification of axotomy-induced alteration of neuropeptide mRNAs in dorsal root ganglion neurons with special reference to neuropeptide Y mRNA and the effects of neonatal capsaicin treatment. J Neurosci Res 1993;35:54–66.
189. Michael GJ, Priestley JV. Differential expression of the mRNA for the vanilloid receptor subtype 1 in cells of the adult rat dorsal root and nodose ganglia and its downregulation by axotomy. J Neurosci 1999;19:1844–1854.
190. Xu XJ, Hao JX, Wiesenfeld-Hallin Z, Hakanson R, Folkers K, Hökfelt T. Spantide II, a novel tachykinin antagonist, and galanin inhibit plasma extravasation induced by antidromic C-fiber stimulation in rat hindpaw. Neuroscience 1991;42:731–737.
191. Green PG, Basbaum AI, Levine JD. Sensory neuropeptide interactions in the production of plasma extravasation in the rat. Neuroscience 1992;50:745–749.
192. Pireau F-K, Sann H, Harti G, Gamse R. Neuropeptides in sensory neurones of pigeon and the insensitivity of avians to capsaicin. In: Schmidt RF, Schaible HG, Vahle-Hinz C, eds. Fine Afferent Fibres and Pain. Weinheim: VCH, 1987; pp. 215–223.
193. Gentle MJ, Hunter LN. Neurogenic inflammation in the chicken (Gallus gallus var domesticus). Comp Biochem Physiol C 1993;105:459–462.

194. Pierau F-K, Sann H, Harti G, Gamse R. Neuropeptides in sensory neurones of pigeon and the insensitivity of avians to capsaicin. In: Schmidt RF, Schaible HG, Vahle-Hinz C, eds. Fine Afferent Fibers and Pain. Weinheim: VCH, 1987; pp. 215–223.
195. Castrignano F, De Stefano ME, Leone F, Mulatero B, Tata AM, Fasolo A et al. Ontogeny of acetylcholinesterase, substance P and calcitonin gene-related peptide-like immunoreactivity in chick dorsal root ganglia. Neuroscience 1990;34:499–510.
196. Harti G, Sharkey KA, Pierau F-K. Effects of capsaicin in rat and pigeon on peripheral nerves containing substance P and calcitonin gene-related peptide. Cell Tissue Res 1989;256:465–474.
197. Ogren SO, Hökfelt T, Kask K, Langel U, Bartfai T. Evidence for a role of the neuropeptide galanin in spatial learning. Neuroscience 1992;51:1–5.
198. Holmberg K, Kuteeva E, Brumovsky P, Kahl U, Kalstrom H, Lucas GA et al. Generation and phenotypic characterization of a galanin overexpressing mouse. Neuroscience 2005;133:59–77.
199. Schmidhuber SM, Santic R, Tam CW, Bauer JW, Kofler B, Brain SD. Galanin-like peptides exert potent vasoactive functions in vivo. J Invest Dermatol 2007;127: 716–721.
200. Ji RR, Zhang X, Zhang Q, Dagerlind A, Nilsson S, Wiesenfeld-Hallin Z et al. Central and peripheral expression of galanin in response to inflammation. Neuroscience 1995;68:563–576.
201. Kofler B, Berger A, Santic R, Moritz K, Almer D, Tuechler C et al. Expression of neuropeptide galanin and receptors in human skin. J Invest Dermatol 2004;122:1050–1053.
202. Baumann MA, Korkmaz Y, Bloch W, Schmidt A, Addicks K, Schroder H. Localization of the neuropeptide galanin in nerve fibers and epithelial keratinocytes of the rat molar gingiva. Eur J Oral Sci 2003;111:175–178.
203. Klein CM, Westlund KN, Coggeshall RE. Percentages of dorsal root axons immunoreactive for galanin are higher than those immunoreactive for calcitonin gene-related peptide in the rat. Brain Res 1990;519:97–101.
204. Heppelmann B, Just S, Pawlak M. Galanin influences the mechanosensitivity of sensory endings in the rat knee joint. Eur J Neurosci 2000;12:1567–1572.
205. Flatters SJ, Fox AJ, Dickenson AH. In vivo and in vitro effects of peripheral galanin on nociceptive transmission in naive and neuropathic states. Neuroscience 2003;116:1005–1012.
206. Kerekes N, Mennicken F, O'Donnell D, Hökfelt T, Hill RH. Galanin increases membrane excitability and enhances Ca(2+) currents in adult, dissociated dorsal root ganglion neurons. Eur J Neurosci 2003;18:2957–2966.
207. Lawson SN. Neuropeptides in morphologically and functionally identified primary afferent neurons in dorsal root ganglia: substance P, CGRP and somatostatin. Prog Brain Res 1995;104:161–173.
208. Kashiba H, Ueda Y, Senba E. Systemic capsaicin in the adult rat differentially affects gene expression for neuropeptides and neurotrophin receptors in primary sensory neurons. Neuroscience 1997;76:299–312.
209. Perry MJ, Lawson SN. Differences in expression of oligosaccharides, neuropeptides, carbonic anhydrase and neurofilament in rat primary afferent neurons retrogradely labelled via skin, muscle or visceral nerves. Neuroscience 1998;85:293–310.

210. Gamse R, Lackner D, Gamse G, Leeman SE. Effect of capsaicin pretreatment on capsaicin-evoked release of immunoreactive somatostatin and substance P from primary sensory neurons. Naunyn Schmiedebergs Arch Pharmacol 1981;316:38–41.
211. Gazelius B, Brodin E, Olgart L, Panopoulos P. Evidence that substance P is a mediator of antidromic vasodilatation using somatostatin as a release inhibitor. Acta Physiol Scand 1981;113:155–159.
212. Lembeck F, Donnerer J, Barthó L. Inhibition of neurogenic vasodilation and plasma extravasation by substance P antagonists, somatostatin and [D-Met2, Pro5]enkephalinamide. Eur J Pharmacol 1982;85:171–176.
213. Gillies HC. Theory and Practice of Counter-irritation. Oxford: MacMillan, 1896.
214. Green PG, Miao FJ, Janig W, Levine JD. Negative feedback neuroendocrine control of the inflammatory response in rats. J Neurosci 1995;15:4678–4686.
215. Pintér E, Szolcsányi J. Systemic anti-inflammatory effect induced by antidromic stimulation of the dorsal roots in the rat. Neurosci Lett 1996;212:33–36.
216. Szolcsányi J, Pintér E, Helyes Z, Oroszi G, Németh J. Systemic anti-inflammatory effect induced by counter-irritation through a local release of somatostatin from nociceptors. Br J Pharmacol 1998;125:916–922.
217. Szolcsányi J, Helyes Z, Oroszi G, Németh J, Pintér E. Release of somatostatin and its role in the mediation of the anti-inflammatory effect induced by antidromic stimulation of sensory fibres of rat sciatic nerve. Br J Pharmacol 1998;123:936–942.
218. Baker JR, Kemmenoe BH, McMartin C, Peters GE. Pharmacokinetics, distribution and elimination of a synthetic octapeptide analogue of somatostatin in the rat. Regul Pept 1984;9:213–226.
219. Schoemaker RC, Burggraaf J, Cohen AF. Assessment of hepatic blood flow using continuous infusion of high clearance drugs. Br J Clin Pharmacol 1998;45:463–469.
220. Gamse R, Jancsó G. Reduced neurogenic inflammation in streptozotocin-diabetic rats due to microvascular changes but not to substance P depletion. Eur J Pharmacol 1985;118:175–180.
221. Brewster WJ, Diemel LT, Leach RM, Tomlinson DR. Reduced sciatic nerve substance P and calcitonin gene-related peptide in rats with short-term diabetes or central hypoxaemia co-exist with normal messenger RNA levels in the lumbar dorsal root ganglia. Neuroscience 1994;58:323–330.
222. Robinson JP, Willars GB, Tomlinson DR, Keen P. Axonal transport and tissue contents of substance P in rats with long-term streptozotocin-diabetes. Effects of the aldose reductase inhibitor 'statil'. Brain Res 1987;426:339–348.
223. Levy DM, Karanth SS, Springall DR, Polak JM. Depletion of cutaneous nerves and neuropeptides in diabetes mellitus: an immunocytochemical study. Diabetologia 1989;32:427–433.
224. Jancsó G, Janka Z. A simple test for topographical diagnosis of sensory nervous system lesions. Eur Neurol 1981;20:84–87.
225. Jancsó G, Husz S, Simon N. Impairment of axon reflex vasodilatation after herpes zoster. Clin Exp Dermatol 1983;8:27–31.
226. Heyer G, Koppert W, Martus P, Handwerker HO. Histamine and cutaneous nociception: histamine-induced responses in patients with atopic eczema, psoriasis and urticaria. Acta Derm Venereol 1998;78:123–126.
227. Heyer G, Hornstein OP, Handwerker HO. Reactions to intradermally injected substance P and topically applied mustard oil in atopic dermatitis patients. Acta Derm Venereol 1991;71:291–295.

228. Westerman RA, Low A, Pratt A, Hutchinson JS, Szolcsányi J, Magerl W et al. Electrically evoked skin vasodilatation: a quantitative test of nociceptor function in man. Clin Exp Neurol 1987;23:81–89.
229. Kramer HH, Schmelz M, Birklein F, Bickel A. Electrically stimulated axon reflexes are diminished in diabetic small fiber neuropathies. Diabetes 2004;53:769–774.
230. Hutchison KJ, Overton TR, Biltek KB, Nixon R, Williams HT. Skin blood flow during histamine flare using the clearance of epicutaneous applied Xenon-133 in diabetic and non-diabetic subjects. Angiology 1983;34:223–230.
231. Stahle-Backdahl M, Hagermark O, Lins LE. The sensitivity of uremic and normal human skin to histamine. Acta Derm Venereol 1988;68:230–235.
232. Jancsó G, Obál F, Jr., Tóth-Kása I, Katona M, Husz S. The modulation of cutaneous inflammatory reactions by peptide-containing sensory nerves. Int J Tissue React 1985;7:449–457.
233. Hou M, Uddman R, Tajti J, Kanje M, Edvinsson L. Capsaicin receptor immunoreactivity in the human trigeminal ganglion. Neurosci Lett 2002;330:223–226.
234. Cortright DN, Crandall M, Sanchez JF, Zou T, Krause JE, White G. The tissue distribution and functional characterization of human VR1. Biochem Biophys Res Commun 2001;281:1183–1189.
235. Maggi CA, Borsini F, Santicioli P, Abelli L, Evangelista S, Manyini S et al. Cutaneous lesions in capsaicin-pretreated rats. A trophic role of capsaicin-sensitive afferents? Naunyn Schmiedebergs Arch Pharmacol 1987;336:538–545.
236. Winkelmann RK. Nerve Endings in Normal and Pathological Skin. Springfield, Illinois: C. C. Thomas, 1960.
237. Stucker M, Hugler P, von Kobyletzki G, Reuther T, Hoffmann K, Laubenthal H et al. Intracutaneous histamine injection can detect damage of cutaneous afferent fibers in postherpetic neuralgia. Dermatology 1997;195:311–316.
238. Morris GC, Gibson SJ, Helme RD. Capsaicin-induced flare and vasodilatation in patients with post-herpetic neuralgia. Pain 1995;63:93–101.
239. Glinski W, Glinska-Ferenz M, Pierozynska-Dubowska M. Neurogenic inflammation induced by capsaicin in patients with psoriasis. Acta Derm Venereol 1991;71:51–54.
240. Pergolizzi S, Vaccaro M, Magaudda L, Mondello MR, Arco A, Bramanti P et al. Immunohistochemical study of epidermal nerve fibres in involved and uninvolved psoriatic skin using confocal laser scanning microscopy. Arch Dermatol Res 1998;290:483–489.
241. Jiang WY, Raychaudhuri SP, Farber EM. Double-labeled immunofluorescence study of cutaneous nerves in psoriasis. Int J Dermatol 1998;37:572–574.
242. Raychaudhuri SP, Dutt S, Raychaudhuri SK, Sanyal M, Farber EM. Severe combined immunodeficiency mouse-human skin chimeras: a unique animal model for the study of psoriasis and cutaneous inflammation. Br J Dermatol 2001;144:931–939.
243. Raychaudhuri SP, Raychaudhuri SK. Role of NGF and neurogenic inflammation in the pathogenesis of psoriasis. Prog Brain Res 2004;146:433–437.
244. Farber EM, Lanigan SW, Boer J. The role of cutaneous sensory nerves in the maintenance of psoriasis. Int J Dermatol 1990;29:418–420.
245. Bernstein JE, Parish LC, Rapaprort M, Rosenbaum MM, Roenigk HH Jr. Effects of topically applied capsaicin on moderate and severe psoriasis vulgaris. J Am Acad Dermatol 1986;15:504–507.

246. Ellis CN, Berberian B, Sulica VI, Dodd WA, Jarratt MT, Katz HI et al. A double-blind evaluation of topical capsaicin in pruritic psoriasis. J Am Acad Dermatol 1993;29:438–442.
247. Liang Y, Marcusson JA, Jacobi HH, Haak-Frendscho M, Johansson O. Histamine- containing mast cells and their relationship to NGFr-immunoreactive nerves in prurigo nodularis: a reappraisal. J Cutan Pathol 1998;25:189–198.
248. Johansson O, Liang Y, Emtestam L. Increased nerve growth factor- and tyrosine kinase A-like immunoreactivities in prurigo nodularis skin – an exploration of the cause of neurohyperplasia. Arch Dermatol Res 2002;293:614–619.
249. Liang Y, Marcusson JA, Johansson O. Light and electron microscopic immunohistochemical observations of p75 nerve growth factor receptor-immunoreactive dermal nerves in prurigo nodularis. Arch Dermatol Res 1999;291:14–21.
250. Husz S, Tóth-Kása I, Obál F, Jancsó G. A possible pathomechanism of the idiopathic cold contact urticaria. Acta Physiol Hung 1991;77:209–215.
251. Ostlere LS, Cowen T, Rustin MH. Neuropeptides in the skin of patients with atopic dermatitis. Clin Exp Dermatol 1995;20:462–467.
252. Toyoda M, Makino T, Kagoura M, Morohashi M. Expression of neuropeptide-degrading enzymes in alopecia areata: an immunohistochemical study. Br J Dermatol 2001;144:46–54.
253. Worrell JT, Cockerell CJ. Histopathology of peripheral nerves in cutaneous herpesvirus infection. Am J Dermatopathol 1997;19:133–137.
254. Springall DR, Karanth SS, Kirkham N, Darley CR, Polak JM. Symptoms of notalgia paresthetica may be explained by increased dermal innervation. J Invest Dermatol 1991;97:555–561.
255. Inaloz HS, Kirtak N, Erguven HG, Karakok M, Inaloz SS. Notalgia paresthetica with a significant increase in the number of intradermal nerves. J Dermatol 2002;29:739–743.
256. Jancsó G, Lynn B. Possible use of capsaicin in pain therapy. Clin J Pain 1987;3:123–126.
257. Walker FO, Lewis SF. Some esthetic and electrophysiologic effects of topical 0.025% capsaicin in man. Reg Anesth 1990;15:61–66.
258. Simone DA, Ochoa J. Early and late effects of prolonged topical capsaicin on cutaneous sensibility and neurogenic vasodilatation in humans. Pain 1991;47:285–294.
259. Orosz K, Jancsó G, Dux M, Erős I. Cutaneous innervation and vascular reactions after topical capsaicin treatment. Eur J Pharm Sci 1998;6(Suppl 1):S55.
260. McMahon SB, Lewin G, Bloom SR. The consequences of long-term topical capsaicin application in the rat. Pain 1991;44:301–310.
261. Carpenter SE, Lynn B. Vascular and sensory responses of human skin to mild injury after topical treatment with capsaicin. Br J Pharmacol 1981;73:755–758.
262. Lynn B, Ye W, Cotsell B. The actions of capsaicin applied topically to the skin of the rat on C-fibre afferents, antidromic vasodilatation and substance P levels. Br J Pharmacol 1992;107:400–406.
263. Bjerring P, Arendt-Nielsen L. Use of a new argon laser technique to evaluate changes in sensory and pain thresholds in human skin following topical capsaicin treatment. Skin Pharmacol 1989;2:162–167.
264. Carrillo P, Camacho M, Manzo J, Martinez-Gomez M, Salas M, Pacheco P. Cutaneous wounds produced by capsaicin treatment of newborn rats are due to trophic disturbances. Neurotoxicol Teratol 1998;20:75–81.

265. Gamse R, Leeman SE, Holzer P, Lembeck F. Differential effects of capsaicin on the content of somatostatin, substance P, and neurotensin in the nervous system of the rat. Naunyn Schmiedebergs Arch Pharmacol 1981;317:140–148.
266. Bernstein JE, Korman NJ, Bickers DR, Dahl MV, Millikan LE. Topical capsaicin treatment of chronic postherpetic neuralgia. J Am Acad Dermatol 1989;21:265–270.
267. Johnson RW, Whitton TL. Management of herpes zoster (shingles) and postherpetic neuralgia. Expert Opin Pharmacother 2004;5:551–559.
268. Szállási Á. Vanilloid receptor ligands: hopes and realities for the future. Drugs Aging 2001;18:561–573.
269. Rains C, Bryson HM. Topical capsaicin. A review of its pharmacological properties and therapeutic potential in post-herpetic neuralgia, diabetic neuropathy and osteoarthritis. Drugs Aging 1995;7:317–328.
270. Mathias BJ, Dillingham TR, Zeigler DN, Chang AS, Belandres PV. Topical capsaicin for chronic neck pain. A pilot study. Am J Phys Med Rehabil 1995;74:39–44.
271. Hautkappe M, Roizen MF, Toledano A, Roth S, Jeffries JA, Ostermeier AM. Review of the effectiveness of capsaicin for painful cutaneous disorders and neural dysfunction. Clin J Pain 1998;14:97–106.
272. Dini D, Bertelli G, Gozza A, Forno GG. Treatment of the post-mastectomy pain syndrome with topical capsaicin. Pain 1993;54:223–226.
273. Watson CP, Evans RJ, Watt VR. The post-mastectomy pain syndrome and the effect of topical capsaicin. Pain 1989;38:177–186.
274. Weintraub M, Golik A, Rubio A. Capsaicin for treatment of post-traumatic amputation stump pain. Lancet 1990;336:1003–1004.
275. Rayner HC, Atkins RC, Westerman RA. Relief of local stump pain by capsaicin cream. Lancet 1989;2:1276–1277.
276. Spruce MC, Potter J, Coppini DV. The pathogenesis and management of painful diabetic neuropathy: a review. Diabet Med 2003;20:88–98.
277. Robbins W. Clinical applications of capsaicinoids. Clin J Pain 2000;16:S86–S89.
278. Tandan R, Lewis GA, Krusinski PB, Badger GB, Fries TJ. Topical capsaicin in painful diabetic neuropathy. Controlled study with long-term follow-up. Diabetes Care 1992;15:8–14.
279. Tandan R, Lewis GA, Badger GB, Fries TJ. Topical capsaicin in painful diabetic neuropathy. Effect on sensory function. Diabetes Care 1992;15:15–18.
280. Scheffler NM, Sheitel PL, Lipton MN. Treatment of painful diabetic neuropathy with capsaicin 0.075%. J Am Podiatr Med Assoc 1991;81:288–293.
281. Levy DM, Abraham RR, Tomlinson DR. Topical capsaicin in the treatment of painful diabetic neuropathy. N Engl J Med 1991;324:776–777.
282. Ross DR, Varipapa RJ. Treatment of painful diabetic neuropathy with topical capsaicin. N Engl J Med 1989;321:474–475.
283. Wallengren J, Klinker M. Successful treatment of notalgia paresthetica with topical capsaicin: vehicle-controlled, double-blind, crossover study. J Am Acad Dermatol 1995;32:287–289.
284. Wist E, Risberg T. Topical capsaicin in treatment of hyperalgesia, allodynia and dysesthetic pain caused by malignant tumour infiltration of the skin. Acta Oncol 1993;32:343.
285. Reimann S, Luger T, Metze D. Topical administration of capsaicin in dermatology for treatment of itching and pain. Hautarzt 2000;51:164–172.

286. Kim DH, Ahn BO, Kim SH, Kim WB. Antipruritic effect of DA-5018, a capsaicin derivative, in mice. Arch Pharmacol Res 1999;22:549–553.
287. Jancsó G, Karcsu S, Kiraly E, Szebeni A, Toth L, Bacsy E et al. Neurotoxin induced nerve cell degeneration: possible involvement of calcium. Brain Res 1984;295:211–216.
288. Sugimoto T, Xiao C, Ichikawa H. Neonatal primary neuronal death induced by capsaicin and axotomy involves an apoptotic mechanism. Brain Res 1998;807:147–154.
289. Hiura A, Nakae Y, Nakagawa H. Cell death of primary afferent nerve cells in neonatal mice treated with capsaicin. Anat Sci Int 2002;77:47–50.
290. Ritter S, Dinh TT. Age-related changes in capsaicin-induced degeneration in rat brain. J Comp Neurol 1992;318:103–116.
291. Ritter S, Dinh TT. Capsaicin-induced neuronal degeneration: silver impregnation of cell bodies, axons, and terminals in the central nervous system of the adult rat. J Comp Neurol 1988;271:79–90.
292. Jancsó G, Kiraly E, Joo F, Such G, Nagy A. Selective degeneration by capsaicin of a subpopulation of primary sensory neurons in the adult rat. Neurosci Lett 1985;59:209–214.
293. Jancsó G, Ambrus A, Török T. Neurotoxic effects of resiniferatoxin on rodent and avian primary sensory neurones. Neuropeptides 1992;22:34.
294. Pan HL, Khan GM, Alloway KD, Chen SR. Resiniferatoxin induces paradoxical changes in thermal and mechanical sensitivities in rats: mechanism of action. J Neurosci 2003;23:2911–2919.
295. Szállási Á, Szállási Z, Blumberg PM. Permanent effects of neonatally administered resiniferatoxin in the rat. Brain Res 1990;537:182–186.
296. Jancsó G, Savay G, Király E. Appearance of histochemically detectable ionic calcium in degenerating primary sensory neurons. Acta Histochem 1978;62:165–169.
297. Jancsó G, Lawson SN. Transganglionic degeneration of capsaicin-sensitive C-fiber primary afferent terminals. Neuroscience 1990;39:501–511.
298. Miller MS, Buck SH, Sipes IG, Burks TF. Altered axoplasmic transport of nerve growth factor by capsaicin. Proc West Pharmacol Soc 1982;25:87–88.
299. Schicho R, Skofitsch G, Donnerer J. Regenerative effect of human recombinant NGF on capsaicin-lesioned sensory neurons in the adult rat. Brain Res 1999;815:60–69.
300. Malmberg AB, Mizisin AP, Calcutt NA, von Stein T, Robbins WR, Bley KR. Reduced heat sensitivity and epidermal nerve fiber immunostaining following single applications of a high-concentration capsaicin patch. Pain 2004;111:360–367.
301. Jancsó G, Domoki F, Sántha P, Varga J, Fischer J, Orosz K et al. Beta-amyloid (1-42) peptide impairs blood-brain barrier function after intracarotid infusion in rats. Neurosci Lett 1998;253:139–141.
302. Klukovits A, Gáspár R, Sántha P, Jancsó G, Falkay G. Role of capsaicin-sensitive nerve fibers in uterine contractility in the rat. Biol Reprod 2004;70:184–190.
303. Avelino A, Cruz F. Peptide immunoreactivity and ultrastructure of rat urinary bladder nerve fibers after topical desensitization by capsaicin or resiniferatoxin. Auton Neurosci 2000;86:37–46.
304. Jancsó G, Ambrus A. Capsaicin sensitivity of primary sensory neurones and its regulation. In: Besson G, Guilbaud G, Ollat H, eds. Peripheral Neurons in Nociception: Physio-Pharmacological Aspects. Paris: John Libbey Eurotex, 1994; pp. 71–87.
305. Jancsó G, Juhász A, Dux M, Sántha P, Domoki F. Axotomy prevents capsaicin-induced sensory ganglion cell degeneration. Prim Sensory Neuron 1997;2:159–165.

306. Hiura A. Neuroanatomical effects of capsaicin on the primary afferent neurons. Arch Histol Cytol 2000;63:199–215.
307. Mason L, Moore RA, Derry S, Edwards JE, McQuay HJ. Systematic review of topical capsaicin for the treatment of chronic pain. BMJ 2004;328:991.
308. Di Marzo V, Blumberg PM, Szállási Á. Endovanilloid signaling in pain. Curr Opin Neurobiol 2002;12:372–379.
309. Pomonis JD, Harrison JE, Mark L, Bristol DR, Valenzano KJ, Walker K. N-(4-Tertiary butylphenyl)-4-(3-cholorphyridin-2-yl)tetrahydropyrazine-1(2H)-carbox-amide (BCTC), a novel, orally effective vanilloid receptor 1 antagonist with analgesic properties: II. in vivo characterization in rat models of inflammatory and neuropathic pain. J Pharmacol Exp Ther 2003;306:387–393.
310. Garcia-Martinez C, Humet M, Planells-Cases R, Gomis A, Caprini M, Viana F et al. Attenuation of thermal nociception and hyperalgesia by VR1 blockers. Proc Natl Acad Sci U S A 2002;99:2374–2379.
311. Huang J, Zhang X, McNaughton PA. Modulation of temperature-sensitive TRP channels. Semin Cell Dev Biol 2006;17:638–645.
312. Price TJ, Jeske NA, Flores CM, Hargreaves KM. Pharmacological interactions between calcium/calmodulin-dependent kinase II alpha and TRPV1 receptors in rat trigeminal sensory neurons. Neurosci Lett 2005;23:1340–1350.

II. NERVES, MEDIATORS, AND RECEPTORS INVOLVED IN NEUROGENIC INFLAMMATION

Neurogenic Inflammation in Health and Disease
Edited by Gábor Jancsó

The Nature and Electrophysiological Properties of the Afferent Neurons Involved in Neurogenic Inflammation in the Skin

BRUCE LYNN

Department of Physiology, University College London, London, UK

ABSTRACT

In all species studied so far, neurogenic inflammation is generated by subclasses of C-fiber nociceptor, although the particular subclass varies between species. In pig and human skin, the key neurons are mechanically insensitive (heat nociceptors and/or silent nociceptors) and polymodal nociceptors are not involved. In the rat and rabbit, a proportion of polymodal nociceptors produce antidromic vasodilatation (ADV) and, in the rat, also plasma extravasation. It is likely that the neurogenic inflammation-producing neurons are from the population that contains the peptides substance P and/or calcitonin gene-related peptide. In the pig, the heat nociceptors that are the only group causing ADV have relatively long duration axonal action potentials. In other species, nociceptive dorsal root ganglion (DRG) neurons all tend to have long duration action potentials, involving significant tetrodotoxin-resistant sodium currents and possibly calcium currents. There is no information on the specific electrophysiological properties of the subpopulation of nociceptive DRG neurons involved in neurogenic inflammation.

1. INTRODUCTION

It has been known for more than 100 years that some dorsal root fibers can cause peripheral vasodilatation [1,2] and the dependence of cutaneous flare on intact sensory – not autonomic – innervation was demonstrated in the 1920s (see Ref. [3]). Lewis in fact thought that the dorsal root fibers involved in flare were not truly sensory, designating them instead "nocifensor" [4], and this idea has had its advocates subsequently [5]. However, the general view has been that neurogenic inflammatory responses represent release of proinflammatory mediators from some sensory neurons. Thus, this group of neurons have a dual function – they act as afferents signaling about the state of the tissue and at the same time act as efferents with neurogenic inflammatory actions.

The response properties of the relevant afferents can be deduced in broad terms from the types of stimuli that generate neurogenic inflammation. Light tactile stimuli and innocuous warming or cooling do not cause inflammatory responses. Therefore, the large numbers of sensitive, specialized mechanoreceptors and thermoreceptors are not involved. Strong heating or mechanical stress are needed to produce neurogenic inflammation. Thus, the relevant afferent group must be the nociceptors. The sections 1-5 of this chapter will consider which subclasses of nociceptor are involved. The section 6 will consider what we know about the electrophysiological properties of these neurons.

2. SUBCLASSES OF NOCICEPTOR

On the basis of their transduction profile, nociceptors can be subdivided into four main classes (Table 1). Note, however, that in any given species or tissue they may not all be present. Mechanical nociceptors were the first to be characterized [6] and were called at that time "high-threshold mechanoreceptors" or HTM, a designation still sometimes used. As their name indicates, these afferents respond to strong mechanical stimuli, notably punctate pressure, at levels well above those that excite specialized mechanoreceptors, although still usually well below levels that cause immediate tissue injury. These afferents typically have small myelinated A-delta (Aδ) axons, although a few mechanical nociceptors with unmyelinated (C) axons are found.

Next to be described were polymodal nociceptors [7]. These respond to a range of noxious stimuli including strong pressure, to heat approaching the noxious range, and to irritant chemicals such as capsaicin or bradykinin. Polymodal nociceptors mostly have C-axons, but significant numbers of nociceptors with Aδ axons that are responsive to pressure and heat have been found in primate skin [8,9].

Two further categories of C-fiber nociceptor both share the characteristic of mechanical insensitivity. One class responds to noxious heating and has been designated heat nociceptor [10–12]. A second class responds only to irritant chemicals and not to physical stimuli. Their lack of response to physical stimuli led to the designation silent nociceptors. Such afferents were first clearly described in joint nerves [13], but similar afferents are found, in variable numbers, in most tissues including the skin. A key feature of these insensitive afferents is that they become much more responsive if the tissues are injured or inflamed. Hence another designation is sleeping nociceptor, indicating that they can be wakened given the correct circumstances. When awake, these afferents have polymodal response properties, developing both heat and pressure sensitivity. Note also that heat nociceptors have a comparable development of pressure responses in inflamed skin and that polymodal nociceptors undergo sensitization, becoming more responsive especially to heat.

Table 1 Classification of cutaneous nociceptors

Class	Alternative designation	Fiber caliber	Responsive to		
			Pressure	Heat	Irritant chemicals
Mechanical	HTM (high-threshold mechanoreceptor)	Aδ (Aβ, C*)	+	0	0
Polymodal	Mechanoheat	C (Aδ†)	+	+	+
Heat		C	0‡	+	+
Sleeping	Silent; CM_iH_i	C	0‡	0‡	+

* A small number of both Aβ and C-fibers have mechanical nociceptor properties.
† Some A-fibres have polymodal properties, notably in primate skin.
‡ In presence of inflammation, respond to pressure and heat.

3. WHICH CUTANEOUS NOCICEPTORS MEDIATE THE FLARE AND ANTIDROMIC VASODILATATION COMPONENTS OF NEUROGENIC INFLAMMATION?

The first clue comes from experiments using graded stimulation of nerve trunks to produce antidromic vasodilatation (ADV). Excitation of Aδ fibers alone can produce a small, relatively short-lasting, vasodilatation in some species [14,15]. However, to get large, prolonged, ADV it is necessary to excite C-fibers. So the mechanical nociceptors with Aδ axons are probably not involved to any significant extent in producing flare or ADV.

To establish the identity of the afferent neurons involved in ADV, it is necessary to stimulate identified single axons while scanning the skin to map increases in blood flow. Results from such experiments in pig skin are shown in Fig. 1. In the pig, only stimulation of C-heat nociceptors gave local blood flow responses [16]. Stimulation of polymodal nociceptors, the other major class of nociceptor in the pig, did not elicit blood flow responses. In the pig there are few C-mechanical nociceptors or silent nociceptors, but of the small number tested, none gave a detectable increase in skin blood flow. Note that the single fiber stimulation gave small zones of vasodilatation that matched closely the extent of afferent receptive fields. This indicates that efferent properties are coextensive with afferent ones, presumably because all terminals play a part in both transduction and secretion. The extent of heat nociceptor receptive fields also matched closely the spread of flare in the same skin area, giving direct confirmation of the axon reflex model.

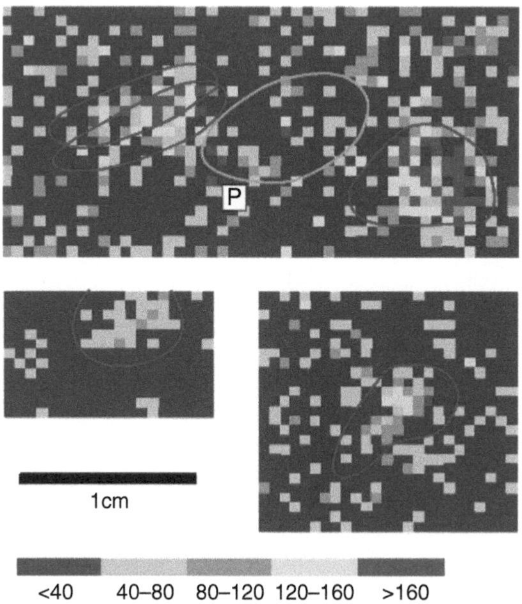

Figure 1. Blood flow images during stimulation of fine nerve filaments containing C-heat nociceptor axons in the pig. Upper field: filament contained three identified heat nociceptors – afferent receptive fields outlined in red and one polymodal nociceptor (P) receptive field outlined in green. Note that only the heat nociceptors gave vasodilator responses. Lower two fields: filaments with one heat nociceptor axon. Note the good relation between afferent receptive field and area of vasodilatation. Images have been scaled so that color levels represent fixed percentages above the average baseline level and have been oriented so that proximal is to the right and anterior at the top. Scans took 1–3 min. Stimulation was 1–2 Hz for the duration of the scans (from Ref. [16]). (See color plate 1).

In smaller animals (rat and rabbit), there are few heat nociceptors in the skin. In these species, a proportion of polymodal nociceptors gave clear blood flow increases on antidromic stimulation (50% in rabbit; 20% in rat) [17]. The two heat nociceptors tested in the rat both also gave increases in skin blood flow within their receptive fields on stimulation. The small numbers of mechanical and silent nociceptors gave no detectable blood flow response, as had been found in the pig. The only distinctive feature of those polymodal nociceptors that gave blood flow responses was a tendency for them to have relatively high mechanical thresholds [17].

In human skin the direct experiment of single fiber stimulation has not been attempted, although it might be possible since intraneural stimulation via tungsten microelectrodes has been used for studies of sensation in humans. However, some indirect evidence has emerged from a comparison of cutaneous stimuli of different area [18]. Polymodal and mechanical nociceptors can be excited readily (<14 mA for 0.2 ms current pulses) by large area (30 mm^2) stimulating electrodes, while mechanically insensitive afferents (heat or sleeping nociceptors) require at least twice as much current. On the other hand, all nociceptors, including the heat and sleeping ones, respond to 10–25 mA stimuli applied via a pointed electrode. As shown in Fig. 2, flare is readily produced by stimulation with the small electrode, but is not produced by stimuli from the 30 mm^2 electrode at intensities sufficient to excite most polymodal nociceptors [18]. It therefore appears that in

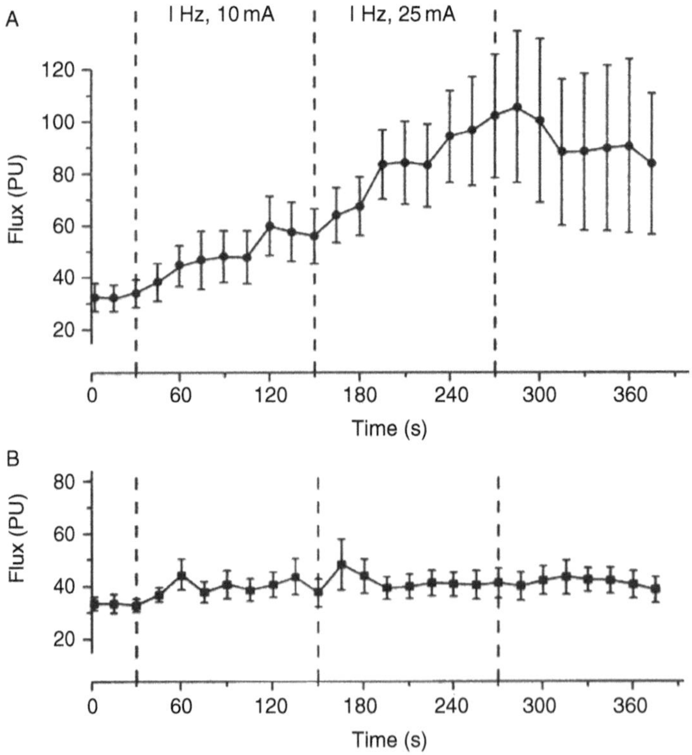

Figure 2. Flare responses in human skin during transcutaneous electrical stimulation of all nociceptors (A) or only mechanoheat (polymodal) nociceptors (B). Only when the mechanically insensitive nociceptors (heat and silent nociceptors) were stimulated did flare appear. Blood flow was monitored by laser Doppler flowmetry 10 mm from the stimulating electrodes. A, point stimulation; B, stimulation via 30 mm^2 electrode (from Ref. [18]).

humans, as in pigs, mechanically insensitive nociceptors play the sole role in producing flare. What cannot be deduced from these experiments is whether only the heat nociceptors are involved, as in pigs, or if the sleeping nociceptors are responsible, either wholly or in part.

4. FIBERS INVOLVED IN NEUROGENIC EDEMA AND CELL INFILTRATION

Plasma extravasation (edema) in response to afferent C-fiber stimulation is readily observed in rats, although it does not appear to occur in humans. In rats, single fiber antidromic stimulation revealed that many polymodal nociceptors could produce plasma extravasation [19]. Thus, the same group of afferents, which are the predominant nociceptor class in this species, produced both vasodilatation and edema.

Afferent nerve stimulation can also trigger inflammatory cell infiltration [20,21]. Because cell infiltration is rather slow (typically experiments look over several hours), no one appears to have attempted the difficult experiment of relating this process to the activity of single nociceptors or single subgroups of nociceptors.

5. EVIDENCE CONCERNING THE ROLE OF BIOCHEMICALLY IDENTIFIED SENSORY AFFERENTS IN NEUROGENIC INFLAMMATION

How far can the work described above on afferent characterization of the neurons involved in neurogenic inflammation be related to the large amount of recent work that has characterized somatosensory afferents in terms of their biochemistry? One obvious correlation is with peptide content. Approximately 50% of DRG neurons innervating the skin of the rat contain calcitonin gene-related peptide (CGRP) [22] and around half of these, 25% overall, also contain substance P (see the review by Lawson [23]). In other species and tissues, the proportions of CGRP and substance P-containing DRG neurons vary, as does the degree of colocalization (see, e.g., Refs [23,24]). These peptidergic neurons are predominantly, but by no means entirely, small neurons with C-fiber axons. Substance P and CGRP both have proinflammatory actions, CGRP mostly acting as a vasodilator, while substance P additionally can increase microvascular permeability [25] and is an attractant for inflammatory cells [21,26]. It is therefore likely that this peptide-containing subpopulation of neurons is the one primarily involved in neurogenic inflammation. This view is reinforced by the finding that neonatal capsaicin treatment, causing loss of this class of neuron (but also many of the nonpeptide-containing neurons), abolishes neurogenic inflammation (see the review by Holzer [27]). Also, antagonists to CGRP can reduce flare or ADV [28,29] while NK1 (substance P receptor) antagonists can greatly reduce neurogenic plasma extravasation [30].

There is an interesting mismatch in the rat between numbers of peptidergic C-afferent neurons (approximately 50%) and the much smaller proportion (<20%) of C-afferents that generated vasodilator responses [17]. It may be a mistake to read too much into negative results from single fiber antidromic experiments. However, this mismatch may relate to another observation on the peptidergic innervation of the skin. This is that many of the terminals of peptidergic sensory neurons are in the epidermis where there are no blood vessels [31–33]. Perhaps some peptidergic afferents act on keratinocytes or immune system cells locally in the epidermis (e.g., Langherans cells) [32,34,35] and do not have short-term vascular actions. These afferents could still be involved in neurogenic inflammation, but would not be detected by methods looking only at immediate microvascular reactions (edema or vasodilatation).

6. ELECTROPHYSIOLOGICAL PROPERTIES OF NEURONS INVOLVED IN NEUROGENIC INFLAMMATION

6.1 Axonal properties

Pig heat nociceptors have action potentials of relatively long duration (Fig. 3) [36]. It is not known what currents underlie the later parts of the spike, but by comparison with cell body spikes, one might suspect the involvement of slow sodium currents and/or calcium currents. Presence of calcium currents in some C-fiber axons has been reported [37]. The presence of a large calcium current might be related to the efferent function since such currents are presumably present in the terminals to trigger peptide release and so might be also found to some extent in the axon.

The recovery of excitability following an action potential, which also shows itself as a slowing of conduction during repetitive firing, is quite different in different functional classes of C-fiber. In rats and humans, the polymodal nociceptors and the sleeping nociceptors – some of which are involved in neurogenic inflammation – have much slower recovery cycles than mechanoreceptor or thermoreceptor C-fibers [38–40]. Recovery cycles for heat nociceptors have not been investigated.

6.2 DRG neuron properties associated with subpopulations likely to be involved in neurogenic inflammation

As discussed above, it is likely that small neurons containing substance P or CGRP will be involved in neurogenic inflammation. Unfortunately, not much is known about the differences

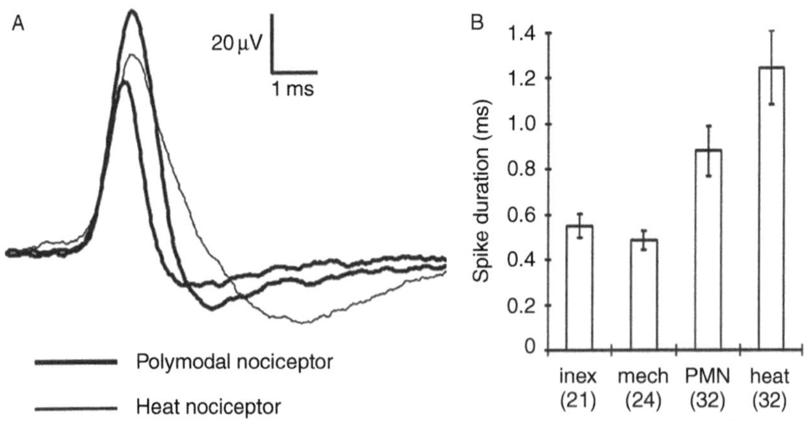

Figure 3. Axonal action potentials for two major classes of nociceptor innervating the pig skin. (A) Examples of three spikes recorded from the same filament. Note that the heat nociceptor has a spike of longer duration than either of the polymodal nociceptors. (B) Average duration of the main peak of the action potential and half-maximum amplitude for different types of C-fiber (mean ± SE). Class abbreviations: inex, unit with no afferent receptive field to pressure or heat; mech, sensitive C-mechanoreceptor; PMN, polymodal nociceptor; heat, heat nociceptor (from Ref. [36]).

in electrophysiological properties of these neurons compared with nonpeptide containing ones. A recent study of the properties of DRG neurons that bind the lectin IB4 showed that these have relatively long duration spikes and that this is due at least in part to the presence of relatively slow tetrodotoxin (TTX)-resistant sodium currents [41]. The IB4 marker has been thought to largely mark the nonpeptide-containing population of DRG neurons [42]. Thus, one might be tempted to conclude that the neurons most likely involved in neurogenic inflammation actually had the spikes with a relatively short duration, thus going against the trend shown by axonal spikes (see above). However, there are a number of complications. In particular recent work has shown that many CGRP-positive DRG neurons do in fact express the IB4 marker, although often at relatively low intensity [43], so it is possible that many of the IB4+ neurons studied by Stucky and Lewin [41] contained inflammatory peptides.

Another approach is to compare the electrophysiological properties of nociceptive DRG neurons with those of nonnociceptive ones. We know that the nonnociceptive neurons are definitely not involved in neurogenic inflammation. What we cannot tell from this type of study is whether particular nociceptor neurons are from the subset that are involved in neurogenic inflammation. Another problem is that the only extensive study has been in the guinea pig and we do not have any data for this species on which classes of nociceptor are actually involved in neurogenic inflammation. There are clear differences in the electrophysiological properties of DRG neurons of different afferent subclasses. Overall, the nociceptive neurons have wider spikes with longer duration after hyperpolarizations [44]. The spikes also overshoot further, i.e., are larger, in nociceptive neurons [45]. Interestingly, the C-nociceptor spikes become shorter in duration if the skin is inflamed for 2–4 days (using a nonneurogenic method, injection of complete Freund's adjuvant) [46].

7. CONCLUSIONS

Neurogenic inflammation is generated by subclasses of C-fiber nociceptor. In the pig skin, the key neurons are heat nociceptors; in the human skin, it appears that heat nociceptors and/or silent nociceptors may be involved. In the rat and rabbit, a proportion of polymodal nociceptors produce ADV and, in the rat, plasma extravasation. Many nociceptors contain CGRP and some also substance P. It is likely that the neurogenic inflammation-producing neurons are from this population. In the pig, the heat nociceptors have relatively long duration axonal action potentials. In other species, nociceptive DRG neurons all tend to have long duration action potentials, involving significant TTX-resistant sodium currents and possibly calcium currents. There is no information on the specific electrophysiological properties of nociceptive DRG neurons involved in neurogenic inflammation.

REFERENCES

1. Stricker S. Sitzb d k Akad d Wissensch Wien 1876;74(3):173.
2. Bayliss WM. On the origin from the spinal cord of the vaso-dilator fibres of the hind-limb, and on the nature of these fibres. J Physiol 1901;26:173–209.
3. Lewis T. The Blood Vessels of the Human Skin and Their Responses. London: Shaw and Son, 1927.
4. Lewis T. The nocifensor system of nerves and its reactions. Br Med J 1937;1:491–494.

5. LaMotte RH. Subpopulations of "nocifensor neurons" contributing to pain and allodynia, itch and alloknesis. APS J 1992;1:115–126.
6. Burgess PR, Perl ER. Myelinated afferent fibres responding specifically to noxious stimulation of the skin. J Physiol 1967;190:541–562.
7. Bessou P, Perl ER. Response of cutaneous sensory units with unmyelinated fibers to noxious stimuli. J Neurophysiol 1969;32:1025–1043.
8. Price DD, Dubner R. Neurons that subserve the sensory-discriminative aspects of pain. Pain 1977;3:307–338.
9. Adriaensen H, Gybels J, Handwerker HO, Van Hees J. Response properties of thin myelinated (A-delta) fibers in human skin nerves. J Neurophysiol 1983;49:111–122.
10. Baumann TK, Simone DA, Shain CN, LaMotte RH. Neurogenic hyperalgesia: the search for the primary cutaneous afferent fibers that contribute to capsaicin-induced pain and hyperalgesia. J Neurophysiol 1991;66:212–227.
11. Schmidt R, Schmelz M, Forster C, Ringkamp M, Torebjork E, Handwerker H. Novel classes of responsive and unresponsive C nociceptors in human skin. J Neurosci 1995;15:333–341.
12. Lynn B, Faulstroh K, Pierau FK. The classification and properties of nociceptive afferent units from the skin of the anaesthetized pig. Eur J Neurosci 1995;7:431–437.
13. Schmidt RF, Schaible HG, Messlinger K, Heppelmann B, Hanesch U, Pawlak M. Silent and active nociceptors: structure, functions and clinical implications. In: Gebhart GF, Hammond DL, Jensen TS, eds. Proc 7th World Congress on Pain. Seattle: IASP, 1994; pp. 213–250.
14. Lynn B, Shakhanbeh J. Neurogenic inflammation in the skin of the rabbit. Agents Actions 1988;25:228–230.
15. Janig W, Lisney SJ. Small diameter myelinated afferents produce vasodilatation but not plasma extravasation in rat skin. J Physiol 1989;415:477–486.
16. Lynn B, Schutterle S, Pierau FK. The vasodilator component of neurogenic inflammation is caused by a special subclass of heat-sensitive nociceptors in the skin of the pig. J Physiol (Lond) 1996;494:587–593.
17. Gee MD, Lynn B, Cotsell B. The relationship between cutaneous C fibre type and antidromic vasodilatation in the rabbit and the rat. J Physiol (Lond) 1997;503:31–44.
18. Schmelz M, Michael K, Weidner C, Schmidt R, Torebjork HE, Handwerker HO. Which nerve fibers mediate the axon reflex flare in human skin? NeuroReport 2000;11: 645–648.
19. Bharali LA, Lisney SJ. The relationship between unmyelinated afferent type and neurogenic plasma extravasation in normal and reinnervated rat skin. Neuroscience 1992;47:703–712.
20. Lynn B, Walsh DA. Cutaneous vasodilatation and macrophage infiltration during prolonged electrical stimulation of the rat saphenous nerve. J Physiol 1998;509P, 163P.
21. Pinter E, Than M, Chu DQ, Fogg C, Brain SD. Interaction between interleukin 1beta and endogenous neurokinin 1 receptor agonists in mediating plasma extravasation and neutrophil accumulation in the cutaneous microvasculature of the rat. Neurosci Lett 2002;318:13–16.
22. O'Brien C, Woolf CJ, Fitzgerald M, Lindsay RM, Molander C. Differences in the chemical expression of rat primary afferent neurons which innervate skin, muscle or joint. Neuroscience 1989;32:493–502.
23. Lawson SN. Morphological and biochemical cell types of sensory neurons. In: Scott SA, ed. Sensory Neurons: Diversity, Development, and Plasticity. New York: Oxford University Press, 1992; pp. 27–59.

24. Lawson SN, Crepps B, Perl ER. Calcitonin gene-related peptide immunoreactivity and afferent receptive properties of dorsal root ganglion neurones in guinea-pigs. J Physiol 2002;540:989–1002.
25. Brain SD. Sensory neuropeptides: their role in inflammation and wound healing. Immunopharmacology 1997;37:133–152.
26. Smith CH, Barker J, Morris RW, MacDonald DM, Lee TH. Neuropeptides induce rapid expression of endothelial cell adhesion molecules and elicit infiltration in human skin. J Immunol 1993;151:3274–3282.
27. Holzer P. Capsaicin: cellular targets, mechanisms of action, and selectivity for thin sensory neurons. Pharmacol Rev 1991;43:143–201.
28. Basile S, Pierau FK, Schutterle S, Ladewig S. Participation of sensory neuropeptides in the flare reaction in pig skin. Proc German Neurosci Assoc 1996;1:167.
29. Habler HJ, Timmermann L, Stegmann JU, Janig W. Involvement of neurokinins in antidromic vasodilatation in hairy and hairless skin of the rat hindlimb. Neuroscience 1999;89:1259–1268.
30. Lembeck F, Donnerer J, Tsuchiya M, Nagahisa A. The non-peptide tachykinin antagonist, CP-96,345, is a potent inhibitor of neurogenic inflammation. Br J Pharmacol 1992;105:527–530.
31. Dalsgaard CJ, Jernbeck J, Stains W, Kjartansson J, Haegerstrand A, Hokfelt T et al. Calcitonin gene-related peptide-like immunoreactivity in nerve fibers in the human skin. Relation to fibers containing substance P-, somatostatin- and vasoactive intestinal polypeptide-like immunoreactivity. Histochemistry 1989;91:35–38.
32. Egan CL, Viglione-Schneck MJ, Walsh LJ, Green B, Trojanowski JQ, Whitaker-Menezes D et al. Characterization of unmyelinated axons uniting epidermal and dermal immune cells in primate and murine skin. J Cutan Pathol 1998;25:20–29.
33. Navarro X, Verdu E, Wendelscafer-Crabb G, Kennedy WR. Innervation of cutaneous structures in the mouse hind paw: a confocal microscopy immunohistochemical study. J Neurosci Res 1995;41:111–120.
34. Song IS, Bunnett NW, Olerud JE, Harten B, Steinhoff M, Brown JR et al. Substance P induction of murine keratinocyte PAM 212 interleukin 1 production is mediated by the neurokinin 2 receptor (NK-2r). Exp Dermatol 2000;9:42–52.
35. Hosoi J, Murphy GF, Egan CL, Lerner EA, Grabbe S, Asahina A et al. Regulation of Langerhans cell function by nerves containing calcitonin gene-related peptide. Nature 1993;363:159–163.
36. Gee MD, Lynn B, Basile S, Pierau FK, Cotsell B. The relationship between axonal spike shape and functional modality in cutaneous C-fibres in the pig and rat. Neuroscience 1999;90:509–518.
37. Mayer C, Quasthoff S, Grafe P. Confocal imaging reveals activity-dependent intracellular Ca^{2+} transients in nociceptive human C fibres. Pain 1999;81:317–322.
38. Thalhammer JG, Raymond SA, Popitz-Bergez FA, Strichartz GR. Modality-dependent modulation of conduction by impulse activity in functionally characterized single cutaneous afferents in the rat. Somatosens Mot Res 1994;11:243–257.
39. Gee MD, Lynn B, Cotsell B. Activity-dependent slowing of conduction velocity provides a method for identifying different functional classes of C-fibre in the rat saphenous nerve. Neuroscience 1996;73:667–675.
40. Serra J, Campero M, Ochoa J, Bostock H. Activity-dependent slowing of conduction differentiates functional subtypes of C fibres innervating human skin [see comments]. J Physiol (Lond) 1999;515:799–811.

41. Stucky CL, Lewin GR. Isolectin B(4)-positive and -negative nociceptors are functionally distinct. J Neurosci 1999;19:6497–6505.
42. Silverman JD, Kruger L. Selective neuronal glycoconjugate expression in sensory and autonomic ganglia: relation of lectin reactivity to peptide and enzyme markers. J Neurocytol 1990;19:789–801.
43. Kashiba H, Uchida Y, Senba E. Difference in binding by isolectin B4 to trkA and c-ret mRNA-expressing neurons in rat sensory ganglia. Brain Res Mol Brain Res 2001;95:18–26.
44. Djouhri L, Bleazard L, Lawson SN. Association of somatic action potential shape with sensory receptive properties in guinea-pig dorsal root ganglion neurones. J Physiol 1998;513:857–872.
45. Djouhri L, Lawson SN. Differences in the size of the somatic action potential overshoot between nociceptive and non-nociceptive dorsal root ganglion neurones in the guinea-pig. Neuroscience 2001;108:479–491.
46. Djouhri L, Lawson SN. Changes in somatic action potential shape in guinea-pig nociceptive primary afferent neurones during inflammation in vivo. J Physiol 1999;520:565–576.

Edited by Gábor Jancsó

Peptidergic Innervation of Blood Vessels

THOMAS M. SCOTT[1] and MICHAEL M. SCOTT[2]

[1]*Schulich School of Medicine and Dentistry, Windsor Program, University of Windsor, Windsor, Ontario, Canada*
[2]*Department of Internal Medicine, Center for Hypothalamic Research, The University of Texas Southwestern Medical Center, Dallas, TX, USA*

ABSTRACT

The vasculature has been shown to be innervated not solely by sympathetic nerve fibers containing catecholamine as was once thought but by nerve fibers containing a variety of peptides. In general, peptidergic perivascular nerve fibers are found in the adventitia of blood vessels with a concentration at the adventitial–medial border. The distribution of nerve fibers varies across species and in different vascular beds. The source of these nerve fibers varies according to the peptides present. Development of perivascular innervation begins before birth, with nerve fiber density and functioning axon terminals being fully established early. In some vascular beds, the perivascular innervation is significantly involved both in the control of the circulation and in disease processes. Perivascular peptidergic innervation is a prime target for therapeutic intervention in hypertension, myocardial infarction, atherosclerosis, migraine, arthritis, and skin disease.

1. INTRODUCTION

The vasculature has been shown to be innervated not solely by sympathetic nerve fibers containing catecholamine as was once thought but by nerve fibers containing a variety of peptides [1]. Such peptidergic perivascular nerve fibers may be important contributors to conditions such as neurogenic inflammation, migraine, hypertension, and coronary vascular disease through release of peptides such as calcitonin gene-related peptide (CGRP), substance P (SP), neurokinin A (NKA), neuropeptide Y (NPY), and vasoactive intestinal polypeptide (VIP), which have profound effects on the vasculature and on the vascularized tissues.

2. THE PERIVASCULAR NEUROPEPTIDES

The neuropeptides are a group of proteins that are widely distributed in the central and peripheral nervous system and that act as neurotransmitters and/or neuromodulators. The first neuropeptide to be discovered was found by von Euler and Gaddum [2] in 1931 when they reported that extracts of equine brain and intestine contained a hypotensive and spasmogenic factor, later determined to be SP. Among the perivascular nerve neuropeptides studied, SP, CGRP, NPY, and VIP are the most widely characterized.

3. THE TACHYKININS

The tachykinins in perivascular nerves are small peptides of 10–13 amino acids with a conserved COOH-terminal sequence and different ionic charges at the NH_2 terminus, critical for receptor binding and affinity. Included in this group are SP, NKA, and neurokinin B (NKB). The C-terminal region is important in the activation of each of the three known mammalian G-protein-coupled tachykinin receptors – NK1, NK2, and NK3 [3,4]. The affinity of these receptors for each tachykinin ligand varies as follows: NK1, SP $\geq$ NKA $>$ NKB; NK2, NKA $>$ NKB $>$ SP; and NK3, NKB $>$ NKA $>$ SP [5,6]. The identification of further tachykinins and expansion of the receptor class have been the subjects of recent reviews [7,8].

The activation of the tachykinin receptors has been implicated in a wide variety of biological actions from smooth muscle contraction, emotional behavior, vasodilation, pain transmission, neurogenic inflammation, hematopoiesis, activation of the immune system, and stimulation of endocrine gland secretion [9]. For example, in the vasculature, SP generally acts as a vasodilator, binding to NK-1 receptors localized on endothelial cells to cause release of nitric oxide and other vasodilator substances [10]. In some circumstances, SP has been shown to constrict mesenteric veins via action at NK3 receptors [11,12]. The importance of NK-1 receptor-mediated neurogenic vasodilation was further demonstrated in mice null for NK-1. Measurement of mouse ear blood flow showed that capsaicin, which acts as an exogenous ligand at the transient receptor potential vanilloid receptor to produce substantial vasodilation, is altered in the absence of NK-1 signaling [13].

4. CGRP

CGRP exists in two forms, as alpha-CGRP generated from a transcript that also encodes calcitonin along with two other small peptides and beta-CGRP, a highly similar peptide transcribed from an independent gene locus [14,15]. Both alpha- and beta-CGRP activate a heteromeric G-protein-coupled receptor [16]. CGRP was demonstrated by Brain et al. [17] to be a potent vasodilator, leading to the suggestion that local extravascular release of CGRP may be involved in the physiological control of blood flow and that circulating CGRP may contribute to hyperemia in certain pathological conditions. CGRP released from perivascular sensory nerve fibers acts on arteriolar smooth muscle to produce profound dilation. Mechanisms of action include both endothelium-dependent and endothelium-independent relaxation of vascular smooth muscle. The vasodilator actions of CGRP have been described in a wide range of arterial beds from coronary [18,19] to cranial blood vessels [20]. In the microcirculation, CGRP is one of the most potent vasodilator substances identified [17]. In keeping with this, the highest density of CGRP binding sites are present in the heart and in blood vessels [21]. In the heart, high-affinity binding sites for CGRP are found in the atria and ventricles [22]. Regardless of the species, the density of the CGRP binding sites in atria invariably exceeds that of ventricles [22]. Autoradiographic studies in the hearts of rats [22], guinea pigs, and humans [23] have shown the highest density of CGRP binding sites in the coronary arteries, coronary veins, and heart valves, whereas a lower density is found in the coronary arterioles and endocardium [21]. These findings attest to the relevance of the CGRP receptor in regulating blood flow as well as inotropic and chronotropic effects in the heart [24,25]. Interestingly, overlapping expression of both alpha- and beta-CGRP suggests a redundancy in CGRP signaling [15]. However, there is conflicting data on the ability of beta-CGRP to compensate for the loss of alpha-CGRP in the induction of vasodilation. Although deletion of the alpha-CGRP/calcitonin transcript resulted in a significant increase in coronary flow rate and basal blood pressure [26], specific deletion of the alpha-CGRP component of the gene locus, leaving calcitonin intact, had no effect on these parameters [27].

5. NPY

NPY has generally been regarded as a vasoconstrictor peptide [28–30] signaling through a variety of G-protein-coupled receptors, designated $Y_{1,2,4,5}$ (reviewed in Ref. [31]), although there is some suggestion that in certain coronary arteries it fails to cause constriction [32]. In isolated ring segments of human middle meningeal and cerebral arteries, NPY caused vasoconstriction but did not potentiate the contractile response of noradrenaline (NA). In the temporal artery, NPY did not induce contraction but potentiated the vasoconstrictor response to NA [33]. Thus, in certain vascular beds, NPY may act in parallel with NA, whereas in other select locations, it may directly affect the release of NA or signaling through adrenergic receptors. Furthermore, NPY may also have a role in modulating the actions of other vasoactive agents [34,35].

Deletion of NPY receptors has allowed for further description of the interaction between NA and NPY in mediating vasoconstriction. Y1 deletion completely abolished the ability of NPY to potentiate NA-dependent vasoconstriction in isolated aorta preparations, leaving basal blood pressure and heart rate unaltered [36]. Deletion of the Y4 receptor subtype revealed the importance of NPY in regulating noradrenergic sympathetic tone, producing a substantial decrease in resting blood pressure and heart rate [37]. Similarly, Y2 deletion was observed to lower basal heart rate while maintaining blood pressure [38].

6. VIP

VIP is a 28 amino acid cleavage product of prepro-VIP, which is the parent molecule also for peptide histidine methionine (PHM), pituitary adenylate cyclase-activating polypeptide (PACAP), and peptide histidine valine (PHV-42), acting at G-protein-coupled VPAC(1) and VPAC(2) receptors (reviewed in Ref. [39]). VIP, found mainly in parasympathetic nerves, is a vasodilator of arteries in various species [32,40–43]. The mechanism of vasodilation includes both endothelium-dependent and endothelium-independent actions [44]. Recently, deletion of the prepro-VIP transcript encoding VIP and peptide histamine (PH) peptides highlighted the role of these molecules in maintaining appropriate pulmonary blood pressure. In the absence of VIP/PH signaling, marked pulmonary arterial hypertension (PAH) is observed, leading to a thickening of the pulmonary arteries and reduced lumen diameter. As expected, treatment with VIP significantly decreased the degree of PAH and associated arterial pathology [45].

7. THE SOURCE OF PERIVASCULAR PEPTIDERGIC NERVE FIBERS

7.1 The ganglia and methods of study

In general, peptidergic perivascular nerve fibers are found in the adventitia of blood vessels with a concentration at the adventitial–medial border. The source of these nerve fibers varies according to the peptides present. Sources include neurons of the dorsal root and trigeminal ganglia (TG), the sympathetic paravertebral ganglia such as the stellate ganglia and the parasympathetic ganglia. Typically, sources have been identified by methods such as immunohistochemical staining, retrograde labeling, and through ganglionectomy. The investigation of peptidergic perivascular innervation of the head and neck will serve as an illustration. Tracing studies of the monkey middle–cerebral artery innervation [46] confirmed earlier studies in rats [47] and cats, showing the superior cervical (SCG) and TG as the primary sources with contralateral involvement in the SCG and TG. In the parasympathetic sphenopalatine ganglion, immunostaining was scarce [46]. The source of cerebrovascular innervation in the gerbil was examined using fluorescence histochemistry and SCG ganglionectomy [48]. Bilateral ganglionectomy resulted in a complete loss of NA-, 5-hydroxytryptamine (5-HT)-, and NPY-containing perivascular nerve fibers in the anterior circulation, whereas a few fibers still persisted in the posterior circulatory bed. Both 5-HT- and NPY-positive cell bodies were shown to be present in the SCG of adult gerbils. The SCG is the primary source of adrenergic fibers for the head and neck and represents the cranial extension of the paravertebral sympathetic chain. In addition to the catecholamines, the most abundant and best characterized neuropeptide in the SCG is NPY. Many sympathetic neurons contain high levels of NPY that also functions as a transmitter or neuromodulator in this system [49–51]. The TG, the main ganglion of the sensory system for cranial structures, is immunopositive for several peptides including SP, CGRP, PACAP, cholecystokinin, SOM, opioid peptides, VIP, galanin, NPY, and nitric oxide synthase [52].

8. THE PEPTIDES AND THEIR SOURCES

8.1 SP and CGRP

Myelinated and unmyelinated small caliber neurons in dorsal root and TG supply most blood vessels with SP- and CGRP-immunoreactive nerve fibers. Many immunohistochemical studies have shown that CGRP and SP are often found together in afferent nerve fibers supplying the mammalian cardiovascular system [53–57]. It has also been demonstrated that the peptides are colocalized in the same secretory vesicles in some perivascular nerve varicosities [58].

8.2 NPY

Most NPY-containing perivascular nerves are derived from sympathetic neurons in the paravertebral ganglia, where numerous NPY-immunoreactive cell bodies have been identified [50,51]. NPY-immunoreactive perivascular fibers innervating the cerebral vasculature are thought to arise from the SCG [59–61].

8.3 VIP

Parasympathetic nerve fibers containing acetylcholinesterase (AChE) and VIP arise mainly from cell bodies located in the sphenopalatine and otic ganglia [35,62–65] for head and neck vasculature and from other parasympathetic ganglia such as those of the vagus nerve for the rest of the body.

9. DEVELOPMENT OF PEPTIDERGIC PERIVASCULAR INNERVATION

Development of perivascular innervation begins before birth, with nerve fiber density and functioning axon terminals being fully established early. Under normal circumstances there is no change once the adult innervation pattern and density have been established. Perivascular innervation is present early during development and is available to contribute to further development of the vasculature, to normal functioning of the tissues and to the response to injury and disease. However, in certain pathological conditions development continues, either as a result of the condition or as a contributor to its cause. Hypertrophy of the perivascular plexus is thought to play a role in the development of hypertension in spontaneously hypertensive rats [66,67]. Furthermore, it has been shown recently that following vascular denervation, perivascular nerves are able to regrow [68]. Treatment with adrenomedullin (1000 ng/h) or Nerve Growth Factor (NGF) restored adrenergic nerve-mediated vasoconstriction and CGRP nerve-mediated vasodilation in the perfused mesenteric artery treated topically with phenol. These results suggest that adrenomedullin, like NGF, has a facilitatory effect on the reinnervation of perivascular nerves. Using immunohistochemistry, it has been shown [69] that adrenomedullin is present in perivascular nerves, colocalized with CGRP in the rat mesenteric artery and that adrenomedullin immunoreactivity and adrenomedullin messenger RNA (mRNA) expression is present in the rat dorsal root ganglia [69]. It is not known to what extent adrenomedullin contributes to normal growth, maintenance, or regrowth following disease or injury.

Scott et al. [70] have followed the development of peptidergic perivascular innervation in the mesenteric arterial bed of the rat. At birth, fibers positive for SP, NPY, and CGRP by immunohistochemistry could be demonstrated although fibers positive for anti-VIP were absent. By 1 week of age, it was found that the jejunal arteries had a higher density of innervation by SP-, NPY-, and CGRP-positive fibers than the superior mesenteric arteries. The fibers showing immunoreactivity to an anti-VIP antibody formed an innervation pattern that was different from that of the other peptides in that an irregular and highly variable plexus was seen on the superior mesenteric artery and vein and on the jejunal artery. The plexus, however, was more regular on the distal jejunal and second- and third-order branches of the jejunal arteries. As development proceeded, the density of innervation as seen at 1 week of age changed little for SP and CGRP although an increase in the density of NPY innervation of jejunal arteries occurred up to 4 weeks of age. It was not clear whether increased branching from existing nerve fibers caused the increase in NPY innervation. At 4 weeks of age, the density of perivascular fibers was not different from that at 8 or 12 weeks of age.

At birth, the innervation of the mesenteric vascular bed was by means of individual fibers, which passed along the length of the vessel giving off occasional branches (Fig. 1). By 1 week of age, SP (Fig. 2), NPY, and CGRP fibers had formed a plexus that changed its form little between 1 and 12 weeks of age. The form of the fiber plexus was similar in the case of SP and CGRP. The fibers forming a superficial plexus in the outer adventitia were smooth-surfaced

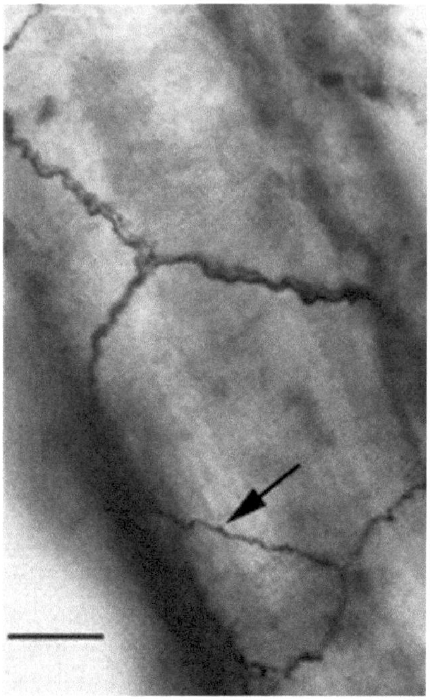

Figure 1. The SP-like immunoreactive nerve fibers (arrow) of a jejunal artery from a 1-day-old rat. Calibration = 20 μm.

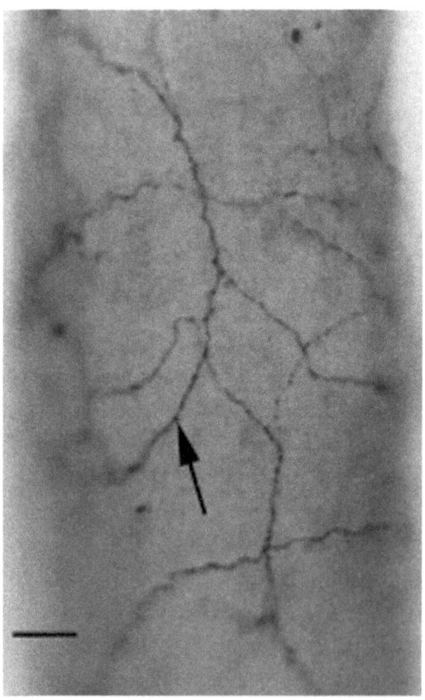

Figure 2. The SP-like immunoreactive nerve plexus (arrow) of a jejunal artery from a 1-week-old rat. Calibration = 20 μm.

fiber bundles that branched to run more deeply. The deep plexus at the adventitial/medial border was formed of beaded fibers, which seemed to run in all directions in the walls of the superior mesenteric and jejunal arteries but with a circular orientation in superior mesenteric veins. No distinguishing features of either the mesh size or shape of the SP or CGRP networks were observed. No particular accumulation of fibers occurred at branch points either in the superior mesenteric artery or in the jejunal arteries. The NPY network was easily identifiable since the mesh was more consistent in both its size and shape particularly in the jejunal arteries. The distribution of NPY fibers in the superior mesenteric artery was more sparse but dense in the superior mesenteric vein. The fibers appeared to be considerably more coarse than those of the SP/CGRP plexus. The change in density of innervation from the superior mesenteric to the jejunal arteries was obvious at the root of the jejunal arteries although again no clear concentration of fibers occurred at the branch point. The VIP plexus on the distal jejunal artery and its branches was somewhat regular but was much less regular on the proximal jejunal and superior mesenteric arteries and on the superior mesenteric vein. Development of a stable density of innervation of the mesenteric vasculature occurs before 1 week of age, and this is maintained during further growth of the tissues although quite dramatic increases in vessel diameter and wall thickness occur.

Luff [71] has carried out an ultrastructural study on the development of the innervation of second-order mesenteric arteries from the ileum region of the rat intestine. It was found that the formation of neuromuscular junctions with mature structural characteristics occurred by 2 weeks postnatal. The plexus of varicose axons developed predominantly between days 4 and 13, which agrees with previous light microscopy studies of catecholaminergic and peptidergic nerves around similar vessels. Before day 4, the axons lacked varicosities and were mainly contained in large bundles located in the outer region of the adventitia. By day 8, there were more axons and most were distributed in smaller bundles. Some had grown through the adventitia to lie at the medial–adventitial border and axon varicosities were also observed. Some varicosities had formed rudimentary neuromuscular contacts. By day 13, there were significantly more contacting varicosities compared to day 8. The ultrastructural study confirms the general pattern of growth of mesenteric perivascular innervation [70] and suggests that functional innervation of the muscle may begin early in postnatal development.

Development of perivascular innervation of the cerebral circulation parallels that of the mesenteric vascular bed. The pre- and postnatal development of trigeminal CGRP [72], AChE [73], NA [72,73], NPY [73], and VIP [73]-containing nerves supplying the cerebral arteries has been studied by immunohistochemistry in rats. It was reported that before birth (embryonic day (E) 18–19), CGRP fibers were present only as one or two longitudinal bundles zigzagging along the anterior cerebral artery and anterior communicating artery. Growth cone-like swellings were found at the terminals of individual fibers [72]. In contrast, at this same prenatal age, NA fibers were present as a meshwork on all cerebral arteries. The density of NA fibers was higher in the rostral than in the caudal parts of the circle of Willis; growth cones were present on individual fibers at the middle segment of the basilar artery and distal parts of major cerebral arteries. At postnatal day 1–2, the outgrowth of CGRP axons extended along the walls of the middle cerebral and internal carotid arteries. These axons were relatively straight and unbranched. At the same time, NA fibers increased in number and density and continued to form the meshwork pattern on all cerebral arteries. At the end of the first postnatal week, all the longitudinal NA bundles on the rostral part of the circle of Willis began to form circular arborizations. At the end of the second postnatal week, the pattern of NA innervation had completely changed, consisting almost entirely of circumferential rather than tangential fibers. Beginning in the first postnatal week, CGRP fibers increased greatly in number and density and began to form a meshwork pattern. At the second postnatal week, the pattern of CGRP innervation, compared to the pattern at fetal and neonatal stages, had changed significantly, consisting predominantly of a meshwork pattern [72]. AChE-positive and VIP-immunoreactive nerves from the internal ethmoidal artery covered the whole internal carotid system during the first postnatal week and projected to the upper basilar artery after the second week, whereas those from the cerebral carotid artery remained limited to the middle cerebral artery throughout development [73]. By 4 weeks after birth, all fiber systems achieved adult densities and patterns. All these nerves entered the cranial cavity along the cerebral, carotid, internal ethmoidal, and vertebral arteries during the early stages of development [73].

Although the timing is different, a similar pattern has been shown in the development of human cerebrovascular perivascular innervation [74]. The development of cerebrovascular nerves containing NA, AChE, NPY, and VIP in the basilar artery was histochemically or immunohistochemically studied in human fetuses of various gestational ages. At the 12th week of gestation, nerve fibers containing NA and AChE appeared. Both types of nerve fibers consisted exclusively of the longitudinal nerve fibers running along the basilar arteries. Subsequently, the circular nerve fibers gradually increased and development was then completed by the 20th week of gestation. NPY- and VIP-containing nerve fibers were detected in human fetal cerebral arteries in the 16th week of gestation with development being completed by the 24th week of gestation.

In conclusion, the development of peptidergic perivascular innervation clearly precedes full development of the vascularized tissues, allowing for a possible role for perivascular innervation in morphological development and functional development of the vasculature and vascularized tissues.

10. DISTRIBUTION OF PEPTIDERGIC PERIVASCULAR INNERVATION

The adult distribution of the peptides appears to vary from bed to bed and from one species to another. The types of fiber present and the density of innervation varies according to vessel diameter and location, with local function playing a role. While in the skin, preterminal arterioles are densely innervated, controlling perfusion [75,76], the pulmonary circulation has a low density of perivascular innervation [77–79], presumably reflecting that local perfusion is determined by alveolar oxygen tension [80]. The range of distribution of the various types of perivascular fiber was outlined in the early eighties in the guinea pig, where practically in every vascular bed peptidergic fibers could be demonstrated. The presence of peptidergic perivascular innervation was, for example, noted in the heart [81], in blood vessels of the urinary and reproductive systems [82], and in the cerebral vascular bed [83]. About this time, many of the noradrenergic nerves supplying the guinea pig cardiovascular system were found also to contain NPY [84]. The literature has since then blossomed with attention being paid to an increasing number of peptides, animal models, and to human vasculature. In some vascular beds, the perivascular innervation is significantly involved in both the control of the circulation and in disease processes. In this light, the perivascular innervation of the cranial/cerebral, coronary, mesenteric and cutaneous vascular beds, and the vasculature of joints will be reviewed further.

11. CRANIAL AND CEREBRAL VASCULAR BEDS

The peptidergic perivascular innervation of cranial and cerebral blood vessels has been studied using morphological, physiological, and pharmacological tools. The early studies mapped out the distribution of the various fiber types in the branches of the external carotid arteries, in particular the superficial temporal artery [85], and the intracranial branches of the internal carotid and basilar. The source and distribution of the cerebral perivascular innervation has been reviewed [86].

11.1 Branches of the external carotid

A sparse to moderate supply of nerve fibers immunoreactive for NPY, VIP, SP, and CGRP was demonstrated in the walls of human middle meningeal arteries [87]. Comparison with similar studies on human cerebral and temporal arteries indicated a similar distribution and density. The peptidergic innervation of the human superficial temporal artery has been investigated by means of immunohistochemical, ultrastructural, and *in vitro* pharmacological techniques [88]. A dense network of nerve fibers was shown to be present in the adventitia. The majority of the nerve fibers displayed immunoreactivity for tyrosine hydroxylase (TH) and NPY. A moderate supply of perivascular nerve fibers showed either AChE activity or immunoreactivity for VIP, PHM, or CGRP. Only a few nerve fibers were SP-, NKA-, and neuropeptide K (NPK)-immunoreactive. In double immunostained preparations, SP immunoreactivity was shown to be colocalized with NPK and CGRP in the same nerve fibers. Ultrastructural studies revealed the presence of numerous axon varicosities at the adventitial–medial border. NPY, VIP, and CGRP immunoreactivities occurred in the same type of large granular vesicles, but in morphologically distinct nerve profiles [88].

11.2 Branches of the internal carotid and basilar arteries

The peptidergic innervation of the guinea pig basilar artery and the posterior, middle, and anterior cerebral arteries was studied by means of immunohistochemical and image analysis techniques using whole mount preparations [89]. In all four cerebral arteries, the majority of nerve fibers possessed NPY and TH immunoreactivity, with VIP, SP, and CGRP detected at lower densities. A pharmacological study performed on small circular segments with an intact endothelium revealed that, in all four cerebral arteries, NPY was a more potent constrictor than NA. The rank order of potency for relaxant agents was CGRP = SP > VIP > ACh in the posterior and middle cerebral arteries, and SP = CGRP > VIP > ACh in the basilar and anterior cerebral arteries [89].

Although the brain has been the target of most studies of the perivascular innervation of the internal carotid artery, the eye is also supplied by branches of the internal carotid artery. The considerable peptidergic innervation of this organ, including the contribution from perivascular nerve fibers, has been reviewed by Troger et al. [90].

A recent emphasis has been on the possible involvement of perivascular innervation in disease processes such as migraine [88] and the value of peptidergic innervation as a target for therapy [91,92]. The significant involvement of CGRP-immunoreactive perivascular nerve fibers in cerebrovascular disease will be discussed later.

12. THE CORONARY VASCULAR BED

The main peptides identified in nerves associated with coronary blood vessels are NPY, VIP, CGRP, and the tachykinins SP and NK [93,94]. Perivascular innervation of the coronary vascular bed has been shown to be important in the function of coronary arteries and veins and may be important in cardiac ischemia and reperfusion and in coronary vasospasm. The peptidergic innervation of proximal and distal regions of human epicardial coronary arteries has been investigated by means of immunohistochemical, chromatographic, radioimmunological, and *in vitro* pharmacological techniques [95]. The use of an antiserum to the general neuronal marker protein gene product 9.5 revealed that the proximal part of epicardial arteries possessed a relatively sparse supply of nerve fibers forming a loose network in the adventitia. In common with most vascular beds, the perivascular network increased in density as the vessels were followed distally. In both proximal and distal regions, the majority of nerve fibers possessed NPY and TH immunoreactivity. VIP, AChE, CGRP-, and SP-immunoreactive nerve fibers were very sparse in the proximal region of the arteries and increased in number distally. Only a few scattered VIP-immunoreactive nerve fibers were detected in both arterial regions. An isolated tissue method demonstrated that NPY did not produce a response in either proximal or distal arterial segments [95,96]. In contrast, CGRP, SP, and VIP all produced a concentration-dependent relaxation of both arterial regions. CGRP and SP were stronger and more potent than VIP. CGRP and SP induced a more potent response in distal compared with proximal regions of the arteries. A recent investigation of perivascular innervation in the rat heart also found a differential distribution of nerve fiber types. While CGRP-immunoreactive perivascular fibers were distributed fairly evenly in the atria and ventricles, NPY-positive fibers were found mostly in the atria, with the most dense innervation being present on smaller diameter vessels [97].

13. THE MESENTERIC VASCULAR BED

The mesenteric vascular bed was an early subject for the investigation of perivascular innervation, probably due to its easy access and extensive vasculature in experimental animals. The bed has acted as a model for various human disease states, such as hypertension, allowing thorough characterization of the peptidergic perivascular innervation.

The mesenteric vascular bed in the rat has been shown to be densely innervated by peptidergic perivascular nerve fibers [70]. SP- and CGRP-immunoreactive fibers are present in high density in the mesenteric artery and mesenteric vein, whereas NPY-immunoreactive fibers show a lower density. In jejunal arteries, SP-, CGRP-, and NPY-immunoreactive fibers are more dense than in the mesenteric artery, with NPY-immunoreactive fibers having the highest density. The fibers showing immunoreactivity to anti-VIP antibody formed an innervation pattern that was different from that of the other peptides in that an irregular and highly variable plexus was seen on the superior mesenteric artery and vein and on the jejunal artery. The form of the fiber plexus was similar in the case of SP and CGRP. The fibers forming a superficial plexus in the outer adventitia were smooth-surfaced fiber bundles that branched to run more deeply. The deep plexus at the adventitial/medial border was formed of beaded fibers, which seemed to run in all directions in the walls of the superior mesenteric and jejunal arteries (Fig. 3), but with a circular orientation in superior mesenteric veins (Fig. 4). No distinguishing features of either the mesh size or

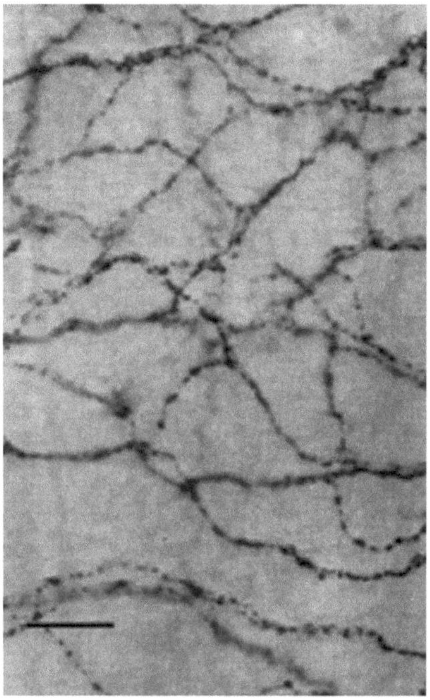

Figure 3. The beaded fibers of the CGRP-like immunoreactive plexus on a jejunal artery of an 8-week-old rat. Calibration = 20 µm.

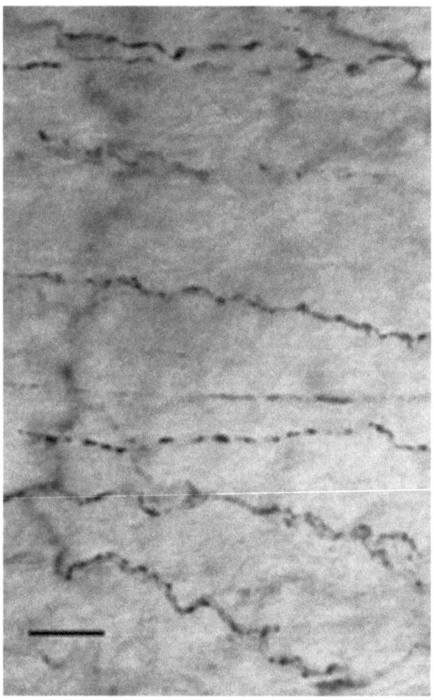

Figure 4. The parallel circular SP-like immunoreactive fibers of a superior mesenteric vein from a 12-week-old rat. Calibration = 20 µm.

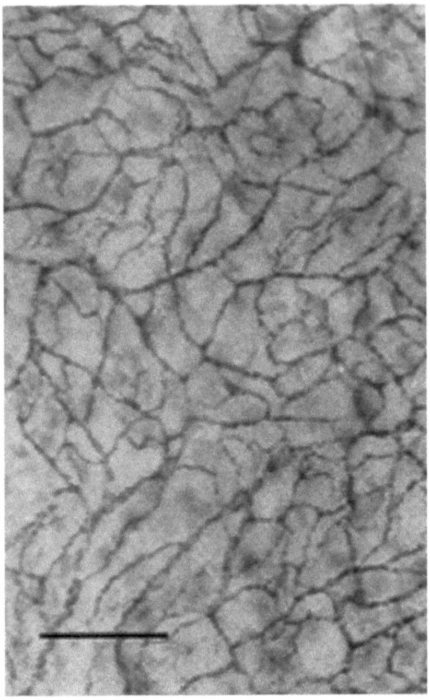

Figure 5. The coarse fibers of the NPY-like immunoreactive plexus of a jejunal artery from an 8-week-old rat. Calibration = 50 μm.

shape of the SP or CGRP networks were observed. No particular accumulation of fibers occurred at branch points either in the superior mesenteric artery or the jejunal arteries. The NPY network, however, was easily identifiable. The mesh was more consistent in both its size and shape particularly in the jejunal arteries (Fig. 5). The distribution of NPY fibers in the superior mesenteric artery was more sparse but dense in the superior mesenteric vein (Fig. 6). The fibers appeared to be considerably more coarse than those of the SP/CGRP plexus. The change in density of innervation from the superior mesenteric to the jejunal arteries was obvious at the root of the jejunal arteries although again no clear concentration of fibers occurred at the branch point. The VIP plexus on the distal jejunal artery and its branches was somewhat regular (Fig. 7) but was much less regular on the proximal jejunal and superior mesenteric arteries and on the superior mesenteric vein.

Most of the reports on the mesenteric vasculature have concerned its value as a model for study of the cardiovascular system. The peptidergic innervation of the mesenteric vascular bed has also been implicated in other disease states. Accumulating evidence indicates that the SP-NK-1 receptor system represents a major immunoregulatory circuit involved in several physiological and pathophysiological gut responses and disease [98].

14. THE SKIN

The skin has an extensive blood supply, reflective of its many functions. The perivascular innervation is involved in both control of the vasculature and in the response of the skin to

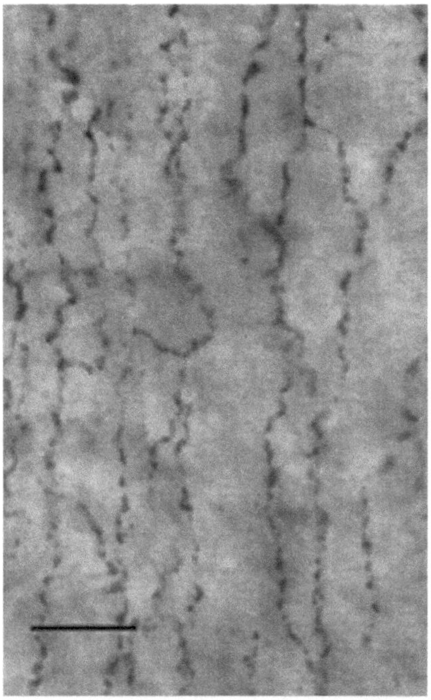

Figure 6. The parallel fibers of the NPY plexus of a vein from a 12-week-old rat. Calibration = 50 μm.

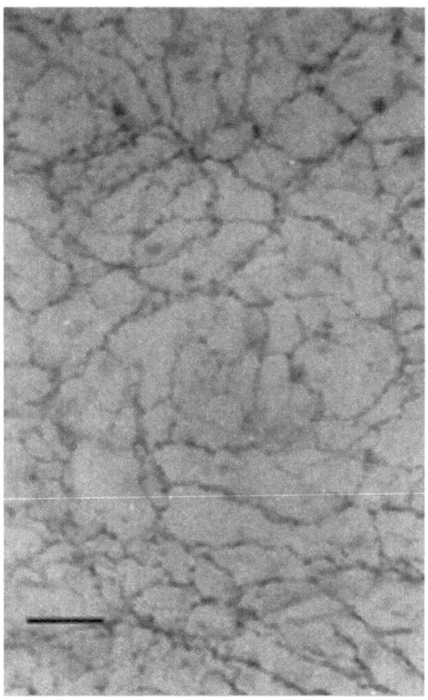

Figure 7. The deep fibers of the VIP-like immunoreactive plexus of a branch from a jejunal artery from a 12-week-old rat. Calibration = 25 μm.

injury and disease. The innervation of cutaneous arterioles is different in smooth (glabrous) or hairy (nonglabrous) areas of human skin. In the glabrous areas such as the palms and soles, the arterioles are innervated only by noradrenergic fibers [99–103]. In nonglabrous areas, cholinergic perivascular nerve fibers are also present [90–104]. Both the catecholaminergic and cholinergic innervation of the cutaneous vasculature appear to be cotransmitter systems [105–107], with NPY being present in catecholaminergic nerve fibers and VIP being released by cholinergic fibers. While the involvement of NPY in vasoconstriction in the skin has been confirmed, there is still discussion on the agent acting with acetylcholine to cause vasodilation [108]. The vasculature is also under local control by a wide range of factors including cutaneous sensory nerves that release vasoactive peptides. The peptidergic innervation of the skin and neuronal skin function have recently been reviewed [109] and will not be considered further.

15. THE JOINTS

The synovial membrane of joints is a highly vascular tissue lining the inner surface of the joint capsule. The high vascularity of the synovium is thought to be responsible for the turnover of synovial fluid [110], nutrition of chondrocytes, and provision of a source of fresh monocytes to replace macrophages in the synovium [111,112]. The synovium is also reported to be richly innervated [113], and many clinical and experimental findings have suggested a contribution to neurogenic inflammation in rheumatoid arthritis (RA) [114,115]. Early studies of human joints suggested a decrease of peptidergic innervation in human synovium in RA [116,117]. Immunohistochemical studies have demonstrated nerve fibers containing neuropeptides, such as SP, CGRP, VIP, NPY, and enkephalin in the synovium of human patients with arthritis [118,119] and of normal as well as adjuvant arthritic rats [120–122]. In the dog immunohistochemical investigations with the general neuronal marker, PGP 9.5 revealed that almost all neuronal structures in the synovium were associated with the arterial tree [123]. The perivascular innervation extended from the main artery to the precapillary arteriole, suggesting an intimate relationship between neuronal regulation and arterial function. Immunostaining for neuropeptides and TH showed the occurrence of SP, CGRP, VIP, NPY and TH immunoreactivities in these synovial nerve fibers. All of these peptides and TH immunoreactivities were demonstrated in varicose nerve fibers around the main artery, primary branches, and secondary or tertiary branches. Around more distal small arteries, SP-, CGRP-, NPY-, and TH-immunoreactive fibers were found, whereas VIP-immunoreactive fibers were no longer seen. At the most peripheral precapillary arterioles, only SP- and CGRP-immunoreactive fibers were found. Based on these findings, Nakayama et al. [123] suggest that the arterial tree receives input from different types of nerve fibers and that the neuronal effects vary according to the branching pattern. It appears that blood flow in the synovial fold in the dog wrist joint is regulated by at least three different nerve systems with regional differences. The proximal parts of the arterial tree are regulated by one vasoconstrictive system of NPY-containing noradrenergic sympathetic fibers and two vasodilative systems of VIP-containing noncatecholaminergic sympathetic fibers and of SP- and CGRP-containing sensory fibers. The more distal parts of the arterial branches are controlled by a vasoconstrictive NPY/noradrenergic sympathetic system and by vasodilatory SP/CGRP perivascular nerve fibers. The most peripheral sublining arterioles appear to be under the influence of SP/CGRP perivascular fibers alone.

The role of perivascular peptidergic innervation in neurogenic inflammation has a long history. Neurogenic inflammation is a process in which capsaicin-sensitive, afferent nerves fire in an antidromic direction to cause the peripheral release of sensory neuropeptides from nerve terminals [124]. In joints, antidromic electrical stimulation of sensory nerves results in protein extravasation and vasodilation of articular blood vessels [125,126], whereas chemical activation of these nerves by local capsaicin treatment causes obvious synovial hyperemia [127,128]. Sensory neuropeptides, which have been shown to be vasodilatory in joints and are therefore considered neurogenic inflammatory mediators, include SP [129–131], CGRP [132,133], and VIP [134]. In acutely inflamed rat knee joints, stimulation of mast cells and leukocytes leading to the secondary release of proinflammatory mediators into the joint subsequently stimulates sensory neuropeptide release from capsaicin-sensitive nerves culminating in vasodilation and increased articular blood flow [135].

16. CLINICAL SIGNIFICANCE AND FUTURE RESEARCH DIRECTIONS

Peptidergic perivascular innervation has been shown to be involved in growth, maintenance, and disease in most tissues and organs in the body. As such it is a prime target for therapeutic intervention. Although the nerves, their peptides, and receptors are well described, further work is still necessary before a full description and understanding is achieved. At this time, however, knowledge of this system forms the basis for many approaches, particularly in the case of the following important conditions.

16.1 Myocardial infarction

An extensive series of studies has shown that CGRP is not only an important mediator in early protection of cardiac ischemic preconditioning [136,137] but also participates in mediation of delayed cardioprotection afforded by other stimuli, such as heat stress [138], intestinal ischemia [139], and some drugs [140,141]. In the cardioprotection of preconditioning, early protection is ascribed to stimulation of CGRP release, whereas delayed protection is related to elevation of the synthesis and release of CGRP. Li and coworkers [142] have shown that cardiac ischemic preconditioning remarkably reduced infarct size and decreased creatine kinase release during ischemia–reperfusion, concomitantly with an increase in concentration of CGRP in plasma and the level of CGRP mRNA in dorsal root ganglion, a major site of CGRP synthesis. The effects of preconditioning were abolished by capsaicin. All indications suggest that delayed cardioprotection induced by various factors is related to the synthesis and release of CGRP.

16.2 Hypertension

Peptidergic perivascular innervation is an important contributor to the development, maintenance, and control of the cardiovascular system. As such it is an important therapeutic target in cardiovascular disease. There is some evidence that changes in peptidergic innervation contribute to the development of hypertension. The sympathetic innervation of the mesenteric vascular bed has been shown to be hypertrophied in the spontaneously hypertensive rat (SHR) [66,67]. Although CGRP and SP have been reported to be involved in hypertension, the experimental results are ambiguous. In addition to its involvement in normal vascular tone, CGRP has been proposed as an antihypertensive agent in that a loss of CGRP-immunoreactive perivascular nerves occurs as hypertension develops in the SHR [143–145]. A reduction in loss of CGRP-immunoreactive perivascular fibers and an increase in the expression of CGRP mRNA in the SHR has been shown as a result of angiotensin II type-I receptor blockade and angiotensin converting enzyme inhibition [145–148]. Lind and Edvinsson [149] have reported enhanced CGRP-induced vasodilation in human subcutaneous arteries in hypertension. The increased relaxant response was present in essential hypertension and renovascular hypertension subgroups. SP-induced vasodilation was not altered by human hypertension. Kawasaki et al. [150] also reported a change in CGRP innervation in hypertension, suggesting that the decreased vasodilator mechanism by CGRP-containing nerves contributes to the development and maintenance of hypertension. Perivascular nerve stimulation of the perfused mesenteric vascular bed evoked an increased release of CGRP-like immunoreactive substance in the perfusate, which was significantly less in 15-week-old SHR than in age-matched Wistar–Kyoto (WKY) rats. Immunohistochemical studies [150] showed an age-related decrease in CGRP-like immunoreactive fibers in the SHR but not in WKY rats. These results suggest that CGRP-containing vasodilator innervation is greatly decreased when the SHR develop and maintain hypertension. The role of SP in hypertension has not been fully determined although increased SP content has been shown in nerve fibers associated with mesenteric veins from deoxycorticosterone acetate-salt hypertensive rats [8].

16.3 Atherosclerosis

The peptides of coronary perivascular nerve fibers may play a role in atherosclerosis. Early models of atherosclerosis revealed loss of perivascular innervation [151]. Subsequent studies showed that perivascular denervation was not responsible for atherogenesis [152,153]. It has been shown that mast cells in the adventitia of coronary arteries are closely related to peptidergic perivascular nerves staining positive for the peptide neurotransmitters SP, VIP, and CGRP, all potentially capable of stimulating mast cells. Contacts between adventitial mast cells and peptidergic sensory fibers were more frequent in atherosclerotic segments of coronary vessels than in the control segments [154]. It is possible that this relationship is the basis for neurogenic inflammation in the heart.

16.4 Migraine

Although the involvement of peptidergic perivascular innervation in migraine has been suspected for many years, recent studies have confirmed the involvement of the trigeminal system [155,156]. As described earlier, the intracranial vessels are innervated by SP, CGRP, and NKA immunopositive fibers [20], but it appears that only the CGRP fibers are involved in migraine [20,157,158]. Migraine research is now focused, among other avenues [150], on prevention through interference with CGRP synthesis, release, and receptor binding.

16.5 Cutaneous disease

There is significant evidence that the cutaneous peripheral nervous system, including the peptidergic perivascular innervation, plays an important role in the extensive catalog of normal skin function and the wide range of skin diseases. Arterial sections of arteriovenous anastomoses, precapillary sphincters of metarterioles, arteries, and capillaries appear to be the most intensely innervated regions. Although sensory nerves are important for vasodilation, neuropeptides from sympathetic neurons such as NPY mediate vasoconstriction supporting an important role for neuropeptides in vascular regulation. Both endothelial cells and smooth muscle cells respond to neuronal modulation during processes such as inflammation, cellular immune responses, neovascularization, wound healing [109], and heating and cooling [108]. A recent review by Roosterman et al. [109] considers the role of innervation in urticaria, psoriasis, atopic dermatitis, immediate and delayed-type hypersensitivity, wound healing, and pruritus, confirming the important role played by peptides in cutaneous disease. Hendrix and Peters [159] have reviewed key studies on bilateral neuroimmune interactions in the skin under both healthy and disease conditions providing a basis for future research. In the skin, "itch" (pruritus), "excessive sweating" (hyperhidrosis), and "flushing" (facial erythema) and many dermatoses such as atopic dermatitis, psoriasis, seborrheic eczema, prurigo nodularis, lichen planus, chronic urticaria, and alopecia areata, can be triggered or aggravated by stress [160], working through the central and peripheral nervous systems. Thus, the peptidergic innervation of the skin can be seen as a novel therapeutic target in the prevention of cutaneous disease in response to stress [161].

16.6 Arthritis

Classical symptoms of both inflammatory and degenerative arthritis may contribute to neurogenic responses like wheal, flare, edema, and pain. RA is an autoimmune disease with a significant immunogenetic component. Neurogenic inflammation has been considered to play an essential role in RA, in part because of the symmetrical and predominant involvement of the most heavily innervated small joints of the hands and the feet. It is possible that the age-related degeneration of the proprioceptive, kinesthetic, and vasoregulatory nerves represents the primary pathogenic events, leading to progressive damage of tissue with extremely poor capacity for self-regeneration [162]. The peptides of the perivascular innervation are involved in the process of RA and represent a target for the development of new therapeutic approaches.

17. CONCLUSION

Peptidergic perivascular innervation is one component of a complex control system that influences the vasculature and the tissues supplied by blood vessels. The system acts both under the control of the central nervous system and independent of it, responding to local conditions through the release of peptides. Peptides released by perivascular nerves not only act on local vascular tissues to alter blood flow but also interact with the components of other body systems resulting in an integrated response. Conversely, dysregulation of peptidergic innervation may be seen as an important contributor to the pathology of a variety of common medical conditions.

REFERENCES

1. Furness JB, Costa M, Papka RE, Della NG, Murphy R. Neuropeptides contained in peripheral cardiovascular nerves. Clin Exp Hypertens A 1984;6(1–2):91–106.
2. von Euler US, Gaddum JH. An unidentified depressor substance in certain tissue extracts. J Physiol (Lond) 1931;72:74–87.

3. Cascieri MA, Huang RR, Fong TM, Cheung AH, Sadowski S, Ber E et al. Determination of the amino acid residues in substance P conferring selectivity and specificity for the rat neurokinin receptors. Mol Pharmacol 1992;41:1096–1099.
4. Maggi CA. The mammalian tachykinin receptors. Gen Pharmacol 1995;26:911–944.
5. Hastrup H, Schwartz TW. Septide neurokinin A are high-affinity ligands on the NK-1 receptor: evidence from homologous versus heterologous binding analysis. FEBS Lett 1996;399:264–266.
6. Lecci A, Maggi CA. Peripheral tachykinin receptors as potential therapeutic targets in visceral diseases. Expert Opin Ther Targets 2003;7:343–362.
7. Page NM. New challenges in the study of the mammalian tachykinins. Peptides 2005;26:1356–1368.
8. Pennefather JN, Leccib A, Candenasc ML, Patakd E, Pintoc FM, Maggi CA. Tachykinins and tachykinin receptors: a growing family. Life Sci 2004;74:1445–1463.
9. Kim HR, Lavielle S, Sagan S. The two NK-1 binding sites are distinguished by one radiolabelled substance P analogue. Biochem Biophys Res Commun 2003;306: 725–729.
10. Brain SD, Grant AD. Vascular actions of calcitonin gene-related peptide and adrenomedullin. Physiol Rev 2004;84:903–934.
11. Galligan JJ, Miller SB, Katki K, Supowit S, DiPette D, Fink GD. Increased substance P content in nerve fibers associated with mesenteric veins from deoxycorticosterone acetate (DOCA)-salt hypertensive rats. Regul Pept 2006;133:97–104.
12. D'Orleans-Juste P, Claing A, Telemaque S, Warner TD, Regoli D. Neurokinins produce selective venocontriction via NK-3 receptors in the rat mesenteric vascular bed. Eur J Pharmacol 1991;204:329–334.
13. Grant AD, Gerard NP, Brain SD. Evidence of a role for NK-1 and CGRP receptors in mediating neurogenic vasodilation in the mouse ear. Br J Pharmacol 2002;135:356–362.
14. Burns DM, Howard GA, Roos BA. An assessment of the anabolic skeletal actions of the common-region peptides derived from the CGRP and calcitonin prohormones. Ann N Y Acad Sci 1992;657:50–62.
15. Amara SG, Arriza JL, Leff SE, Swanson LW, Evans RM, Rosenfeld MG. Expression in brain of a messenger RNA encoding a novel neuropeptide homologous to calcitonin gene-related peptide. Science 1985;229:1094–1097.
16. Mclatchie LM, Fraser NJ, Main MJ, Wise A, Brown J, Thompson N et al. RAMPs regulate the transport and ligand specificity of the calcitonin-receptor-like-receptor. Nature 1998;393:333–339.
17. Brain SD, Williams TJ, Tippins JR, Morris HR, MacIntyre I. Calcitonin gene-related peptide is a potent vasodilator. Nature 1985;313:54–56.
18. Saetrum Opgaard O, Gulbenkian S, Bergdahl A, Barroso CP, Costa Andrade N, Polak JM et al. Innervation of human epicardial coronary veins: immunohistochemistry and vasomotility. Cardiovasc Res 1995;29:463–468.
19. Gulbenkian S, Saetrum Opgaard O, Ekman R, Costa Andrade N, Wharton J, Polak JM et al. Peptidergic innervation of human epicardial coronary arteries. Circ Res 1993;73:579–588.
20. Edvinsson L. Blockade of CGRP receptors in the intracranial vasculature: a new target in the treatment of headache. Cephalalgia 2004;24:611–622.
21. Wimalawansa SJ. Blood pressure and cardiovascular tone: role of CGRP family of peptides. Scientific World 2001;1:32.

22. Chang Y, Stover SR, Hoover DB. Regional localization and abundance of calcitonin gene-related peptide receptors in guinea pig heart. J Mol Cell Cardiol 2001;33:745–754.
23. Coupe MO, Mak JC, Yacoub M, Oldershaw PJ, Barnes PJ. Autoradiographic mapping of calcitonin gene-related peptide receptors in human and guinea pig hearts. Circulation 1990;81:741–747.
24. Poyner DR, Sexton PM, Marshall I, Smith DM, Quirion R, Born W et al. International Union of Pharmacology: XXXII. The mammalian calcitonin gene-related peptides, adrenomedullin, amylin, and calcitonin receptors. Pharmacol Rev 2002;54:233–246.
25. Saetrum Opgaard O, Hasbak P, de Vries R, Saxena PR, Edvinsson L. Positive inotropy mediated via CGRP receptors in isolated human myocardial trabeculae. Eur J Pharmacol 2000;397:373–382.
26. Gangula PRR, Zhao H, Supowit SC, Wimalawansa SJ, Dipette DJ, Westlund KN et al. Increased blood pressure in alpha-calcitonin gene-related peptide/calcitonin gene knockout mice. Hypertension 2000;35:470–475.
27. Lu JT, Son Y-J, Lee J, Jetton TJ, Shiota M, Moscoso L et al. Mice lacking alpha-calcitonin gene-related peptide exhibit normal cardiovascular regulation and neuromuscular development. Mol Cell Neurosci 1999;14:99–120.
28. Edvinsson L, Hfakanson R, Wahlestedt C, Uddman R. Effects of neuropeptide Y on the cardiovascular system. Trends Pharmacol Sci 1987;8:231–235.
29. Edvinsson L, Ekman R, Hedner P, Valdermarsson S. Congestive heart failure: involvement of perivascular peptides reflecting activity in sympathetic, parasympathetic and afferent fibres. Eur J Clin Invest 1990;20:85–89.
30. Tseng C-J, Robertson D, Light RT, Atkinson JR, Robertson RM. Neuropeptide Y is a vasoconstrictor of human coronary arteries. Am J Med Sci 1988;296:11–16.
31. Brain SD, Cox HM. Neuropeptides and their receptors: innovative science providing novel therapeutic targets. Br J Pharmacol 2006;147:S202–S211.
32. Gulbenkian S, Saetrum Opgaard O, Ekman R, Costa Andrade N, Wharton J, Polak JM et al. Peptidergic innervation of human epicardial coronary arteries. Circ Res 1993;73:579–588.
33. Jansen I, Uddman R, Ekman R, Olesen J, Ottosson A, Edvinsson L. Distribution and effects of neuropeptide Y, vasoactive intestinal peptide, substance P, and calcitonin gene-related peptide in human middle meningeal arteries: comparison with cerebral and temporal arteries. Peptides 1992;13(3):527–536.
34. Fallgren B, Ekblad E, Edvinsson L. Co-existence of neuropeptides and differential inhibition of vasodilator responses by neuropeptide Y in guinea pig uterine arteries. Neurosci Lett 1989;100:71–76.
35. Gulbenkian S, Edvinsson L, Saetrum Opgaard O, Valenqa A, Wharton J, Polak JM. Neuropeptide Y modulates the action of vasodilator agents in guinea pig epicardial coronary arteries. Regul Pept 1992;40:351–362.
36. Pedrazzini T, Seydoux J, Kunster P, Aubert J-F, Grouzmann E, Beermann F et al. Cardiovascular response, feeding behavior and locomotor activity in mice lacking the NPY Y1 receptor. Nature 1998;4(6):722–726.
37. Smith-White MA, Herzog H, Potter EK. Cardiac function in neuropeptide Y Y_4 receptor-knockout mice. Regul Pept 2002;110:47–54.
38. Naveilhan P, Hassani H, Canals JM, Ekstrand AJ, Larefalk A, Chhajlani V et al. Normal feeding behavior, body weight and leptin response require the neuropeptide Y Y_2 receptor. Nature 1999;5(10):1188–1193.

39. Laburthe M, Couvineau A, Marie JC. VPAC receptors for VIP and PACAP. Recept Channels 2002;8(3–4):137–153.
40. Franco-Cereceda A, Rudehill A. Capsaicin-induced vasodilation of human coronary arteries in vitro is mediated by calcitonin gene-related peptide rather than substance P or neurokinin. Acta Physiol Scand 1989;136:575–580.
41. Popma JJ, Smitherman TC, Bedotto JB, Eichhorn EJ, Said SI, Dehmer GJ. Direct coronary vasodilation induced by intracoronary vasoactive intestinal peptide. J Cardiovasc Pharmacol 1990;16:1000–1006.
42. Ignarro LJ, Byrns RE, Buga GM, Wood KS. Mechanisms of endothelium-dependent vascular smooth muscle relaxation elicited by bradykinin and VIP. Am J Physiol 1987;253:H1074–H1082.
43. Sata T, Linden J, Liu L-W, Kubota E, Said SI. Vasoactive intestinal peptide evokes endothelium-dependent relaxation and cyclic AMP accumulation in rat aorta. Peptides 1988;9:853–858.
44. Pernow J, Lundberg JM. Neuropeptide Y induces potent contraction of arterial vascular smooth muscle via an endothelium independent mechanism. Acta Physiol Scand 1988;134:157–158.
45. Said SI, Hamidi A, Dickman KG, Szema AM, Lyubsky S, Lin RZ et al. Moderate pulmonary arterial hypertension in male mice lacking the vasoactive intestinal peptide gene. Circulation 2007;115:1260–1268.
46. Edvinsson L, Uddman R, Juul R. Peptidergic innervation of the cerebral circulation. Role in subarachnoid hemorrhage in man. Neurosurg Rev 1990;13(4):265–272.
47. Edvinsson L, Hara H, Uddman R. Retrograde tracing of nerve fibers to the rat middle cerebral artery with true blue: colocalization with different peptides. J Cereb Blood Flow Metab 1989;9(2):212–218.
48. Alafaci C, Tomasello F, Burnstock G. Histochemical and immunohistochemical study of noradrenergic, serotonergic and peptidergic innervation of the cerebral circulation. Funct Neurol 1987;2(4):417–420.
49. Hökfelt T, Johansson O, Ljungdahl A, Lundberg JM, Schultzberg M. Peptidergic neurons. Nature 1980;284:515–521.
50. Lundberg JM, Terenius L, Hökfelt T, Martling CR, Tatemoto K, Mutt V et al. Neuropeptide Y (NPY)-like immunoreactivity in peripheral noradrenergic neurons and effects of NPY on sympathetic function. Acta Physiol Scand 1982;116: 477–480.
51. Lundberg JM, Franco-Cereceda A, Hokfelt T, Terenius L, Goldstein M. Differential effects of reserpine and 6-hydroxydopamine on neuropeptide Y (NPY) and noradrenaline in peripheral neurones. Naunyn Schmiedebergs Arch Pharmacol 1985;328:331–340.
52. Lazarov NE. Comparative analysis of the chemical neuroanatomy of the mammalian trigeminal ganglion and mesencephalic trigeminal nucleus. Prog Neurobiol 2002;66:19–59.
53. Gulbenkian S, Edvinsson L, Saetrum Opgaard O, Wharton J, Polak JM et al. Peptide-containing nerve fibres in guinea-pig coronary arteries: immunohistochemistry, ultrastructure and vasomotility. J Auton Nerv Syst 1990;31:153–168.
54. Wharton J, Gulbenkian S, Mulderry PK, Ghatei MA, MacGregor GP, Bloom SR et al. Capsaicin induces a depletion of calcitonin gene-related peptide (CGRP)-immunoreactive nerves in the cardiovascular system of the guinea pig and rat. J Auton Nerv Syst 1986;16:289–309.

55. Franco-Cereceda A, Henke H, Lundberg JM, Petermann JB, Hökfelt T, Fischer JA. Calcitonin gene-related peptide (CGRP) in capsaicin-sensitive substance P-immunoreactive sensory neurons in animals and man: distribution and release by capsaicin. Peptides 1987;8:399–410.
56. Lundberg JM, Franco-Cereceda A, Hua X, Hökfelt T, Fischer JA. Co-existence of substance P and calcitonin gene-related peptide-like immunoreactivities in sensory nerves in relation to cardiovascular and bronchoconstrictor effects of capsaicin. Eur J Pharmacol 1985;108:315–319.
57. Scott T M, Robinson J, Foote J. Two fibre plexuses of the rat mesenteric vascular bed. J Anat 1987;152:262.
58. Gulbenkian S, Merighi A, Wharton J, Varndell IM, Polak JM. Ultrastructural evidence for the coexistence of calcitonin gene-related peptide and substance P in secretory vesicles of peripheral nerves in the guinea pig. J Neurocytol 1986;15: 535–542.
59. Edvinsson L, Emson P, McCulloch J, Tatemoto K, Uddman R., Neuropeptide Y. Cerebrovascular innervation and vasomotor effects in the cat. Neurosci Lett 1983;43:79–84.
60. Edvinsson L, Copeland JR, Emson PC, McCulloch J, Uddman R. Nerve fibres containing neuropeptide Y in the cerebrovascular bed: immunocytochemistry, radioimmunoassay, and vasomotor effects. J Cereb Blood Flow Metab 1987;7:45–57.
61. Nielsen KC, Owman C. Adrenergic innervation of pial arteries related to the circle of Willis in the cat. Brain Res 1967;6:773–776.
62. Edvinsson L, Hara H, Uddman R. Retrograde tracing of nerve fibres to the rat middle cerebral artery with True Blue. Colocalization with different peptides. J Cereb Blood Flow Metabol 1989;9:212–228.
63. Edvinsson L, Nielsen KC, Owman C, Sporrong B. Cholinergic mechanisms in pial vessels. Histochemistry, electron microscopy and pharmacology. Z Zellforsch 1972;311:311–325.
64. Hamel E, Edvinsson L, MacKenzie ET. Heterogeneous vasomotor responses of anatomically distinct feline cerebral arteries. Br J Pharmacol 1988;94:423–436.
65. Larsson LI, Edvinsson L, Fahrenkrug J, Hanson R, Owman C, Schaffalitzky D et al. Immunohistochemical localization of a vasodilatory polypeptide (VIP) in cerebrovascular nerves. Brain Res 1976;113:400–404.
66. Luff SE, Young SB, McLachlan EM. Hyperinnervation of mesenteric arteries in spontaneously hypertensive rats by sympathetic but not primary afferent axons. J Vasc Res 2005;42(4):348–358.
67. Scott TM, Pang SC. The correlation between the development of sympathetic innervation and the development of medial hypertrophy in jejunal arteries in normotensive and spontaneously hypertensive rats. J Auton Nerv Syst 1983;8:25–32.
68. Hobara N, Goda AM, Kitamura AY, Sendou BT, Gomitac C, Kawasakia H. Adrenomedullin facilitates reinnervation of phenol-injured perivascular nerves in the rat mesenteric resistance artery. Neuroscience 2007;144:721–730.
69. Hobara N, Nakamura A, Ohtsuka A, Narasaki M, Shibata K, Gomoita Y et al. Distribution of adrenomedullin-containing perivascular nerves in the rat mesenteric artery. Peptides 2004;25:589–599.
70. Scott TM, Robinson J, Foote J. The peptidergic innervation of the developing mesenteric vascular bed in the rat. J Anat 1989;62:177–183.
71. Luff SE. Development of neuromuscular junctions on small mesenteric arteries of the rat. J Neurocytol 1999;28(1):47–62.

72. Tsai SH, Tew JM, Shipley MT. Cerebral arterial innervation: II. Development of calcitonin-gene-related peptide and norepinephrine in the rat. J Comp Neurol 1989;279(1):1–12.
73. Ando K, Ishikawa A, Kawakami H, Sakuma Y. Development of AChE-positive, NA-containing and VIP- and NPY-immunoreactive nerves in the major cerebral arteries of the rat. Anat Embryol (Berl) 1991;184(1):25–32.
74. Kawamura K, Takebayashi S. The development of noradrenaline acetylcholinesterase-, neuropeptide Y- and vasoactive intestinal polypeptide-containing nerves in human cerebral arteries. Neurosci Lett 1994;175(1–2):1–4.
75. Gibbins IL, Morris JL. Sympathetic noradrenergic neurons containing dynorphin but not neuropeptide Y innervate small cutaneous blood vessels of guinea-pigs. J Auton Nerv Syst 1990;29:137–150.
76. Janig W, Baron R. Complex regional pain syndrome is a disease of the central nervous system. Clin Auton Res 2002;12:150–164.
77. Cech S. Adrenergic innervation of blood vessels in the lung of some mammals. Acta Anat 1969;74:169–182.
78. O'Donnell SR, Saar N, Wood LJ. The density of adrenergic nerves at various levels in the guinea pig lung. Clin Exp Pharmacol Physiol 1978;5:325–332.
79. Haberberger R, Schemann M, Sann H, Kummer W. Innervation pattern of guinea pig pulmonary vasculature depends on vascular diameter. J Appl Physiol 1997;82:426–434.
80. West JB. Blood flow to the lung and gas exchange. Anesthesiology 1974;41:124–138.
81. Wharton J, Polak JM, Mcgregor GP, Bishop AE, Bloom SR. The distribution of substance P-like immunoreactive nerves in the guinea-pig heart. Neuroscience 1981;6:2193–2204.
82. Wharton J, Polak JM, Probert J, De May GP, Mcgregor GP, Bryant MG et al. Peptide containing nerves in the ureter of the guinea-pig and cat. Neuroscience 1981;6:969–982.
83. Wanaka A, Matsuyama T, Yoneda S, Kimura K, Kamada T, Girgis S et al. Origins and distribution of calcitonin gene-related peptide-containing nerves in the walls of the cerebral arteries of the guinea-pig with special reference to the coexistence with substance P. Brain Res 1986;369:185–192.
84. Uddman R, Ekblad E, Edvinsson L, Hakanson R, Sundler F. Neuropeptide Y-like immunoreactivity in perivascular nerve fibres of the guinea-pig. Regul Pept 1985;10:243–257.
85. Uddman R, Edvinsson L, Hara H. Axonal tracing of autonomic nerve fibers to the superficial temporal artery in the rat. Cell Tissue Res 1989;256:559–565.
86. Edvinsson L, Hamel E. Perivascular nerves in brain vessels. In: Edvinsson L, Krause DN, eds. Cerebral Blood Flow and Metabolism 2nd ed. Philadelphia, PA: Lippincott Williams & Wilkins, 2002; pp. 43–67.
87. Jansen I, Uddman R, Ekman R, Olesen J, Ottosson A, Edvinsson L. Distribution and effects of neuropeptide Y, vasoactive intestinal peptide, substance P, and calcitonin gene-related peptide in human middle meningeal arteries: comparison with cerebral and temporal arteries. Peptides 1992;13(3):527–536.
88. Olesen IJ, Gulbenkian S, Valenca A, Antunes JL, Wharton J, Polak JM et al. The peptidergic innervation of the human superficial temporal artery: immunocytochemistry, ultrastructure, and vasomotilin. Peptides 1995;16(2):275–287.
89. You J, Gulbenkian S, Jansen Olesen I, Marron K, Wharton J, Barroso CP et al. Peptidergic innervation of guinea-pig brain vessels: comparison with immunohistochemistry and in vitro pharmacology in rostrally and caudally located arteries. J Auton Nerv Syst 1995;55(3):179–188.

90. Troger J, Kieselbach G, Teuchner B, Kralinger M, Nguyen OA, Haas G et al. Peptidergic nerves in the eye, their source and potential pathophysiological relevance. Brain Res Rev 2007;53:39–62.
91. Edvinsson L. Calcitonin gene-related peptide (CGRP) in cerebrovascular disease. Sci World J 2002;30(2):1484–1490.
92. Geppetti P, Trevisani M, Nicoletti P, Capone JG. Cephalic pain. Neurol Sci 2006;27:S111–S114.
93. Nobin A, Owman C, Arenklo-Nobin B eds. Neuronal Messengers in Vascular Function. Amsterdam: Elsevier Science Publishing Co, Inc, 1987; p. 570.
94. Wharton J, Gulbenkian S. Peptides in the mammalian cardiovascular system. Experientia 1987;43:821–832.
95. Gulbenkian S, Saetrum Opgaard O, Ekman R, Costa Andrade N, Wharton J, Polak JM et al. peptidergic innervation of human epicardial coronary arteries. Circ Res 1993;73:579–588.
96. Saetrum Opgaard O, Gulbenkian S, Bergdahl A, Barroso CP, Costa Andrade N, Polak JM et al. Innervation of human epicardial coronary veins: immunohistochemistry and vasomotility. Cardiovasc Res 1995;29:463–468.
97. Sequeiraa IM, Haberbergera RV, Kummera W. Atrial and ventricular rat coronary arteries are differently supplied by noradrenergic, cholinergic and nitrergic, but not sensory nerve fibres. Ann Anat 2005;187:345–355.
98. Koon HW, Pothoulakisi C. Immunomodulatory properties of substance P: the gastrointestinal system as a model. Ann NY Acad Sci 2006;1088:23–40.
99. Fox RH, Edholm OG. Nervous control of the cutaneous circulation. Br Med Bull 1963;19:110–114.
100. Johnson JM. Nonthermoregulatory control of human skin blood flow. J Appl Physiol 1986;61:1613–1622.
101. Johnson JM, Brengelmann GL, Hales JRS, Vanhoutte PM, Wenger CB. Regulation of the cutaneous circulation. Fed Proc 1986;45:2841–2850.
102. Johnson JM, Proppe DW. Cardiovascular adjustments to heat stress. In: Fregley MJ, Blatteis CM, eds. Handbook of Physiology. Environmental Physiology. Bethesda, MD: Am. Physiol Soc., 1996; pp. 215–243.
103. Rowell LB. Reflex control of the cutaneous vasculature. J Invest Dermatol 1977;69:154–166.
104. Charkoudian N, Johnson JM. Reflex control of cutaneous vasoconstrictor system is reset by exogenous female reproductive hormones. J Appl Physiol 1999;87:381–385.
105. Stephens DP, Aoki K, Kosiba WA, Johnson JM. Nonnoradrenergic mechanism of reflex cutaneous vasoconstriction in men. Am J Physiol Heart Circ Physiol 2001;280:H1496–H1504.
106. Thompson CS, Kenney WL. Altered neurotransmitter control of reflex vasoconstriction in aged human skin. J Physiol 2004;558:697–704.
107. Kellogg DL, Pergola PE, Kosiba WA, Grossmann M, Johnson JM. Cutaneous active vasodilation in humans is mediated by cholinergic nerve co-transmission. Circ Res 1995;77:1222–1228.
108. Kellogg DL. In vivo mechanisms of cutaneous vasodilation and vasoconstriction in humans during thermoregulatory challenges. J Appl Physiol 2006;100:1709–1718.
109. Roosterman D, Goerge T, Schneider SW, Bunnett NW, Steinhoff M. Neuronal control of skin function: the skin as a neuroimmunoendocrine organ. Physiol Rev 2006;86: 1309–1379.

110. Levick JR. An investigation into the validity of subatmospheric pressure recordings from synovial fluid and their dependence on joint angle. J Physiol 1979;289:55–67.
111. Edwards JCW. Functions of synovial lining. In: Henderson B, Edwards JCW, eds. The Synovial Lining in Health and Disease. London: Chapmann & Hall, 1987; 41–74.
112. Wilkinson LS, Edwards JCW. Microvascular distribution in normal human synovium. J Anat 1989;167:129–136.
113. Kennedy JC, Alexander J, Hayer KC. Nerve supply of the human knee and its functional importance. Am J Sports Med 1982;10:329–335.
114. Levine JD, Moskowitz MA, Basbaum AT. The contribution of neurogenic inflammation in experimental arthritis. J Immunol 1985;135:843–847.
115. Kidd BL, Mapp PI, Gibson SJ, Polak JM, O'Higgins F, Buckland-Wright, JC. Hypothesis. A neurogenic mechanism for symmetrical arthritis. Lancet 1989;2:1128–1130.
116. Pereira da Silva JA, Carmo-Fonseca M. Peptide containing nerves in human synovium: immunohistochemical evidence for decreased innervation in rheumatoid arthritis. J Rheumatol 1990;17(12):1592–1599.
117. Mapp PI, Kidd BL, Gibson SJ, Terry JM, Revell PA, Ibrahim NB et al. Substance P-, calcitonin gene-related peptide- and C-flanking peptide of neuropeptide Y-immunoreactive fibres are present in normal synovium but depleted in patients with rheumatoid arthritis. Neuroscience 1990;37(1):143–153.
118. Gronblad R, Konttinen YT, Korkala O, Liesi P, Hukkanen M, Polak JM. Neuropeptides in synovium of patients with rheumatoid arthritis and osteoarthritis. J Rheum 1988;16:1807–1810.
119. Silva JAP, Carmo-Fonseca M. Peptide containing nerves in human synovium: Immunohistochemical evidence for decreased innervation in rheumatoid arthritis. J Rheum 1990;17:1592–1599.
120. Bjurholm A, Kreicbergs A, Ahmed M, Schultzberg M. Noradrenergic and peptidergic nerves in the synovial membrane of the Sprague-Dawley rat. Arthritis Rheum 1990;33:859–865.
121. Konttinen YT, Rees R, Hukkanen M, Gronblad M, Tolvanen E, Gibson SJ et al. Nerves in inflammatory synovium: immunohistochemical observations on the adjuvant arthritic rat model. J Rheum 1990;17:1586–1591.
122. Hukkanen M, Konttinen YT, Rees RG, Santavirta S, Terenghi G, Polak JM. Distribution of nerve endings and sensory neuropeptides in rat synovium, meniscus and bone. Int J Tissue React 1992;14:1–10.
123. Nakayama K, Masuko S, Tsuruta T, Kutsuna T, Watanabe H. Vascularization and innervation of the canine wrist Joint synovial membrane. Tohoku J Exp Med 1995;175:195–209.
124. Jancso G, Jancso-Gabor A, Szolcsanyi J. Direct evidence for neurogenic inflammation and its prevention by denervation and pretreatment with capsaicin. Br J Pharmacol Chemother 1967;31:138–151.
125. Ferrell WR, Russell NJW. Extravasation in the knee induced by antidromic stimulation of articular C fibre afferents of the anaesthetised cat. J Physiol 1986;379:407–416.
126. Ferrell WR, Cant R. Vasodilation of articular blood vessels induced by antidromic electrical stimulation of joint C fibres. In: Schmidt RF, Schaible H-G, Vahle-Hinz C. eds. Fine Afferent Nerve Fibres and Pain. Weinheim: VCH Verlagsgesellschaft, 1987; 187–192.
127. Karimian SM, Mcdougall JJ, Ferrell WR. Neuropeptidergic and autonomic control of the vasculature of the rat knee joint revealed by laser Doppler perfusion imaging. Exp Physiol 1995;80:341–348.

128. Varga A, Nemeth J, Szabo A, Mcdougall JJ, Zhang C, Elekes K et al. Effects of the novel TRPV1 receptor antagonist SB366791 *in vitro* and *in vivo* in the rat. Neurosci Lett 2005;398:135–138.
129. Lam FY, Ferrell WR. Specific neurokinin receptors mediate plasma extravasation in the rat knee joint. Br J Pharmacol 1991;103:1263–1267.
130. Lam FY, Ferrell WR. Effects of interactions of naturally-occurring neuropeptides on blood flow in the rat knee joint. Br J Pharmacol 1993;8:694–699.
131. Mcdougall JJ, Karimian SM, Ferrell WR. Prolonged alteration of sympathetic vasoconstrictor and peptidergic vasodilator responses in rat knee joints by adjuvant-induced arthritis. Exp Physiol 1995;80:349–357.
132. Ferrell WR, Mcdougall JJ, Bray RC. Spatial heterogeneity in the effects of calcitonin gene-related peptide (CGRP) on the microvasculature of ligaments in the rabbit knee joint. Br J Pharmacol 1997;121:1397–1405.
133. Mcmurdo L, Lockhart JC, Ferrell WR. Modulation of synovial blood flow by the calcitonin gene-related peptide (CGRP) receptor antagonist, $CGRP_{(8-37)}$. Br. J Pharmacol 1997;121:1075–1080.
134. Mcdougall JJ, Barin AK. The role of joint nerves and mast cells in the alteration of vasoactive intestinal peptide (VIP) sensitivity during inflammation progression in rats. Br J Pharmacol 2005;145:104–113.
135. Zhang C, McDougall JJ. Stimulation of sensory neuropeptide release by nociceptin/orphanin FQ leads to hyperaemia in acutely inflamed rat knees. Br J Pharm 2006;148:938–946.
136. Ouyang W, Qian XX, Li ZL, Fu XY, Wang SH. The roles of endogenous calcitonin gene-related peptide on myocardial ischemic preconditioning in intact rat. China J Arterioscler 1999;7:24–26.
137. Lu R, Li YJ, Deng HW. Evidence for calcitonin gene-related peptide-mediated ischemic preconditioning in the rat heart. Regul Pept 1999;30:53–57.
138. Song QJ, Li YJ, Deng HW. Early and delayed cardioprotection by heat stress is mediated by calcitonin gene-related peptide. Naunyn Schmiedeberg's Arch Pharmacol 1999;359:477–483.
139. Tang ZL, Dai W, Li YJ, Deng HW. Involvement of capsaicin sensitive sensory nerves in early and delayed cardioprotection induced by a brief ischaemia of the small intestine. Naunyn Schmiedeberg's Arch Pharmacol 1999;359:243–247.
140. He SY, Deng HW, Li YJ. Monophosphoryl lipid A-induced delayed preconditioning is mediated by calcitonin gene-related peptide. Eur J Pharmacol 2001;420:143–149.
141. Zhou ZH, Peng J, Ye F, Li NS, Deng HW, Li YJ. Delayed cardioprotection induced by nitroglycerin is mediated by alphacalcitonin gene-related peptide. Naunyn Schmiedeberg's Arch Pharmacol 2002;365:253–259.
142. Luo D, Deng PY, Ye F, Peng WJ, Deng HW, Li YJ. Delayed preconditioning by cardiac ischemia involves endogenous calcitonin gene-related peptide via the nitric oxide pathway. Eur J Pharm 2004;502:135–141.
143. Kawasaki H, Saito A, Takasaki K. Age-related decrease of calcitonin gene-related peptide-containing vasodilator innervation in the mesenteric resistance vessel of the spontaneously hypertensive rat. Circ Res 1990;67:733–743.
144. Kawasaki H, Takasaki K. Age-related decrease of neurogenic release of calcitonin gene-related peptide from perivascular nerves in spontaneously hypertensive rats. Clin Exp Hypertens 1992;14:989–1001.

145. Kawasaki H, Nuki Y, Yamaga N, Kurosaki Y, Taguchi T. Decreased depressor response mediated by calcitonin gene-related peptide (CGRP)-containing vasodilator nerves to spinal cord stimulation and levels of CGRP mRNA of the dorsal root ganglia in spontaneously hypertensive rats. Hypertens Res 2001;23:693–699.
146. Kawasaki H, Okazaki M, Nakatsuma A, Mimaki Y, Araki H, Gomita Y. Long-term treatment with angiotensin converting enzyme inhibitor restores reduced calcitonin gene-related peptide-containing vasodilator nerve function in mesenteric artery of spontaneously hypertensive rats. Jpn J Pharmacol 1999;79:221–229.
147. Kawasaki H, Inaizumi K, Nakamura A, Hobara N, Kurosaki Y. Chronic angiotensin II inhibition increases level of calcitonin gene related peptide mRNA of the dorsal root ganglia in spontaneously hypertensive rats. Hypertens Res 2003;26:257–263.
148. Hobara N, Goda M, Kitamura Y, Takayama F, Kawasaki H. Innervation and functional changes in mesenteric perivascular calcitonin gene-related peptide- and neuropeptide Y-containing nerves following topical phenol treatment. Neuroscience 2006;141:1087–1099.
149. Lind H, Edvinsson L. Enhanced vasodilator responses to calcitonin gene-related peptide (CGRP) in subcutaneous arteries in human hypertension. J Human Hyp 2002;16:53–59.
150. Kawasaki H, Saito A, Takasaki K. Age-related decrease of neurogenic release of calcitonin gene-related peptide from perivascular nerves in spontaneously hypertensive rats. Clin Exp Hypertens A 1992;14(6):989–1001.
151. Scott TM, Honey AC, Martin JF, Booth RF. Perivascular innervation is lost in experimental atherosclerosis. Cardioscience 1992;3(3):145–153.
152. Butt RD, Scott TM. Vascular innervation in atherogenesis. Artery 1997;22(6):336–345.
153. De Meyer GR, Van Put DJ, Kockx MM, Van Schil P, Bosmans R, Bult H et al. Possible mechanisms of collar-induced intimal thickening. Arterioscler Thromb Vasc Biol 1997;17(10):1924–1930.
154. Laine P, Naukkarinen A, Heikkilä L, Penttilä A, Kovanen PT. Adventitial mast cells connect with sensory nerve fibers in atherosclerotic coronary arteries. Circulation 2000;101:1665–1669.
155. De Vries P, Villalón CM, Saxena PR. Pharmacological aspects of experimental headache models in relation to acute antimigraine therapy. Eur J Pharmacol 1999;375:61–74.
156. Villalón CM, Centurion D, Valdivia LF, De Vries P, Saxena PR. An introduction to migraine: from ancient treatment to functional pharmacology and antimigraine therapy. Proc West Pharmacol Soc 2002;45:199–210.
157. Goadsby PJ, Lipton RB, Ferrari MD. Migraine – current understanding and treatment. N Engl J Med 2002;346:257–270.
158. Arulmania U, VanDenBrinka AM, Villalon CM, Saxenaa PR. Calcitonin gene-related peptide and its role in migraine pathophysiology. Eur J Pharm 2004;500:315–330.
159. Hendrix S, Peters EMJ. Neuronal plasticity and neuroregeneration in the skin: the role of inflammation. J Neuroimmunol 2007;184(1–2):113–126.
160. Webster EL, Barrientos RM, Contoreggi C, Isaac MG, Ligier S, Gabry KE et al. Corticotropin releasing hormone (CRH) antagonist attenuates adjuvant induced arthritis: role of CRH in peripheral inflammation. J Rheumatol 2002;29:1252–1261.
161. Paus R, Theoharides TC, Arck PC. Neuroimmunoendocrine circuitry of the "brain-skin connection". Trends Immunol 2006;27:32–39.
162. Niissalo S, Hukkanen M, Imai S, Törnwall J, Konttinen YT. Neuropeptides in experimental and degenerative arthritis. Ann NY Acad Sci 2002;966:384–399.

Neurogenic Inflammation in Health and Disease
Edited by Gábor Jancsó

Molecular Mechanisms of TRPV1-Mediated Pain

ISTVAN NAGY, CLEOPER C. PAULE, and JOHN P.M. WHITE

Department of Anaesthetics and Intensive Care, Imperial College, Faculty of Medicine, Chelsea and Westminster Hospital, London, UK

ABSTRACT

The transient receptor potential vanilloid type 1 receptor ion channel (TRPV1) (formerly known as the "capsaicin receptor") plays a crucial role in the excitation of the large subpopulation of nociceptive primary sensory neurons which are sensitive to noxious heat and to a large variety of other pain-inducing stimuli. These ion channels are the important principal mediators of the pain sensations associated with peripheral inflammation which is by far the most important source of pathological pain in humans. Since the cloning of TRPV1, numerous investigations have resulted in the accumulation of substantial amounts of data concerning this ion channel's molecular, biophysical, and pharmacological properties. Moreover, the mechanisms through which TRPV1 activity may initiate the perception of pain are also beginning to be elucidated. This chapter offers a summary of developments in the rapidly evolving field of TRPV1 research, and considers how TRPV1 activity may contribute to the development of various acute and chronic pain sensations.

1. INTRODUCTION

Certain plants and animals produce substances which either deter their predators or help them to catch their prey. The identification of some of these agents has assisted in the discovery of several receptors and ion channels in different tissues, especially in the nervous system. Capsaicin is one of the best known deterrent compounds produced by plants. It is also one of the most widely used experimental tools of plant origin.

Capsaicin is a highly lipophilic molecule. Its synthesis in different quantities is responsible for the varying burning hot sensations associated with ingestion of different peppers [1]. Chemically, capsaicin belongs to the vanilloids [2]. Capsaicin acts as a deterrent to its ingestion by animals by inducing a burning pain sensation when it comes into contact with mucous membranes or gains entry into tissues. Remarkably, capsaicin induces these effects only in mammals, and not in birds. The evolutionary advantage for hot chili plants in capsaicin production appears to reside in this burning hot flavor which discourages mammals from consuming the chilies and thereby prevents an inefficient method of seed dispersal by those animals, while favoring seed dispersal by more effective methods, such as dispersal by birds [3].

Systemic application of capsaicin induces acute effects only in mammals, which is consistent with the finding that topical application of capsaicin serves only as a deterrent to mammals. These acute effects include the Bezold–Jarisch reflex (hypotension, bradycardia), bronchoconstriction, apnea, increased bladder motility, hypothermia, heat loss, increased O_2 consumption, and pain-related behavior [4–15]. The majority of the chronic effects induced by systemic capsaicin application are the opposite of those evoked acutely. For example, acute hypothermia is followed by hyperthermia; or acute pain-related behavior is followed by reduced responses to noxious stimuli [5,13–18].

That capsaicin mediates its effects through a specific receptor was suggested by the early findings that (i) capsaicin activates and induces degeneration in a well-defined group of neurons [16,17]; (ii) the inorganic dye, ruthenium red, and the synthetic compound, capsazepine, selectively block capsaicin-induced responses [19–20]; and (iii) the ultrapotent capsaicin analog, resiniferatoxin (RTX), specifically binds to the cytoplasmic membranes of a subpopulation of neurons [21]. Electrophysiological recordings from capsaicin-responsive primary sensory neurons indicated that the receptor for capsaicin must be a ligand-gated, nonselective, cationic channel with a high permeability for Ca^{2+} [22,23].

In 1997, David Julius and his coworkers announced the cloning of a rat protein, denominated vanilloid receptor 1 (VR1), which responded to capsaicin [24]. This receptor was subsequently identified as belonging to the transient receptor potential (TRP) family of ion channels and it is now denominated "transient receptor potential vanilloid type 1 receptor (TRPV1)" [25]. A number of homologs, as well as TRPV1 in other species, including humans, have now been cloned [26–33]. However, TRPV1 remains the only molecule thus far identified that displays sensitivity to capsaicin and other vanilloids.

About 40% of the total neuronal population of primary sensory neurons express TRPV1 [34]. TRPV1 is expressed in the perikarya, as well as in both the central and peripheral terminals, of primary sensory neurons. In peripheral tissues, TRPV1-expressing sensory fibers can be found in the dermis, along the epidermal/dermal junction, epidermis, and also in Meissner's corpuscles in the skin [34,35]. In the viscera, TRPV1 immunopositive fibers are found in the mucous membrane, submucous, and muscular layers [36,37]. TRPV1-expressing fibers also innervate the pancreas [38,39]. They are also found in the synovial membrane of certain joints [40]. Many neurons in the central nervous system also express TRPV1 [41–43]. Moreover, some nonneuronal cells have also been shown to express TRPV1 [44,45].

The central terminals of TRPV1-expressing primary sensory neurons terminate primarily in the superficial dorsal horn of the spinal cord, which is involved in nociceptive processing [46]. In addition, some TRPV1-expressing fibers can also be found in the deep dorsal horn and around the central canal [47]. Thus, TRPV1 is expressed in cells and locations which are highly relevant for detecting, transducing, transmitting, and processing noxious stimuli in the peripheral nervous system and in the spinal cord. This expression pattern of TRPV1 is coupled with its capacity to respond to a range of noxious stimuli, including heat above approximately 43°C, protons, and pain-inducing exogenous and endogenous agents [24]. In consequence, TRPV1 is regarded as being one of the most important molecules responsible for initiating and maintaining pain-related behavior in animals and pain experience in humans [24,30,46].

The molecular basis of this TRPV1-mediated nociception is the subject of this chapter. The rat and mouse receptors, on which the majority of the data has been obtained, will be described to introduce the mechanism involved in capsaicin receptor-mediated nociception, as the basic characteristics of TRPV1 in different mammalian species are the same. Significant differences between TRPV1 in humans and these animals will then be noted.

2. THE STRUCTURE OF TRPV1

The term "TRPV1" is generally applied to describe the TRPV1 ion channel. That ion channel is believed to be comprised of four TRPV1 molecules, referred to as "subunits." In this chapter, the term, "TRPV1" is employed to refer to the ion channel, while the term "TRPV1 molecule" is used to refer to channel subunits.

2.1 TRPV1 is a member of the transient receptor potential family of ion channels

The TPRV1 molecule is an approximately 95 kDa protein consisting of 838 (human: 839) amino acids. The amino acids are arranged in six transmembrane domains and in an additional short hydrophobic loop connecting the fifth and sixth transmembrane domains (Fig. 1). It is thought that this additional loop is involved in the formation of the channel pore. Both the N- and C-termini of TRPV1 are intracellular. The N-terminus is rich in proline and has three ankaryn repeat domains. The molecule has several phosphorylation, glycosylation, and nucleotide-binding sites [24,30]. The human TRPV1 ion channel (hTRPV1) expresses several naturally occurring polymorphisms (mutations). P91S, I315M, T469I, T505A, and V585I are missense single nucleotide polymorphisms which result in a change in the amino acid [30,48]. The I315M and P91S variants elicit a greater response to capsaicin than do wild-type hTRPV1 receptors [49]. The TRPV1 gene is localized on chromosome 17, p13 in humans [30].

The amino acid sequence of individual TRPV1 molecules, and the structure of the TRPV1 ion channel, are related to the TRP family of cationic channels [50,51]. These channels represent a major route for Ca^{2+} entry into many cell types, including neurons. Thus, the TRPV1 ion channel has been classified as a member of the TRP family and, within that family, it is a member of the vanilloid (TRPV) subfamily [25]. The structure of TRP channel proteins suggests that several TRPV1 molecules must coassemble to form a functioning ion channel. Indeed, the functional TPRV1 channel is a multimer both in dorsal root ganglion neurons and in heterologous expression systems, with a tetramer as the predominant form [52,53].

TRPV1 molecules, in addition to being able to form homomers, may also be assembled in heteromers, with other members of the TRPV subfamily and with TRPV1 splice variants. The

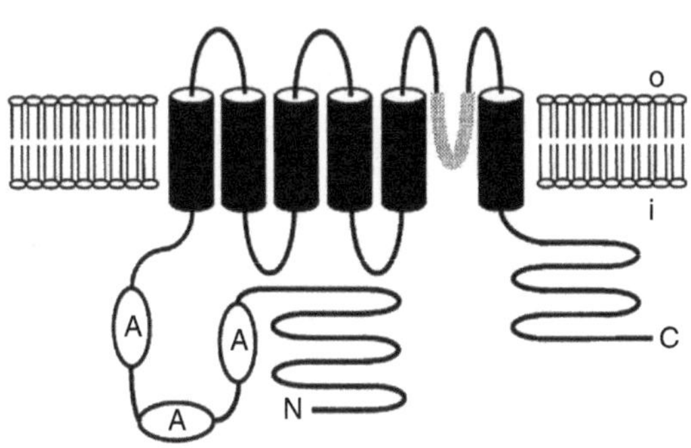

Figure 1. Predicted membrane topology of the TRPV1 molecule. TRPV1, like other members of the TRP family, has six transmembrane domains. Both the N- and C-termini are intracellular. The hydrophobic loop connecting transmembrane domains five and six is believed to be part of the channel. o, extracellular; i, intracellular (reprinted by permission from Ref. [24]).

assembly of TRPV1 with TRPV3 may be of importance in relation to nociception, because TRPV3 is coexpressed with TRPV1 in primary sensory neurons. TRPV3 responds to noxious heat with a threshold of about 39°C without being responsive to capsaicin or other vanilloids. When expressed heterologously, TRPV3 is able to assemble with TRPV1 and may modulate its responses [28].

The assembly of TRPV1 with a recently identified splice variant – the dominant negative TRPV1beta – may also be of importance in relation to nociception [54]. The TRPV1beta protein, which is expressed in mouse, lacks 10 amino acids, which are coded by a portion of exon 7 of the TRPV1 gene. TRPV1beta is not functional when expressed alone. However, when coexpressed with the TRPV1 molecule it has an inhibitory effect on TRPV1 channel function. A similar splice variant – called TRPV1b – has been found in humans [55]. The TRPV1b molecule lacks all the 60 amino acids coded by exon 7. When expressed alone, TRPV1b is functional. It responds to heat with a threshold of approximately 47°C. However, it does not respond to capsaicin. Like the TRPV1beta molecule, the TRPV1b molecule also has an inhibitory effect on TRPV1 when they are coexpressed. Our unpublished data show that the expression of the TRPV1 and TRPV1b molecules may be differentially regulated in inflammatory conditions. These findings raise the possibility that changes in the composition of the TRPV1 ion channel may also contribute to the sensitivity and degree of activation of this ion channel.

2.2 The TRPV1 ion channel is linked to other proteins in transducisomes

TRP channels, such as those involved in the phototransduction cascade in *Drosophila*, form molecular complexes called *transducisomes* or *signalplexes* [51]. These complexes contain a TRP channel, a scaffolding protein, and several other molecules, including phospholipase C (PLC) and protein kinase C (PKC) [51]. Chuang and colleagues [56] demonstrated that the capsaicin receptor is also a transducisome containing TRPV1 molecules and probably several other molecules (Fig. 2). Some of the auxiliary proteins in the TRPV1 signalplex may contribute to

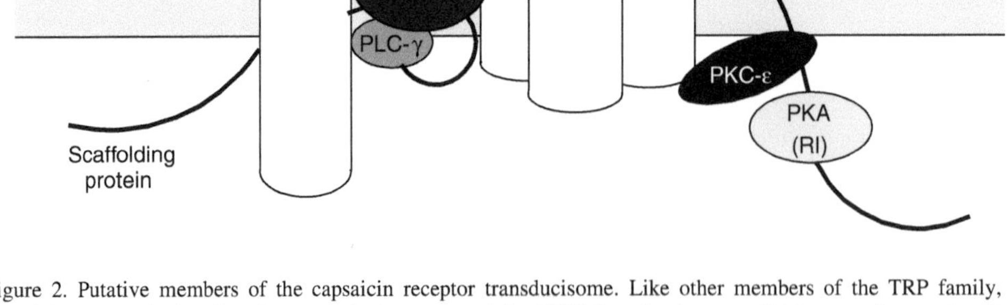

Figure 2. Putative members of the capsaicin receptor transducisome. Like other members of the TRP family, the capsaicin receptor is thought to be a transducisome, containing TRPV1 molecules, a scaffolding protein, tyrosine kinase A (TrkA), phospholipase C-γ (PLC-γ), phosphatidylinositol-4,5-biphosphate (PtdIns(4,5)P_2), protein kinase C-ε (PKC-ε), and the regulatory subunit (RI) of protein kinase A (PKA).

TRPV1-mediated responses of nociceptive primary sensory neurons. Of these proteins, snapin and synaptotagmin IX may be important in the regulation of the membrane expression of TRPV1. These proteins have been shown to associate and colocalize with TRPV1 in cytoplasmic vesicles. While they do not affect the channel function of TRPV1, they inhibit TRPV1 translocation from the cytoplasm to the plasma membrane [57]. Regulation of the expression of these proteins may affect TRPV1-mediated cellular responses since pathological conditions induce upregulation of TRPV1 expression and translocation from the cytoplasm to the membrane [57–59].

Unlike snapin and synaptotagmin IX, the PLC-γ molecule, as a member of the TRPV1 signaling complex, does modulate TRPV1 channel function. PLC-γ, together with TRPV1, is precipitated from lysates of nonneuronal cells transfected with TRPV1 cDNA. PLC-γ hydrolyzes phosphatidylinositol-4,5-bisphosphate (PtdIns(4,5)P_2) that binds to, and has an inhibitory effect on, TRPV1. PLC-γ is activated by at least two receptors, which are coexpressed with TRPV1 in primary sensory neurons, namely: the tyrosine receptor kinase A (TrkA), which is the high-affinity receptor for nerve growth factor (NGF); and the bradykinin B2 receptor [56]. Both NGF and bradykinin are inflammatory mediators, which are released during inflammation or tissue injury.

Protein kinases may also form part of the TRPV1 signaling complex, at least transiently. Rathee and coworkers (2002) reported that, following protein kinase A (PKA) activation, the catalytic subunits of this protein kinase translocate to the membrane [59A]. Similar translocation has been found following the activation of PKC [60]. Both PKA and PKC mediate TRPV1 phosphorylation which results in increased channel activity. These protein kinases can be activated by receptors of inflammatory mediators, including bradykinin and prostaglandin E2 (PGE2) [61–63].

3. TRPV1 ACTIVATORS

TRPV1 is responsive to a variety of activators. These include, in addition to capsaicin, and related vanilloids, protons and some other positive charge carriers, heat above approximately 43°C, and depolarization [24,46,64]. The sensitivity of TRPV1 to protons, heat, and capsaicin is consistent with the common experience that hot chilies and acids induce a similar sensation to that induced by moderately hot temperatures. These activators interact with TRPV1 itself and may, therefore, be referred as direct activators of this ion channel.

3.1 Ligands

Capsaicin, and other vanilloids, such as the ultrapotent RTX, directly activate TRPV1 by binding to the channel's ligand-binding site. Capsaicin binding induces conformational changes in the TRPV1 ion channel, which differ from the conformational changes which are induced by proton binding or by thermal activation. The fact that this difference exists is consistent with there being independent activating pathways of TRPV1 for capsaicin, protons, and heat [65].

Some of the agents released from inflamed or injured tissues, can also directly activate TRPV1. Quite incongruously, these agents are known as "endogenous vanilloids" or "endovanilloids." These endogenous agonists include anandamide [66], *N*-arachidonoyl-dopamine (NADA) [67], *N*-oleoyldopamine (OLDA) [68], lipoxygenase products, such as 12-or 15-(*S*)-HPETE [69], and unsaturated C18 *N*-acylethanolamines (NAEs) [70]. All of these agonists have been shown to compete for the same binding site [71,72]. Of major interest is the finding that anandamide and NADA, in addition to being agonists at the TRPV1 ion channel, also activate the cannabinoid 1 (CB1) receptor [66,73].

Anandamide is the best characterized of the endogenous ligands of TRPV1. Anandamide is a member of the group of bioactive lipids known as "long-chain C18 NAEs." Several endogenous NAEs, many of which are more abundant than anandamide in rat tissues, are capable of activating TRPV1 and may therefore play a role as endogenous TRPV1 modulators [70]. While other exogenous and endogenous TRPV1 activators induce desensitization and tachyphylaxis [74,75], i.e., reduction in the response evoked by long-lasting or repeated application, anandamide has been found not to induce these effects in any conditions thus far investigated [76].

The ability of anandamide to activate TRPV1 in normal physiological conditions is very limited. However, when TRPV1 is activated by other stimuli, such as inflammatory mediators, anandamide becomes a powerful activator of TRPV1 [77]. Its limited capacity to activate TRPV1 in physiological conditions is essential to prevent unwarranted activation of TRPV1 which would result in signaling pain in the absence of a relevant pain-inducing stimulus. Since anandamide, and the other endogenous activators, only activate TRPV1 in certain conditions, these agents are appropriately described as "conditional activators" of TRPV1.

3.2 Heat

Sensitivity to noxious heat is the most distinctive feature of TRPV1. Heat contributes to TRPV1 activation in two distinct ways. First, heat reduces the threshold for the activation of TRPV1 by all other TRPV1 activators. Second, heat above approximately 43°C independently activates TRPV1 [46]. At less than 43°C, TRPV1 openings are few and brief. However, raising the ambient temperature rapidly increases the frequency of channel openings [78]. There is also some evidence from studies on heterologously expressed human TRPV1 that the presence of either reducing or oxidizing agents results in an increased response to heat by TRPV1 channels [79].

3.3 Protons

The pH threshold for proton-evoked TRPV1 activation is approximately 6.5. At lower pH values, protons can themselves activate TRPV1. Capsaicin binding, the temperature threshold, and channel gating are all affected by pH. Lowering pH enhances the apparent binding affinity of capsaicin, and lowers the heat threshold for activation of the channel. It also promotes the occurrence of long openings and short closures, and stabilizes at least one of the open conformations of the channel. Moreover, capsaicin binding and protonation of the channel interact allosterically, where each of these agonists can alter the effect of the other [80].

3.4 Depolarization

TRPV1 is a voltage-gated ion channel [64,81]. The voltage-activity curve of TRPV1 is much wider than that of other voltage-activated channels, including voltage-gated Na^+, K^+, or Ca^{2+} channels [64]. Thus, while voltage-activated Na^+, K^+, or Ca^{2+} channels reach their maximal opening probability from zero within a narrow voltage range ($\sim$50 mV), full TRPV1 activation occurs within a much wider voltage range ($\sim$150 mV) [64].

3.5 Cations

In addition to protons, excess extracellular positive charges carried by various ions are also able to activate TRPV1. Ahern and colleagues [82] showed that extracellular Na^+, Mg^{2+}, and Ca^{2+} can directly gate the TPRV1 ion channel under extreme or pathophysiological concentrations. However, even discrete pathology may result in the generation of extracellular microenvironments which readily achieve concentrations of cations which are sufficient to directly gate TRPV1. For example, in bone, the Ca^{2+} concentration surrounding resorbing osteoclasts approaches 20 mM [83]. A yet more significant effect of extracellular cations may occur under normal physiological conditions where millimolar increases in cation concentrations (1–5 mM) are sufficient to sensitize TRPV1 to various ligands, including capsaicin, anandamide, and NADA [82].

3.6 Indirect activators of TRPV1

While ligands, heat, protons, and depolarization induce TRPV1 opening directly, gating of this ion channel can also be evoked indirectly. The indirect activators include various inflammatory mediators, which are produced and released during inflammation in tissues. These agents, through activating their own target receptors, which are also expressed on the nociceptive primary sensory neuron expressing TRPV1, induce activity in the neuron's intracellular second messenger system. That, in turn, results in posttranslational modification of TRPV1. Hydrolysis of phosphotidylinositol 4,5-bisphosphate ($PtdIns(4,5)P_2$) in the cytoplasmic membrane, and PKA- and PKC-mediated phosphorylation of the capsaicin receptor, have been shown to induce opening of the TRPV1 ion channel [56,84]. Hydrolysis of $PtdIns(4,5)P_2$ can be initiated by the activity of either TrkA or bradykinin B2 receptors, while PKC-mediated phosphorylation can be induced by activating the bradykinin B2 receptor [56]. Activation of PGE2 receptors induces PKA-mediated phosphorylation of TRPV1. TrkA, bradykinin B2, and EP receptors are all expressed on capsaicin-sensitive primary sensory neurons [85–87].

4. MOLECULAR BASIS FOR TRPV1 ACTIVATION

4.1 Different domains of TRPV1 are responsible for the sensitivity of the receptor to the different activators

The evidence from functional studies suggests that all of the individual activators each mediate their respective effects independently of the others. Thus, ligand binding, protonation, electrical energy, thermal energy, and posttranslational changes can each control the gating of the channel. Results of mutagenesis studies confirm that different TRPV1 activators activate the molecule at different activation sites. Studies by Jordt and Julius [27] suggest that a segment of rat TRPV1 in, and adjacent to, the third transmembrane domain is responsible for the capsaicin sensitivity of the ion channel. Mutations of amino acids in this region indicate that tyrosine at position 511 is specifically responsible for capsaicin binding. According to the predicted membrane topology of TRPV1, residue 511 is located intracellularly, which is in agreement with previous assumptions that capsaicin binds to the cytoplasmic side of the receptor [88]. Interestingly, binding of RTX depends on residues tyrosine 550 and methionine 547 [71]. However, the capsaicin binding site is also the locus for binding of competitive antagonists, such as capsazepine [19,27]. Recent data suggest that extracellular sites may also be involved in vanilloid binding to TRPV1 [89,90]. Moreover, data from several sources support the suggestion that TRPV1 may bind capsaicin at multiple sites [91–93].

Unlike capsaicin, protons interact with TRPV1 at the extracellular side of the receptor [46]. Mutations generated at residues of the extracellular loops reveal that the charge of the amino acid at position 600 determines the proton-induced increase in responses to capsaicin or heat, while the amino acid at position 648 seems to be responsible for proton-induced opening [94].

A series of other thermo-sensitive ion channels have been identified more recently [95]. As expected, the main structure of these ion channels is very similar, and they appear to exhibit a "modular" C-terminus [96]. Brauchi and colleagues [96] prepared chimera from TRPV1 and the cold-sensing TRP channel, TRPM8, by altering their respective C-termini. They found that the TRPV1, when altered to have the C-terminus of TRPM8, became cold-sensitive, while the TRPM8 with the C-terminus of TRPV1 became sensitive to heat. These findings suggest that the thermosensor resides in the C-terminus. However, the amino acids which govern the sensitivity of the channel to heat remain unknown.

There are several sites in TRPV1 which are targets for transcriptional changes. PKA and PKC phosphorylate serine and threonine residues at S116, T144, T370, S502, S774, S800 [97–99] and at S502, T704, S744, S800, S820 [100,101], respectively. Tyrosine kinase-mediated phosphorylation by Src at Y199 (Y200 in humans) controls TRPV1 trafficking [102]. The segment of the C-terminus cytoplasmic domain of TRPV1 comprising residues from 777 to 820 has been identified as the PtdIns(4,5)P_2 binding site [103].

There are no data on the voltage sensor for TRPV1, although the voltage sensor of various other voltage-gated ion channels, and of the thermo-sensitive TRP channel, TRPM8, have been mapped [104–107]. However, since the voltage sensor seems to be highly conserved in all voltage-gated ion channels, including TRPM8, it is anticipated that this sensor in TRPV1 will also be found to be localized around the fourth transmembrane segment.

4.2 TRPV1 is a stimulus integrator for different exogenous stimuli

Various activators of TRPV1 can increase the effect evoked by another, or others, of its activators. For example, a dish made with chili pepper produces a more intense burning sensation when it is warm than when it is cold. It seems that this "cooperation" of various ligands is synergistic, rather than additive, in activating TRPV1 [24,46,108]. The findings that various ligands potentiate each other's effect on TRPV1 indicate that the different activator sites are coupled in TRPV1. It has been suggested that the TRPV1 ion channel acts as a "stimulus integrator" given the polymodal nature of its activation, and the potentiating effect of each of the activating stimuli on the effect produced by another activator of TRPV1 [46].

The mechanism of stimulus integration remains unknown. However, the evidence available suggests that the heat sensor may play a central role in stimulus integration and channel gating. Tominaga and coworkers [46,109] found that protons reduce the heat threshold for activation of TRPV1. Vyklicky and his team [110] reported that capsaicin also significantly reduces the heat threshold. Liang and colleagues [111] proposed that a reduced heat threshold is responsible for bradykinin-evoked TRPV1 activation. Subsequently, Sugiura and coworkers [62] have provided evidence that bradykinin indeed reduces the heat threshold of TRPV1 well below body temperature. If indeed temperature is crucial in TRPV1 activation, then neither capsaicin nor protons, nor any other TRPV1 activator should open the channel at low temperatures. Indeed,

Babes and colleagues [112] reported that cooling the superfusate to approximately 10°C inhibits capsaicin-induced currents in primary sensory neurons. However, more recently, Voets and colleagues [64] have suggested that all activators mediate their effects through changing the voltage-evoked activation properties of TRPV1. This proposal is based on the finding, first, that temperature sensitivity is modulated by the transmembrane voltage, and, second, that changes in ambient temperature result in graded shifts of the voltage dependence of channel activation [64]. These investigators have also suggested that the heat and voltage sensors are directly coupled. However, Brauchi and colleagues [113] proposed a model in which the voltage- and temperature-dependent activation of the channel can occur independently, although the sensors are coupled allosterically. A definitive identification of the mechanisms involved in stimulus integration is important because the integration of different stimuli seems to be pivotal in the regulation of channel activity.

5. ROLE OF TRPV1 IN PHYSIOLOGICAL AND PATHOLOGICAL PAIN SENSATION

The first evidence suggesting a significant role for the capsaicin receptor in nociception was provided by Porszasz and Jancso [17] who found that capsaicin selectively activates nociceptive primary sensory neurons. Many years later, the elucidation of the expression pattern and responsiveness of cloned TRPV1 provided increased support for the putative role of TRPV1 in nociception. The most compelling evidence, however, of the involvement of TRPV1 in mediating certain physiological and pathological pain conditions, has been provided by studies performed in TRPV1 "knockout" mice (TRPV1$^{-/-}$), i.e., in mice in which the transcription and translation of the TRPV1 gene has been disrupted [114,115].

Acute application of capsaicin to TRPV1$^{-/-}$ animals, or to primary sensory neurons obtained from TRPV1$^{-/-}$ mice, does not induce pain-related behavior or membrane responses in either case [116], indicating that TRPV1 is the only receptor on primary sensory neurons that responds to capsaicin, and the only receptor which, when activated by capsaicin, results in pain sensation [115]. Noxious heat below approximately 50°C also fails to produce heat-induced currents in primary sensory neurons of the TRPV1 knockout animals. However, noxious heat below 50°C does induce pain-related behavior in TRPV1$^{-/-}$ mice *in vivo*, indicating that another, capsaicin-insensitive, receptor may also be involved in the detection of heat between approximately 42°C and 50°C [114–115]. Experiments in TRPV1$^{-/-}$ animals also show that TRPV1 is not the only ion channel which responds to protons. The proton-activated responses in neurons of TRPV1$^{-/-}$ mice are reduced but are not completely eliminated, indicating the existence of acid-sensing ion channels on capsaicin-sensitive primary sensory neurons [116].

Studies on TRPV1 knockout mice have also confirmed that TRPV1 plays an important role in the development of pathological pain sensations, such as enhanced pain to noxious heat stimuli (known as heat *hyperalgesia*), and pain produced by otherwise innocuous thermal stimuli (known as heat *allodynia*) [24,114]. Pathological responses to thermal stimuli usually develop after prolonged, or chronic, inflammation of peripheral tissues, or, after physical, or metabolic, injury to peripheral nerves. Inflammatory heat hyperalgesia substantially fails to develop in TRPV1 knockout mice, but pathological thermal sensations after peripheral nerve injury remain the same in both wild-type and TRPV1$^{-/-}$ animals [114,115]. Both Caterina and colleagues [114] and Davis and colleagues [115] concluded that these findings demonstrate that TRPV1 is an essential component in the development of pathological responses to thermal stimuli, but probably limited to pain sensations of inflammatory origin.

There is, however, evidence that TRPV1 could be involved in the development of heat hyperalgesia and allodynia in certain forms of neuropathy. Michael and Priestley [86] found that axotomy results in the downregulation of TRPV1 mRNA in dorsal root ganglion cells. Subsequently, Hudson and colleagues [117] confirmed that after total, or partial, sciatic nerve transection, or spinal nerve ligation, TRPV1 immunoreactivity is significantly reduced in the somata of damaged dorsal root ganglion neurons. However, they also found that the position is otherwise in the somata of *undamaged* dorsal root ganglion neurons where TRPV1 expression is actually increased. Moreover, the perikarya of large, noninjured, nonnociceptive neurons change their phenotype and also express TRPV1; and this expression persists in the damaged axons proximal to the injury. In cases of peripheral neuropathy, the central terminals of nonnociceptive primary sensory neurons may sprout into the termination area of the nociceptive cells [118]. Moreover, the excitability of the neuronal membrane is increased in those instances where a neuroma develops after axotomy. Thus, alterations in TRPV1 expression may well have significance in the development of certain neuropathic pain conditions. This is supported by the finding that TRPV1 antiserum can significantly reduce both heat hyperalgesia and allodynia in diabetic neuropathy [119]. After sciatic nerve ligation, intrathecal application of capsazepine, a TRPV1 antagonist, blocks A-delta fiber-evoked responses in the dorsal horn neurons of rats [120]. There is also evidence that the contribution made by TRPV1 to neuropathic pain conditions may even be species dependent. Capsazepine reverses mechanical hyperalgesia in a guinea pig model of partial sciatic nerve ligation, but has no effect in rat or mouse models of neuropathic pain [121]. Recently, Christoph and colleagues [122] reported that in an *in vivo* rat model of spinal nerve ligation, both intravenous application of the TRPV1 antagonist thioxo-BCTC and intrathecal administration of the antisense oligonucleotide against TRPV1 reduce mechanical hypersensitivity in a similar manner, evidencing the involvement of TRPV1 in such neuropathic pain conditions in rat. Interestingly, intrathecal injection of an siRNA against TRPV1 reduces cold allodynia of mononeuropathic rats by more than 50% over a period of approximately 5 days [123].

6. MECHANISMS INVOLVED IN THE DEVELOPMENT OF TRPV1-MEDIATED CHRONIC PAIN SENSATION IN INFLAMMATORY CONDITIONS

Obviously, TRPV1 must be in a state of continuous activation in any TRPV1-dependent pain state. This sustained TRPV1 activity depends upon several processes, including (i) the production and release of agents which induce upregulation in TRPV1 expression and TRPV1 sensitivity; (ii) the mechanisms involved in the upregulation of TRPV1 expression by these agents; and (iii) the mechanisms involved in the ongoing activation of TRPV1.

6.1 Production and release of "inflammatory mediators," i.e., agents which induce upregulation of TRPV1expression and TRPV1 sensitivity

A variety of agents, commonly described as "inflammatory mediators," are produced, released, or activated, in the inflammatory process which results from tissue injury or disease. These "inflammatory mediators" are able to induce upregulation of TRPV1 expression and activation. They include bradykinin, ecosanoids, various neurotrophic factors, anandamide, and protons and cations.

Bradykinin is a nonapeptide and is a powerful inflammatory mediator which is produced at sites of tissue injury. The proteolytic enzyme kallikrein acts on kininogen, a plasma alpha-globulin, to form either bradykinin or kallidin. Kallidin is a peptide which is almost identical with bradykinin, but comprises an additional amino acid residue. The kallikrein enzyme is derived from prekallikrein, which is an inactive precursor present in plasma. Prekallikrein can be converted to kallikrein in several ways, the most significant of which is by reaction with Hageman factor (which is factor XII of the clotting sequence). Hageman factor is normally in an inactive form in the plasma and is activated by contact with surfaces having a negative charge, including collagen, basement membrane, and bacterial lipopolysaccharides. The increased vascular permeability which accompanies inflammation results in leakage of plasma, containing prekallikrein, Hageman factor and kininogen, which leads to bradykinin formation. [124]

The *ecosanoids* comprise three principal classes of inflammatory mediators, namely, prostanoids, leukotrienes, and platelet activating factor. Ecosanoids are not constitutively expressed in tissues but are instead generated *de novo* from phospholipids, mainly from arachidonic acid which is found esterified in phospholipids. Many stimuli, including general cell damage, can result in the release of arachidonic acid from phospholipid either by an action of phospholipase A2 (PLA2) or an action of PLC followed by an action of diacylglycerol lipase [124]. Among the ecosanoids, prostanoids play a major role in increasing the sensitivity of TRPV1 in inflammation.

Prostanoids are a group of bioactive lipids working as local mediators and comprise the Prostaglandins and thromboxanes. They are the cyclooxygenase (COX) metabolites of arachidonic acid which are synthesized and released upon cell stimulation. They act on cells in the vicinity of their cell of synthesis to exert their actions. The prostanoids include PGE2, PGI2, PGD2, PGF2alpha, and thromboxane A2. The COX enzymes, COX-1 and COX-2, initiate the biosynthesis of both PGs and thromboxanes from free arachidonic acid within the cell.

NGF was initially identified as a molecule that promotes the survival and differentiation of sensory and sympathetic neurons. Its functions in neural development have been characterized extensively, but recent studies indicate that its role in development is only one of several biological actions of NGF. In particular, NGF constitutes an important inflammatory mediator and is an important component in the process of sensitization of TRPV1 ion channels in primary sensory neurons. NGF is produced by a variety of cells in the context of inflammation. Monocytes, produce, store, and release NGF [125]; as do eosinophils [126]. Mast cells also synthesize, store, and release NGF [127]. Histamine – itself an inflammatory mediator – is a potent stimulator of NGF production. Histamine is released by mast cells. The main signaling pathway involved in the stimulation of NGF production by histamine, includes activation of histamine H(1)-receptor, stimulation of Ca^{2+}-dependent PKC, and mitogen-activated protein kinase. The same signaling pathway is involved in the interactions between histamine, interleukin-1beta (IL-1beta), and interleukin-6 (IL-6), where NGF secretion is amplified. Whereas histamine and IL-1beta have an additive stimulatory effect on NGF secretion, interaction between histamine and IL-6 causes a long-term synergism [128]. The proinflammatory cytokines, IL-1beta and tumor necrosis factor-alpha (TNF-alpha), stimulate the expression of NGF by human intervertebral discs in which NGF is constitutively expressed in annulus fibrosus cells and nucleus pulposus cells [129]. It appears likely that other tissues constitutively express NGF and that proinflammatory cytokines have a broader role in stimulating its production. Thus, for example, TNF-alpha stimulates the synthesis and secretion of biologically active NGF in quiescent fibroblasts and also in glioblastoma cells [130]. Inflammation induced by intraplantar injection of complete Freund's adjuvant (CFA) in adult rats results in significant elevation in the levels of TNF alpha, IL-1beta, and NGF in the inflamed paw [131].

Artemin, neurturin, and *glial cell line-derived neurotrophic factor (GDNF)* are members of the GDNF family and are potent regulators of TRPV1 function. Members of the GDNF family are produced in the form of a "prepro" precursor. The signal sequence is cleaved on secretion, and activation of the "pro" derivative is suggested to occur by proteolytic cleavage. GDNF family members appear to bind to heparin sulfate side chains of extracellular matrix proteoglygans, but little is known of the gene regulation and mechanisms of secretion of these proteins, or of the activation of their precursor [132]. Artemin mRNA (but not GDNF) is substantially upregulated during cutaneous inflammation evoked by hind paw injection of CFA, suggesting that artemin, in particular, enhances TRPV1 signaling in response to inflammatory injury [133].

Stimulus-evoked *anandamide (N-arachidonylethanolamine)* production in a subgroup of nociceptive primary sensory neurons is one of the most recently identified factors implicated in controlling the activity and excitability of these neurons [134,135] In animal tissues, anandamide is principally formed, together with other NAEs, from glycerophospholipid by two successive enzymatic reactions: (i) *N*-acylation of phosphatidylethanolamine to generate *N*-acylphosphatidylethanolamine (NAPE) by Ca^{2+}-dependent *N*-acyltransferase; and (ii) release of NAE from NAPE by a phosphodiesterase of the phospholipase D type (NAPE-PLD) [136].

An increase in the *local hydrogen ion concentration* is a common accompaniment of inflammation consequent upon tissue damage or disease. It has long been known that low pH (down to 4.7) is commonly found in inflamed tissues and that acidic solutions are particularly painful when injected into the skin [137–139]. Protons selectively induce lasting excitation and sensitization to mechanical stimulation of nociceptors in rat skin *in vitro* [140]. *In vivo*, a decrease in pH occurs immediately after plantar incision in rats and is sustained for at least 4 days. During the period of decreased tissue pH, pain behaviors are evident, with these diminishing when the tissue pH returns to normal. The fall in pH is localized at the incision site and does not extend to areas surrounding the incision. However, the magnitude of the pH change varies between tissues [141].

6.2 Mechanisms involved in the upregulation of TRPV1 expression by inflammatory mediators

It is likely that the half-life of nonactivated TRPV1 ion channels is approximately 1 day, in common with other ionotropic receptor subunits, such as the subunits of ionotropic glutamate receptors. It is also very likely that this period is significantly reduced when the TRPV1 ion channel is activated [142]. Thus, in pathological conditions, chronic TRPV1-mediated heat hyperalgesia and/or allodynia can be maintained only if TRPV1 expression in capsaicin-sensitive primary sensory neurons is upregulated. Indeed, TRPV1 expression has been shown to be upregulated in primary sensory neurons in inflammatory conditions [58,59]. The upregulation in inflammation occurs presumably in neurons which also express the ion channel in naive conditions. This is in contrast with the upregulation of TRPV1 expression in neuropathy, described by Hudson and colleagues [117], because it also occurs in nonnociceptive cells, which normally do not express TRPV1.

Neurotrophic factors, such as NGF and members of the GDNF family have been implicated in regulating TRPV1 transcription in primary sensory neurons [143–145]. The role of NGF in inducing upregulation of TRPV1 expression *in vivo* in inflammation is supported by the fact that NGF production is increased in this condition [146]. The NGF-evoked upregulation involves the small GTPase, Ras, and the mitogen-activated protein kinase p38 [58,147].

Artemin, a member of the GDNF family, is found in TRPV1-expressing primary sensory neurons and its level in the skin appears to modulate gene expression and response properties of afferents that project to the skin, with the resulting changes leading to behavioral sensitivity to both hot and cold stimuli. Overexpression of artemin in the skin of transgenic mice enhances the expression of TRPV1 in cutaneous sensory neurons and leads to behavioral sensitivity to heat and cold [148].

However, the extent of pathology-induced upregulation of the expression of the various TRPV1 channel subunits may be different. Inflammatory conditions seem to induce greater upregulation of TRPV1 expression than they do of the expression of the dominant-negative TRPV1 splice variant, TRPV1b (Nagy et al., unpublished observations). At present, it is not known whether similar differential upregulation occurs in other pathological conditions. Furthermore, the nature of the mechanisms which regulate that differential increase in the expression of the various subunits also remains unknown.

6.3 Mechanisms involved in ongoing TRPV1 activation

Of the mediators released from inflamed or injured tissues, protons seem to be the only activators which can directly activate TRPV1. All other agents of tissue origin seem to increase TRPV1 activity through a process known as "sensitization." Sensitization is frequently used by various parts of the nervous system for the purpose of adaptation. It results in increased responsiveness of cells and/or molecules. Sensitization involves posttranslational modifications of molecules. It is very likely that, in the case of TRPV1 in primary sensory neurons, sensitization involves both an increase of the responsiveness of the already responsive ion channels, and the recruitment of TRPV1 ion channels which are not responsive to activators in normal physiological conditions.

Regarding recruitment of "silent TRPV1," a comparison of the number of primary sensory neurons in intact dorsal root ganglia which express TRPV1 [86] with the number of those neurons that actually respond to challenge by a TRPV1 activator [149] indicates that, in naive conditions, only about 20% of the TRPV1-expressing neurons can respond to such activators. The fact that the number of capsaicin-responsive cells increases in inflammation suggests that the inflammatory process recruits TRPV1 ion channels which are nonresponsive in normal physiological conditions [150]. This recruitment is probably the result of similar posttranslational modification of TRPV1 which sensitizes the already responding receptor. GDNF was previously considered to be unimportant in enhancing TRPV1 activity, on the assumption that GDNF does not induce heat hyperalgesia or allodynia [151,152]. However, recent data show that the GDNF family members, artemin and neurturin, as well as GDNF, are capable of potentiating TRPV1 signaling and of inducing behavioral hyperalgesia. This potentiation seems to include the recruitment of silent TRPV1. Thus, following their superfusion with GDNF family members, primary sensory neurons, which did not respond to capsaicin prior to the superfusion, start to respond to capsaicin. Artemin mRNA is profoundly upregulated in response to inflammation induced by hind paw injection of CFA. Interestingly, while NGF expression only doubles by day 7, artemin expression shows a 10-fold increase 1 day after the CFA injection. However, no increase occurs in neurturin or GDNF expression.

NGF, unlike members of the GDNF family, appears to be involved in increasing the activity of already responding TRPV1 channels (Fig. 3) [163]. At present, there are conflicting views on the mechanisms by which NGF produces this effect. PKA and the phosphatidylinositol-3 kinase (PI3K) have been reported to be members of the signaling pathway activated by the binding of NGF to the NGF receptor, TrkA, resulting in TRPV1 sensitization [153,154]. A competing hypothesis is that activation of PLC-γ by TrkA leads to hydrolysis of PtdIns(4,5)-P_2 and release

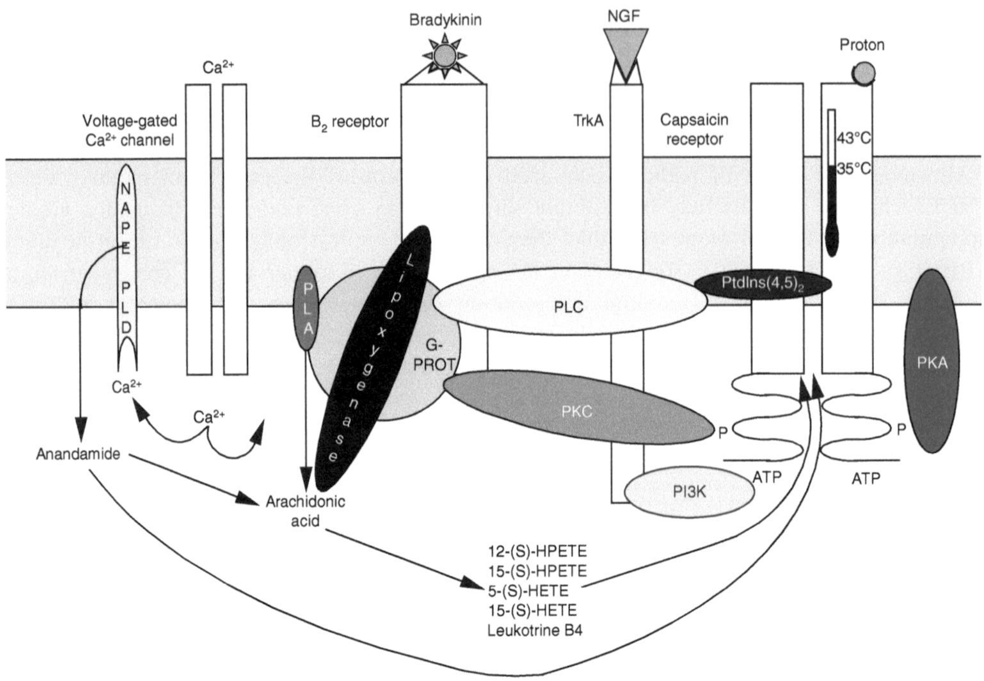

Figure 3. Molecular mechanisms involved in the activation of TRPV1 in pathological conditions. Ligand-binding (protons, anandamide, and lipoxygenase products) and posttranslational modifications (removal of phosphatidylinositol-4,5-biphosphate (PtdIns(4,5)P_2) and protein kinase C (PKC)-mediated phosphorylation) can reduce the heat threshold of TRPV1 below body temperature. Protein kinase A (PKA)-mediated phosphorylation reduces desensitization. Phosphatidylinositol 3-kinase is involved in nerve growth factor (NGF)-induced sensitization. ATP binding also potentiates responses evoked by ligands.

of TRPV1 from inhibition by endogenous PtdIns(4,5)-P_2 [56]. Zhang and colleagues have proposed that NGF, acting on the TrkA receptor, activates a signaling pathway in which PI3K plays a crucial early role, with Src kinase as the downstream element which binds to, and phosphorylates, TRPV1. Phosphorylation of TRPV1 at a single tyrosine residue, Y200, followed by insertion of TRPV1 channels into the surface membrane, is claimed to account for most of the rapid sensitizing action of NGF [102]. More recently, Stein and coworkers have proposed a model for NGF-mediated sensitization in which physical coupling of TRPV1 and PI3K in a signal transduction complex facilitates trafficking of TRPV1 to the plasma membrane [155]. These investigators propose a model in which PtdIns(4,5)-P_2 binding to TRPV1 sensitizes the activity of the channel, while PI3K activity is required for NGF-mediated sensitization, with that sensitization consisting of an increase in the number of channels in the plasma membrane [155]. This is in contrast with the results of Chuang and colleagues that PtdIns(4,5)-P_2 inhibits the activity of TRPV1 [56].

Other activators released in the inflammatory response also seem to increase the responsiveness of those TRPV1s which are already responsive (Fig. 3). Thus, protons which bind to TRPV1 have two effects. First, they directly open the ion channel, resulting in depolarization and an increase in the intracellular Ca^{2+} concentration. Second, they reduce the threshold for activation for both ligands and heat, thereby increasing the responsiveness of the ion channel.

Bradykinin and PGE2 activate TRPV1-expressing primary sensory neurons by binding to their own respective receptors on the cells. This activation (i) induces Ca^{2+} influx through voltage-gated Ca^{2+} channels [156], which results in depolarization; and (ii) activates various components of the intracellular second messenger system, such as PKA, PKC, or PLC [63,157]. Depolarization, in turn, reduces the heat threshold of TRPV1 [64]. Calcium influx through voltage-gated Ca^{2+} channels initiates anandamide production [158], and stimulates other members of the second messenger system, such as Ca^{2+}/calmodulin-dependent protein kinases [159], the activity of which may be essential for keeping TRPV1 ion channels active. Anandamide, PKA-, and PKC-mediated phosphorylation, and PLC-mediated hydrolysis of PtdIns(4,5)-P_2 can open the TRPV1 channel, probably by reducing the heat threshold of the receptor below body temperature (Fig. 3) [56,61,87,97,160].

In addition to direct binding to TRPV1, anandamide can also activate TRPV1 through its products. Anandamide is hydrolyzed by fatty acid amide hydrolase into arachidonic acid, which is the substrate of lipoxygenases [161,162]. Lipoxygenases are activated by bradykinin B2 receptor activation and the products, in turn, activate TRPV1 [162]. Again, it is most likely that lipoxygenase products activate TRPV1 by reducing the heat threshold of the molecule below body temperature. The findings that bradykinin induces Ca^{2+} influx, reduces the heat threshold through different mechanisms (PKC-mediated phosphorylation, PLC-mediated PtdIns(4,5)-P_2 hydrolyses) and activates TRPV1 by lipoxygenase products, indicate that bradykinin may play a particularly important role in the indirect activation of the capsaicin receptor (Fig. 3).

This polymodal nature of TRPV1 activation in pathological conditions is possibly the most intriguing feature of this ion channel. The effects of inflammatory mediators, involving ligand binding, protonation, posttranslational modifications, and possible depolarization of primary sensory neurons, are combined by TRPV1 into channel gating. While some of the activators, including protons and ligands, can open the channel on their own, all activators seem to ultimately induce sensitization. Interestingly, the sensitization produced by the majority, if not all of the activators, induces the reduction of the heat threshold of TRPV1. Therefore, the mechanisms which set the heat threshold may well be pivotal in the activation of TRPV1 in pathological conditions.

7. CONCLUSIONS

Overwhelming evidence supports the hypothesis that the TRPV1 ion channel plays a central role in the development of heat hyperalgesia and heat allodynia in pathological conditions, particularly in inflammation of the peripheral tissues. Recent findings indicate that the mechanisms involved in the development of TRPV1-mediated heat hyperalgesia and allodynia are complex, involving induction of various posttranslational changes of TRPV1, induction of endovanilloid synthesis, and upregulation of TRPV1 transcription and translation. These events result in sustained TRPV1-mediated excitation of a major subpopulation of primary sensory neurons and transmitter release both from the peripheral and central terminals of the neurons. While transmitter release from the peripheral terminals contributes to the neurogenic inflammatory response, transmitter release from central terminals initiates spinal nociceptive information processing. Thus, inhibition of the events leading to the sustained excitation of TPRVI-expressing primary sensory neurons may provide a significant reduction in pathological pain sensation, especially where the pain is of inflammatory origin.

REFERENCES

1. Scoville WL. Note on capsicum. J Am Pharm Assoc 1912;1:453.
2. Neson EK. The constitution of capsaicin – the pungent principle of capsaicum. J Am Chem Assoc 1919;41:1115–1117.
3. Tewksbury JJ, Nabhan GP. Seed dispersal. Directed deterrence by capsaicin in chilies. Nature 2001;412:403–404.
4. Cheng CL, Liu JC, Chang SY, Ma CP, De Groat WC. Effect of capsaicin on the micturition reflex in normal and chronic spinal cord-injured cats. Am J Physiol 1999;277:R786–R794.
5. Masters DB, Komisaruk BR. Neonatal capsaicin treatment attenuates sensory-induced analgesia and nociception. Physiol Behav 1991;50:901–906.
6. Szolcsanyi J, Bartho L, Petho G. Capsaicin-sensitive bronchopulmonary receptors with dual sensory-efferent function: mode of action of capsaicin antagonists. Acta Physiol Hung 1991;77:293–304.
7. Kawada T, Watanabe T, Takaishi T, Tanaka T, Iwai K. Capsaicin-induced beta-adrenergic action on energy metabolism in rats: influence of capsaicin on oxygen consumption, the respiratory quotient, and substrate utilization. Proc Soc Exp Biol Med 1986;183:250–256.
8. Green JF, Schmidt ND, Schultz HD, Roberts AM, Coleridge HM, Coleridge JC. Pulmonary C-fibers evoke both apnea and tachypnea of pulmonary chemoreflex. J Appl Physiol 1984;57:562–567.
9. Donnerer J, Lembeck F. Heat loss reaction to capsaicin through a peripheral site of action. Br J Pharmacol 1983;79:719–723.
10. Donnerer J, Lembeck F. Analysis of the effects of intravenously injected capsaicin in the rat. Naunyn Schmiedebergs Arch Pharmacol 1982;320:54–57.
11. Szikszay M, Obal FJ, Obal F. Dose-response relationships in the thermoregulatory effects of capsaicin. Naunyn Schmiedebergs Arch Pharmacol 1982;320:97–100.
12. Russell JA, Lai-Fook SJ. Reflex bronchoconstriction induced by capsaicin in the dog. J Appl Physiol 1979;47:961–967.
13. Szolcsanyi JA. Pharmacological approach to elucidation of the role of different nerve fibres and receptor endings in mediation of pain. J Physiol (Paris) 1977;73:251–259.
14. Jancso-Gabor A, Szolcsanyi J, Jancso N. Irreversible impairment of thermoregulation induced by capsaicin and similar pungent substances in rats and guinea-pigs. J Physiol 1970;206:495–507.
15. Jancso-Gabor A, Szolcsanyi J, Jancso N. Stimulation and desensitization of the hypothalamic heat-sensitive structures by capsaicin in rats. J Physiol 1970;208:449–459.
16. Jancso G, Kiraly E, Jancso-Gabor A. Pharmacologically induced selective degeneration of chemosensitive primary sensory neurones. Nature 1977;741:741–743.
17. Porszasz R, Jancso N. Studies on the action potentials of sensory nerves in animals desensitized with capsazepine. Acta Physiol Acad Sci Hung 1959;16:299–306.
18. Hogyes E. Adatok a capsicum annuum (paprika) alkatreszeinek elletani hatasahoz. Orv Hetil 1878;18:181–188.
19. Bevan S, Hothi S, Hughes G, James I-F, Rang H-P, Shah K et al. Capsazepine: a competitive antagonist of the sensory neurone excitant capsaicin. Br J Pharmacol 1992;107:544–552.
20. Maggi C-A, Santicioli P, Geppetti P, Parlani M, Astolfi M, Pradelles P et al. The antagonism induced by ruthenium red of the actions of capsaicin on the peripheral terminals of sensory neurons: further studies. Eur J Pharmacol 1988;154:1–10.

21. Szallasi A, Blumberg P-M. Resiniferatoxin and its analogs provide novel insights into the pharmacology of the vanilloid (capsaicin) receptor. Life Sci 1990;47:1399–1408.
22. Marsh S-J, Stansfeld C-E, Brown D-A, Davey R, McCarthy D. The mechanism of action of capsaicin on sensory C-type neurons and their axons in vitro. Neuroscience 1987;23:275–289.
23. Heyman I, Rang H-P. Depolarizing responses to capsaicin in a subpopulation of rat dorsal root ganglion cells. Neurosci Lett 1985;56:69–75.
24. Caterina MJ, Schumacher MA, Tominaga M, Rosen TA, Levine JD, Julius D. The capsaicin receptor: a heat-activated ion channel in the pain pathway. Nature 1997;389:816–824.
25. Gunthorpe MJ, Benham CD, Randall A, Davis JB. The diversity in the vanilloid (TRPV) receptor family of ion channels. Trends Pharmacol Sci 2002;23:183–191.
26. Savidge J, Davis C, Shah K, Colley S, Phillips E, Ranasinghe S et al. Cloning and functional characterization of the guinea pig vanilloid receptor 1. Neuropharmacology 2002;43:450.
27. Jordt SE, Julius D. Molecular basis for species-specific sensitivity to "hot" chili peppers. Cell 2002;108:421–430.
28. Smith GD, Gunthorpe MJ, Kelsell RE, Hayes PD, Reilly P, Facer P et al. TRPV3 is a temperature-sensitive vanilloid receptor-like protein. Nature 2002;418:186–190.
29. Delany NS, Hurle M, Facer P, Alnadaf T, Plumpton C, Kinghorn I et al. Identification and characterization of a novel human vanilloid receptor-like protein, VRL-2. Physiol Genomics 2001;4:165–174.
30. Hayes P, Meadows HJ, Gunthorpe MJ, Harries MH, Duckworth DM, Cairns W et al. Cloning and functional expression of a human orthologue of rat vanilloid receptor-1. Pain 2000;88:205–215.
31. Schumacher MA, Jong BE, Frey SL, Sudanagunta SP, Capra NF, Levine JD. The stretch-inactivated channel, a vanilloid receptor variant, is expressed in small-diameter sensory neurons in the rat. Neurosci Lett 2000;287:215–218.
32. Schumacher MA, Moff I, Sudanagunta SP, Levine JD. Molecular cloning of an N-terminal splice variant of the capsaicin receptor. Loss of N-terminal domain suggests functional divergence among capsaicin receptor subtypes. J Biol Chem 2000;275:2756–2762.
33. Caterina MJ, Rosen TA, Tominaga M, Brake AJ, Julius DA. Capsaicin-receptor homologue with a high threshold for noxious heat. Nature 1999;398:436–441.
34. Guo A, Vulchanova L, Wang J, Li X, Elde R. Immunocytochemical localisation of the vanilloid receptor (VR1): relationship to neuropeptides, the P2X3 purinoceptor and IB4 binding sites. Eur J Neurosci 1999;11:946–958.
35. Pare M, Elde R, Mazurkiewicz JE, Smith AM, Rice FL. The Meissner corpuscle revised: a multiafferented mechanoreceptor with nociceptor immunochemical properties. J Neurosci 2001;21:7236–7246.
36. Avelino A, Cruz C, Nagy I, Cruz F. Vanilloid receptor 1 expression in the rat urinary tract. Neuroscience 2002;109:787–798.
37. Ward SM, Bayguinov J, Won KJ, Grundy D, Berthoud HR. Distribution of the vanilloid receptor (VR1) in the gastrointestinal tract. J Comp Neurol 2003;465:121–135.
38. Gram DX, Ahren B, Nagy I, Olsen UB, Brand CL, Sundler F et al. Capsaicin-sensitive sensory fibers in the islets of Langerhans contribute to defective insulin secretion in Zucker diabetic rat, an animal model for some aspects of human type 2 diabetes. Eur J Neurosci 2007;25:213–223.

39. Razavi Y, Chan F, Afifiyan X, Liu X, Wan J, Yantha H et al. TRPV1+ sensory neurons control beta cell stress and islet inflammation in autoimmune diabetes. Cell 2006;127:1123–1135.
40. Sato J, Segami N, Yoshitake Y, Kaneyama K, Abe A, Yoshimura H et al. Expression of capsaicin receptor TRPV-1 in synovial tissues of patients with symptomatic internal derangement of the temporomandibular joint and joint pain. Oral Surg Oral Med Oral Pathol Oral Radiol Endod 2005;100:674–681.
41. Mezey E, Toth ZE, Cortright DN, Arzubi MK, Krause JE, Elde R et al. Distribution of mRNA for vanilloid receptor subtype 1 (VR1), and VR1-like immunoreactivity, in the central nervous system of the rat and human. Proc Natl Acad Sci U S A 2000;97:3655–3660.
42. Cristino L, De Petrocellis L, Pryce G, Baker D, Guglielmotti V, Ci Marzo V. Immunohistochemical localisazation of cannabinoid type 1 and vanilloid transient receptor potential vanilloid type 1 receptors in the mouse brain. Neuroscience 2006;139:1405–1415.
43. Roberts JC, Davis JB, Benham CK. [3H]resiniferatoxin autoradiography in the CNS of wild-type and TRPV1 null mice defines TRPV1 (VR-1) protein distribution. Brain Res 2004;995:176–183.
44. Balaban CD. Type 1 vanilloid receptor expression by mammalian inner ear ganglion cells. Hear Res 2003;175:165–170.
45. Zheng Dai JC, Steyger PS, Kim Y, Vass Z, Ren T et al. Vanilloid receptors in hearing: altered cochlear sensitivity by vanilloids and expression of TRPV1 in the organ of corti. J Neurophysiol 2003;90:444–455.
46. Tominaga M, Caterina MJ, Malmberg AB, Rosen TA, Gilbert H, Skinner K et al. The cloned capsaicin receptor integrates multiple pain-producing stimuli. Neuron 1998;21:531–543.
47. Valtschanoff JG, Rustioni A, Guo A, Hwang SJ. Vanilloid receptor VR1 is both presynaptic and postsynaptic in the superficial laminae of the rat dorsal horn. J Comp Neurol 2001;436:225–235.
48. Kim H, Neubert JK, San Miguel A, Xu K, Krishnaraju RK, Iadarola MJ et al. Genetic influence on variability in human acute experimental pain sensitivity associated with gender, ethnicity and psychological temperament. Pain 2004;109:488–489.
49. Xu H, Tian W, Fu Y, Oyama TT, Anderson S, Cohen DM. Functional effects of synonymous polymorphisms in the human TRPV1 gene. Am J Physiol Renal Physiol 2007;293(6):F1865–F1876.
50. Montell C, Birnbaumer L, Flockerzi V. The TRP channels, a remarkably functional family. Cell 2002;108:595–598.
51. Vennekens R, Voets T, Bindels RJ, Droogmans G, Nilius B. Current understanding of mammalian TRP homologues. Cell Calcium 2002;31:253–264.
52. Kedei N, Szabo T, Lile JD, Treanor JJ, Olah Z, Iadarola MJ et al. Analysis of the native quaternary structure of vanilloid receptor 1. J Biol Chem 2001;276:28613–28619.
53. Kuzhikandathil EV, Wang H, Szabo T, Morozova N, Blumberg PM, Oxford GS. Functional analysis of capsaicin receptor (vanilloid receptor subtype 1) multimerization and agonist respsonsiveness using a dominant negative mutation. J Neurosci 2001;21:8697–8706.
54. Wang C, Hu HZ, Colton CK, Wood JD, Zhu MX. An alternative splicing product of the murine TRPV1 gene dominant negatively modulates the activity of TRPV1 channels. J Biol Chem 2004;279:37423–37430.

55. Lu G, Henderson D, Liu L, Reinhart PH, Simon SA. TRPV1b, a functional human vanilloid receptor splice variant. Mol Pharmacol 2005;67:1119–1127.
56. Chuang HH, Prescott ED, Kong H, Shields S, Jordt SE, Basbaum AI et al. Bradykinin and nerve growth factor release the capsaicin receptor from PtdIns(4,5)P2-mediated inhibition. Nature 2001;411:957–962.
57. Morenilla-Palao C, Planells-Cases R, Garcia-Sanz N, Ferrer-Montiel A. Regulated exocytosis contributes to protein kinase C potentiation of vanilloid receptor activity. J Biol Chem 2004;279:25665–25672.
57A. Rathee PK, Distler C, Obreja O, Neuhuber W, Wang GK, Wang SY, Nau C, Kress M. PKA/AKAP/VR-1 module A common link of Gs – mediated signalling to thermal hyperalgesia. J Neurosci 2002;222:4740–4745.
58. Ji R, Samad T, Jin S, Schmoll R, Woolf C. p38 MAPK activation by NGF in primary sensory neurons after inflammation increases TRPV1 levels and maintains heat hyperalgesia. Neuron 2002;36:57–68.
59. Tohda C, Sasaki M, Konemura T, Sasamura T, Itoh M, Kuraishi Y. Axonal transport of VR1 capsaicin receptor mRNA in primary afferents and its participation in inflammation-induced increase in capsaicin sensitivity. J Neurochem 2001;76:1628–1635.
60. Cesare P, Dekker LV, Sardini A, Parker PJ, McNaughton PA. Specific involvement of PKC-e in sensitization of the neuronal response to painful heat. Neuron 1999;23:617–624.
61. Premkumar LS, Ahern GP. Induction of vanilloid receptor channel activity by protein kinase C. Nature 2000;408:985–990.
62. Sugiura T, Tominaga M, Katsuya H, Mizumura K. Bradykinin lowers the threshold temperature for heat activation by vanilloid receptor 1. J Neurophysiol 2002;88:544–548.
63. Fischer MJ, Reeh PW. Sensitization to heat through G-protein-coupled receptor pathways in the isolated sciatic mouse nerve. Eur J Neurosci 2007;25:3570–3575.
64. Voets T, Droogmans G, Wissenbach U, Janssens A, Flockerzi V, Nilius B. The principle of temperature-dependent gating in cold- and heat-sensitive TRP channels. Nature 2004;430:748–754.
65. Welch JM, Simon SA, Reinhart PH. The activation mechanism of rat vanilloid receptor 1 by capsaicin involves the pore domain and differs from the activation by either acid or heat. Proc Natl Acad Sci U S A 2000;97:13889–13894.
66. Zygmunt PM, Peterson J, Anderson DA, Chuang H, Sorgard M, Di Marzo V. Vanilloid receptors on sensory nerves mediate the vasodilator action of anandamide. Nature 1999;400:452–457.
67. Huang SM, Bisogno T, Trevisani M, Al-Hayani A, De Petrocellis L, Fezza F et al. An endogenous capsaicin-like substance with high potency at recombinant and native vanilloid VR1 receptors. Proc Natl Acad Sci U S A 2002;99:8400–8405.
68. Szolcsanyi J, Sandor Z, Petho G, Varga A, Bolcskei K, Almasi R et al. Direct evidence for activation and desensitization of the capsaicin receptor by N-oleoyldopamine on TRPV1-transfected cell line, in gene deleted mice and in the rat. Neurosci Lett 2004;361:155–158.
69. Hwang SW, Cho H, Kwak J, Lee SY, Kang CJ, Jung J et al. Direct activation of capsaicin receptors by products of lipoxygenases: endogenous capsaicin-like substances. Proc Natl Acad Sci U S A 2000;97:6155–6160.
70. Movahed P, Jonsson BAG, Birnir B, Wingstrand JA, Jorgensen TK, Ermund A et al. Endogenous unsaturated C18 N-acylethanolamines are vanilloid receptor (TRPV1) agonists. J Biol Chem 2005;280:38496–38504.

71. Gavva NR, Klinonsky L, Qu Y, Shi L, Tamir R, Edenson S et al. Molecular determinants of vanilloid sensitivity in TRPV1. J Biol Chem 2004;279:20283–20295.
72. Phillips E, Reeve A, Bevan S, McIntyre P. Identification of species-specific determinants of the action of the antagonist capsazepine and the agonist PPAHV on TRPV1. J Biol Chem 2004;279:17165–17172.
73. Bisogno T, Melck D, Bobrov MYu, Gretskaya NM, Bezuglov VV, De Petrocellis L et al. N-acyl-dopamines: novel synthetic CB(1) cannabinoid-receptor ligands and inhibitors of anandamide inactivation with cannabimimetic activity in vitro and in vivo. Biochem J 2000;351:817–824.
74. Liu B, Zhang C, Qin F. Functional recovery from desensitization of vanilloid receptor TRPV1 requires resynthesis of phosphatidylinositol 4,5-bisphosphate. J Neurosci 2005;25:4835–4843.
75. Ferreira J, da Silva GL, Calixto JB. Contribution of vanilloid receptors to the overt nociception induced by B2 kinin receptor activation in mice. Br J Pharmacol 2004;141:787–794.
76. Dinis P, Charrua A, Avelino A, Yaqoob M, Bevan S, Nagy I et al. Anandamide-evoked activation of vanilloid receptor 1 contributes to the development of bladder hyperreflexia and nociceptive transmission to spinal dorsal horn neurons in cystitis. J Neurosci 2004;24:11253–11263.
77. Olah Z, Karai L, Iadarola MJ. Anandamide activates vanilloid receptor 1 (VR1) at acidic pH in dorsal root ganglia neurons and cells ectopically expressing VR1. J Biol Chem 2001;276:31163–31170.
78. Liu B, Hui K, Qin F. Thermodynamics of heat activation of single capsaicin ion channels VR1. Biophysical J 2003;85:2988–3006.
79. Susankova K, Tousova K, Vyklicky L, Teisinger J, Vlachova V. Reducing and oxidizing agents sensitize heat-activated vanilloid receptor (TRPV1) current. Mol Pharmacol 2006;70:383–394.
80. Ryu S, Liu B, Qin F. Low pH potentiates both capsaicin binding and channel gating of VR1 receptors. J Gen Physiol 2003;122:45–61.
81. Gunthorpe MJ, Harries MH, Prinjha RK, Davis JB, Randall A. Voltage- and time-dependent properties of the recombinant rat vanilloid receptor (rVR1). J Physiol 2000;525:747–759.
82. Ahern GP, Brooks IM, Miyares RL, Wang X. Extracellular cations sensitise and gate capsaicin receptor TRPV1 modulating pain signalling. J Neurosci 2005;25: 5109–5116.
83. Silver IA, Murrills RJ, Etherington DJ. Microelectrode studies on the acid microenvironment beneath adherent macrophages and osteoclasts. Exp Cell Res 1988;175:266–276.
84. Premkumar LS. Interaction between vanilloid receptors and purinergic metabotropic receptors: pain perception and beyond. Proc Natl Acad Sci U S A 2001;98:6537–6539.
85. Lee YJ, Zachrisson O, Tonge DA, McNaughton PA. Upregulation of bradykinin B2 receptor expression by neurotrophic factors and nerve injury in mouse sensory neurons. Mol Cell Neurosci 2002;19:186–200.
86. Michael GJ, Priestley JV. Differential expression of the mRNA for the vanilloid receptor subtype 1 in cells of the adult rat dorsal root and nodose ganglia and its downregulation by axotomy. J Neurosci 1999;19:1844–1854.
87. Dray A, Perkins MN. Bradykinin activates peripheral capsaicin-sensitive fibres via a second messenger system. Agents Actions 1988;25:214–215.

88. Jung J, Hwang SW, Kwak J, Lee SY, Kang CJ, Kim WB et al. Capsaicin binds to the intracellular domain of the capsaicin-activated ion channel. J Neurosci 1999;19:529–538.
89. Rami HK, Thompson M, Whyman P, Jerman JC, Egerton J, Brough S et al. Discovery of small molecule antagonists of TRPV1. Bioorg Med Chem Lett 2004;14:3631–3634.
90. Vyklicky L, Lyfenko A, Kuffler DP, Vlachova V. Vanilloid receptor TRPV1 is not activated by vanilloids applied intracellularly. NeuroReport 2003;14:1061–1065.
91. Hui K, Liu B, Qin F. Capsaicin activation of the pain receptor, VR1: multiple open states from both partial and full binding. Biophysical J 2003;84:2957–2968.
92. Jung J, Lee SY, Hwang SW, Cho H, Shin J, Kang YS et al. Agonist recognition sites in the cytosolic tails of vanilloid receptor 1. J Biol Chem 2002;277:44448–44454.
93. Vlachova V, Teisinger J, Susankova K, Lyfenko A, Ettrich R, Vyklicky L. Functional role of C-terminal cytoplasmic tail of rat vanilloid receptor 1. J Neurosci 2003;23:1340–1350.
94. Jordt SE, Tominaga M, Julius D. Acid potentiation of the capsaicin receptor determined by a key extracellular site. Proc Natl Acad Sci U S A 2000;97:8134–8139.
95. Huang J, Zhang Z, McNaughton PA. Modulation of temperature-sensitive TRP channels. Semin Cell Biol 2006;17:638–645.
96. Brauchi S, Orta G, Salazar M, Rosenmann E, Latorre R. A hot-sensing cold receptor: C-terminal domain determines thermosensation in transient receptor potential channels. J Neurosci 2006;26:4835–4840.
97. Bhave G, Zhu W, Wang H, Brasier DJ, Gereau RW. cAMP-dependent protein kinase regulates desensitization of the capsaicin receptor (VR1) by direct phosphorylation. Neuron 2002;35:721–731.
98. Mohapatra DP, Nau C. Regulation of Ca++-dependent desensitization in the vanilloid receptor TRPV1 by calcineurin and cAMP-dependent protein kinase. J Biol Chem 2005;280:13424–13432.
99. Jeske NA, Patwardhan AM, Gamper N, Price TJ, Akopian AN, Hargreaves KM. Cannabinoid WIN 55,212-2 regulates TRPV1 phosphorylation in sensory neurons. J Biol Chem 2006;281:32879–32890.
100. Numazaki M, Tominaga T, Toyooka H, Tominaga M. Direct phosphorylation of capsaicin receptor VR1 by protein kinase ce and identification of two target serine residues. J Biol Chem 2002;277:13375–13378.
101. Bhave G, Hu HJ, Glauner KS, Zhu W, Wang H, Brasier DJ. Protein kinase C phosphorylation sensitizes but does not activate the capsaicin receptor transient receptor potential vanilloid 1 (TRPV1). Proc Natl Acad Sci U S A 2003;100:12480–12485.
102. Zhang X, Huang J, McNaughton PA. NGF rapidly increases membrane expression of TPV1 heat-gated ion channels. EMBO J 2005;24:4211–4223.
103. Prescott ED, Julius D. A modular PIP2 binding site as a determinant of capsaicin receptor sensitivity. Science 2003;300:1284–1288.
104. Latorre R, Brauchi S, Orta G, Zaelzer C, Vargas G. ThermoTRP channels as modular proteins with allosteric gating. Cell Calcium 2007;42:427–438.
105. Voets T, Owsianik G, Janssens A, Talavera K, Nilius B. TRPM8 voltage sensor mutants reveal a mechanism for integrating thermal and chemical stimuli. Nat Chem Biol 2007;3:174–182.
106. Stuhmer W, Conti F, Suzuki H, Wang XD, Noda M, Yahagi N et al. Structural parts involved in activation and inactivation of the sodium channel. Nature 1989;39:597–603.
107. Tombola F, Pathak MM, Isachoff EY. How does voltage open an ion channel? Annu Rev Cell Dev Biol 2006;22:23–252.

108. Kress M, Fetzer S, Reeh PW, Vyklicky L. Low pH facilitates capsaicin responses in isolated sensory neurons of the rat. Neurosci Lett 1996;211:5–8.
109. Tominaga M, Julius D. Capsaicin receptor in the pain pathway. Jpn J Pharmacol 2000;83:20–24.
110. Vyklicky L, Vlachova V, Vitaskova Z, Dittert I, Kabat M, Orkand RK. Temperature coefficient of membrane currents induced by noxious heat in sensory neurones in the rat. J Physiol 1999;517:181–192.
111. Liang YF, Haake B, Reeh PW. Sustained sensitization and recruitment of rat cutaneous nociceptors by bradykinin and a novel theory of its excitatory action. J Physiol 2001;523:229–239.
112. Babes A, Amuzescu B, Krause U, Scholz A, Flonta ML, Reid G. Cooling inhibits capsaicin-induced currents in cultured rat dorsal root ganglion neurones. Neurosci Lett 2002;317:131–134.
113. Brauchi S, Orio P, Latorre R. Clues to understanding cold sensation: thermodynamics and electrophysiological analysis of the cold receptor TRPM8. Proc Natl Acad Sci U S A 2004;101:15494–15499.
114. Caterina MJ, Leffler A, Malmberg AB, Martin WJ, Trafton J, Petersen-Zeitz KR et al. Impaired nociception and pain sensation in mice lacking the capsaicin receptor. Science 2000;288:306–313.
115. Davis JB, Gray J, Gunthorpe MJ, Hatcher JP, Davey PT, Overend P et al. Vanilloid receptor-1 is essential for inflammatory thermal hyperalgesia. Nature 2000;405:183–187.
116. Chen CC, England S, Akopian AN, Wood JN. A sensory neuron-specific, proton-gated ion channel. Proc Natl Acad Sci U S A 2002;95:10240–10245.
117. Hudson LJ, Bevan S, Wotherspoon G, Gentry C, Fox A, Winter J. VR1 protein expression increases in undamaged DRG neurons after partial nerve injury. Eur J Neurosci 2001;13:2105–2114.
118. Woolf CJ, Shortland P, Coggeshall RE. Peripheral nerve injury triggers central sprouting of myelinated afferents. Nature 1992;355:75–78.
119. Kamei J, Zushida K, Morita K, Sasaki M, Tanaka S. Role of vanilloid VR1 receptor in thermal allodynia and hyperalgesia in diabetic mice. Eur J Pharmacol 2001;422:83–86.
120. Kelly S, Chapman V. Spinal administration of capsazepine inhibits noxious evoked responses of dorsal horn neurons in non-inflamed and carrageenan inflamed rats. Brain Res 2002;935:103–108.
121. Walker K, Urban L, Medhurst SJ, Patel S, Panesar M, Fox AJ et al. The VR1 antagonist, capsazepine, reverses hyperalgesia in models of inflammatory and neuropathic pain. J Pharmacol Exp Ther 2003;304:56–62.
122. Christoph T, Gillen C, Mika J, Grunweller A, Schater MKH, Schiene K et al. Antinociceptive effect of antisense oligonucleotides against the vanilloid receptor VR1/TRPV1. Neurochem Int 2007;50:281–290.
123. Christoph T, Grunweller A, Mika J, Schafter MK-H, Wade EJ, Weihe E et al. Silencing of vanilloid receptor TRPV1 by RNAi reduces neuropathic and visceral pain in vivo. Biochem Biophys Res Commun 2006;350:238–243.
124. Rang HP, Dale MM, Ritter JM, Moore PK. Local hormones, inflammation, and immune reactions. In: Rang HP, Dale MM, Ritter JM, Moore PK, eds. Pharmacology, 5 ed. Edinburgh: Churchill Livingstone, 2003; pp. 217–243.

125. Rost B, Hanf G, Ohnemus U, Otto-Knapp R, Groneberg DA et al. Monocytes of allergics and non-allergics produce, store and release the neurotrophins NGF, BDNF and NT-3. Regul Pept 2005;124:19–25.
126. Noga O, Englmann C, Hanf G, Grutzkau A, Seybold J, Kunkel G. The production, storage and release of the neurotrophins nerve growth factor, brain-derived neurotrophic factor and neurotrophin-3 by human peripheral eosinophils in allergics and non-allergics. Clin Exp Allergy 2003;33:649–654.
127. Leon A, Buriani A, Dal Toso R, Fabris M, Romanello S, Aloe L et al. Mast cells synthesize, store, and release nerve growth factor. Proc Natl Acad Sci U S A 1994;91:3739–3743.
128. Lipnik-Stangelij M. Multiple role of histamine H1-receptor-PKC-MAPK signalling pathway in histamine-stimulated nerve growth factor synthesis and secretion. Biochem Pharmacol 2006;72:1375–1381.
129. Abe Y, Akeda K, An HS, Aoki Y, Pichika R, Muehleman C et al. Proinflammatory cytokines stimulate the expression of nerve growth factor by human intervertebral disc cells. Spine 2007;32:635–642.
130. Hattori A, Tanaka E, Murase K, Ishida N, Chatani Y, Tsujimoto M et al. Tumor necrosis factor stimulates the synthesis and secretion of biologically active nerve growth factor in non-neuronal cells. J Biol Chem 1993;268:2577–2582.
131. Woolf CJ, Allchorne A, Safieh-Garabedian B, Poole S. Cytokines, nerve growth factor and inflammatory hyperalgesia: the contribution of tumour necrosis factor alpha. Br J Pharmacol 1997;121:417–424.
132. Airaksinen MS, Saarma M. The GDNF family: signalling, biological functions and therapeutic value. Nat Rev Neurosci 2002;3:383–394.
133. Malin SA, Molliver DC, Koerber HR, Cornuet P, Frye R, Albers KM et al. Glial cell line-derived neurotrophic factor family members sensitize nociceptors in vitro and produce thermal hyperalgesia in vivo. J Neurosci 2006;26:8588–8599.
134. Ahluwalia J, Yaqoob M, Urban L, Bevan S, Nagy I. Activation of capsaicin-sensitive primary sensory neurones induces anandamide production and release. J Neurochem 2003;84:585–591.
135. van der Stelt M, Trevisani M, Vellani V, De Petrocellis L, Schiano Moriello A, Campi B et al. Anandamide acts as an intracellular messenger amplifying Ca^{2+} influx via TRPV1 channels. EMBO J 2005;24:3026–3037.
136. Okamoto Y, Wang J, Morishita J, Ueda N. Biosynthetic pathways of the endocannabinoid anandamide. Chem Biodivers 2007;4:1842–1857.
137. Habler C. Uber den K+ und Ca++ gehalt von eiter and exsudaten und seine beziehungen zum entzundungsschmerz. Klin Wochenschr 1929;8:1569–1572.
138. von Gaza W, Brandi B. Beziehungen zwischen wasserstoffionenkonzentration und schmerzempfindung. Klin Wochenschr 1926;5:1123–1127.
139. Steen KH, Steen AE, Kreysel H-W, Reeh PW. Inflammatory mediators potentiate pain induced by experimental tissue acidosis. Pain 1996;66:163–170.
140. Steen KH, Reeh PW, Anton F, Handwerder HO. Protons selectively induce lasting excitation and sensitization to mechanical stimulation of nociceptors in rat skin, in vitro. J Neurosci 1992;12:86–95.
141. Woo YC, Park SS, Subleta AR, Brennan TJ. Changes in tissue pH and temperature after incision indicate acidosis May contribute to postoperative pain. Anesthesiology 2004;101:468–475.

142. O'Brien RJ, Kamboj S, Ehlers MD, Rosen KR, Fischbach GD, Huganir RL. Activity-dependent modulation of synaptic AMPA receptor accumulation. Neuron 1998;21:1067–1078.
143. Ogun-Muyiwa P, Helliwell R, McIntyre P, Winter J. Glial cell line derived neurotrophic factor (GDNF) regulates VR1 and substance P in cultured sensory neurons. NeuroReport 1999;10:2107–2111.
144. Winston J, Toma H, Shenoy M, Pasricha PJ. Nerve growth factor regulates VR-1 mRNA levels in cultures of adult dorsal root ganglion neurons. Pain 2001;89:181–186.
145. Anand U, Otto WR, Casula MA, Day NC, Davis JB, Bountra C et al. The effect of neurotrophic factors on morphology, TRPV1 expression and capsaicin responses of cultured human DRG sensory neurons. Neurosci Lett 2006;399:51–56.
146. Woolf CJ, Safieh-Garabedian B, Ma QP, Crilly P, Winter J. Nerve growth factor contributes to the generation of inflammatory sensory hypersensitivity. Neuroscience 1994;62:327–331.
147. Bron R, Klesse LJ, Shah K, Parada LF, Winter J. Activation of Ras is necessary and sufficient for upregulation of vanilloid receptor type 1 in sensory neurons by neurotrophic factors. Mol Cell Neurosci 2003;22:118–132.
148. Elitt CM, McIlwrath SL, Lawson JJ, Malin SA, Molliver DC, Cornuet PK et al. Artemin overexpression in skin enhances expression of TRPV1 and TRPA1 in cutaneous sensory neurons and leads to behavioral sensitivity to heat and cold. J Neurosci 2006;26:8578–8587.
149. Nagy I, Pabla R, Matesz C, Dray A, Woolf CJ, Urban L. Cobalt uptake enables identification of capsaicin- and bradykinin-sensitive subpopulations of rat dorsal root ganglion cells in vitro. Neuroscience 1993;56:241–246.
150. Nicholas RS, Winter J, Wren P, Bergmann R, Woolf CJ. Peripheral inflammation increases the capsaicin sensitivity of dorsal root ganglion neurons in a nerve growth factor-dependent manner. Neuroscience 1999;91:1425–1433.
151. Boucher TJ, Okuse K, Bennett DL, Munson JB, Wood JN, McMahon SB. Potent analgesic effects of GDNF in neuropathic pain states. Science 2000;290:124–127.
152. Shembalkar PK, Till S, Boettger MK, Terenghi G, Tate S, Bountra C et al. Increased sodium channel SNS/PN3 immunoreactivity in a causalgic finger. Eur J Pain 2001;5:319–323.
153. Shu X, Mendell LM. Acute sensitization by NGF of the response of small-diameter sensory neurones to capsaicin. J Neurphysiol 2001;86:2931–2938.
154. Bonnington JK, McNaughton PA. Signalling pathways involved in the sensitisation of mouse nocicpetive neurones by nerve growth factor. J Physiol 2003;551:433–446.
155. Stein AT, Ufret-Vincenty CA, Hua L, Santana LF, Gordon SE. Phosphoinositide 3-kinase binds to TRPV1 and mediates NGF-stimulated TRPV1 trafficking to the plasma membrane. J Gen Physiol 2006;128:509–522.
156. Linhart O, Obreja O, Kress M. The inflammatory mediators serotonin, prostaglandin E2 and bradykinin evoke calcium influx in rat sensory neurons. Neuroscience 2003;118:69–74.
157. Kopp UC, Cicha MZ, Smith LA. PGE(2) increases release of substance P from renal sensory nerves by activating the cAMP-PKA transduction cascade. Am J Physiol Regul Integr Comp Physiol 2002;282:R1618–1627.
158. Ahluwalia J, Urban L, Capogna M, Bevan S, Nagy I Cannabinoid 1 receptors are expressed. Neuroscience 2000;100:685–688.

159. Novakova-Tousova K, Vyklicky L, Susankova K, Benedikt J, Samad A, Teisinger J et al. Functional changes in the vanilloid receptor subtype 1 channel during and after acute desensitization. Neuroscience 2007;149:144–154.
160. Singh Tahim A, Santha P, Nagy I. Inflammatory mediators convert anandamide into a potent activator of the vanilloid type 1 transient receptor potential receptor in nociceptive primary sensory neurons. Neuroscience 2005;136:539–548.
161. Hillard CJ. Biochemistry and pharmacology of the endocannabinoids arachidonylethanolamide and 2-arachidonylglycerol. Prostaglandins Other Lipid Mediat 2000;61:3–18.
162. Shin J, Cho H, Hwang SW, Jung J, Shin CY, Lee SY et al. Bradykinin-12-lipoxygenase-VR1 signaling pathway for inflammatory hyperalgesia. Proc Natl Acad Sci U S A 2002;99:10150–10255.
163. Galoyan SM, Petruska JC, Mendell LM. Mechanisms of sensitization of the response of single dorsal root ganglion cells from adult rat to noxious heat. Eur J Neurosci 2003;18:535–541.

Edited by Gábor Jancsó

The Role of the Vanilloid and Related Receptors in Nociceptor Function and Neuroimmune Regulation

DANIEL N. CORTRIGHT[1] and ARPAD SZALLASI[2]

[1] *Department of Biochemistry and Molecular Biology, Neurogen Corporation, Branford, CT, USA*
[2] *Department of Pathology, Monmouth Medical Center, NJ, USA*

ABSTRACT

The contribution of sensory nerve fibers to inflammatory responses and immune regulation has been the subject of intense study for many years. Capsaicin has been particularly useful as a tool to probe this connection as it dissects a specific subset of sensory nerve fibers, namely those with unmyelinated fibers that carry polymodal nociceptors. These nerves are sites of release for various proinflammatory neuropeptides that initiate the cascade of neurogenic inflammation, and they convey pain to the central nervous system. The capsaicin receptor, also known as the vanilloid VR1 or TRPV1 receptor, plays a pivotal role in the hyperalgesic response that follows inflammatory challenge in rodent models of pain. TRPV1 is activated by noxious heat, and endogenous proinflammatory molecules, such as arachidonic acid metabolites, nerve growth factor, bradykinin, and protons (tissue acidification), act in concert to reduce the heat activation threshold of this receptor below body temperature.

TRPV1 protein levels are up-regulated in experimental models of inflammatory hyperalgesia and in a number of human disorders (e.g., inflammatory bowel disease). Hence, the development of novel TRPV1 antagonists could bring about new therapies for the treatment of inflammatory diseases.

The interaction between sensory nerves and immune cells is, however, even more complex than previously thought. For example, functional TRPV1 receptors are expressed on a broad array of cells (e.g., mast cells, lymphocytes, dendritic cells) involved in immune responses, and, in turn, mediators released from these cells (e.g., mast cell tryptase via proteinase-activated receptor 2 and protein kinase C) can activate TRPV1 in sensory nerves.

A better understanding of the complex cross talk between sensory nerves and immune cells may identify novel indications for pharmacological intervention.

1. INTRODUCTION

Cross talk between the nervous and immune systems is manifest at many levels. One of the more interesting areas of interaction is at the sensory neuron fiber terminal. These terminals are distributed throughout the body but have been the subject of intense study in the skin, the mouth, the gastrointestinal tract, the bladder, and the lung. It is widely reported (as recently reviewed in Refs [1–6]) that immune system mediators, such as monoamines, prostaglandins, and cytokines, can activate these fibers resulting in sensations ranging from mild discomfort (e.g., itchiness) to pain (e.g., burns). Sensory neurons, in turn, can release via their peripheral terminals a variety of substances, including substance P (SP) and calcitonin gene-related peptide (CGRP) [7–9]. In addition to exerting potent vasoregulatory actions, these substances can profoundly affect the intensity of immune responses [10–13]. In fact, the presence of receptors for SP (neurokinin NK1 receptors) in lymphocytes has been reported [14].

Vanilloids, including capsaicin and resiniferatoxin (RTX), have been widely used as tools to investigate the function of sensory neurons in modulating immune and neurogenic inflammatory responses [15–17]. These compounds are potent activators of SP and CGRP release from sensory neurons [18,19] by interacting at a specific membrane recognition site, referred to as the vanilloid receptor TRPV1 [20]. Vanilloid receptors mediate large calcium currents in cultured sensory neurons and dramatically increase neuron-firing rate *in vivo* [15,19,21,22]. Capsaicin, best known as the agent responsible for the piquancy of red peppers [20], is a noxious agent that initiates dramatic hyperalgesia via TRPV1 in rodents [23] and humans [24,25]. Interestingly, large dose administration of capsaicin results in the development of a state in which the previously excited neurons not only do not respond to subsequent capsaicin challenge but are also unresponsive to various noxious stimuli, both chemical and physical, including heat and acids [18–20]. This effect, first described by the late Jancsó half a century ago [26,27], is traditionally referred to as desensitization. The mechanisms underlying this desensitization are only beginning to be understood. Both the acute activating and long-term desensitization effects of vanilloids have been exploited to study sensory neuron–immune interactions [15–20]. A recently emerging area is the use of vanilloid compounds for direct immunomodulation. For example, both mast cells [28–30] and lymphocytes (in specific, CD4+ as well as double-negative T cells residing in the thymic cortex [31]) are reported to possess vanilloid receptors. The implications of these new findings are detailed below.

2. THE VANILLOID (CAPSAICIN) RECEPTOR VR1 (TRPV1)

Capsaicin and other vanilloids mediate their effects via the vanilloid receptor subtype-1, abbreviated as TRPV1. TRPV1 is a nonselective cation channel first cloned by Julius' group at UCSF in 1997 [32] and is a member of the transient release potential (TRP) channel superfamily [33]. VR1 (also known as TRPV1) helps define a subclass of these channels known as TRPV channels, which contain three conserved ankyrin domains [32]. Several conserved protein kinase A (PKA) and protein kinase C (PKC) phosphorylation sites are also observed in TRPV1 [32], which have important roles in the regulation of TRPV1 functions.

The physiology of TRPV1 [3,6,20,34] and related TRP channels [35] has recently been subject to exhausting reviews. It is important to emphasize that TRPV1 is the only known "vanilloid receptor" defined as a receptor activated by vanilloid compounds. Hence, reports of non-TRPV1-mediated effects of vanilloids, such as that vanilloids may modulate T-cell activation independent of TRPV1 receptors by blocking voltage-gated K^+ channels as well as the store-operated Ca^{2+} channel responsible for capacitative Ca^{2+} entry [36] must be viewed with caution.

Interestingly, mustard oil, which has also been used to study neurogenic inflammation and sensory nerve function [37], does not activate TRPV1 but a distantly related channel referred to as TRPA1 [38,39]. The substantial colocalization of TRPA1 in TRPV1-expressing fibers explains the earlier findings that mustard oil effects are blunted by capsaicin pretreatment [37].

2.1 Other VR1 ligands

In addition to its vanilloid sensitivity, TRPV1 is activated by noxious heat (>43°C), by low pH (<6.5), by a variety of inflammatory lipid metabolites, and by phosphorylation [40,41]. Therefore, the emerging concept is that TRPV1 functions as a general sensor of noxious stimuli (a nociceptor) rather than a specific membrane recognition site for compounds carrying a vanillyl moiety [3,20]. As a matter of fact, this concept is nothing new because this is exactly how Jancsó, who almost single handedly transformed capsaicin from a pharmacological oddity to a widely used tool to study a specific subset of sensory neurons, had defined the capsaicin receptor [42].

Among the newly discovered vanilloids, anandamide (*N*-arachidonoyl-ethanolamine) is no doubt the most controversial [43]. Anandamide is an endogenous eicosanoid, which preferentially acts on cannabinoid CB1 receptors: that is, an "endocannabinoid" [44]. Anandamide is formed "on demand" from the hydrolysis of phospholipid precursors, catalyzed by phospholipase D. Anandamide was reported to act as a full agonist at both heterologously expressed and native rat [43,45] and human [46] vanilloid receptors. This was, however, an apparently low-affinity interaction, resulting in understandable reluctance in accepting anandamide as a possible "endovanilloid" (for further details, see Ref. [47]). New insights, however, enhance the possibility of anandamide being a potent "endovanilloid" [6]. Now it is clear that the vanilloid binding site on VR1 is intracellular [48]. Anandamide is transported through the membrane via a specific transporter [49]. The potency of anandamide at TRPV1 is enhanced by coapplication of other TRPV1 agonists such as low pH and increased temperature [50–52]. Agents, like nitric oxide, that stimulate the anandamide membrane transporter enhance the apparent potency of anandamide at TRPV1 [53]. This recognition has exciting implications. Cannabinoid CB1 and vanilloid TRPV1 receptors are coexpressed on a subset of sensory neurons [54]. Generally speaking, agents that activate TRPV1 are excitatory, whereas those acting on CB1 inhibit neuronal activity. The binding site for anandamide on TRPV1 is intracellular [48]. The anandamide recognition domain on CB1 is, in contrast, extracellular [55]. Thus, anandamide may have opposing actions on the very same nerve terminal depending on the functional state of its membrane transporter. Indeed, both excitation (via TRPV1) and inhibition (via CB1) of primary sensory neurons by anandamide have been described (reviewed in Ref. [6]). A novel SP-like peptide named hemokinin was identified in immune cells. Hemokinin is from a different precursor protein and gene than is SP [56].

There is also good evidence that some *in vivo* anandamide actions are mediated by TRPV1. For example, it has been shown that capsazepine inhibits the proinflammatory effects of exogenous anandamide or 2-acylglycerol injected into the rat ileum lumen [57]. Subsequent to the findings of endocannabinoid effects on TRPV1, other potent endogenous capsaicin-like substances, the so-called endovanilloids were identified [6]. The first example is *N*-arachidonoyl-dopamine (NADA), a brain substance, which is similar to capsaicin not only structurally, but also with regard to its potency at TRPV1 [58]. Certain lipoxygenase metabolites, such as 12-hydroperoxy-eicosatetraenoic acid (12-HPETE), have also been identified as putative endogenous TRPV1 agonists [59]. These findings may provide a mechanistic link between TRPV1 and the 5-lipoxygenase products that are important mediators of airway inflammation [60].

Most recently, ethanol has been added to the list of controversial vanilloids [61]. High concentrations of ethanol activate C-fibers innervating the esophagus and the skin, evoking neuropeptide release and resultant neurogenic inflammation [61]. Ethanol also excites dorsal root ganglion neurons possessing native vanilloid receptors as well as HEK cells transfected with TRPV1 [60]. This excitation by ethanol is abolished by the coadministration of capsazepine [61]. Even more interesting is that ethanol lowers the heat threshold of TRPV1 activation from 42°C to 34°C [60].

There is mounting evidence that TRPV1 is balanced on the edge between open and closed states [6]. Agents that promote the open state are nociceptive. By contrast, agents that shift TRPV1 toward the closed state are antinociceptive. As of today, phosphatidylinositol(4,5)-bis-phosphate is the only known endogenous molecule that helps keeping TRPV1 in the closed state [62]. Agents that favor the open state are, however, numerous and can be divided into two major categories: (i) agents that increase the probability of channel opening directly and (ii) agents that release TRPV1 under the inhibitory control of phosphatidylinositol(4,5)-bis-phosphate. The first category is exemplified by the enzymes PKC and PKA (reviewed in Refs [6,34, 40]). Factors that stimulate phospholipase C, the enzyme that cleaves phosphatidylinositol(4,5)-bis-phosphate, fall into the second category [63]. Bradykinin (acting on B2 receptors) has a dual effect on TRPV1: it stimulates PKC by promoting diacylglycerol formation, and at the same time, it degrades phosphatidylinositol(4,5)-bis-phosphate by facilitating phospholipase C [62]. It is tempting to speculate that this dual action on TRPV1 gives an important contribution to the profound algesic activity of bradykinin. Nerve growth factor, a key player in inflammatory hyperalgesia [63], also releases TRPV1 under the inhibitory control of phosphatidylinositol(4,5)-bis-phosphate [62].

2.2 Tissue and cellular expression of TRPV1 and its splice variants

TRPV1 is expressed in small, unmyelinated sensory nerve fibers called C-fibers, which conduct very slow action potentials and contain various neuropeptides including CGRP and SP [9,64]. These TRPV1 immunoreactivity positive fibers have been observed innervating the skin [65], the bladder [30,66,67], and the gastric mucosa [68], just to cite a few examples. TRPV1 RNA and immunoreactivity are also found in the central nervous system (reviewed in Ref. [69]), albeit at much lower levels than found in dorsal root ganglia.

There is good evidence for TRPV1 gene expression in other non-neural tissues as well, including the human skin (keratinocytes, dermal blood vessels, hair follicles, sebocytes, and sweat glands) [29], bladder (urothelium, smooth muscle, mast cells, and capillaries) [30,70], and brain microvasculature endothelium [71]. Importantly, TRPV1 in these tissues is functional, lending experimental foundation to the novel concept that both squamous and transitional epithelium can play sensory roles [72]. In animals, the tissue distribution of TRPV1 is even broader, encompassing heart, liver, kidney, spleen, and lung (reviewed in Ref. [34]). Of note, rat cardiomyocytes express TRPV1 in neonates but not in adults, suggesting that TRPV1 expression is developmentally regulated, at least in the heart [73]. The identification of central nervous system (CNS) and peripheral nonneuronal sites of VR1 expression negates one of the central dogmas of the field (i.e., vanilloid receptor expression is a functional signature of primary sensory neurons) and should revolutionize the ways we are thinking about vanilloid receptors.

As reviewed elsewhere [34,40,74], several lines of evidence demonstrate that TRPV1 protein levels are regulated. For instance, increased TRPV1 immunoreactive fiber innervation has been observed in inflamed human skin [29,75] and gastrointestinal tract, both in esophagus [76] and in large intestine [77]. This phenomenon is believed to contribute to the pathogenesis of various diseases, such as reflux esophagitis (also known as gastroesophageal reflux disease) [76], inflammatory bowel disease (both Crohn's disease and ulcerative colitis) [76], irritable bowel syndrome [78], vulvar allodynia [75], and prurigo nodularis [29]. A new finding with far-reaching implications is the demonstration of "aberrant" TRPV1 expression in cervical carcinoma cells [79], implying a future role for TRPV1 agonists as adjuvant chemotherapeutic drugs in cancer treatment regimens.

Several splice variants of TRPV1 have been reported [80–84]. One, called the rat TRPV1 5′ splice variant (VR1 5′sv), differs from TRPV1 only in the amino terminal tail region [81]. A similar, but not identical, variant has also been isolated from humans, called TRPV1b [82]. Neither VR1 5′sv nor TRPV1b are sensitive to capsaicin or low pH; interestingly, TRPV1b may be activated by high temperature [81]. Given the hypothesis that TRPV1 is a tetramer, it is possible that these splice variants function as dominant negative regulators [83]. VR1 5′sv RNA is expressed at very low levels in sensory neurons relative to TRPV1 but is comparable to TRPV1 gene expression in the CNS [81]. The existence of SIC (stretch inhibitable channel) [84] remains controversial as several studies failed to confirm the original observation.

2.3 The role of vanilloid receptors in nociception

The pivotal role that TRPV1-expressing nerves play in nociception has been the subject of several excellent recent reviews [3,20,35,85–87]. Here it suffices to mention that there are three main lines of evidence to implicate TRPV1 directly in nociception. First and foremost, TRPV1 receptor knockout mice show reduced inflammatory hyperalgesia and thermal nociception [88,89]. Second, anti-TRPV1 antiserum ameliorates thermal allodynia in diabetic mice [90]. Third, TRPV1 antagonists such as capsazepine exert beneficial effects in animal models of neuropathic pain [23].

3. VANILLOIDS AND NEUROIMMUNE REGULATION

3.1 Evidence that SP released from TRPV1-expressing nerves plays an important role in the neural control of immunity

Primary afferent unmyelinated fibers possessing TRPV1 modulate the inflammatory response to injury by initiating the release of a cascade of proinflammatory mediators, known as neurogenic inflammation. Indeed, rats treated as newborns with capsaicin to destroy these nerves permanently have a markedly diminished capacity to mount a neurogenic inflammatory response [91,92]. A pivotal role for SP in the initiation of the neurogenic inflammatory response has been postulated [93,94]. In accord, edema formation is suppressed in NK1 receptor knock-out mice in response to scald injury [95]. The connection between capsaicin-sensitive nerves and neurogenic inflammation is well-known and has been reviewed elsewhere [7–9,15–20]. Less recognized is the fact that capsaicin-treated rats also have altered humoral and cellular immune responses. For instance, these animals show reduced ability to mount a primary antibody response to sheep red blood cell antigen [96]. T-cell proliferation in response to mitogen or interleukin-2 administration is also attenuated in capsaicin-treated rats [97]. In contrast, allergic response to ovalbumin is enhanced in rats treated with capsaicin as neonates [98]. The mechanisms underlying this altered immunoregulation remain to be understood. A subset of thymic T cells express TRPV1 [31], therefore the possibility that capsaicin may kill lymphocytes directly cannot be discounted. This possibility not withstanding, most effects by capsaicin treatment on immunoregulation appear to be secondary to the depletion of SP by capsaicin from afferent nerves in that exogenous SP is able to fully restore antibody response to sheep erythrocyte antigens [97] and prevent apoptosis in T cells [99].

SP augments the proliferation of CD4-positive T lymphocytes in the presence of the mitogen Con A [100]. This effect is abolished by the coadministration of the nonpeptide NK1 receptor antagonist SR140333 [100]. This is in accord with the demonstration of mRNA for NK1 receptor in human lymphocytes [14]. Parenthetically, these cells also produce SP [14], implying an autocrine regulatory role for SP in lymphocytes. Of course, lymphocytes are not the only targets cells for SP. For example, SP directly up-regulates the expression of cellular adhesion molecules such as VCAM-1 on endothelial cells in the microvasculature of the human skin [101]. In turn, VCAM-1 plays a key role in the extravasation of inflammatory cells [101].

Here we mention briefly that there is good evidence to suggest an important role for up-regulated SP expression in the pathogenesis of inflammatory conditions. In the nematode-infected rat, there is marked increase in SP-like immunoreactivity not only in the gut wall but also in dorsal root ganglia and the dorsal horn of the spinal cord [102]. This is in keeping with the demonstration of elevated SP receptor (NK1 receptor) levels in patients with inflammatory bowel disease [103]. Interestingly, patients with inflammatory bowel disease (IBD) also show a higher density of TRPV1-like immunoreactivity in their bowels [77,78].

3.2 New insights into the participation of TRPV1-expressing nerves in the control of cystitis and enteritis

As discussed above, agents that directly activate TRPV1 now range from vanilloids to heat to protons to lipid metabolites to PKC-mediated phosphorylation. Of relevance to enteritis is the finding that *Clostridium difficile* toxin A-induced SP release in the rat ileum can be prevented by pretreatment with the TRPV1 antagonist capsazepine [104]. Although capsazepine is not entirely specific to TRPV1, this observation was interpreted to suggest a direct involvement of TRPV1 in pseudomembranous enteritis.

There is an increasing recognition that the participation of TRPV1-expressing nerves in bladder and bowel inflammation is far more complex than thought previously. One focus of attention is now proteinase-activated receptor 2 (PAR-2). PAR-2 is highly expressed both in colon [105] and in bladder [106] and is unique in that it needs to be cleaved first by proteases to be activated by tethered ligands. As discussed above, mast cells express TRPV1 receptors, and mast cell tryptase is a protease that is known to activate PAR-2. In turn, activated PAR-2 can sensitize TRPV1, via PKC-dependent phosphorylation [107]. TRPV1 and PAR-2 are colocalized in bladder nerve fibers, with an up-regulation in PAR-2 expression during experimental cystitis [106]. During colitis, TRPV1 expression has also been shown to increase [77]. In the skin, injection of PAR-2 agonists evokes thermal hyperalgesia that is prevented by TRPV1 antagonists [107]. Taken together, these findings imply a powerful proinflammatory cascade in which both PAR-2 and TRPV1 are up-regulated during inflammation: TRPV1 contributes to PAR-2 activation by releasing mast cell tryptase (as well as other proteases), and PAR-2 further sensitizes TRPV1 via PKC phosphorylation.

In the stomach, PAR-2 induces mucus secretion [108]. This action is partially inhibited by the TRPV1 receptor antagonist capsazepine [108]. This is an interesting link to explore given the protective action of TRPV1 activation by capsaicin and RTX against gastric ulcer formation (reviewed in Ref. [20]).

To make this regulatory loop even more complex, PAR-2 is now recognized to exert an anti-inflammatory signal for colonic lamina propria lymphocytes [105]. In a mouse model of colitis, a synthetic peptide corresponding to the murine tethered ligand dose dependently attenuates the symptoms of 2,4,6-trinitrobenzene sulfuric acid (TNBS) induced colitis [105]. This protective mechanism is, however, defunct in mice that underwent neonatal capsaicin treatment. As yet, it is not clear how one can reconcile the apparently contradictory findings obtained with PAR-2 activators in skin (thermal hyperalgesia) [107] and bowel (attenuation of colitis) [105].

3.3 Capsaicin-sensitive nerves protect against allergic and contact sensitivity reactions

In ovalbumin-immunized rats, capsaicin administration inhibits the synthesis of IgG but enhances the level of serum IgA [109]. Even more intriguing is the observation that pulmonary neutrophil recruitment by ovalbumin challenge is exacerbated following capsaicin treatment [98,110]. These findings imply a protective role for unmyelinated fibers against allergic reactions. Capsaicin-treated mice also exhibit enhanced contact sensitivity reactions to various haptens [111,112] that cannot be transferred to naïve mice by effector T cells [111]. Interleukin-1 release is significantly (by over 20-fold) higher in the capsaicin-treated animals than their similarly sensitized littermates [112]. Furthermore, UV light exposure inhibits contact hypersensitivity reactions in control but not in capsaicin-treated mice [113]. Collectively, these results indicate that C-fibers, presumably via neuropeptide release, control cutaneous, delayed-type hypersensitivity reactions in the mouse. Surprisingly, CGRP, but not tachykinin, receptor antagonists mimic the effects of capsaicin denervation in this experimental paradigm [113]. Conversely, capsaicin-treated and NK1 antagonist-treated mice exhibit reduced delayed-type hypersensitivity responses [111,114,115].

3.4 Contribution of capsaicin-sensitive nerves to airway inflammation

The mechanistic role for C-fibers in airway inflammatory diseases has been extensively studied (for reviews, see Refs [9,15,16,116]). In general, sensory fiber activation is thought to mediate neurogenic inflammation, typified by increased plasma protein extravasation. Other consequences of sensory fiber neuropeptides may include recruitment/retention of dendritic cells to the airway after antigen challenge [117]. The involvement of TRPV1, specifically, in mediating airway sensory responses in the pathophysiological state is supported by the finding that capsazepine can block low pH [118] and bradykinin [119] activation of airway sensory nerve endings. Additionally, particulate matter air pollutants have been shown to provoke airway inflammation by interacting at sensory nerve fibers [120]. Such pollutants may also activate TRPV1 expressed in airway epithelial cells [121]. Of note, NK1 receptor knockouts show no plasma extravasation response in an immune-complex airway inflammation model, implicating SP as a significant mediator of airway inflammatory response [122].

3.5 Capsaicin-sensitive nerves contribute to the symptoms of arthritis

As discussed above, intact capsaicin-sensitive nerves protect against allergic and contact hypersensitivity reactions. In animal models of these conditions, capsaicin pretreatment exacerbates symptoms. By contrast, defunctionalization of C-fibers by high-dose capsaicin is beneficial in disease states with a significant component of neurogenic inflammation. A well-characterized example of this phenomenon is the adjuvant-induced experimental arthritis. In a guinea pig model, capsaicin pretreatment resulted in a significant attenuation of clinical (e.g., tissue swelling) and radiological scores [123]. Importantly, the synovial T-cell infiltration was also selectively reduced in these animals [124].

4. THE COMPLEX INTERACTION BETWEEN MAST CELLS AND TRPV1-EXPRESSING NERVES

As has repeatedly been alluded to above, there is an intimate anatomical and physiological relationship between mast cells and capsaicin-sensitive nerves possessing TRPV1 [29,30,124]. Several lines of evidence suggest that these nerves exert a trophic effect on mast cell density. For example, a diminished synovial mast cell density was found in rats whose C-fibers were destroyed by high-dose capsaicin administration [125]. In keeping with this, 28% fewer intestinal mucosal mast cells are present in the small bowel of capsaicin-treated rats compared with littermate controls [126]. This change is similar to that caused by truncal vagotomy (a 25% decrease in jejunal mast cell density) [126], implying that the component of the capsaicin-sensitive innervation of the gut, which is associated with mast cells, is vagal in origin. The

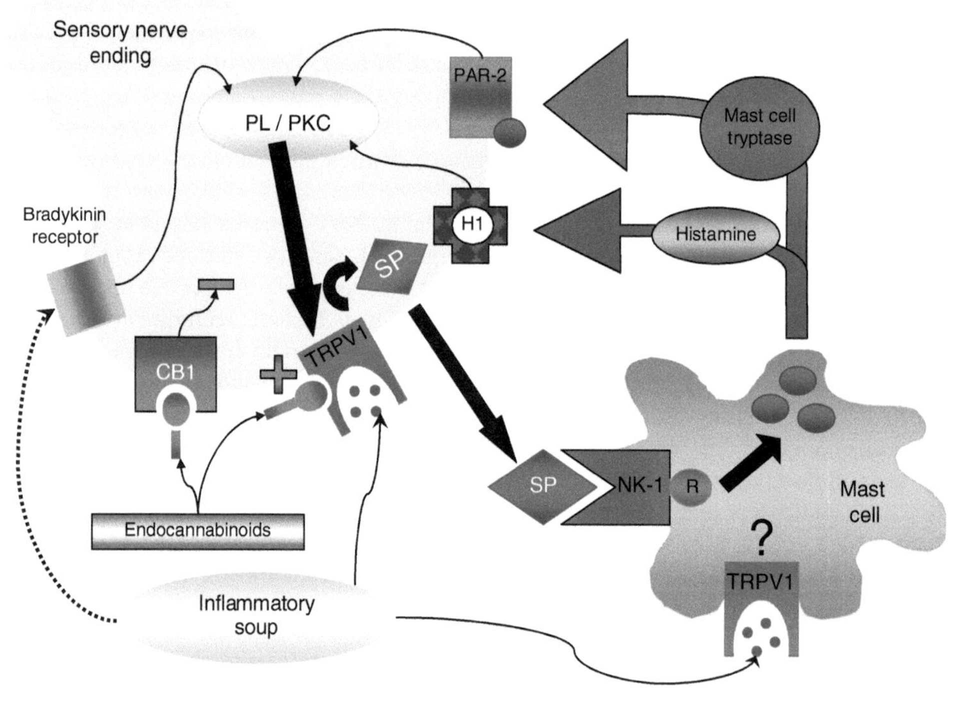

Figure 1. Hypothetical interaction of TRPV1-expressing sensory afferents with mast cells. During inflammation, numerous mediators are generated, including bradykinin, prostaglandin, and protons. This "inflammatory soup" can directly activate TRPV1 (e.g., low pH) or indirectly sensitize TRPV1 (e.g., bradykinin through its receptor, bradykinin R). Activation of TRPV1 in sensory afferents results in the release of neuropeptides, including substance P (SP). SP via NK1 can activate mast cells resulting in the release of mast cell tryptase – an activator of PAR2. Sensitization of TRPV1 via PAR2 and/or histamine receptor (H1) would complete a positive regulatory loop between mast cells and sensory afferents that drives neurogenic inflammation. Endocannabinoids, such as anandamide, may activate sensory neurons through direct interaction with TRPV1(+) or decrease sensory neuron activity via cannabinoid receptor, CB1(−). PL, phospholipase; PKC, protein kinase C. (See color Plate 2.)

mediator of this trophic action on mast cells is not known but SP is a good candidate to play this role. A trophic action by SP on other cell types, like keratinocytes, has been described [127], and mast cells are known to carry both NK1 receptors [128] and a second receptor that recognizes the C-terminal domain of SP [129].

Capsaicin evokes the degranulation of mast cells in the human skin via release of sensory neuropeptides [130]. It has been suggested that mast cells express either TRPV1 or a closely related receptor [28], and TRPV1 immunoreactivity has been observed in resident mast cells of human skin [29] and bladder [30]. Capsaicin itself, however, does not release histamine from mast cells [130]; therefore the physiological role of TRPV1 in these cells is unclear. In any case, mast cell activation results in the release tryptase, a protease that can activate PAR-2, and histamine, which can activate H1 receptors [131]. Each of these receptors can sensitize TRPV1 [107]. This exacerbates neuropeptide release further and maintains a vicious circle in which more and more proinflammatory mediators are generated (Fig. 1).

5. CONCLUDING REMARKS

The essential contribution of capsaicin-sensitive nerves to nociception and neurogenic inflammation was firmly established decades ago [7–9,15–20]. A receptor for capsaicin, termed the vanilloid receptor TRPV1, was cloned in 1997, heralding a new era in sensory pharmacology [32]. TRPV1 integrates effects of noxious heat, protons, arachidonic acid metabolites, and vanilloid ligands [3,34,35,40]. Considerable progress has been made in delineating the contribution of endovanilloid signaling to neuropathic pain and neuroimmune regulation. Endogenous TRPV1 agonists (endovanilloids) include anandamide, 12-HPETE, leukotriene B4, and NADA [6]. PKs sensitize TRPV1 to agonists by enhancing the probability of channel opening [6] and provide a mechanism by which proinflammatory mediators, such as bradykinin, mast cell tryptase, and histamine, activate sensory nerve afferents. As a consequence of the anatomical and biochemical interaction between sensory nerve afferents and immune cells, TRPV1 is a key molecular player in neuro–immune interactions. A better understanding of endovanilloid signaling via TRPV1 is expected to provide important clues to the development of innovative analgesic and anti-inflammatory drugs.

ACKNOWLEDGMENTS

The authors thank Marianne Flynn for preparing Fig. 1.

REFERENCES

1. Shu XQ, Mendell LM. Neurotrophins and hyperalgesia. Proc Natl Acad Sci USA 1999;96:7693–7696.
2. Boddeke EW. Involvement of chemokines in chronic pain. Eur J Pharmacol 2001;429:115–119.
3. Julius D, Basbaum AI. Molecular mechanisms of pain. Nature 2001;413:203–210.
4. Kidd BL, Urban LA. Mechanisms of inflammatory pain. Br J Anaesth 2001;87:3–11.
5. Curran DR, Walsh MT, Costello RW. Interactions between inflammatory cells and nerves. Curr Opin Pharmacol 2002;2:243–248.

6. Di Marzo V, Blumberg PM, Szallasi A. Endovanilloid signaling in pain. Curr Opin Neurobiol 2002;12:372–380.
7. Szolcsányi J. Capsaicin and neurogenic inflammation: history and early findings. In: Chahl LA, Szolcsányi J, Lembeck F, eds. Antidromic Vasodilatation and Neurogenic Inflammation. Budapest: Akadèmiai Kiadó, 1984; pp. 2–25.
8. Foreman JC. Peptides and neurogenic inflammation. Br Med Bull 1987;43:386–400.
9. Holzer P. Local effector functions of capsaicin-sensitive sensory nerve endings: involvement of tachykinins, calcitonin gene-related peptide and other neuropeptides. Neuroscience 1988;24:739–768.
10. Rameshwar P. Substance P: a regulatory neuropeptide for hematopoiesis and immune functions. Clin Immunol Immunopathol 1997;85:129–133.
11. Holzer P. Implications for tachykinins and calcitonin gene-related peptide in inflammatory bowel disease. Digestion 1998;59:269–283.
12. Scholzen T, Armstrong CA, Bunnett NW, Hyer TA, Olerud JE, Ansel JC. Neuropeptides in the skin: interactions between the neuroendocrine and the skin immune systems. Exp Dermatol 1998;7:81–96.
13. Levite M, Chowers Y. Nerve-driven immunity: neuropeptides regulate cytokine secretion of T cells and intestinal epithelial cells in a direct, powerful, and contextual manner. Ann Oncol 2001;12(Suppl. 2):S19–S25.
14. Lai JP, Douglas SD, Ho WZ. Human lymphocytes express substance P and its receptor. J Neuroimmunol 1998;86:80–86.
15. Wood J (ed.). Capsaicin in the Study of Pain. San Diego, CA: Academic Press, 1993.
16. Geppetti P, Holzer P (eds). Neurogenic Inflammation. Boca Raton, FL: CRC Press, 1996.
17. Szallasi A, Blumberg PM. Mechanisms and therapeutic potential of vanilloids (capsaicin-like molecules). In: August TJ, Anders MV, Murad F, eds. Advances in Pharmacology. San Diego, CA: Academic Press, 1993; pp. 123–155.
18. Buck SH, Burks TF. The neuropharmacology of capsaicin: a review of some recent observations. Pharmacol Rev 1986;38:179–226.
19. Holzer P. Capsaicin: cellular targets, mechanisms of action, and selectivity for thin sensory neurons. Pharmacol Rev 1991;43:143–201.
20. Szallasi A, Blumberg PM. Vanilloid (capsaicin) receptors and mechanisms. Pharmacol Rev 1999;51:159–211.
21. Wood JN, Winter J, James IF, Rang HP, Yeats J, Bevan S. Capsaicin-induced ion fluxes in dorsal root ganglion cells in culture. J Neurosci 1988;8:3208–3220.
22. Liu L, Simon SA. A rapid capsaicin-activated current in rat trigeminal ganglion neurons. Proc Natl Acad Sci USA 1994;91:738–741.
23. Kwak JY, Jung JY, Hwang SW, Lee WT, Oh U. A capsaicin receptor antagonist, capsazepine, reduces inflammation-induced hyperalgesia in the rat: evidence for an endogenous capsaicin-like substance. Neuroscience 1998;86:619–625.
24. Liu M, Max MB, Robinowitz E, Gracely RH, Bennett GJ. The human capsaicin model of allodynia and hyperalgesia: sources of variability and methods of reduction. J Pain Symptom Manage 1998;16:10–20.
25. Baron R. Capsaicin and nociception: from basic mechanisms to novel drugs. Lancet 2000;356:785–787.
26. Jancsó M and Jancsó M. Desensitization of sensory nerve endings (in Hungarian). Kísèrletes Orvostudomány [Experimental Medicine] 1949;2(Suppl.):15(English translation in ref 20).

27. Jancsó G. Histamine, capsaicin, and neurogenic inflammation. A historical note on the contribution of Miklós (Nicholas) Jancsó (1903–1966) to sensory pharmacology. In: Bérczi I, Szelényi J, eds. Advances in Psychoneuroimmunology. New York, NY: Plenum Press, 1994; pp. 17–23.
28. Bíró T, Maurer M, Modarres S, Lewin NE, Brodie C, Ács G, Ács P, Paus P, Blumberg PM. Characterization of functional vanilloid receptors expressed by mast cells. Blood 1998;91:1332–1340.
29. Stander S, Moormann C, Schumacher M, Buddenkotte J, Artuc M, Shpakovitch V, Brzoska T, Lippert U, Henz BM, Luger TA, Metze D, Steinhoff M. Expression of vanilloid receptor subtype-1 in cutaneous sensory nerve fibers, mast cells, and epithelial cells of appendage structure. Exp Dermatol 2004;13:129–139.
30. Lazzeri M, Vannucchi MG, Zardo C, Spinelli M, Beneforti P, Turini D, Faussone-Pellegrini M-S. Immunohistochemical evidence of vanilloid receptor 1 in normal human urinary bladder. Eur Urol 2004;46:792–798.
31. Amantini C, Mosca M, Lucciarini R, Perfumi M, Morrone S, Piccoli M, Santoni G. Distinct thymocyte subsets express the vanilloid receptor VR1 that mediates capsaicin-induced apoptotic cell death. Cell Death Differ 2004;11:1342–1356.
32. Caterina MJ, Schumacher MA, Tominaga M, Rosen TA, Levine JD, Julius D. The capsaicin receptor: a heat-activated channel in the pain pathway. Nature 1997;389:816–824.
33. Montell C, Birnbaumer L, Flockerzi V, Bindels RJ, Bruford EA, Caterina MJ, Clapham DE, Harteneck C, Heller C, Julius DE, Kogima I, Mori Y, Penner R, Prawitt D, Scharenberg AM, Schultz G, Shimizu N, Zhu MX. A unified nomenclature for the superfamily of TRP channels. Mol Cell 2002;9:229–231.
34. Nagy I, Sántha P, Jancsó G, Urbán L. The role of the vanilloid (capsaicin) receptor (TRPV1) in physiology and pathology. Eur J Pharmacol 2004;500:351–369.
35. Krause JE, Chenard BL, Cortright DN. Transient receptor potential ion channels as targets for discovery of pain therapeutics. Curr Opin Investig Drugs 2005;6:48–57.
36. Fischer BS, Qin D, Kim K, McDonald TV. Capsaicin inhibits Jurkat cell activation by blocking calcium entry current I(CRAC). J Pharmacol Exp Ther 2001;299:238–246.
37. Jancsó N, Jancsó-Gábor A, Takáts J. Pain and inflammation induced by nicotine, acetylcholine, and structurally-related compounds and their prevention by desensitizating agents. Acta Physiol Acad Sci Hung 1961;19:113–131.
38. Jordt SE, Bautista DM, Chuang HH, McKemy DD, Zygmunt PM, Hogestatt ED, Meng ID, Julius D. Mustard oils and cannabinoids excite sensory nerve fibers through the TRP channel ANKTM1. Nature 2004;427:260–265.
39. Bandell M, Story GM, Hwang SW, Viswanath V, Eid SR, Petrus MJ, Earley TJ, Patapoutian A. Noxious cold ion channel TRPA1 is activated by pungent compounds and bradykinin. Neuron 2004;41:849–857.
40. Cortright DN, Szallasi A. Biochemical pharmacology of the vanilloid receptor TRPV1. An update. Eur J Biochem 2004;271:1814–1819.
41. Jin X, Morsey N, Winston J, Pasricha PJ, Garrett K, Akbarali H. Modulation of TRPV1 by nonreceptor tyrosine kinase, c-Src kinase. Am J Physiol Cell Physiol 2004;287: C558–C563.
42. Jancsó N. Desensitization with capsaicin and related acylamides as a tool for studying the function of pain receptors. In: Lin K, Armstrong D, Pardo ED, eds. Pharmacology of Pain. Oxford: Pergamon Press, 1968; pp. 33–55.

43. Zygmunt PM, Petersson J, Andersson DA, Chuang H, Sørgård M, Di Marzo V, Julius D, Högestätt ED. Vanilloid receptors on sensory nerves mediate the vasodilator action of anandamide. Nature 1999;400:452–457.
44. Pertwee RG. Pharmacology of cannabinoid CB1 and CB2 receptors. Pharmacol Ther 1997;74:129–180.
45. Tognetto M, Amadesi S, Harrison S, Creminon C, Trevisani M, Carreras M, Matera M, Geppetti P, Bianchi A. Anandamide excites central terminals of dorsal root ganglion neurons via vanilloid receptor-1. J Neurosci 2001;21:1104–1109.
46. Smart D, Gunthorpe MJ, Jerman JC, Nasir S, Gray J, Muir AI, Chambers JK, Randall AD, Davis JB. The endogenous lipid anandamide is a full agonist at the human vanilloid receptor (hVR1). Br J Pharmacol 2000;129:227–230.
47. Di Marzo V, Bisogno T, De Petrocellis L. Anandamide: some like it hot. Trends Pharmacol Sci 2001;22:346–349.
48. Gavva NR, Klionsky L, Qu Y, Shi L, Tamir R, Edenson S, Zhang TJ, Viswanadhan VN, Toth A, Pearce LV, Vanderah TW, Porreca F, Blumberg PM, Lile J, Sun Y, Wild K, Louis J-C, Treanor JJS. Molecular determinants of vanilloid sensitivity in TRPV1. J Biol Chem 2004;279:20283–20295.
49. Hillard CJ, Jarrahian A. The movement of N-arachidonoylethanolamine (anandamide) across the cell membrane. Chem Phys Lipids 2000;108:123–134.
50. Premkumar LS, Ahern GP. Induction of vanilloid receptor channel activity by protein kinase C. Nature 2000;408:985–990.
51. Vellani V, Mapplebeck S, Moriondo A, Davis JB, McNaughton PA. Protein kinase C activation potentiates gating of the vanilloid receptor VR1 by capsaicin, protons, heat and anandamide. J Physiol 2001;534:813–825.
52. Olah Z, Karai I, Iadarola MJ. Anandamide activates vanilloid receptor 1 (VR1) at acidic pH in dorsal root ganglia and cells ectopically expressing VR1. J Biol Chem 2001;276:31163–31170.
53. De Petrocellis L, Bisogno T, Maccarrone M, Davis JB, Finazzi-Agro A, Di Marzo V. The activity of anandamide at vanilloid VR1 receptors requires facilitated transport across the cell membrane and is limited by intracellular metabolism. J Biol Chem 2001;276:12856–12863.
54. Ahluwalia J, Urban L, Capogna M, Bevan S, Nagy I. Cannabinoid 1 receptors are expressed on nociceptive primary sensory neurons. Neuroscience 2000;100:685–688.
55. Song ZH, Bonner TI. A lysine residue of the cannabinoid receptor is critical for receptor recognition by several agonists but not WIN55212-2. Mol Pharmacol 1996;49:891–896.
56. Metwali A, Blum AM, Elliott DE, Seiawan T, Weinstock JV. Cutting edge: hemokinin has substance P-like function and expression in inflammation. J Immunol 2004;172:6528–6532.
57. McVey DC, Schmid PC, Schmid HH, Vigna SR. Endocannabinoids induce ileitis in rats via the capsaicin receptor (VR1). J Pharmacol Exp Ther 2003;304:713–722.
58. Huang SM, Bisogno T, Trevisani M, Al-Hayani A, De Petrocellis L, Fezza M, Krey JF, Petros TJ, Chu CJ, Miller JD, Davies SN, Geppetti P, Walker M, Di Marzo V. An endogenous capsaicin-like substance with high potency at recombinant at native vanilloid VR1 receptors. Proc Natl Acad Sci USA 2002;99:8400–8405.
59. Hwang SW, Cho H, Kwak J, Lee SY, Kang CJ, Jung J, Cho S, Min KH, Suh YG, Kim D, Oh U. Direct activation of capsaicin receptors by products of lipoxygenases: endogenous capsaicin-like substances. Proc Natl Acad Sci USA 2000;97:6155–6160.

60. Hwang SW, Oh U. Hot channels in airways: pharmacology of the vanilloid receptor. Curr Opin Pharmacol 2002;2:235–242.
61. Trevisani M, Smart D, Gunthorpe MJ, Tognetto M, Barbieri M, Campi B, Amadesi S, Grey J, Jerman JC, Brough SJ, Owen D, Smith GD, Randall AD, Harrison S, Bianchi A, Davis JB, Geppetti P. Ethanol elicits and potentiates nociceptor responses via the vanilloid receptor-1. Nat Neurosci 2002;5:546–551.
62. Chuang HH, Prescott ED, Kong H, Shields S, Jordt SE, Basbaum A, Chao MV, Julius D. Bradykinin and nerve growth factor release the capsaicin receptor from PtdIns(4,5) P2-mediated inhibition. Nature 2001;411:957–962.
63. Shu XQ, Mendell LM. Neurotrophins and pain. Proc Natl Acad Sci USA 1999;96:7693–7696.
64. Szallasi A. Vanilloid-sensitive neurons: a fundamental subdivision of the peripheral nervous system. J Peripher Nerv Syst 1996;1:6–18.
65. Guo A, Vulchanova L, Wang J, Li X, Elde R. Immunocytochemical localization of the vanilloid receptor 1 (VR1): relationship to neuropeptides, the P2X3 purinoceptor and IB4 binding sites. Eur J Neurosci 1999;11:946–958.
66. Avelino A, Cruz C, Nagy I, Cruz F. Vanilloid receptor 1 expression in rat urinary tract. Neuroscience 2002;109:787–798.
67. Yiangou Y, Facer P, Ford A, Brady C, Wiseman O, Fowler CJ, Anand P. Capsaicin receptor VR1 and ATP-gated ion channel P2X3 in human urinary bladder. BJU Int 2001;87:774–779.
68. Nozawa Y, Nishihara K, Yamamoto A, Nakano M, Ajioka H, Matsuura N. Distribution and characterization of vanilloid receptors in the rat stomach. Neurosci Lett 2001;309:33–36.
69. Szallasi A, Di Marzo V. New perspectives on enigmatic vanilloid receptors. Trends Neurosci 2000;23:491–497.
70. Kim JC, Beckel JM, Birder LA, Kiss S, Washabaugh C, Kanai A, Reynolds I, Dineley K, de Groat WC, Caterina MJ. Identification of functional vanilloid receptors in human bladder urothelial cells using a nitric oxide microsensor technique and reverse transcriptase polymerase chain reaction. J Urol 2001;165 (Suppl.):34.
71. Golech SA, McCarron RM, Chen Y, Bembry J, Lenz F, Mechoulam R, Shohami E, Spatz M. Human brain endothelium: coexpression and function of vanilloid and endocannabinoid receptors. Brain Res Mol Brain Res 2004;132:87–92.
72. De Groat WC. The urothelium: passive bystander or active participant? Urology 2004;64(Suppl. 6):7–11.
73. Dvorakova M, Kummer W. Transient expression of vanilloid receptor subtype 1 in rat cardiomyocytes during development. Histochem Cell Biol 2001;116:223–225.
74. Szallasi A. Vanilloid (capsaicin) receptors in health and disease. Am J Clin Pathol 2002;118:110–121.
75. Tympanidis P, Casula MA, Yiangou Y, Terenghi G, Dowd P, Anand P. Increased vanilloid receptor VR1 innervation in vulvodynia. Eur J Pain 2004;8:129–133.
76. Matthews PJ, Aziz Q, Facer P, Davis JB, Thompson DG, Anand P. Increased capsaicin receptor TRPV1 nerve fibres in the inflamed human oesophagus. Eur J Gastroenterol Hepatol 2004;16:897–902.
77. Yiangou Y, Facer P, Dyer NHC, Knowles C, Williams NS, Anand P. Vanilloid receptor 1 immunoreactivity in inflamed human bowel. Lancet 2001;357:1338–1339.

78. Chan CL, Facer P, Davis JB, Smith GD, Egerton J, Bountra C, Williams NS, Anand P. Sensory fibres expressing capsaicin receptor TRPV1 in patients with rectal hypersensitivity and faecal urgency. Lancet 2003;361:385–391.
79. Contassot E, Tenan E, Schnuriger V, Pelte MF, Dietrich PT. Arachidonyl ethanolamine induces apoptosis of uterine cervix cancer cells via aberrantly expressed vanilloid receptor-1. Gynecol Oncol 2004;93:182–188.
80. Schumacher MA, Moff I, Sudanagunta SP, Levine JD. Molecular cloning of an N-terminal splice variant of the capsaicin receptor. J Biol Chem 2000;275:2756–2762.
81. Sanchez JF, Krause JE, Cortright DN. The distribution and regulation of VR1 and VR1 5′ splice variant RNA expression in the rat. Neuroscience 2001;107:373–381.
82. Lu G, Henderson D, Liu L, Reinhart PH, Simon SA. TRPV1b: a functional human vanilloid receptor splice variant. Mol Pharmacol 2005;67:1119–1127.
83. Wang C, Hu HZ, Colton CK, Wood JD, Zhu MX. An alternative splicing product of the murine trpv1 gene dominant negatively modulates the activity of TRPV1 channels. J Biol Chem 2004;279:37423–37430.
84. Suzuki M, Sato J, Kutsuwada K, Ooki G, Imai M. Cloning of a stretch-inhibitable nonselective cation channel. J Biol Chem 1999;274:6330–6335.
85. Caterina MJ, Julius D. The vanilloid receptor: a molecular gateway to the pain pathway. Annu Rev Neurosci 1999;24:487–517.
86. Alvarez FJ, Fyffe RE. Nociceptors for the 21th century. Curr Opin Pain 2000;4:451–458.
87. Reeh PW, Kress M. Molecular physiology of proton transduction in nociceptors. Curr Opin Pharmacol 2001;4:45–51.
88. Caterina MJ, Leffler A, Malmberg AB, Martin WJ, Trafton J, Petersen-Zeitz KR, Koltzenburg M, Basbaum AI, Julius D. Impaired nociception and pain sensation in mice lacking the capsaicin receptor. Science 2000;288:306–313.
89. Davis JB, Gray J, Gunthorpe MJ, Hatcher JP, Davey PT, Overend P, Harries MH, Latcham J, Clapham C, Atkinson K, Hughes SA, Rance K, Grau E, Harper AJ, Pugh PL, Rogers DC, Bingham S, Randall A, Sheardown SA. Vanilloid receptor-1 is essential for inflammatory thermal hyperalgesia. Nature 2000;405:183–187.
90. Kamei J, Zushida K, Morita K, Sasaki M, Tanaka S. Role of vanilloid receptors in thermal allodynia and hyperalgesia in diabetic mice. Eur J Pharmacol 2001;422:83–86.
91. Lundberg JM, Saria A. Capsaicin-induced desensitization of airway mucosa to cigarette smoke, mechanical and chemical irritants. Nature 1983;302:251–253.
92. Jancsó G, Obál F, Jr., Tóth-Kása I, Katona M, Husz S. The modulation of cutaneous inflammatory reactions by peptide-containing sensory nerves. Int J Tissue React 1985;7:449–457.
93. Pernow B. Role of tachykinins in neurogenic inflammation. J Immunol 1985;135: 812–815.
94. Krause JE, Takeda Y, Hershey AD. Structure, functions, and mechanisms of substance P receptor actions. J Invest Dermatol 1992;98(Suppl. 6):2S–7S.
95. Cao T, Gerard NP, Brain SD. Use of NK1 knockout mice to analyze substance P-induced edema formation. Am J Physiol 1999;277:R476–R481.
96. Helme RD, Eglezos A, Dandie GW, Andrews PV, Boyd RL. The effect of substance P on the regional lymph node antibody response to antigenic stimulation in capsaicin-pretreated rats. J Immunol 1987;139:3470–3473.
97. Santoni G, Perfumi M, Bressan AM, Piccoli M. Capsaicin-induced inhibition of mitogen and interleukin 2-stimulated T cell proliferation: its reversal by in vivo substance P administration. J Neuroimmunol 1996;68:131–138.

98. Franco-Penteado CF, Desouza IA, Camargo EA, Teixeira SA, Muscara MN, De Nucci G, Antunes E. Mechanisms involved in the enhancement of allergic airways neutrophil influx by permanent C-fibers degeneration in rats. J Pharmacol Exp Ther 2005;313:440–448.
99. Santoni G, Amantini C, Lucciarini R, Perfumi M, Pompei P, Piccoli M. Neonatal capsaicin treatment affects rat thymocyte proliferation and cell death by modulating substance P and neurokinin-1 receptor expression. Neuroimmunomodulation 2004;11:160–172.
100. Santoni G, Perfumi MC, Spreghini E, Romagnoli S, Piccoli M. Neurokinin type-1 receptor antagonist inhibits enhancement of T cell functions by substance P in normal and neuromanipulated capsaicin-treated rats. J Neuroimmunol 1999;93:15–25.
101. Quinlan KL, Song IS, Naik SM, Letran EL, Olerud JE, Bunnett NW, Armstrong CA, Caughman SW, Ansel JC. VCAM-1 expression on human dermal microvascular endothelial cells in directly and specifically up-regulated by substance P. J Immunol 1999;162:1656–1661.
102. De Giorgio R, Blennerhassett P, Wang L, Stanghellini V, Corinaldesi R, Collins SM, Tougas G. Intestinal inflammation and activation of sensory nerve pathways: a functional and morphological study in the nematode infected rat. Gut 2001;49: 822–827.
103. Mantyh CR, Gates TS, Zimmerman RP, Welton ML, Passaro EP, Vigna SR, Maggio JE, Kruger L, Mantyh PW. Receptor binding sites for substance P, but not substance K or neuromedin K, are expressed in high concentrations by arterioles, venules, and lymph nodules in surgical specimens obtained from patients with ulcerative colitis or Crohn disease. Proc Natl Acad Sci USA 1988;85:3235–3239.
104. McVey DC, Vigna SR. The capsaicin VR1 receptor mediates substance P release in toxin A-induced enteritis in rats. Peptides 2001;22:1439–1446.
105. Fiorucci S, Mencarelli A, Palazzetti B, Distrutti E, Vergnolle N, Hollenberg MD, Wallace JL, Morelli A, Cirino G. Proteinase-activated receptor 2 is an anti- inflammatory signal for colonic lamina propria lymphocytes in a mouse model of colitis. Proc Natl Acad Sci USA 2001;98:13936–13941.
106. Dattilio A, Vizzard MA. Up-regulation of protease activated receptors in bladder after cyclophosphamide induced cystitis and colocalization with capsaicin receptor (VR1) in bladder nerve fibers. J Urol 2005;173:635–639.
107. Amadesi S, Nie J, Vergnolle N, Cottrell GS, Grady EF, Trevisani M, Manni C, Geppetti P, McRoberts JA, Ennes H, Davis JB, Mayer EA, Bunnett NW. Protease activated receptor-2 sensitizes the capsaicin receptor transient receptor potential vanilloid receptor 1 to induce hyperalgesia. J Neurosci 2004;24:4300–4312.
108. Katawaba A, Kinoshita M, Kuroda R, Kakehi K. Capsazepine partially inhibits neurally mediated gastric mucus secretion following activation of proteinase-activated receptor 2. Clin Exp Pharmacol Physiol 2002;29:360–361.
109. Nilsson G, Alving K, Ahlstedt S. Effects of immune responses in rats after neuromanipulation with capsaicin. Int J Immunopharmacol 1991;13:21–26.
110. Medeiros MV, Macedo-Soares MF, De Luca IM, Hyslop S, De Nucci G, Antunes E. Contribution of C-fibers to leukocyte recruitment in bronchoalveolar lavage fluid and pleural cavity in the rat. Eur J Pharmacol 2001;421:133–140.
111. Girolomoni G, Tigelaar RE. Capsaicin-sensitive primary sensory neurons are potent modulators of murine delayed-type hypersensitivity reactions. J Immunol 1990;145: 1105–1112.

112. Veronesi B, Williams WC, Smialowicz RJ, Sailstad DM, Doerfler D, Selgrade MJ. Neuropeptide denervation alters both the elicitation and induction phases of contact hypersensitivity in mice. Toxicol Appl Pharmacol 1998;153:243–249.
113. Garssen J, Buckley TL, Van Loveren H. A role for neuropeptides in UBV-induced systemic immunosuppression. Photochem Photobiol 1998;68:205–210.
114. Buckley TL, Nijkamps FP. Airways hyperreactivity and cellular accumulation in a delayed type hypersensitivity reaction in the mouse: modulation by capsaicin-sensitive nerves. Am J Respir Crit Care Med 1994;149:400–407.
115. Kraneveld AD, Buckley TL, van Heuven-Nolsen D, van Schaik Y, Koster AS, Nijkamp FP. Delayed type hypersensitivity-induced increase in vascular permeability in the mouse small intestine: inhibition by depletion of sensory neuropeptides and NK1 receptor blockade. Br J Pharmacol 1995;114:1483–1489.
116. Belvisi M. Sensory nerves and airway inflammation: role of A delta and C fibers. Pulm Pharmacol Ther 2003;16:1–7.
117. Kradin R, MacLean J, Duckett S, Schneeberger EE, Waeber C, Pinto C. Pulmonary response to inhaled antigen: neuroimmune interactions promote the recruitment of dendritic cells to the lung and the cell response to inhaled antigen. Am J Pathol 1997;150:1735–1743.
118. Fox AJ, Urban L, Barnes PJ, Dray A. Effects of capsazepine against capsaicin- and proton-evoked excitation of single airway C-fibers and vagus nerves from guinea pig. Neuroscience 1995;57:741–752.
119. Carr MJ, Kollarik M, Meeker SN, Undem BJ. A role for TRPV1 for bradykinin induced excitation of vagal airway afferent nerve terminals. J Pharmacol Exp Ther 2003;304:1275–1279.
120. Agopyan N, Head J, Yu S, Simon SA. TRPV1 receptors mediate particular matter-induced apoptosis. Am J Physiol Lung Cell Mol Physiol 2004;286:L563–L572.
121. Veronesi B, Oortgiesen M. Neurogenic inflammation and particulate matter (PM) air pollutants. Neurotoxicol 2001;22:795–810.
122. Boric CR, Lu B, Hopken UE, Gerard C, Gerard NP. Neurogenic amplification of immune complex inflammation. Science 1996;273:1722–1725.
123. Hood VC, Cruwys SC, Urban L, Kidd BL. The neurogenic contribution to synovial leukocyte infiltration and other outcome measures in a guinea pig model of arthritis. Neurosci Lett 2001;299:201–204.
124. Nilsson G, Alving K, Ahlstedt S, Hökfelt T, Lundberg JM. Peptidergic innervation of rat lymphoid tissue and lung: relation to mast cells and sensitivity to capsaicin and immunization. Cell Tissue Res 1990;262:125–133.
125. Levine JD, Coderre TJ, Covinsky K, Basbaum AI. Neural influences on synovial mast cell density in rat. J Neurosci Res 1990;26:301–307.
126. Gottwald T, Lhotak S, Stead RH. Effect of truncal vagotomy and capsaicin on mast cells and IgA-positive plasma cells in rat jejunal mucosa. Neurogastroenterol Motil 1997;9:25–32.
127. Paus R, Heinzelmann T, Robicsek S, Czarnetzki BM, Maurer M. Substance P stimulates murine epidermal keratinocyte proliferation and dermal mast cell degranulation in situ. Arch Dermatol Res 1995;287:500–502.
128. Krumins SA, Broomfield CA. Evidence for NK1 and NK2 tachykinin receptors and their involvement in histamine release in a murine mast cell line. Neuropeptides 1992;21:65–72.

129. Krumins SA, Broomfield CA. C-terminal substance P fragments elicit histamine release from a murine mast cell line. Neuropeptides 1993;24:5–10.
130. Bunker CB, Cerio R, Bull HA, Evans J, Dowd PM, Foreman JC. The effect of capsaicin application on mast cells in normal human skin. Agents Actions 1991;33:195–196.
131. Kim BM, Lee SH, Shim WS, Oh U. Histamine-induced Ca^{2+} influx via the PLA_2/lipoxygenase/TRPV1 pathway in rat sensory neurons. Neurosci Lett 2004;361:159–162.

III. MODULATION AND PHYSIOLOGICAL SIGNIFICANCE OF NEUROGENIC INFLAMMATION

Neurogenic Inflammation in Health and Disease
Edited by Gábor Jancsó

Somatostatin as an Anti-Inflammatory Neuropeptide: From Physiological Basis to Drug Development

ERIKA PINTÉR, ZSUZSANNA HELYES, JÓZSEF NÉMETH, and JÁNOS SZOLCSÁNYI

Department of Pharmacology and Pharmacotherapy, Faculty of Medicine, University of Pécs, Pécs, Hungary

ABSTRACT

This chapter demonstrates a promising opportunity for anti-inflammatory drug therapy. The theoretical background of the development of these novel-type compounds is an original observation derived from our laboratory. According to our experimental data, somatostatin is released from the capsaicin-sensitive sensory nerve endings by different stimuli. It reaches distant parts of the body by the circulation and exerts systemic anti-inflammatory effects via binding to G-protein-coupled membrane receptors. This receptor family consists of five subtypes (sst_1–sst_5). Presumably, sst_1 and sst_4 are responsible for the mediation of anti-inflammatory actions. Somatostatin influences neurogenic inflammation by presynaptic and postsynaptic actions. It inhibits the release of proinflammatory neuropeptides from the sensory nerve endings and also acts on receptors of vascular endothelial, inflammatory, and immune cells. As somatostatin itself is not a suitable lead molecule for drug development because of its wide spectrum of actions and very short elimination half-life, stable, selective, well-tolerated somatostatin agonists with anti-inflammatory activity have been synthesized and investigated. TT-232, a heptapeptide anti-inflammatory analog, is under clinical investigation and several nonpeptide analogs have been preclinically tested. Somatostatin analog anti-inflammatory agents would mean a novel chance in the pharmacotherapy of chronic inflammatory and immune diseases.

1. PHYSIOLOGICAL ACTIONS OF SOMATOSTATIN

Somatostatin (SST) is a peptide widely distributed in the body, in both the central nervous system (CNS) and the peripheral tissues. Somatostatin or somatotropin release-inhibiting factor (SRIF) was first discovered in hypothalamic extracts as an activity that inhibited growth hormone secretion from cultured anterior pituitaries [1]. Brazeu et al. [2] characterized this factor as a cyclic peptide consisting of 14 amino acids (SST14). Several years later, a second bioactive form, an NH_2-terminally extended somatostatin molecule consisting of 28 amino acids (SST28) was isolated and characterized [3]. Somatostatin has an inhibitory action on a variety of physiological functions in different organ systems, including the hypothalamus, the anterior pituitary gland, the gastrointestinal tract, exocrine and endocrine pancreas, thyroid, kidney, adrenals, reproductive organs, and immune cells [4]. SST acts as a neurotransmitter in the CNS with effects on locomotor activity and cognitive function [5]. In the peripheral nervous system, it is found in sympathetic and sensory neurons [6] and may have an inhibitory effect on nociception [7]. Somatostatin has also been shown to possess antiproliferative effects *in vitro* [8].

The physiological actions of SST are initiated with its binding to membrane receptors. Five human somatostatin receptors (sst) have been cloned and characterized and referred to as sst_{1-5} receptors using the nomenclature suggested by Hoyer et al. [9]. Structurally, somatostatin receptors are seven transmembrane domain glycoproteins, comprised of seven membrane-spanning α-helical domains connected by short loops, an N-terminal extracellular domain and a C-terminal intracellular domain. On the basis of binding studies using synthetic somatostatin analogs, sst receptors can be divided into two different subgroups: $SRIF_1$ group comprising sst_2, sst_3, and sst_5 are able to bind octapeptide analogs, whereas $SRIF_2$ group comprising sst_1 and sst_4 have negligible affinity for these compounds. None of the peptide analogs bind exclusively to only one of the sst subtypes although new approaches might yield subtype-selective agonists and antagonists [4–16]. Somatostatin receptors are linked via G-proteins to multiple cellular effector systems. They mediate the inhibition of adenylate cyclase activity [4,17], reduce the conductance of voltage-dependent Ca^{++} channels [4,18], and activate K^+ channels [19–21]. Somatostatin receptors also mediate the stimulation of tyrosine phosphatase activity and induce a reduction of cell proliferation and the inhibition of an Na^+/H^+ exchanger (NHE1) [4,22,23]. sst receptors represent a major class of inhibitory receptors that play an important role in modulating higher brain functions, secretory processes, cell proliferation, and apoptosis. A great deal remains to be learned about the biology of sst dysfunction in neurological, gastroenterological, and immunological disorders [12]. Although the molecular cloning of many different sst subtypes has greatly improved our knowledge of their physiological role, several questions in this field still have not been answered. This apparent problem may be because of the fact that each receptor may activate multiple intracellular pathways and the same intracellular pathway may be activated by different sst subtypes, often expressed on the same cells [11]. This chapter focuses on the anti-inflammatory role of somatostatin.

2. SOURCES OF SOMATOSTATIN

Many tissues in the body are innervated by peptidergic sensory nerves of unmyelinated C-fibers and small myelinated Aδ-fibers [14,16]. Proinflammatory and anti-inflammatory neuropeptides are released by the activation of these sensory nerve endings. Neuronal somatostatin acting on sst receptors exerts anti-inflammatory action. The inhibition of neurogenic inflammation can be realized by the reduction of the release of proinflammatory neuropeptides. In endocrine organs and the gastrointestinal tract, somatostatin is produced by specialized neuroendocrine cells. In addition, immune cells themselves have been shown to contain somatostatin [24]. SST produced by immune cells may act as an autocrine or paracrine regulator in the local environment. Activated synovial cells can also produce the peptide [25].

3. SOMATOSTATIN RECEPTORS IN INFLAMED TISSUES

SST receptor autoradiography, in which tissue sections are incubated with radioisotope-labeled somatostatin or somatostatin analogs, has shown the presence or even upregulation of somatostatin-binding sites in infectious diseases (e.g., tuberculosis), autoimmune diseases (Graves ophthalmopathy), and other immune-mediated diseases [16]. sst receptors could also be visualized in biopsies of inflamed tissue from patients with sarcoidosis, rheumatoid arthritis, and inflammatory bowel disease. In sarcoidosis, the granulomas were labeled, whereas in rheumatoid arthritis and inflammatory bowel disease, mainly the venules were labeled [25–27]. Binding sites were also detected in human and mouse lymphoid organs, where sst receptors were identified on normal and activated leukocytes [28,29].

4. ANTI-INFLAMMATORY EFFECT OF EXOGENOUS SOMATOSTATIN

It has been established that exogenous somatostatin inhibited vasodilatation and plasma extravasation induced by saphenous nerve stimulation [30] and mustard oil-induced plasma leakage in the skin of the rat hind paw [31]. Repeated SST treatments attenuated adjuvant-induced chronic arthritis and mechanical allodynia in the rat [32], as well as endotoxin-evoked subacute pulmonary inflammation and consequent bronchial hyperreactivity [33]. SST also inhibited the release of substance P (SP) from central and peripheral nerve endings [34]. Vascular somatostatin receptors play a pivotal role in inflammatory skin conditions and are particularly important in pathomechanism of conditions such as dermatitis and psoriasis [35]. Inhibition of lymphoid cell proliferation is the best-documented effect of somatostatin on immune function [28]. Besides the inhibitory effect on proliferation, it has been shown to inhibit T-lymphocyte function, such as cytokine production [36]. Somatostatin reduced the secretion of interferon-γ (IFN-γ) from human peripheral blood mononuclear cells [37]. In several studies, SST has been shown to inhibit immunoglobulin production by B lymphocytes [38]. Somatostatin dose dependently reduced the IgA secretion from a murine plasmocytoma cell line [39] and inhibited the primary antibody response to antigenic stimulation in the rat [40]. SST has been demonstrated to influence a number of monocyte–macrophage functions [41,42]. Chemotaxis of human peripheral blood monocytes and murine macrophage cell lines was inhibited by this peptide, but dose-dependency was not observed [43,44]. There is evidence that, besides endothelial and immune cells, somatostatin acts on different cell types that are able to produce proinflammatory mediators. Intestinal epithelial cells secrete proinflammatory cytokines and chemokines that are crucial in mucosal defense. SST caused marked inhibition of spontaneous and tumor necrosis factor α (TNF-α)-induced secretion of interleukin-8 (IL-8) and IL-1β from the intestinal epithelial cells [45]. These data suggest that somatostatin plays an active role in regulating the mucosal inflammatory response of intestinal epithelial cells in pathological circumstances such as inflammatory bowel diseases or bacterial invasion (Table 1) [46].

5. EVIDENCE FOR THE EXISTENCE OF NEUROGENIC ANTI-INFLAMMATORY SUBSTANCES

Direct evidence for neurogenic inflammation evoked by antidromic stimulation of the sensory fibers was presented more than 30 years ago by Jancsó et al. [47]. These findings suggested that beyond the well-known phenomenon of arteriolar "antidromic vasodilatation" and "flare reaction" [48,49], venular plasma leakage forms a cardinal sign of neurogenic inflammation. Subsequently, it was realized that these local efferent responses are mediated by the release of proinflammatory neuropeptides such as SP, neurokinin A (NKA), and calcitonin gene-related peptide (CGRP) [50,51]. An unexpected observation made by Pintér and Szolcsányi [52] raised

Table 1 Anti-inflammatory effects of exogenous somatostatin

Inhibition of antidromic vasodilatation and plasma extravasation in the rat	[30]
Inhibition of the release of substance P from central and peripheral nerve endings	[34]
Inhibition of immunoglobulin production of B cells	[38–40]
Reduction of the secretion of reactive oxygen species from macrophages	[42]
Inhibition of SP-induced enhancement of the secretion of cytokines, IL-1, IL-6, TNF-α from LPS-stimulated murine peritoneal macrophages	[41]
Inhibition of mustard oil-induced plasma extravasation in the rat	[58]
Regulation of T-cell function. Inhibition of IFN-γ release from T cells	[36]
Marked inhibition of the spontaneous and TNF-α-induced secretion of IL-8 and IL-1β from intestinal epithelial cells	[45]
Significant attenuation of adjuvant-induced chronic arthritis and consequent mechanical allodynia in the rat	[32]
Inhibition of endotoxin-evoked subacute airway inflammation and inflammatory bronchial hyperresponsiveness	[78]

the possibility that mediators released from the capsaicin-sensitive nerve terminals possess not only local but also systemic neurohumoral effects. Electrical stimulation with C-fiber strength (20 V, 0.5 ms, 5 Hz for 5 min) of the peripheral stump of cut dorsal roots (lumbar 4–6) evoked intensive plasma extravasation in the skin of the rat hind paw and intravenously given Evans blue dye bound to the plasma protein accumulated in the inflamed tissues. In this series of experiments, a search was made to map internal organs in which neurogenic inflammation can be evoked; therefore, after the first stimulation had been completed a second stimulation was applied to another dorsal root. To our surprise, subsequent antidromic stimulation of the contralateral stump ($\Delta t = 5$ min) caused a reduced effect, with the intensity of the secondary plasma extravasation diminished by 50% (Fig. 1Aa). The difference in responses disappeared when the second train of stimulation was applied 60 min after the first one. Bilateral simultaneous stimulation of the cut dorsal roots evoked identical responses in both sides. According to these early findings, it was presumed that besides proinflammatory peptides, anti-inflammatory mediators are also released from the sensory nerve endings and reach the distant parts of the body through the circulation exerting a short-lasting systemic anti-inflammatory action [52]. Inflammation evoked by chemical agents with or without the intervention of sensory fibers was also inhibited by antidromic stimulation of dorsal roots. In a subsequent publication, Pintér and Szolcsányi [53] showed in 1996 that antidromic stimulation of the dorsal roots inhibited a subsequent inflammatory response because of subplantar carrageenin injection or instillation of capsaicin solution into the eye. Carrageenin, a sulfated polysaccharide from seaweed, is a commonly used agent for the induction of experimental edema formation in animal models. Complement activation plays a minor role in the carrageenin-induced edema, but the involvement of neutrophils and the kinin system has been demonstrated [54–57]. In addition, there is evidence for a neurogenic component mediated by tachykinins [57].

Initiation of dorsal root stimulation-induced anti-inflammatory responses was prevented by the degeneration of the capsaicin-sensitive afferents after perineural capsaicin-pretreatment but not by bilateral adrenalectomy (Fig. 1Ab,c). Taken together, these data provided evidence that the presumed anti-inflammatory substances are released from the capsaicin-sensitive nerves and act not only on neurogenic inflammation but also on mixed-type inflammatory reactions such as carrageenin-induced edema formation. Significant action of corticosteroids can be excluded, because bilateral adrenalectomy did not affect this anti-inflammatory response. In further

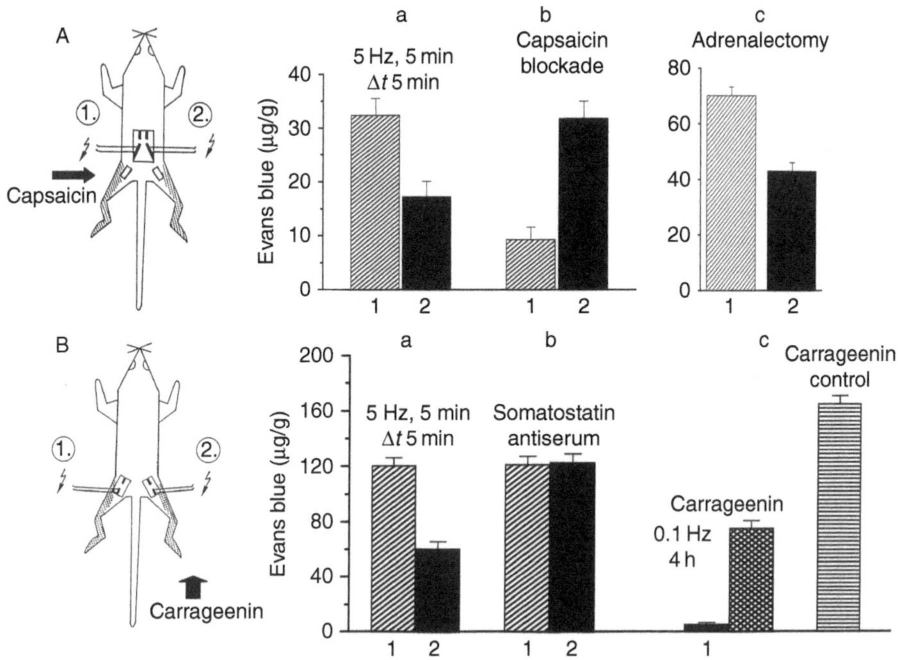

Figure 1. Antidromic stimulation of the peripheral stumps of cut lumbar (L_{4-6}) dorsal roots caused plasma extravasation in the skin of the innervated area, which was determined quantitatively by accumulation of Evans blue dye. The effect of the second stimulation was reduced by 50% compared with the first response (Aa). Perineural capsaicin pretreatment of the sciatic nerve not only prevented the plasma extravasation by ipsilateral dorsal root stimulation but abolished the contralateral antiplasma leakage effect as well (Ab). Systemic anti-inflammatory effect of dorsal root stimulation was not affected by bilateral adrenalectomy (Ac). The systemic anti-inflammatory effect has also been observed after antidromic sciatic nerve stimulation (Ba), but this phenomenon was prevented by somatostatin antiserum pretreatment (Bb). Although long lasting stimulation on very low frequency (0.1 Hz for 4 h) did not induce plasma extravasation on the ipsilateral hindpaw skin, it significantly reduced the carrageenin-induced intraplantar plasma extravasation on the contralateral side (Bc). (① first electrical stimulation, ② second electrical stimulation).

experiments, it was proven that besides dorsal root stimulation, antidromic excitation of a peripheral nerve also elicited the inhibitory action. Szolcsányi et al. [58] published in 1998 that antidromic stimulation of the sciatic nerve at a frequency that primarily stimulates C-fibers (20 V, 0.5 ms, 5 Hz for 5 min) caused neurogenic plasma leakage in the innervated area and inhibited the development of a subsequent Evans blue accumulation by 50% in response to similar stimulation of the contralateral sciatic nerve, which was abolished by polyclonal somatostatin antiserum pretreatment that suggested the potential mediator role of somatostatin (Fig. 1Ba,b). In these studies, the sympathetic neuron blocker guanethidine (8 mg/kg i.p.) was injected 1 h before the nerve stimulation to abolish the vascular effects of admixed sympathetic fibers, and pipecuronium bromide (200 μg/kg i.v.) was given to block the neuromuscular transmission. Under these conditions, nerve stimulation at 0.1 Hz for 4 h elicited no local plasma extravasation in the stimulated hind leg but still reduced the carrageenin-induced edema by 52% in the paw of the contralateral side (Fig. 1Bc). On the basis of this observation, we concluded that the actual development of inflammatory processes was not required to induce the anti-inflammatory effect. Consequently, the inhibitory substance(s) was supposed to be released directly from the nerve fibers and not from the inflamed tissues. Chemical activation of the capsaicin-sensitive sensory nerve terminals is also suitable to release sufficient amount of the anti-inflammatory mediator for evoking actions in distant parts of the body.

Neurogenic plasma leakage evoked by topical application of 1% mustard oil (allylisothiocyanate) on the skin of the acutely denervated hind leg (primary reaction) inhibited the progression of a secondary oil-induced plasma extravasation in the skin of the contralateral hind leg by 49.3% and in the conjunctival mucosa due to 0.1% capsaicin instillation by 33.5% in the rat. In this study, acute denervation of the hind leg was performed by cutting both the sciatic and the saphenous nerves 30 min before the experiments to avoid the interference of autonomic reflexes [31]. The primary reaction also inhibited the nonneurogenic hind paw edema evoked by s.c. injection of 5% dextran into the chronically denervated hind paw by 48%.

Chemical agents without neurogenic component of inflammation do not evoke this anti-inflammatory action. Pure non-neurogenic inflammation can be produced by dextran, which acts via activation of mast cells [59]. To exclude neurogenic inflammation, the hind legs were denervated 5 days prior to the experiments to induce degeneration of the nerve fibers. After chronic denervation, subplantar injection of 5% dextran elicited non-neurogenic edema formation without causing any systemic anti-inflammatory effect. This observation provided further evidence that the source of inhibitory substances must be the capsaicin-sensitive sensory nerves. As in the case of sciatic nerve stimulation, bilateral adrenalectomy did not inhibit the mustard oil-induced anti-inflammatory effect in the contralateral leg.

The most important goal of the subsequent studies was to elucidate the mediator background of the neurogenic anti-inflammatory action. On the basis of earlier data, anti-inflammatory sensory neuropeptides with a putative neurohumoral role include somatostatin [30,60,61], opioid peptides [30,62], and galanin [63].

6. EVIDENCE FOR THE ANTI-INFLAMMATORY ROLE OF NEURONAL ENDOGENOUS SOMATOSTATIN

Although the mediator role of somatostatin could be assumed by previous data, our later systemic studies presented clear evidence that SST released from the activated capsaicin-sensitive sensory afferent fibers is responsible for the systemic anti-inflammatory effect [31,58,64]. Pretreatment with polyclonal somatostatin antiserum (0.5 ml/rat i.v.), or the selective somatostatin depleting agent cysteamine (280 mg/kg, s.c.), prevented the anti-inflammatory effect of sciatic nerve stimulation or mustard oil application on a subsequent neurogenic plasma extravasation of the contralateral paw skin. The inhibitory effect of antidromic sciatic nerve excitation on plasma extravasation in response to vagal nerve stimulation was also prevented by somatostatin antiserum pretreatment. Somatostatin-like immunoreactivity (SST-LI) in the plasma increased more than fourfold after the stimulation of both sciatic nerves (5 Hz, 5 min), but the stimulus-evoked elevation was not observed in cysteamine (280 mg/kg, s.c.) pretreated rats [58]. Furthermore, inflammation-induced significant increase of plasma SST-LI has been shown in the adjuvant-induced chronic arthritis model of the rat [32] and also in the endotoxin-evoked subacute airway inflammation model of the mouse [33]. The functional significance of somatostatin released from the capsaicin-sensitive sensory nerve endings has been described with the help of cyclo-somatostatin, which antagonizes all the five somatostatin receptors. Repeated cyclo-somatostatin injections significantly increased edema formation and inflammatory changes in the rat joint [32], as well as histopathological alterations, inflammatory cytokine production, and hyperresponsiveness in the mouse lung [33].

As these functional data provide clear evidence for the pivotal role of somatostatin in the systemic anti-inflammatory action of sensory nerve stimulation, later studies were devoted to finding somatostatin analogs with anti-inflammatory properties.

7. SOMATOSTATIN, SUBSTANCE P, AND CALCITONIN GENE-RELATED PEPTIDE IN THE INFLAMMATORY MECHANISMS

SST counteracts the effect of SP in neurogenic inflammation because of the inhibition of proinflammatory neuropeptide release or opposing receptor actions [16]. SST acting on its own receptors inhibits neurokinin 1 receptor (NK_1)-mediated SP actions on vasodilatation in arterioles as well as on plasma protein extravasation and leukocyte accumulation from postcapillary venules and cytokine production of immune and other stimulated inflammatory cells. There is evidence that NK_1 and sst receptors also exist on inflammatory and immune cells where they mediate antagonistic effects [16]. CGRP acting on $CGRP_1$ receptors, causing vasodilatation and potentiating the SP effect on edema formation by increasing the blood supply, acts against SST in vascular inflammatory events. However, synergistic actions can also be shown between these peptides, especially in the cellular phases of the inflammatory process. Similar to SST, CGRP inhibits the $^{\bullet}O_2^-$ production, adhesion, and accumulation of neutrophil granulocytes. Both peptides also inhibit B and T lymphocyte proliferation and IgG production, as well as proinflammatory cytokine release from macrophages. CGRP was also found to upregulate the production of the anti-inflammatory cytokine IL-10 from macrophages [65]. These interactions demonstrate the existence of numerous binding sites for neuropeptides in the regulatory pathways of inflammatory and immune processes.

8. THERAPEUTIC VALUES OF ANTI-INFLAMMATORY SOMATOSTATIN ANALOGS

Because of its wide range of physiological functions, somatostatin could be a useful tool in the treatment of numerous human diseases. However, the clinical value of SST has been limited because of some disadvantages of the endogenous peptide, such as its short half-life in the circulation (<3 min) and its lack of selectivity. To overcome these problems, hundreds of peptide analogs and recently nonpeptide mimetics have been synthesized [15]. The octapeptide analog, octreotide binding predominantly to sst_2, sst_3, and sst_5, which is effectively used in the treatment of hormone-secreting tumors [66], had no influence on neurogenic inflammation and dextran-induced edema. However, there is some evidence that octreotide inhibits carrageenin-induced edema and leukocyte accumulation [60] as well as formalin-induced nociception [67]. Recently, a series of novel, potent, stable analogs have been synthesized in our laboratories to study the relative importance of specific substitutions in selectivity between these receptor subtypes. An analog with a cyclopenta-ring structure – D-Phe-Cys-Tyr-D-Thr-Lys-Cys-Thr-NH_2, called TT-232 – was found promising, in that it had no endocrine activity. This analog failed to inhibit growth hormone release or gastrin secretion *in vivo* but had a strong antiproliferative and apoptotic effect on tumor cells *in vivo* and *in vitro* [68,69]. Binding studies and functional agonist assays performed by Würster and Engström (Juvantia Pharma, Turku, Finland) demonstrated that TT-232 shows the greatest affinity to sst_4 and in a lesser extent to sst_1 receptors and behaves as an agonist. It proved to be an effective anti-inflammatory compound, which in μg/kg dose range exerts significant inhibitory effects in both *in vitro* and *in vivo* models (Table 2)

Table 2 Anti-inflammatory effect of TT-232

Animal model	Dose range and form of administration	Inhibitory effect
Rat models		
Plasma extravasation evoked by mustard oil on the paw skin	10 μg/kg i.p.	63.62%
Non-neurogenic dextran–edema of the chronically denervated hindpaw	10 μg/kg i.p.	51.92%
Bradykinin-induced plasma extravasation in the skin of the chronically denervated hindpaw	10 μg/kg i.v.	32.23%
Freund adjuvant-induced arthritis	2 × 500 μg/kg per day s.c. throughout the period of 18 days	42.22% (on the 19th day)
Carrageenin-induced paw edema for 3 h	3 × 2.5–20 μg/kg i.v.	19.5–85.5%
Plasma extravasation induced by intra-articular bradykinin injection	5–20 μg/kg i.v.	29.48–37.46%
Cutaneous neutrophil accumulation over a 3 h period after i.d. injection of carrageenin	3 × 80 μg/kg i.v.	47.2%
Cutaneous neutrophil accumulation over a 3-h period after i.d. injection of IL-1β	3 × 80 μg/kg i.v.	84.2%
SP, CGRP, and somatostatin release from the isolated trachea induced by electrical field stimulation (40 V, 0.1 ms, 10 Hz, 120 s) or by capsaicin (10^{-7} M),	500 nM	79 (SP), 48 (CGRP) and 74% (SST) elicited by capsaicin 36 (SP), 48 (CGRP) and 42% (SST) in response to electrical field stimulation
Mouse model		
Edema formation induced by topically applied capsaicin on the ear	5–20 μg/kg i.v.	62.4–84.6%
Endotoxin-induced subacute airway inflammation and consequent hyperreactivity	500 μg/kg i.p. repeatedly (3 times/day)	28.5–47.8% (histological scores of the inflammatory parameters) 33.2–54.1% (bronchoconstriction)
Ovalbumin-evoked chronic airway inflammation and bronchial hyperresponsiveness	500 μg/kg i.p. repeatedly (3 times/day throughout 3 days of the elicitation phase)	31.5–50.8% (histological scores of the inflammatory parameters) 20.2–74.3% (bronchoconstriction)

(for review, see Ref. [70]). In the rat, low doses of TT-232 inhibited the mustard oil-induced plasma extravasation in the skin of the hind paw, the dextran-evoked paw edema and bradykinin-induced plasma extravasation in the paw skin or ankle joint (5–10 μg/kg), and the carrageenin-induced paw edema (3 × 5 μg/kg). Neutrophil leukocyte accumulation after carrageenin or IL-1β injection was only inhibited by higher doses (3 × 80 μg/kg). Both the acute vascular and the longer lasting cellular inflammatory responses were inhibited. In the mouse, TT-232 has proven to be a potent anti-inflammatory agent against the capsaicin-induced neurogenic edema formation in the ear. Also, capsaicin and electrical field stimulation evoked a significant increase in SP, CGRP, and SST release from isolated rat tracheae in control samples, which was significantly inhibited by TT-232 (500 nM). TT-232 itself did not influence the basal, nonstimulated peptide release [71,72].

In rat lung tissue, the mRNA of two somatostatin receptor subtypes were identified. Predominantly, the sst_4 and, to a lesser extent, sst_1 subtype were expressed [73]. In addition, there are studies that have established that in the rat adjuvant arthritis model, sst_3 and sst_4 receptor subtypes are over-expressed in the immune system. They found that systemic administration of octreotide, which has a high affinity for sst_2 and sst_5, did not influence either the incidence or the severity of adjuvant arthritis in the rat [74]. Our previous findings also demonstrated the lack of efficacy of octreotide in mustard oil-induced neurogenic and in dextran-induced non-neurogenic inflammatory models in the rat, where TT-232 significantly inhibited the neurogenic plasma extravasation or paw swelling [75]. More recently, we have shown that repeated TT-232 treatment markedly attenuated adjuvant-induced chronic arthritis and mechanical allodynia in the rat [32]. It also inhibited endotoxin-provoked pulmonary inflammation and ovalbumin-evoked allergic airway inflammation and hyperreactivity in the mouse [76]. Therefore, it is tempting to assume that sst_4, and probably sst_1, receptors are responsible for the anti-inflammatory effect of SST and TT-232 on the neurogenic inflammation. The agonistic effect of TT-232 on G-protein-coupled somatostatin membrane receptors was also established by our data as pertussis toxin prevented the inhibitory effect of the heptapeptide SST analog on the release of neuropeptides from the rat isolated trachea preparation [72]. Presumably, inhibition of tyrosine kinases by TT-232 [71,72] does not play a pivotal role in its anti-inflammatory effect since a potent tyrosine kinase inhibitor, genistein, failed to influence the stimulated neuropeptide release. Extremely high doses of TT-232 (up to 5 mg/kg) did not induce damage of the gastrointestinal mucosa in the rat. As each of the present steroidal anti-inflammatory drugs (SAIDs) and nonsteroidal anti-inflammatory drugs (NSAIDs) cause harmful actions on the gastric and duodenal mucosa, clinical application of TT-232 suggests a possible new route to eliminate the most frequent, serious unwanted effects of anti-inflammatory therapy. Findings summarized in Table 2 established a broad spectrum of anti-inflammatory effect of TT-232 (for review, see Ref. [70]).

A novel sulfonamido-peptidomimetic compound, J-2156 [(1′S,2S)-4-amino-*N*-(1′-carbamoyl-2′-phenylethyl)-2-(4″-methyl-1″-naphthalenesulfonamino)butanamide], synthesized at Juvantia Pharma (Turku, Finland), belongs to a chemically novel class of somatostatin receptor ligands. J-2156 possesses nanomolar affinity for the human somatostatin sst_4 receptors and it is 400-fold more selective for this receptor than any other human sst receptor subtype [77]. In a cyclic adenosine monophoshate (cAMP) assay, the compound acted as a full agonist similar to native SST-14 or SST-28. In a [^{35}S]guanosine-5′-*O*-(3-thio)triphosphate functional assay, J-2156 elicited a two to three times larger response than native somatostatin, and increasing concentrations of somatostatin-14 caused a concentration-dependent rightward shift of the concentration-response curves of J-2156 without affecting its maximal effect. J-2156 exerted greater agonism on the sst_4 receptor than its endogenous ligands, and it has been defined as a selective, high affinity agonist [77]. We have recently published that J-2156 exerted potent anti-inflammatory in several acute rat and mouse models [76,78]. Furthermore, it also significantly diminished IL-1β release in response to stimulation in peritoneal macrophage culture [76].

Therefore, these compounds might be effective in diseases where, besides neurogenic inflammation, several other mediators produce vascular and cellular inflammatory responses. Bronchial asthma, rheumatoid arthritis, and psoriasis are examples of chronic diseases where involvement of sensory neuropeptides and neurogenic inflammation probably play important roles [79] and where treatment with parenteral corticoids or cytotoxic anticancer drugs is required in severe cases. Other studies suggest the therapeutic use of somatostatin in these diseases [80,81]. Therefore, sst_4-selective somatostatin analogs are proposed as candidates for the treatment of bronchial asthma, psoriasis, eczema, dermatitis, or rheumatoid arthritis.

9. CONCLUSIONS

This chapter demonstrates the potent anti-inflammatory actions of SST on different *in vitro* and *in vivo* models. Somatostatin acting on the sensory nerve endings inhibits not only its own release (negative feedback) but also the release of proinflammatory neuropeptides such as SP or CGRP. Besides this presynaptic action, somatostatin also interacts with these peptides at the level of effector tissues. During the vascular phases of inflammation, CGRP causes long lasting arteriolar vasodilatation, whereas SST inhibits it. SP induces venular plasma leakage via activation of endothelial NK_1 receptors, and again this is inhibited by somatostatin binding to its own receptors.

A recent review discussing the effects of SP and CGRP on inflammatory and immune cells clearly demonstrates that SST and SP exert opposing effects throughout the inflammatory and immune response, whereas CGRP can be synergistic with SST in several aspects [16].

Somatostatin itself is not suitable as an anti-inflammatory drug because of its short duration of action and broad-spectrum effects. Modification of fragments of SST has generated more stable, potent, and selective analogs. Octreotide was the first somatostatin analog to be used as a therapeutic agent and is still the most widely used peptide for the treatment of acromegaly, adenomas and pancreatic, breast and prostate tumors [10]. Recent evidence suggests that the antihormonal and antimitotic actions of somatostatin are mediated via sst_2 and sst_5 receptor subtypes, and the $SRIF_1$-selective analog, octreotide, proved to be effective in these cases. Octreotide has not shown significant inhibition in our experiments of neurogenic and non-neurogenic plasma extravasation. On the other hand, our most effective heptapeptide analog, TT-232, exerted considerable inhibitory actions in both vascular and cellular inflammatory models [71,72]. On the basis of these data, we presume that the $SRIF_2$ receptor family, sst_1 or sst_4 or both, are responsible for the anti-inflammatory effects of somatostatin.

The present SAIDs and NSAIDs, which are clinically available, possess harmful actions against the gastrointestinal mucosa to a greater or lesser extent. These unwanted effects can be particularly severe during treatment of chronic inflammatory conditions, such as rheumatoid arthritis, dermatitis, psoriasis, asthma, inflammatory bowel diseases, etc. The proper therapy of these diseases is still an open question. We suggest that the introduction of sst_4-selective somatostatin analogs with significant anti-inflammatory potency would provide novel perspectives for drug development.

REFERENCES

1. Krulich L, Dhariwal APS, McCann SM. Stimulatory and inhibitory effects of purified hypothalamic extracts on growth hormone release from rat pituitary in vitro. Endocrinolology 1968;83:787–790.
2. Brazeu P. Somatostatin: a peptide with unexpected physiologic activities. Am J Med 1986;81(Suppl. 6B):8–13.
3. Pradayrol L, Jornwall H, Mutt V, Ribet A. N-terminally extended somatostatin: the primary structure of somatostatin-28. FEBS Lett 1980;109:55–58.
4. Patel YC, Greenwood MT, Panetta L, Demchyshyn L, Niznik H, Srikant CB. Minireview: the somatostatin receptor family. Life Sci 1995;57:1249–1265.
5. Reisine T, Bell GI. Molecular properties of somatostatin receptors. Neuroscience 1995;67:777–790.

6. Reichlin S. SRIF. New Engl J Med 1983;309:1495–1563.
7. Schindler M, Holloway S, Hathway G, Wolf CJ, Humphrey PP, Emson PC. Identification of somatostatin sst2(a) receptor expressing neurones in central regions involved in nociception. Brain Res 1998;798:25–35.
8. Schally AV. Oncological application of somatostatin analogues. Cancer Res 1988;48:6977–6985.
9. Hoyer D, Bell GI, Berelowitz M, Epelbaum J, Ferniuk W, Humphrey PA et al. Classification and nomenclature of somatostatin receptors. Trends Pharmacol Sci 1995;16:86–88.
10. Hofland LJ, Visser-Wisselaar HA, Lamberts SWJ. Somatostatin analogs: clinical application in relation to human somatostatin receptor subtypes. Biochem Pharmacol 1995;50:287–297.
11. Florio T, Schettini G. Multiple intracellular effectors modulate physiological functions of the cloned somatostatin receptors. J Mol Endocrinol 1996;17:89–100.
12. Patel YC. Molecular pharmacology of somatostatin receptor subtypes. J Endocrinol Invest 1997;20:348–367.
13. Meyerhof W. The elucidation of somatostatin receptor functions: a current view. Rev Physiol Biochem Pharmacol 1998;133:55–108.
14. Selmer IS, Schindler M, Allen JP, Humphrey PPA, Emson PC. Advances in understanding neuronal somatostatin receptors. Regul Pept 2000;90:1–18.
15. Janecka A, Zubrzycka M, Janecki T. Somatostatin analogs. J Pept Res 2001;58:91–107.
16. ten Bokum AMC, Hofland LJ, van Hagen PM. Somatostatin and somatostatin receptors in the immune system: a review. Eur Cytokine Netw 2000;11:161–176.
17. Jakobs K, Aktories K, Schultz G. A nucleotide regulatory site for SRIF inhibition of adenylyl cyclase in S49 lymphoma cells. Nature 1983;303:177–178.
18. Ikeda S, Schofield G. SRIF blocks a ca++ current in rat sympathetic ganglion neurones. J Physiol 1989;409:221–240.
19. Moore S, Madamba S, Joels M, Siggins G. Somatostatin augments the M-current in hippocampal neurons. Science 1988;239:278–280.
20. Mihara S, North R, Suprenant A. SRIF increases and inwardly rectifying potassium conductance in guinea pig submucous plexus neurones. J Physiol 1987;390:335–355.
21. Wang H, Bogen C, Reisine Dichter TM. SRIF-14 and SRIF-28 induce opposite effects on potassium currents in rat neocortical neurons. Proc Natl Acad Sci USA 1989;86:9616–9620.
22. Barber D, Mc Guire M, Ganz M. Beta-adrenergic and SRIF receptor regulate na^{+}-H^{+} exchange independent of cAMP. J Biol Chem 1989;264:21038–21042.
23. Buscail L, Delesque N, Esteve JP, Saint-Laurent N, Prats H, Clerc P et al. Stimulation of tyrosine phosphatase and inhibition of cell proloferation by SRIF analogues: mediation by human SRIF receptor subtypes SSTR1 and SSTR2. Proc Natl Acad Sci USA 1994;91:2315–2319.
24. Weinstock JV. Production of neuropeptides by inflammatory cells within the granulomas of murine schistosomiasis mansoni. Eur J Clin Invest 1991;21:145–153.
25. van Hagen PM, Markusse HM, Lamberts SW, Kwekkeboom DJ, Reubi JC, Krenning EP. Somatostatin receptor imaging. The presence of somatostatin receptors in rheumatoid arthritis. Arthritis Rheum 1994;37:1521–1527.
26. Reubi JC, Waser B, Markusse HM, Krenning EP, van Hagen M, Laissue JA. Vascular somatostatin receptors in synovium from patients with rheumatoid arthritis. Eur J Pharmacol 1994;271:371–378.

27. Reubi JC, Mazzucchelli L, Laissue JA. Intestinal vessels express a high density of somatostatin receptors in human inflammatory bowel disease. Gastroenterology 1994;106:951–959.
28. van Hagen PM, Krenning EP, Kwekkeboom DJ, Reubi JC, Anker-Lugtenburg PJ, Löwenberg B et al. Somatostatin and the immune and haematopoetic system; a review. Eur J Clin Invest 1994;24:91–99.
29. Sreeedharan S, Kodama K, Peterson K, Goetzl E. Distinct subsets of somatostatin receptors on cultured human lymphocytes. J Biol Chem 1989;264:949–952.
30. Lembeck F, Donnerer J, Barthó L. Inhibition of neurogenic vasodilatation by substance P antagonists, somatostatin and (D-met^2, pro^5)enkephalinamide. Eur J Pharmacol 1982;85:171–176.
31. Szolcsányi J, Pintér E, Helyes Oroszi G, Németh J. Systemic anti-inflammatory effect induced by counter-irritation through a local release of somatostatin from nociceptors. Br J Pharmacol 1998;125:916–922.
32. Helyes Zs, Szabó Á, Németh J, Jakab B, Pintér E, Bánvölgyi Á et al. Anti-inflammatory and analgesic effects of somatostatin released from capsaicin-sensitive sensory nerve terminals in Freund's adjuvant induced chronic arthritis model of the rat. Arthritis Rheum 2004;50:1677–1685.
33. Helyes Zs, Elekes K, Németh J, Pozsgai G, Sándor K, Kereskai L et al. Role of transient receptor potential vanilloid 1 receptors in endotoxin-induced airway inflammation in the mouse. Am J Physiol Lung Cell Mol Physiol 2007;292(5):L1173–L1181.
34. Sicuteri F, Raino L, Geppetti P. Substance P and endogenous opioids: how and where they could play a role in cluster headache. Cephalalgia 1983;3(Suppl. 1):143–145.
35. Misery L. Skin, immunity and the nervous system. Br J Dermatol 1997;137:843–850.
36. Elliott DE, Li J, Blum AM, Metwali A, Patel YC, Weinstock JV. SSTR2A is the dominant somatostatin receptor subtype by inflammatory cells, is widely expressed and directly regulates T cell IFN-γ release. Eur J Immunol 1999;29:2454–2463.
37. Muscettola G, Grasso G. Somatostatin and vasoactive intestinal peptide reduce interferon gamma production by human peripheral blood mononuclear cells. Immunobiology 1990;180:419–430.
38. Blum AM, Andrews PV, Metwali A, Mathew RC, Elliott D, Weinstock JV. Substance P and somatostatin can modulate the amount of IgG2a secreted in response to schistosome egg antigens in murine schistosomiasis mansoni. J Immunol 1993;151:6994–7004.
39. Scicchitano R, Dazin P, Bienenstock J, Payan DG, Stanisz AM. The murine IgA-secreting plasmacytoma MOPPC-315 expresses somatostatin receptors. J Immunol 1988;141:937–941.
40. Eglezos A, Andrews PV, Helme RD. In vivo inhibition of the rat primary antibody response to antigenic stimulation by somatostatin. Immunol Cell Biol 1993;71:125–129.
41. Berman AS, Chancellor-Freeland C, Zhu G, Black PH. Substance P primes murine peritoneal macrophages for an augmented pro-inflammatory cytokine response to lipopolysaccharide. Neuroimmunomodulation 1996;3:141–149.
42. Niedermühl-Bichler M, Wiedermann CJ. Suppression of superoxide release from human monocyte by somatostatin-related peptides. Regul Pept 1992;41:39–45.
43. Wiedermann CJ, Reinisch N, Braunsteiner H. Stimulation of monocyte chemotaxis by human growth hormone and its deactivation by somatostatin. Blood 1993;82:954–960.
44. Kolasinski SL, Haines KA, Siegel EL, Cronstein BN, Abramson SB. Neuropeptides and inflammation. A somatostatin analog as a selective antagonist of neutrophil activation by substance P. Arthritis Rheum 1992;35:369–375.

45. Chowers Y, Cahalon L, Lahav M, Schor H, Tal R, Bal-Meir S et al. Somatostatin through its specific receptor inhibits spontaneous and TNF-α and bacteria-. Induced IL-8 and IL-1β secretion from intestinal epithelial cells. J Immunol 2000;165:2955–2961.
46. Pintér E, Helyes Zs, Szolcsányi J. Inhibitory effect of somatostatin on inflammation and nociception. Pharmacol Ther 2006;112:440–456.
47. Jancsó M, Jancsó-Gábor A, Szolcsányi J. Direct evidence for neurogenic inflammation and its prevention by denervation and by pretreatment with capsaicin. Br J Pharmacol 1967;31:138–151.
48. Bayliss WM. The Vaso-Motor System. London: Longmans, Green and Co, 1923.
49. Lewis T. The Blood Vessels of the Human Skin and their Responses. London: Shaw, 1927.
50. Maggi CA. Tachykinins and calcitonin gene-related peptide (CGRP) as co-transmitters released from peripheral endings of sensory nerves. Prog Neurobiol 1995;45:1–98.
51. Lundberg JM. Pharmacology and co-transmission in the autonomic nervous system: integrative aspects on amines, neuropeptides, adenosine triphosphate, amino acids and nitric oxide. Pharmacol Rev 1996;48:113–178.
52. Pintér E, Szolcsányi J. Inflammatory and anti-inflammatory effects of antidromic stimulation of dorsal roots in the rat. Agents Actions 1988;25:240–242.
53. Pintér E, Szolcsányi J. Systemic anti-inflammatory effect induced by antidromic stimulation of the dorsal roots in the rat. Neurosci Lett 1996;212:33–36.
54. Winter CA, Risley EA, Nuss GW. Carrageenin-induced oedema in hind paw of the rat as an assay for anti-inflammatory drugs. Proc Soc Exp Biol Med 1962;111:544.
55. Vinegar RJ, Truax F, Selph JL. Quantitative studies of the pathway to acute carrageenin inflammation. Fed Proc 1976;35:2447.
56. Damas J, Remacle-Volon G. Influence of a long-acting bradykinin antagonist, hoe 140, on some acute inflammatory reactions in the rat. Eur J Pharmacol 1992;211:81.
57. Cao T, Pintér E, Al-Rashed S, Gerard N, Brain SD. Neurokinin-1 receptor agonists are involved in mediating neutrophil accumulation in the inflamed, but normal cutaneous microvasculature: an in vivo study using neurokinin-1 receptor knockout mice. J Immunol 2000;164:5424–5429.
58. Szolcsányi J, Helyes Oroszi G, Németh J, Pintér E. Release of somatostatin and its role in the mediation of the anti-inflammatory effect induced by antidromic stimulation of sensory fibres of rat sciatic nerve. Br J Pharmacol 1998;123:936–942.
59. Selye H. The Mast Cell. Washington: Butterworths, 1965.
60. Karalis K, Mastokaros G, Chrousos GP, Tolis G. Somatostatin analogues suppress the inflammatory reaction in vivo. J Clin Invest 1994;93:2000–2006.
61. Fioravanti A, Govoni M, La Montagna G, Perpignano G, Tirri G, Trotta F et al. Somatostatin-14 and joint inflammation: evidence for intra-articular efficacy of prolonged administration in rheumatoid arthritis. Drugs Exp Clin Res 1995;21:97–103.
62. Barthó L, Szolcsányi J. Opiate agonists inhibit neurogenic plasma extravasation in the rat. Eur J Pharmacol 1981;73:101–104.
63. Xu XJ, Hao JX, Wiesenfeld-Hallin Zs, Hakanson R, Folkers K, Hökfelt T et al. A novel tachykinin antagonist and galanin inhibit plasma extravasation induced by antidromic C-fibre stimulation in the rat hindpaw. Neuroscience 1991;42:731–737.
64. Pintér E, Helyes Németh J, Oroszi G, Szolcsányi J. Somatostatin, as anti-inflammatory neuromediator: in vivo and in vitro evidence. Naunyn-Schmiedeberg's Arch Pharmacol 1997;356(Suppl. 1):R-44.

65. Grant A. Leukocytes and neurogenic inflammation. Immunopharmacology 2002;9: 403–420.
66. Pincus T, Smith S, Oates JA. Somatostatin analogue octreotide is effective in revealing musculoskeletal pain associated with carcinoid syndrome. Arthritis Rheum 1989;32(Suppl. 4):S71.
67. Carlton SM, Du Davidson JE, Zhou S, Coggeshall RE. Somatostatin receptors on peripheral primary afferent terminals: inhibition on sensitised nociceptors. Pain 2001;90:233–244.
68. Kéri Gy, Mező I, Vadász Zs, Horváth A, Idei M, Vántus Á et al. Structure-activity relationship studies of novel somatostatin analogs with antitumor activity. Pept Res 1993;6:281–288.
69. Kéri Gy, Érchegyi J, Horváth A, Mező I, Idei M, Vántus T et al. A tumor-selective somatostatin analog (TT-232) with strong in vitro and in vivo anti-tumor activity. Proc Natl Acad Sci USA 1996;93:12513–12518.
70. Helyes Zs, Pintér E, Szolcsányi J. TT-232. Somatostatin sstr1/sstr4 agonist. Treatment of neuropathic pain. Treatment of inflammation. Drug Fut 2005;30:558–566.
71. Helyes Zs, Pintér E, Németh J, Kéri Gy, Thán M, Oroszi G et al. Anti-inflammatory effect of synthetic somatostatin analogues in the rat. Br J Pharmacol 2001;134:1571–1579.
72. Pintér E, Helyes Zs, Németh J, Pórszász R, Pethő G, Thán M et al. Pharmacological characterisation of the somatostatin analogue TT-232: effects on neurogenic and non-neurogenic inflammation and neuropathic hyperalgesia. Naunyn-Schmiedeberg's Arch Pharmacol 2002;366:142–150.
73. Schloos J, Raulf F, Hoyer D, Bruns C. Identification and pharmacological. Characterization of somatostatin receptors in rat lung. Br J Pharmacol 1997;121:963–971.
74. ten Bokum AMC, Lichtenauer-Kaligis EGR, Melief MJ, van Koetsveld PM, Bruns C, van Hagen PM et al. Somatostatin receptor subtype. Expression in cells of the rat immune system during adjuvant arthritis. J Endocrinol 1999;61:167–175.
75. Helyes Zs, Pintér E, Szolcsányi J, Horváth J. Anti-inflammatory and antinociceptive effect of different somatostatin analogs. Neurobiology 1996;4:115–117.
76. Elekes K, Helyes Zs, Kereskai L, Sándor K, Pintér E, Pozsgai G et al. Inhibitory effects of synthetic somatostatin receptor subtype 4 agonists on acute and chronic airway inflammation and hyperreactivity in the mouse. Eur J Pharmacol 2008;578:313–322.
77. Engström M, Tomperi J, El Darwish K, Ahman M, Savola JM, Würster S. Superagonism at the human somatostatin receptor subtype 4. J Pharmacol Exp Ther 2005;312:332–338.
78. Helyes Zs, Pintér E, Németh J, Sándor K, Elekes K, Szabó Á et al. Effects of the somatostatin receptor subtype 4 selective agonist J-2156 on sensory neuropeptide release and inflammatory reactions in rodents. Br J Pharmacol 2006;149:405–415.
79. Berczi I, Chalmers IM, Nagy E, Warrington RJ. The immune effects of neuro-Peptides. Baillieres Clin Rheumatol 1996;10:227–257.
80. Scholzen T, Amstrong CA, Bunnett NW, Luger TA, Olerud JE, Ansel JC. Neuropeptides in the skin: interactions between the neuroendocrine and the skin immune systems. Exp Dermatol 1998;7:81–96.
81. Sanake T, Suzuki N. The role of somatostatin in the pathophysiology of rheumatoid arthritis. Clin Exp Rheumatol 1998;16:745–749.

Neurogenic Inflammation in Health and Disease
Edited by Gábor Jancsó

Effects of Bradykinin on Nociceptors

GÁBOR PETHŐ[1] and PETER W. REEH[2]

[1] *Department of Pharmacology and Pharmacotherapy, Faculty of Medicine, University of Pécs, Pécs, Hungary*
[2] *Institute for Physiology and Pathophysiology, University of Erlangen-Nürnberg, Erlangen, Germany*

ABSTRACT

Bradykinin is a potent mediator formed upon tissue damage and inflammation. It can both excite and sensitize nociceptors to heat, mechanical, and chemical stimuli. Two types of bradykinin receptors (B_1 and B_2) have been identified of which the constitutive B_2 receptors mediate most of the acute effects of bradykinin in uninflamed tissues while the B_1 receptors are induced and become activated during inflammation. Both receptor subtypes utilize similar signaling pathways including activation of protein kinase C (PKC), elevation of intracellular Ca^{2+} concentration, and release of arachidonic acid. PKC activation is the major mechanism underlying the neuronal excitatory and heat-sensitizing actions of bradykinin while Ca^{2+} accumulation induces formation of nitric oxide within sensory neurons which is involved – together with receptor downregulation – in the development of tachyphylaxis of B_2 receptor-mediated effects of bradykinin. Nitric oxide, however, may also contribute to the excitatory and sensitizing actions of bradykinin. Cyclooxygenase metabolites of arachidonic acid (prostanoids) may also be involved in both the excitatory and the sensitizing effects of bradykinin. Recently, a new signaling mechanism has been revealed for bradykinin which involves activation of the capsaicin TRPV1 receptor through PKC activation and formation of 12-lipoxygenase products of arachidonic acid. According to a novel hypothesis, the neuronal excitatory action of bradykinin is not a separate effect but in fact a heat response as a result of a massive heat sensitization with a threshold drop below the ambient temperature. Prostanoids can also sensitize nociceptors to heat, mechanical, and chemical stimuli predominantly via the cyclic adenosine 3′,5′-monophosphate–protein kinase A pathway that modulates various membrane channels including Ca^{2+}-dependent or voltage-gated K^+ channels, tetrodotoxin-resistant Na^+ channels as well as ligand-gated or noxious heat-gated ion channels.

1. INTRODUCTION

A huge array of chemical substances either preformed or *de novo* synthesized are released during tissue trauma or inflammation. These agents can contribute to the development and maintenance of inflammatory pain via two principal mechanisms. Activation or excitation of nociceptors implicates generation of action potentials by these substances which then travel to the central nervous system. Sensitization of nociceptors refers to their increased responsiveness to heat, mechanical, or chemical stimuli. According to our present knowledge, both mechanisms contribute to pain experienced in inflammatory disease states. Bradykinin is one of the most potent mediators produced under inflammatory conditions, and a multitude of its excitatory and sensitizing actions on nociceptors have been described in various experimental models. This review primarily aimed at giving a comprehensive and up-to-date account of the excitatory and sensitizing actions of bradykinin on the afferent function of peripheral terminals of nociceptive primary sensory neurons under inflammatory conditions. The efferent tissue responses secondary to bradykinin-induced release of neuropeptides from peptidergic nociceptors and effects within the central nervous system will not be discussed.

2. SYNTHESIS AND BREAKDOWN OF BRADYKININ

The nonapeptide bradykinin and the related decapeptide kallidin (lysyl-bradykinin) collectively termed kinins can be synthesized both within the vessels in the blood and extravascularly in tissues. These peptides are cleaved from their protein precursors called kininogens by proteolytic enzymes, the kallikreins which are formed from their precursors, the prekallikreins. The plasma prekallikrein is converted to kallikrein when the clotting factor XII (Hageman) is activated by contact with the negatively charged surfaces. Plasma kallikrein acts on high-molecular-weight kininogen producing bradykinin and kallidin. The tissue prekallikrein is transformed to kallikrein in response to inflammation or tissue damage. The substrates for tissue kallikrein are high- and low-molecular-weight kininogens. The kinins are metabolized rapidly by kininases I and II having a half-life of less than 1 min. Kininase I cleaves only the C-terminal arginine residue yielding des-Arg^9-bradykinin and des-Arg^{10}-kallidin which retain biological activity (on B_1 receptors, see below), while kininase II – which is identical to angiotensin-converting enzyme – cleaves two amino acids from the C-terminal resulting in inactive fragments. Kininase activity is reduced in an acidic environment which can be a factor contributing to higher levels of bradykinin during inflammation [1,2]. Recently, a modulation of bradykinin signaling by plasma membrane-bound endopeptidases was revealed [3]. In cultured rat trigeminal sensory neurons, the endopeptidases 24.15, 24.16, and bradykinin B_2 receptors were shown to associate with lipid rafts. Inhibition of endopeptidases 24.15/16 led to a facilitation of bradykinin-induced inositol phosphate accumulation suggesting that these plasma membrane-associated enzymes can breakdown bradykinin and thereby regulate the degree of B_2 receptor activation.

3. BRADYKININ RECEPTOR EXPRESSION

3.1 Types of bradykinin receptors: general features

Two types of bradykinin or kinin receptors termed B_1 and B_2 have been cloned and characterized [4–8]. Bradykinin and kallidin act preferentially at the B_2 receptors while des-Arg^9-bradykinin and des-Arg^{10}-kallidin produced by kininase I from bradykinin and kallidin, respectively, are selective agonists of the B_1 receptors. The B_2 receptors are constitutive, being present at a relatively constant density on various cells including nociceptive afferent neurons, smooth muscle, endothelium, macrophages, and postganglionic sympathetic fibers [7,8]. In contrast, the B_1 receptors are considered to be inducible, found only at a very low level under resting conditions but largely upregulated upon tissue trauma or inflammation [9–11]. Of the natural extracellular mediators, various cytokines have been shown to induce B_1 receptors with most evidence having been accumulated for interleukin (IL)-1β [12–16]. The induction of B_1 receptors involves activation of various protein kinase enzymes [protein kinase C (PKC), tyrosine kinase, and mitogen-activated protein kinase] as well as the transcription nuclear factor-κB [16]. B_1 receptor induction is diminished by glucocorticoid pretreatment [17,18]; moreover, endogenous glucocorticoid hormones exert a tonic inhibitory control on receptor expression [19]. In addition to inducibility, another striking difference between the two bradykinin receptors subtypes is that while B_2 receptors undergo a rapid desensitization (see later), B_1 receptors do not exhibit any notable tachyphylaxis [20,21].

In uninflamed tissues, most effects of bradykinin are mediated by B_2 receptors as shown by studies employing selective peptide (HOE 140 also known as icatibant, NPC 18521) and nonpeptide B_2 receptor antagonists (WIN 64338, FR 173657, NPC 18884, and bradyzide) as well as B_2 receptor knock-out mice [22–49]. Consistent with this, selective B_1 receptor agonists do not activate or sensitize nociceptors under normal conditions [14,22,24,26,30,31,35, 41–43,48,50–59].

Although B_1 receptors are generally considered not to participate in the acute effects of bradykinin, some studies provided evidence against this. The B_1 receptor antagonist des-Arg^9[Leu]^8-bradykinin reduced both phases of the nociceptive response to formalin in mice and rats [34,60–62]. Both phases of formalin-induced nociception and the algogenic effect of capsaicin were reduced in B_1 receptor knock-out mice [63]. The late phase of formalin-evoked nocifension was reduced by a novel nonpeptide B_1 receptor antagonist [64]. Another novel nonpeptide B_1 receptor antagonist was antinociceptive in both phases of the formalin test in rats, in the second phase in the mouse, and also in the acetic acid-induced writhing assay [65]. In a modified version of the formalin test, employing injection of the agent into the upper lip of the rat, activation of the B_1 receptor was revealed in the second phase, whereas that of the B_2 receptor was revealed in both phases [66]. As these nociceptive responses develop within 5–30 min, the results suggest that B_1 receptors can also be involved in some acute/subacute pronociceptive effects of bradykinin, and the basis for this may be their constitutive presence in uninflamed tissue (see below) and/or a very rapid induction process. It is worth mentioning that an involvement of B_2 receptor activation in both phases of the formalin-induced nociception was found in some [33,61,66] but not other studies [34,62].

3.2 Localization of B_1 receptors

Concerning the localization of B_1 receptors, conflicting data have been obtained. In a study employing both binding and functional assays before and after development of inflammation or treatment with IL-1β failed to detect either a constitutive expression or an induction of B_1 receptors in sensory neurons [53]. Therefore, an indirect mechanism was proposed for the B_1 receptor-mediated hyperalgesia according to which activation of the induced B_1 receptors on non-neuronal cells leads to release of mediators, e.g., cytokines and prostaglandins from them, which in turn sensitize the adjacent nociceptors [67]. No constitutive B_1 receptor expression was detected in cultured rat or mouse sensory neurons using the gold-labeling technique, and northern blot analysis failed to detect B_1 receptor mRNA in these cells [68,69]. Similar results were obtained in the mouse [70]. In contrast, with the more sensitive reverse transcriptase–polymerase chain reaction, B_1 receptor mRNA was detected in dorsal root ganglion (DRG) neurons of the mouse and the rat [35,71], and using the gold-labeling technique, B_1 receptor induction was revealed in rat sensory neurons [72]. In a recent study, B_1 receptor mRNA was detected in the dorsal root ganglia in monkeys as well as in plantar skin and sciatic nerve in mice [73,74].

Moreover, constitutive B_1 receptor expression in both peptidergic and nonpeptidergic rat sensory neurons giving rise to C- and Aδ-fibers has been shown with immunocytochemistry which would allow a direct effect of B_1 receptor agonists on nociceptors [75–77]. In support of this, a low level of B_1 receptor mRNA expression and B_1 receptor agonist-induced translocation of PKC subtype ε was revealed in freshly isolated rat and mouse sensory neurons expressing the TRPV1 capsaicin receptor (see later) and the lectin IB_4, but not neuropeptides such as substance P (SP) or calcitonin gene-related peptide (CGRP) [78]. Both B_1 receptor expression and B_1 receptor-mediated PKCε translocation were strongly upregulated by glial cell-derived neurotrophic factor (GDNF), but not nerve growth factor (NGF), and after GDNF treatment, a B_1 receptor agonist induced a much more sustained facilitation of the heat-gated membrane current than that produced by B_2 receptor activation.

4. SIGNAL TRANSDUCTION MECHANISMS OF BRADYKININ RECEPTORS

4.1 Classical intracellular signaling mechanisms of bradykinin receptors

Both the B_1 and the B_2 receptors belong to the family of G-protein-coupled receptors containing seven transmembrane regions. Generally, they appear to utilize similar signal transduction mechanisms. As the B_2 receptors mediate predominantly the acute effects of bradykinin and kallidin, most of our knowledge concerning the biochemical background of the actions of the kinins refers to signaling mechanisms of the B_2 receptors. These data were obtained mostly in somata of cultured primary sensory neurons. The excitatory effect of bradykinin on cultured sensory neurons was shown to be mediated by a pertussis toxin-insensitive G-protein [26,50, 79–81]. This G-protein subsequently identified as a $G_{q/11}$ results in the activation of phospholipase C (PLC). This enzyme produces two second messengers: inositol-1,4,5-trisphosphate (IP_3) and diacylglycerol (DAG). Both of them have been shown to be increased in DRG neurons in response to bradykinin [45,50,82]. DAG mediates most of the excitatory effects of bradykinin by activating PKC which can phosphorylate various target proteins. The membrane depolarization underlying the sensory stimulant (spike-generating) effect of bradykinin is due to an inward current carried mainly by Na^+, which probably results from opening by PKC-mediated phosphorylation of an ion channel permeable to both Na^+ and K^+ [50,80]. The identity of this cation channel is still unclear (for the possible candidates see Section 4.2). The fundamental role of PKC in the excitatory effect of bradykinin is indicated by considerable evidence. Staurosporine, an inhibitor of protein kinases, attenuated the excitatory effect of bradykinin in the neonatal rat spinal cord–tail preparation *in vitro* [83], in cultured dorsal root or trigeminal ganglion cells [50,81], and in the canine testis–spermatic nerve preparation *in vitro* [84]. The involvement of PKC in bradykinin-induced prostaglandin E_2 (PGE_2) release from cultured rat trigeminal neurons was also revealed by using a selective inhibitor [42]. Downregulation of PKC inhibited bradykinin-induced release of SP and CGRP from cultured rat sensory neurons [85]. Phorbol esters that activate PKC mimicked most effects of bradykinin in the experimental systems mentioned above, reinforcing the involvement of PKC in neuronal excitation by bradykinin [26,50,85].

In addition to PKC activation, bradykinin causes an elevation of the intracellular-free Ca^{2+} concentration as well. This response is mainly due to release of Ca^{2+} from intracellular stores via IP_3 formation as it was still observed in the absence of extracellular Ca^{2+} and could be inhibited by depletion of intracellular Ca^{2+} stores [82,86–88]. The importance of intracellular Ca^{2+} release in the excitatory effect of bradykinin is shown by the fact that ryanodine which depletes intracellular Ca^{2+} stores strongly inhibited bradykinin's depolarizing action whereas blockers of the voltage-sensitive Ca^{2+} channels had no effect [26,81]. Bradykinin is, however, also able to cause an influx of extracellular Ca^{2+} through voltage-sensitive Ca^{2+} channels, which seems to be important for generation of cyclic guanosine monophosphate (cGMP) and the consequent tachyphylaxis to bradykinin (see below), rather than for excitation of the target neurons [50,89]. In a recent study on cultured rat sensory neurons, however, the bradykinin-induced elevation of the intracellular Ca^{2+} concentration was absent in a Ca^{2+}-free medium and also in the presence of the nonselective Ca^{2+} channel blocker Cd^{2+} pointing to an exclusive extracellular source of Ca^{2+} [90].

Bradykinin can also lead to release of arachidonic acid in cultured sensory neurons [50,82]. This response was shown to depend on influx of extracellular Ca^{2+} [91]. The mechanism of arachidonic acid release can be metabolism of DAG by DAG lipase to arachidonic acid and monoacylglycerol, the latter being converted by monoacylglycerol lipase to arachidonic acid and glycerol [91,92]. Alternatively, activation of phospholipase A_2 (PLA_2) probably through the G-protein is another mechanism resulting in arachidonic acid production in sensory neurons [42,91,93] similar to non-neuronal cells [94]. Arachidonic acid release may lead to production of prostaglandins by cyclooxygenase (COX) and subsequently other enzymes. The role of prostaglandins in the effects of bradykinin will be discussed later. In one study, a lipoxygenase (LOX) inhibitor reduced the depolarizing effect of bradykinin on sensory neurons while a COX inhibitor was ineffective [81]. Recent data suggest a major involvement of 12-LOX products in bradykinin signaling (see Section 4.2). The signaling mechanisms of B_1 receptors were less extensively studied, but they appear to involve similar pathways as revealed for B_2 receptors [21,95].

4.2 The role of TRP channels in bradykinin signaling

Recently, a novel signal transduction pathway has been outlined for bradykinin which involves the pharmacological receptor for capsaicin, the pungent agent of the red pepper, cloned in 1997 [96]. This receptor was initially called vanilloid receptor type 1 (VR1), reflecting the vanilloid structure of capsaicin but subsequently renamed to TRPV1 since it was the first identified member of the vanilloid subgroup of the so-called transient receptor potential (TRP) family of ion channels (for review see Refs [97–99]). TRPV1 is a nonselective cation channel located on polymodal nociceptive primary afferent neurons and it is also activated by other chemical stimuli including resiniferatoxin, low pH, and anandamide as well as by noxious heat (>43°C) stimuli [100]. Activation of the TRPV1 receptors results in membrane depolarization due to mainly Na^+ influx and release of neuropeptides (SP, CGRP, and somatostatin) as a result of Ca^{2+} influx through the channel. Higher concentrations of, and/or longer exposures to, capsaicin result in a decreased responsiveness of the sensory nerve terminals to both heat, chemical, and mechanical stimuli often referred to as capsaicin desensitization [101]. Capsazepine and iodo-resiniferatoxin (I-RTX) antagonize the effect of capsaicin at the TRPV1 receptor in a competitive manner, whereas ruthenium red is a functional antagonist blocking the ion channel part of the receptor.

Capsaicin has been shown to activate the TRPV1 receptor via an interaction with its cytosolic domains raising the possibility that putative endogenous TRPV1 receptor agonists also act intracellularly [102,103]. Certain LOX products, namely the 12- and 15-[S]-hydroperoxy-eicosatetraenoic acids (HPETEs), 5- and 15-[S]-hydroxy-eicosatetraenoic acids (HETEs) as well as leukotriene B_4, directly activated TRPV1 channels from the cytoplasmic side in both sensory neurons and TRPV1-transfected host cells in a capsazepine-sensitive manner [104]. Furthermore, arachidonic acid, albeit less, was also efficacious at TRPV1 receptors, and its effect was reduced by inhibitors of the 5- and 12-LOXs suggesting that arachidonic acid is converted to HPETEs and/or HETEs within sensory neurons. Considering the ability of bradykinin to release arachidonic acid in sensory neurons [50,82,91], a hypothesis was put forward that algogenic and sensitizing substances such as bradykinin may act on nociceptors by releasing arachidonic acid intracellularly which is converted to LOX products activating the TRPV1 receptor via an interaction with its cytosolic domains [104]. A direct support for this hypothesis was provided soon by a study showing that bradykinin could activate TRPV1 receptors via an intracellular pathway involving mobilization of arachidonic acid by PLA_2 and subsequent generation of LOX products [93]. In addition, bradykinin stimulated the production of the TRPV1 receptor agonist 12-HPETE in sensory neurons via the 12-LOX pathway. This study employing several methodological approaches revealed a major contribution of the PLA_2–12-LOX–TRPV1 pathway to the excitatory and heat-sensitizing effects of bradykinin. A further support for this hypothesis was provided by a study on guinea pig vagal afferents in which the action potential-generating effect of bradykinin was diminished by capsazepine or ruthenium red as well as by 5- or 12-LOX inhibition [105]. The effect of bradykinin, however, was not altered by PLA_2 inhibition similarly to a previous work [26] pointing to a PLA_2-independent liberation of arachidonic acid in vagal nociceptive nerve endings. A subsequent work from the same laboratory provided conflicting results in that bradykinin was able to induce action potentials in bronchopulmonary vagal afferents of TRPV1 knock-out mice, albeit the response was less persistent compared to wild-type animals [57]. In acutely isolated cardiac DRG neurons, bradykinin increased the firing frequency and lowered the threshold for action potential generation that was inhibited by the TRPV1 receptor antagonist I-RTX, a selective 12-LOX inhibitor, an IP_3 antagonist, and depletion of intracellular Ca^{2+} stores [88]. It is worth mentioning that bradykinin is able to stimulate 12-LOX and cause formation of 12-[S]-HETE in rat skin *in vivo* [106]. Furthermore, 12-LOX also occurs in platelets, and activated platelets excite nociceptors and induce hyperalgesia [107,108].

Parallel to the above mentioned studies, another bradykinin-activated intracellular pathway leading to TRPV1 receptor stimulation through PKC activation has been elucidated. Phorbol ester that activates PKC-induced TRPV1 channel activity in both TRPV1-transfected host cells and sensory neurons and bradykinin itself was able to activate TRPV1 receptors in a PKC-dependent manner [109]. Bradykinin was shown to lower the heat threshold of the TRPV1 receptor in host cells transfected with both TRPV1 and B_2 receptors through activation of PKCε [110]. Consistent with this, PKC activation by phorbol esters potentiated the gating of the TRPV1 receptor by capsaicin, heat, protons, and anandamide [111]. In another study, phorbol ester-induced PKC activation led to sensitization of the TRPV1 receptor to capsaicin, heat, and protons [112]. Adenosine 5′-triphosphate (ATP) increased the TRPV1 currents evoked by capsaicin or protons through activation of $P2Y_1$ purinergic receptors in a PKC-dependent manner and lowered the heat threshold of TRPV1 below normal body (core) temperature

[113]. Direct phosphorylation of the TRPV1 receptor by PKCε on exposure to the phorbol ester phorbol 12-myristate 13-acetate (PMA) was shown and two target serine residues were identified [114]. The TRPV1 receptor having point mutations introduced at these sites failed to be sensitized to either capsaicin or heat in response to PMA or ATP. Recently, using an *in vivo* phosphorylation method, bradykinin was shown to phosphorylate the TRPV1 receptor via activation of PKC [115].

The data mentioned above suggest that two signal transduction pathways of bradykinin, namely PKC activation and arachidonic acid mobilization, may finally converge on a common target, i.e., on TRPV1 receptors. In a recent study, evidence has been provided that the TRPV1 receptor is essential for the development of the various sensitizing effects of bradykinin [116]. Bradykinin failed to induce heat hyperalgesia in TRPV1 receptor knock-out mice, and using host cells expressing B_2 with or without TRPV1 receptors, bradykinin has been shown to potentiate the effect of low pH or capsaicin by modifying the gating of TRPV1 channels. Moreover, this study demonstrated in TRPV1-transfected cells that a reduction in the phosphatidylinositol-4,5-bisphosphate (PIP_2) level of the membrane by a specific antibody was able to activate the TRPV1 receptor, to mimic the sensitizing effects of bradykinin on acid- or capsaicin-induced responses, and to increase the heat responsiveness of TRPV1 receptors. This antibody also stimulated basal currents and augmented capsaicin-induced responses in DRG cells. These data suggest that endogenous PIP_2 functions as a physiological break on TRPV1 receptors, and agents activating PLC such as bradykinin may act by removing this break by hydrolysis of PIP_2 [116]. This would mean that PLC is involved in the intracellular signaling mechanisms in sensory neurons in a twofold way: by reducing the level of its substrate PIP_2 in the membrane and by forming its product, DAG which in turn activates PKC.

Several further studies support the involvement of TRPV1 receptors in various actions of bradykinin. Bradykinin-induced overt nociception in mice was shown to be mediated by the TRPV1 receptor and activation of PLC, PKC, PLA_2, and 5-LOX, providing substantial *in vivo* evidence for the recently discovered signaling mechanisms of bradykinin mentioned above [43]. I-RTX and ruthenium red inhibited the excitatory effect of bradykinin as well as ischemia on cardiac sympathetic afferents in the anesthetized ferret [117]. The spike-generating effect of bradykinin on jejunal afferents was inhibited by capsazepine [118]. In an *ex vivo* innervated trachea/bronchus preparation of the guinea pig, the bradykinin-induced action potential discharges in C-fibers was diminished by I-RTX [119]. The bradykinin-induced Ca^{2+} uptake in trigeminal sensory neurons was inhibited by ruthenium red and attenuated in TRPV1 knock-out mice compared to wild-type animals [120]. Sensitizing effects of bradykinin on responses evoked by the TRPV1 receptor agonist capsaicin were observed in various models of TRPV1 activation including the nocifensive reflex in the neonatal rat spinal cord–tail preparation [121], action potential firing in single vagal sensory C-fibers innervating the trachea [122], nociceptive behavior in the mouse [43], and Ca^{2+} concentration increase in rat primary sensory neurons [45]. In neonatal rat, DRG neurons bradykinin increased the number of neurons responding to capsaicin or low pH [123]. A sustained (3 h) pretreatment with bradykinin enhanced the SP release induced by capsaicin in rat DRG neurons [124,125]. Bradykinin had a sensitizing effect on capsaicin- or anandamide-evoked cobalt uptake that was inhibited by capsazepine in rat primary sensory neurons [126].

It should be remembered, however, that in several other studies the excitatory effect of bradykinin was not inhibited by TRPV1 receptor antagonists. Ruthenium red was ineffective against bradykinin evoking a depressor reflex in the isolated perfused, innervated rabbit ear [127]. The bradykinin-induced release of CGRP from guinea pig atria was unaffected by ruthenium red [128]. Similarly, ruthenium red failed to inhibit the bradykinin-evoked rise of Ca^{2+} in sensory neurons or activation of peripheral nociceptors in the neonatal rat spinal cord–tail preparation *in vitro* [129]. In the latter model, capsazepine was also ineffective against bradykinin [130]. I-RTX failed to inhibit the sympathoexcitatory reflex evoked by epicardial application of bradykinin in rats [131]. The spike-generating effect of bradykinin on jejunal afferents was not different in TRPV1 receptor knock-out and wild-type mice although the bradykinin effect was inhibited by capsazepine in wild-type animals [118]. In vagal sensory neurons of the guinea pig, the B_2 receptor activation-induced membrane depolarization was not attenuated by TRPV1 receptor antagonism [44]. Spinal neuron activation induced by intrapericardial administration of bradykinin was unaffected by a similar pretreatment with capsazepine; however, the effect of capsaicin was attenuated [132]. The bradykinin-induced increased gene transcription in cultured DRG neurons was not inhibited by capsazepine [87]. Furthermore, the reported major contribution of LOX products to bradykinin signaling is difficult to reconcile with previous studies concluding on a high degree of prostaglandin involvement in bradykinin's actions (see Section 5.1).

Recently, bradykinin was shown to inversely affect TRPV1 and TRPM8 (the cool-sensing, menthol-activated [133] channels in both heterologous expression system and DRG neurons) [134]. Bradykinin, in a PKC-dependent manner, enhanced the function of TRPV1 (in accord with previous studies, see above) and diminished the activity of TRPM8. Both phenomena appear to be mediated by alteration of channel phosphorylation state: TRPV1 is phosphorylated, TRPM8 is dephosphorylated by PKC activation (the latter action was proposed to be mediated by a PKC-activated phosphatase). The down-regulation of TRPM8 function might have pathophysiological relevance because TRPM8 activation by intraplantarly applied menthol was shown to reduce the nocifensive effect of coapplied capsaicin. Therefore, the bradykinin-induced downregulation of TRPM8 function may aggravate the pronociceptive action of the bradykinin-evoked enhanced TRPV1 activity. It is worth mentioning that a reciprocal modulation of TRPV1 (tonic inhibition, see above) and TRPM8 (activation) by PIP_2 was also shown [116,135]. In a very recent study, bradykinin inhibited the effect of cooling in cultured sensory neurons by reducing the amplitude of response to cooling and decreasing the threshold temperature for activation via stimulation of PKC [136].

Also recently, an involvement of TRPA1, the noxious cold-detecting ion channel [137], was revealed in the excitatory action of bradykinin [138]. TRPA1-expressing Chinese hamster ovary cells transiently transfected with the B_2 bradykinin receptor showed current responses when exposed to bradykinin, and this effect was blocked by PLC inhibition. The role of PLC was also shown by the fact that a membrane permeable analog of DAG was also effective. In addition, bradykinin increased Ca^{2+} levels in a great majority of DRG neurons activated by cinnamaldehyde, an established agonist of TRPA1 receptor. A study on TRPA1 knock-out mice provided further evidence for involvement of this channel in actions of bradykinin [120]. The bradykinin-induced Ca^{2+} uptake in trigeminal sensory neurons and the bradykinin-evoked heat hyperalgesia were attenuated and absent, respectively, in TRPA1 deficient mice compared to wild-type animals. As mentioned above, the former response was also diminished in trigeminal neurons from TRPV1-deficient mice and by ruthenium red. Complete Freund's adjuvant (CFA), however, induced similar degree of thermal hyperalgesia in TRPA1 knock-out and wild-type mice showing that TRPA1, unlike TRPV1, is not an absolute prerequisite for development of inflammatory thermal hyperalgesia. A model was proposed by the authors according to which TRPV1 acts upstream of TRPA1 in bradykinin's action. B_2 receptor activation causes PLC stimulation leading to PIP_2 hydrolysis, PKC activation, and IP_3-mediated release of Ca^{2+} from intracellular stores. The consequent modest TRPV1 activation (see above the mechanisms) also results in elevation of Ca^{2+} levels through entry from the extracellular space. The increase in intracellular Ca^{2+} then activates TRPA1 [139], which contributes to the bulk of the excitation. The model is supported by the coexpression of TRPV1 and TRPA1 [137].

4.3 Other recently identified mechanisms of bradykinin receptor signaling

The bradykinin-induced PGE_2 release from rat trigeminal sensory neurons was shown to involve activation of mitogen-activated protein kinase [42]. SP release evoked by a sustained (3 h) bradykinin application from rat DRG neurons was shown to involve activation of mitogen-activated protein kinase, extracellular signal-regulated kinase, and *de novo* COX-2 expression but not IP_3 receptors [125]. The bradykinin-induced overt nociception (hind paw licking/flinching) was shorter, mechanical hyperalgesia smaller, and the heat hyperalgesia lacking in $Na_v1.9$ gene-deficient mice compared to wild-type animals pointing to a role of this subtype of voltage-gated Na^+ channels in both excitatory and sensitizing actions of bradykinin [140]. Very recently, evidence has been provided that bradykinin can alter gene expression in cultured DRG neurons [87]. B_2 receptor activation-induced increase in intracellular Ca^{2+} caused nuclear translocation of the transcription factor nuclear factor of activated T-cells NFAT-4 with resultant increase in COX-2 mRNA level. In vagal sensory neurons, bradykinin was shown to inhibit the slow after-hyperpolarization (AHP) that is mediated by a Ca^{2+}-dependent K^+ current [141,142]. In a recent study on rat DRG neurons, bradykinin shortened the AHP by accelerating Ca^{2+} efflux via PKC-dependent facilitation of the plasma membrane pump isoform 4 [143]. The shortened AHP may be involved in the neuronal excitatory action of bradykinin because the AHP plays an important role in slowing of firing rate in sensory neurons [144]. Also in vagal sensory neurons of the guinea pig, the B_2 receptor activation-induced membrane depolarization was independent of TRPV1 but was due to a decrease in resting K^+ conductance and increase in Ca^{2+}-activated Cl^- conductance [44]. The latter mechanism, mediated by Ca^{2+}-activated Cl^- channels, can also contribute to membrane depolarization because primary afferent neurons have elevated intracellular Cl^- concentrations owing to the activity of the Na^+–K^+–Cl^- cotransporter [145,146]. In an *ex vivo* innervated trachea/bronchus preparation of the guinea pig, the bradykinin-induced action potential discharges in C-fibers was diminished by either I-RTX or the Ca^{2+}-activated Cl^- channel inhibitor niflumic acid [119]. The combination of both inhibitors abolished the bradykinin effect showing the parallel involvement of TRPV1 receptor and Ca^{2+}-activated Cl^- channels in the excitatory action of bradykinin in vagal afferents. In this study, γ-aminobutyric acid (GABA) $GABA_A$ receptor activation evoked action potentials in bradykinin-sensitive jugular ganglion neurons indicating that the reversal potential for Cl^- in these cells is in fact more positive than the action potential threshold owing to elevated intracellular Cl^- concentration. The hypothesis that Cl^- efflux is a major component of bradykinin-induced action potentials in vagal afferent C-fiber terminals is analogous to odorant receptor potentials in the dendrites of olfactory neurons [147].

Finally, an unorthodox way of activation of bradykinin receptors must be mentioned: the endogenous opioid peptide dynorphin A was reported to activate bradykinin receptors in hybridoma cells made of embryonic DRG neurons and mouse neuroblastoma cells [148]. In this system, dynorphin A_{2-13}, having no affinity for opioid receptors, induced an increase in intracellular Ca^{2+} that was abolished by the B_2 receptor antagonist icatibant (HOE 140). This effect involved activation of protein kinase A (PKA) and L as well as P/Q-type voltage-gated Ca^{2+} channels but not the typical signaling mechanism such as production of IP_3. Binding studies showed that dynorphin A_{2-13} binds to both B_1 and B_2 receptors with equal affinity that is lower than that of bradykinin.

4.4 Molecular mechanisms of the tachyphylaxis to bradykinin

The B_2 receptor-mediated excitatory actions of bradykinin, unlike those through B_1 receptor activation [21], showed a rapid tachyphylaxis or desensitization in several studies [50,82,149–152]. The signal transduction pathways involved in the tachyphylaxis of B_2 receptors are different from those responsible for the excitatory actions of bradykinin. Bradykinin was shown to elevate cGMP but not cyclic adenosine monophosphate (cAMP) levels in cultured sensory neurons [89]. This response was dependent on Ca^{2+} influx and was ascribed to activation of guanylyl cyclase. A role for cGMP increase in reducing the excitatory action of bradykinin was also proposed by this study. It is worth mentioning that the B_1 receptor agonist des-Arg^9-bradykinin failed to alter cGMP levels in sensory neurons [89]. Studied also in cultured sensory neurons it was shown that B_2 receptor stimulation induced Ca^{2+} influx which activated nitric oxide (NO) synthase; the resulting NO stimulated the soluble guanylyl cyclase, and the cGMP produced led to desensitization of the B_2 receptors at the level of the receptor or the G-protein [153,154]. The role of the NO–cGMP pathway in reducing bradykinin sensitivity was shown in the *in vitro* neonatal rat spinal cord–tail preparation as well [26,155]. In this model after development of desensitization to bradykinin, phorbol ester was still capable of activating PKC suggesting that tachyphylaxis took place up-stream of PKC activation [26]. Consistent with these findings, NO synthase-like immunoreactivity was shown in DRG neurons [156–158]. Although these studies have provided clear evidence for the role of the Ca^{2+}–NO–cGMP pathway in B_2 receptor desensitization, they have also shown that this is not the sole mechanism involved. In a further study in which the effect of the NO–cGMP axis on the bradykinin-induced IP_3 formation and on the number of bradykinin binding sites was simultaneously examined, repeated bradykinin applications led to a reduction in both parameters [159]. Activation of the NO–cGMP pathway also reduced bradykinin-induced IP_3 formation but failed to alter binding of bradykinin, and inhibition of NO synthesis prevented desensitization to bradykinin as assessed by measuring IP_3 formation. These results indicate that the NO–cGMP pathway is involved in the functional uncoupling of the B_2 receptors that occurs downstream of bradykinin binding and up-stream of IP_3 formation but it does not contribute to reduction of receptor density (receptor downregulation).

Concerning bradykinin receptor downregulation, a rapid internalization of the B_2 receptors dependent on phosphorylation processes has been described [21,160]. Bradykinin stimulation of heterologously expressed B_2 receptors induced colocalization of the agonist-activated receptor with β-arrestin into endosomes [161]. Following agonist removal, β-arrestin rapidly dissociates from the receptor in the endosomes and the receptors recycle to plasma membrane causing resensitization. It was shown that the C-terminal of the B_2 receptor is responsible for regulation of interaction with β-arrestin. Bradykinin is also capable of evoking heterologous desensitization, i.e., reduced responsiveness to other agents acting on nociceptors. A mutual cross-desensitization between bradykinin and neuropeptide Y was revealed in cultured primary sensory neurons [159]. In cultured neuronal hybrid cells, bradykinin pretreatment diminished responsiveness to carbachol and ATP, and this interaction was shown to be due to depletion or alteration of an intracellular Ca^{2+} store from which Ca^{2+} is mobilized by these agonists [162].

5. THE PRONOCICEPTIVE ACTIONS OF BRADYKININ

5.1 The neural excitatory action of bradykinin

The excitatory spike-generating effect of bradykinin on nociceptive primary sensory neurons has been studied in various experimental models including patch clamp examinations on cultured DRG neurons, single-unit recording from nociceptors *in vivo* and *in vitro*, behavioral studies in animals and humans as well as measuring stimulated neuropeptide release. There are two extensively used *in vitro* models which have provided a large amount of data about the features of bradykinin's effects on nociceptors: the rat skin–saphenous nerve preparation and the canine testis–spermatic nerve preparation, suitable for studying the cutaneous and visceral nociceptors, respectively [149,163]. Of the various types of nociceptive fibers, bradykinin acts primarily on the mechano-heat-sensitive or polymodal nociceptors including both the unmyelinated C- and thinly myelinated Aδ-units as studied in the rabbit, rat, or human skin [150,164,165], dog testis [166], guinea pig trachea and bronchus [58], rat oral cavity [167], temporomandibular joint [168,169], and periodontium [170] as well as on cat cardiac sympathetic [54,171,172], cat abdominal visceral [173], rat renal [174,175], rat vagal cardiac [176], and murine mesenteric afferents [177]. In the skin of the cat, rat, and monkey, about half of the C-fibers are excited by bradykinin [56,150,178]. In a study on the rabbit ear, practically all cutaneous C-polymodal nociceptors but not other C-afferents were activated by close arterial injection of low doses of bradykinin [164]. The incidence of bradykinin sensitivity is significantly higher in deep tissues: 100% in nodose, 84% in jugular ganglion vagal C-fibers innervating the pulmonary system of the guinea pig [179], 96% in the temporomandibular joint of the rat [168], 93% in dog testis [166], 92% in cat knee joint [180], 87% in dog muscle [181], 83% in cat cardiovascular system [171], 76% in rat ankle joint [182], 73% in cat abdominal organs [183], 67% of splanchnic colonic afferents in the cat [184], 66% of splanchnic, and 11% of pelvic colonic afferents in mice [46]. Bradykinin excited cardiac sympathetic afferents in cats responding to ischemia with the effect being higher in mechanically insensitive fibers compared to mechanically sensitive ones [172]. In the human forearm, bradykinin elicited pain when applied to the blister base or intradermally [22,151,165,185,186], and this pain was absent after topical capsaicin pretreatment [187]. In most studies, a strong tachyphylaxis to the excitatory effect of bradykinin was noted [50,82,149–152].

There exists plenty of experimental data demonstrating that the excitatory effect of bradykinin involves formation and action of COX products, i.e., prostaglandins. Although numerous studies revealed the involvement of COX products in various manifestations of the bradykinin-induced excitation of sensory neurons, only few of them identified the prostaglandin(s) involved. In most cases, the complexity of the system studied did not allow identification of the source(s) of prostaglandins which can be the sensory neurons and/or other, adjacent cells. The pronociceptive effect of bradykinin was inhibited by the COX inhibitor indomethacin in the isolated perfused, innervated rabbit ear [188]. In this model, bradykinin evoked PGE_1 release which was not reduced following chronic denervation suggesting that the bulk of prostaglandins originated from non-neuronal cells [189]. Acetylsalicylic acid (aspirin) was reported to diminish the excitatory effect of bradykinin in the *in vitro* dog testis–spermatic nerve preparation and its effect was largely restored by the exogenously applied PGE_2 [84]. In the neonatal rat spinal cord–tail preparation *in vitro*, activation of nociceptors by low concentrations of bradykinin was diminished by COX inhibition, whereas the effect of high concentrations remained unaltered [26,121]. The reflex response evoked by excitation of perivascular afferents by bradykinin in the dog was reduced by COX inhibition [190]. Bradykinin-induced activation of renal afferents in the rat was inhibited by a COX blocker [174]. Activation of cardiac ischemia-sensitive afferents by bradykinin was reduced by indomethacin [54]. Epicardial administration of bradykinin in dogs causes an increase in renal sympathetic nerve activity which is mediated mainly by NO. In dogs with heart failure, this effect of bradykinin is augmented and it predominantly depends on prostaglandin synthesis [191]. Activation of abdominal ischemia-sensitive visceral afferents in the cat by intraarterial bradykinin injection was inhibited by COX blockade demonstrating that prostaglandins were involved [173]. Excitation of serosal afferents of the rat jejunum by bradykinin was markedly reduced by COX blockade and fully restored by the exogenously applied PGE_2 [39]. Bradykinin-induced release of CGRP from sensory nerves in the guinea pig atria and in the isolated rat trachea was abolished by COX inhibition [52,128]. Bradykinin was shown to promote exocytotic release of noradrenaline from guinea pig heart synaptosomes which is, at least in part, mediated by SP and CGRP released from capsaicin-sensitive nerve endings and acting on sympathetic terminals. This response was inhibited by COX blockade pointing to a role of prostaglandins [192]. In cultured sensory neurons, the bradykinin-induced neuropeptide release was attenuated by indomethacin pretreatment [193]. The inhibitory effect of bradykinin on the slow AHP was abolished by pretreatment with a COX inhibitor in rabbit vagal sensory neurons [141]. Similarly, in acutely isolated nodose sensory neurons of the guinea pig, the slow AHP-reducing effect of bradykinin was prevented by inhibition of either COX or PGI_2 synthase; furthermore, bradykinin selectively enhanced PGI_2 release from these neurons [194]. Consistent with this, COX activity was revealed in somata of sensory neurons [193,195]. Subsequent studies have confirmed that bradykinin is able to release prostaglandins from sensory neurons. Bradykinin was shown to release PGE_2 from cultured rat trigeminal neurons [42]. A short-term (30 min) exposure to bradykinin led to a small PGE_2 release from cultured rat DRG neurons and it was reduced predominantly by COX-1 inhibition [196]. Conversely, a long-term (3 h) exposure to bradykinin induced a massive PGE_2 release that was abolished by indomethacin or a selective COX-2 inhibitor. COX-2 mRNA and protein was shown to be induced by sustained (3 h) B_2, but not B_1, receptor activation. SP release evoked by a sustained (3 h) bradykinin application from rat DRG neurons was reduced by either COX-1 or COX-2 inhibition, showing involvement of both isoenzymes in the effect [197]. This prolonged bradykinin effect was also shown to involve *de novo* COX-2 expression [125].

In other studies, however, a lack of involvement of prostaglandins was revealed. Bradykinin-evoked release of CGRP from the isolated guinea pig heart was not influenced by indomethacin [198]. The depolarizing action of bradykinin on cultured rat sensory neurons remained unaltered by indomethacin pretreatment but was inhibited by LOX blockade [81]. Sensory C-fibers of the vagus nerve in the guinea pig airways were also activated by bradykinin and this response was not sensitive to ibuprofen [51]. The spike-generating effect of bradykinin in the isolated guinea pig trachea–vagal nerve and in the rat skin–saphenous nerve preparation was unaffected by COX blockade [58,93,199]. The bradykinin-evoked overt nociception [200] was also unaffected by COX blockade in the mouse [43]. In vagal sensory neurons of the guinea pig, the B_2 receptor activation-induced membrane depolarization did not include formation of COX products [44].

5.2 The sensitizing action of bradykinin to heat stimuli

The features and molecular mechanisms of the heat-sensitizing effect of bradykinin have been studied in great detail and the results obtained suggest that probably this action of bradykinin is of primary importance with regard to inflammatory pain. The facilitatory effect of bradykinin on the heat sensitivity of nociceptors was revealed in various animal and human models. *In vitro*, first evidence for the heat-sensitizing effect of bradykinin was obtained in the isolated rat skin–saphenous nerve preparation that also provided the reverse finding that heat stimulation could markedly facilitate a subsequent bradykinin response even when tachyphylaxis had developed [150]. In the canine testis–spermatic nerve preparation, the peptide enhanced the noxious heat response (number of spikes evoked by heat stimulation) of the Aδ polymodal nociceptors [201]. This augmenting effect could be induced by 100 times lower concentrations than required for inducing discharges of nociceptors; it was short-lived out-lasting the bradykinin superfusion for no more than 10 min; it was suppressed by a B_2 receptor antagonist but appeared resistant against acetylsalicylic acid (aspirin) in very high concentration. Although there was a strong correlation between the magnitude of the heat-sensitizing and the spike-generating (excitatory) effects of bradykinin, a significant augmentation of the heat response could be observed also in the absence of bradykinin-induced nociceptor discharges.

Injection of bradykinin into the human skin induced heat hyperalgesia in addition to overt pain observed at higher dosage but no alteration of the nociceptive mechanical threshold as assessed by subjective ratings [151]. In this study, the algogenic effect exhibited a near-complete tachyphylaxis upon repeated applications, whereas the heat-sensitizing action, albeit reduced, remained still significant. In contrast, bradykinin applied iontophoretically into the human skin *in vivo* failed to decrease the heat threshold of mechano-heat-sensitive C-fibers [165]. Bradykinin caused a sensitization to heat but not to mechanical stimuli in identified nociceptors in the hairy skin of the monkey [56]. This action comprised a decrease in the heat threshold and an augmentation of the heat response. Interestingly, both B_1 and B_2 receptor-selective agonists had similar sensitizing effects. In a study conducted in the isolated rat skin–saphenous nerve preparation, bradykinin induced a heat sensitization of mechano-heat-sensitive polymodal C-fibers which was characterized by a drop of the heat threshold, an increase in the number of spikes evoked by the heat stimulus as well as a leftward shift and increased slope of the stimulus–response function [202]. Like in a previous study using the same model [150], no alteration of the von Frey threshold of C-fibers was observed following bradykinin application [202]. This study provided unequivocal evidence that, unlike bradykinin-induced mechanical hyperalgesia in rats (see below), the heat-sensitizing effect of the peptide does not involve activation of sympathetic postganglionic fibers as it was unaffected by sympathectomy. This finding was reinforced by two other studies investigating bradykinin-induced heat sensitization in humans and rats [30,203]. In the neonatal rat spinal cord–tail preparation *in vitro*, bradykinin also induced an augmentation of the heat response that was not reduced by inhibition of either COX or LOX enzymes suggesting a lack of involvement of eicosanoids [121]. In contrast, in a behavioral study employing the plantar test in the rat, heat sensitization induced by intraplantar bradykinin injection was strongly reduced by COX blockade indicating that prostaglandins contribute to the response [30]. In a recent study performed in the isolated rat skin–saphenous nerve preparation, the excitatory and heat-sensitizing actions of bradykinin were compared with regard to prevalence and susceptibility to tachyphylaxis [152]. A 5-min exposure to 10 μM bradykinin excited and sensitized to heat 40% of the mechano-heat-sensitive polymodal C-fibers, whereas in the remaining 60% of fibers, not excited by bradykinin, heat sensitization was observed in 3/4 of the units (altogether 85%). The heat-sensitizing effect could be repeated several times – by repeated bradykinin applications – without any decrease in the magnitude of the sensitization while the excitatory effect showed a profound tachyphylaxis. The high-threshold mechanosensitive C-fibers were not excited by bradykinin but 50% of them gained transient heat sensitivity following bradykinin treatment indicating a *de novo* recruitment of heat responsiveness by the compound.

A novel heat-activated ionic current mediated by a nonselective cation channel was discovered in the somata of primary sensory neurons that was sensitized by a short (20 s) exposure to bradykinin or the PKC activator phorbol ester PMA [204]. The sensitizing effect manifested itself as a drop of the heat threshold and an increase in the depolarizing cation conductance; it was prolonged by the phosphatase inhibitor calyculin A while the PMA effect was blocked by the nonselective protein kinase inhibitor staurosporine. These results provided evidence that the heat-sensitizing effect of bradykinin was due to a PKC-mediated phosphorylation in this model. In a subsequent study, the Ca^{2+}-independent PKCε was identified as the molecular entity responsible for this action [205]. The Ca^{2+}-independent nature of the heat-sensitizing effect of bradykinin was also evident from the lack of effect of BAPTA-AM, a membrane permeant Ca^{2+} buffer, on the heat sensitization of polymodal nociceptors in the rat skin–saphenous nerve preparation [206]. In a recent study, bradykinin massively lowered the threshold temperature of the noxious heat-sensitive TRPV1 channel in transfected cells and in DRG neurons in a PKC-dependent manner [110]. The crucial role of PKC activation (by phorbol esters) in the heat-sensitizing effect of bradykinin has previously been suggested by recordings from canine testicular nociceptors [207]. Noxious heat was also shown to release CGRP from the isolated rat skin, and this heat response was facilitated by bradykinin or PMA indicating PKC involvement [208]. A further confirmation of the role of PKC in heat sensitization was provided by two recent studies in which PKC activation by phorbol esters facilitated the heat-induced activation of the TRPV1 channel in transfected cells and DRG neurons [111,112]. In isolated desheathed axons of the mouse, bradykinin applied for 10 min enhanced the heat-evoked CGRP release, and this effect was abolished by PKC inhibition [209]. It is worth mentioning that PKCε is also involved in the long-term modulation of nociception as it mediates a chronic hypersensitivity for inflammatory nociceptor sensitization [210].

Although PKC activation is of crucial importance in the immediate heat sensitization induced by bradykinin, its role is not exclusive. In a study conducted in the isolated rat skin–saphenous nerve preparation, the heat sensitization evoked by a more sustained (5 min) application of bradykinin was abolished by the active, but not the inactive, enantiomer of the COX inhibitor flurbiprofen, and the effect of the active isomer was largely restored by the exogenously applied PGE_2 [199]. As it is widely accepted that the heat-sensitizing action of PGE_2 is predominantly mediated by the cAMP–PKA cascade [211–214], these results suggest that PKA activation may also be involved in the heat-sensitizing effect of bradykinin. It was proposed that the early phase of heat sensitization by bradykinin predominantly depends on PKC activation while in the sustained or after-effects, the COX products and the cAMP–PKA signaling mechanism gain increasing importance [199]. This hypothesis may explain why a major contribution of PKC activation was revealed in studies employing short bradykinin exposures, whereas a predominant role of COX products became evident after a prolonged bradykinin superfusion. It should be kept in mind, however, that a cross-talk between the PKC and cAMP–PKA pathways has been described in PC-12 cells following bradykinin exposure [215], and phorbol ester-induced activation of PKC led to activation of adenylyl cyclase in various cell types [216–218]. In a recent study employing also the isolated rat skin–saphenous nerve preparation, the heat-sensitizing effect of bradykinin on mechano-heat-sensitive polymodal C-fibers was reduced by either COX-1 or COX-2 inhibition showing involvement of both COX isoenzymes in the response [219]. Bradykinin-evoked PGE_2 release from the skin was also attenuated by both specific COX inhibitors. COX-1 immunoreactivity was present in nerve branches, nerve endings in the skin as well as in non-neuronal elements such as mast cells. COX-2 immunoreactivity was weaker but showed similar neuronal and extraneuronal localization [219]. The cutaneous source of prostaglandins sustaining the heat-sensitizing effect of bradykinin can be the nociceptors and/or various non-neuronal cell types in the vicinity of them as bradykinin receptors have been identified not only on nerve endings but also on endothelial, epidermal, and mast cells in the skin [7]. Although COX activity is present in the somata of sensory neurons [193,195] and bradykinin was shown to release PGE_2 from cultured rat trigeminal or DRG neurons [42,196], the peripheral COX products originate from non-neuronal cells because the bradykinin-induced PGE_2 release remained unaltered following chronic denervation of the rat skin [220].

The final step in the bradykinin-induced heat sensitization, not accompanied by mechanical hypersensitivity, is most probably a facilitation of the heat-sensitive ion channels located in the peripheral nerve endings and/or a recruitment of heat transducers hardly activated by thermal stimuli under normal conditions. Four types of ion channels sensitive to heat have been identified so far: TRPV1, VRL-1, VRL-3, and VRL-2 [96,221,222]. They are located on primary afferent neurons and belong to the vanilloid subgroup of the TRP family of ion channels and therefore are denoted also as TRPV1, TRPV2, TRPV3, and TRPV4 [97–99]. Recent evidence suggests that TRPV1 is a likely candidate for being a target for the heat-sensitizing effect of bradykinin as bradykinin lowered the heat threshold of TRPV1 both in TRPV1-transfected cells and in DRG neurons [110]. The role of TRPV1 is also supported by a finding that in TRPV1 receptor knock-out mice, intraplantar injection of bradykinin failed to evoke thermal hyperalgesia whereas it was effective in wild-type animals [116]. The heat-sensitizing effect of bradykinin and PGE_2 combination on sciatic nerve axons was absent in TRPV1 knock-out mice [209]. The role of TRPV2, TRPV3, and TRPV4 channels in the bradykinin-induced heat sensitization is not yet established.

5.3 The sensitizing action of bradykinin to mechanical stimuli

Bradykinin's sensitizing action to mechanical stimuli was extensively investigated in behavioral studies in which the mechanonociceptive threshold of the rat hindpaw was measured using the Randall–Selitto paw withdrawal method [223–225]. In this model, bradykinin applied intradermally was shown to induce a mechanical hyperalgesia which was absent following guanethidine pretreatment implicating that an intact noradrenergic neurotransmission from sympathetic postganglionic fibers is a prerequisite of the phenomenon [224]. As the bradykinin-induced decrease of the mechanonociceptive threshold had previously been shown to be inhibited by COX blockade [223], a hypothesis was put forward that bradykinin may exert its mechanical sensitizing effect indirectly by releasing from the sympathetic postganglionic fibers prostaglandins that eventually sensitize nociceptors to mechanical stimuli [224]. In accord with this, PGE_2 and PGI_2, but not PGD_2 or $PGF_{2\alpha}$, were shown to induce mechanical hyperalgesia and some sensitization in behavioral and cutaneous single-unit recording studies, respectively [223,226–230] and production of PGE_2 and PGI_2 by sympathetic postganglionic neurons was also reported [231]. In subsequent studies, the bradykinin-induced mechanical hyperalgesia was suggested to involve PLA_2-dependent synthesis of PGE_2 [227,232]. Later, bradykinin was shown to depolarize somata of postganglionic sympathetic neurons in both the rat and the mouse through activation of B_2 receptors [233,234]. With regard to receptor background of bradykinin-induced hyperalgesia, it was revealed that in the normal skin the response was mediated by B_2 receptors while in a sustained inflammatory state induced by CFA both B_2 and B_1 receptors were involved [235]. As reported by another group, the bradykinin-induced mechanical hyperalgesia was diminished by COX blockade both in the rat and the mouse [28,49]. Other *in vivo* models in which bradykinin's sensitizing action to mechanical stimuli became apparent include single-unit recordings from afferent fibers supplying the gastrocnemius-soleus muscle or the knee joint of the cat and the ankle joint of the rat [182,236–238]. Bradykinin enhanced the mechanical response of splanchnic but not pelvic colonic afferents in mice, *in vitro* [46]. Bradykinin lowered the mechanical threshold of cardiac vagal polymodal C- and Aδ-nociceptors [176].

The bradykinin-induced mechanical hyperalgesia was abolished by a local administration of tumor necrosis factor alpha (TNF-α) antiserum and partially inhibited by antisera against IL-1β, IL-6, or IL-8 [27,28]. The bradykinin effect was inhibited by B_2 and B_1 receptor antagonist [38]. The bradykinin effect was also inhibited by COX blockade or an adrenergic β_1 receptor antagonist. Interestingly, the B_1 receptor agonist-evoked response also involved both prostaglandins and sympathetic amines [38]. Evidence was also provided that TNF-α can activate both the IL6–IL-1β–prostaglandin and the IL-8–sympathetic pathways [239,240]. It turned out that rats do not produce IL-8 but synthesize cytokine-induced neutrophil chemoattractant 1 (CINC-1) that can produce mechanical hyperalgesia via release of sympathomimetic amines but not prostaglandins [241]. Carrageenan or lipopolysaccharide whose hyperalgesic action involves endogenous kinins were shown to induce CINC-1 formation. These data suggest that bradykinin-induced mechanical hyperalgesia involves first release of TNF-α which can initiate two pathways: IL-6, then IL-1β that produces prostaglandins as well as CINC-1 releasing sympathetic amines. In a very recent study in mice, evidence has been provided that B_2 and B_1 receptors can induce mechanical hyperalgesia via different mechanisms [49]. In naïve mice, bradykinin evoked hyperalgesia through the B_2 receptor which triggered production/release of prostaglandins and sympathomimetic amines but not cytokines. Conversely, in LPS-treated mice, B_1 receptors mediated the hyperalgesic effect of bradykinin depending on synthesis of TNF-α and IL-1β which then induce production/release of prostaglandins and sympathomimetic amines.

In other models, however, bradykinin failed to sensitize nociceptors to mechanical stimuli. In the isolated rat skin–saphenous nerve preparation, bradykinin – even co-applied with histamine, serotonin, and PGE_2 – was unable to produce a mechanical nociceptor sensitization [150,208]. In humans intradermal injection of bradykinin evoked sensitization to heat but not to mechanical stimuli [151,165]. Also, no development of muscular mechanical hyperalgesia was found in humans by intramuscular injection of bradykinin alone although the combination with serotonin was reported effective [242–244]. The site of bradykinin application was reported to determine whether mechanical hyperalgesia develops or not as intradermal, but not subcutaneous, injection of bradykinin produced mechanical hyperalgesia in the rat [235].

5.4 The sensitizing action of bradykinin to chemical stimuli

The sensitizing effect of bradykinin to other chemical stimuli has been studied far less extensively as compared to its action on heat and mechanical responsiveness of nociceptors. This difference refers to both the number of studies conducted and identification of the mechanisms involved. Several studies have revealed an enhancing action of bradykinin on capsaicin-evoked responses. These have already been mentioned shortly (see Section 4.2.) as data supporting the involvement of the TRPV1 receptor in the signaling mechanisms of bradykinin. Here a more detailed description including molecular mechanisms is provided. In host cells transfected with both B_2 and TRPV1 receptors, bradykinin was shown to potentiate the effect of capsaicin or low pH by modifying TRPV1 channel gating [116]. Low concentrations of bradykinin, lacking an excitatory action of their own, enhanced the effect of the TRPV1 receptor agonist capsaicin on the nocifensive reflex in the neonatal rat spinal cord–tail preparation [121]. This interaction was unaffected by inhibitors of either the COX or LOX arguing against a role of eicosanoids. Firing responses to capsaicin of single vagal sensory C-fibers innervating the trachea were increased following application of bradykinin to their receptive fields at concentrations that failed to cause discharges on their own [122]. Bradykinin was shown to increase the proportion of neonatal rat DRG neurons responding to capsaicin and to protons [123]. This response was unaffected by inhibition of COX or LOX. Bradykinin, at a concentration at which it was ineffective by itself, enhanced the increase in intracellular Ca^{2+} concentration evoked by the TRPV1 agonist capsaicin in rat DRG neurons [45]. This facilitatory action was abolished by a B_2 receptor antagonist and by inhibition of PLC or COX-1 but not COX-2. Interestingly, the bradykinin-induced elevation in IP_3 levels was also inhibited by COX-1 blockade suggesting that COX products are involved in the intracellular liberation of IP_3. A sustained (3 h) pretreatment with bradykinin enhanced the SP release induced by capsaicin in rat DRG neurons, and this facilitatory effect was shown to depend on PKA activation [124]. As COX-2 expression was also induced by bradykinin, its potentiating effect was probably involving prostaglandin formation with subsequent PKA-mediated sensitization of TRPV1 possibly by direct phosphorylation. This response was shown to involve IP_3-dependent Ca^{2+} release as well [125]. Bradykinin had a sensitizing effect on capsaicin- or anandamide-evoked cobalt uptake that was inhibited by capsazepine in rat primary sensory neurons [126]. Capsazepine also reduced the facilitatory effect of bradykinin on the action of anandamide. In another study, bradykinin, at a dose in which it did not evoke nociception by itself, enabled a subthreshold dose of capsaicin or low pH to induce nociceptive behavior in the mouse [43].

Bradykinin sensitized rat cutaneous nociceptors to PGE_1 [245]. In cultured rat sensory neurons, bradykinin alone or more effectively in combination with histamine and PGE_2 resulted in a significant enhancement of the sustained ionic current evoked by low pH [246]. In the isolated rat skin–saphenous nerve preparation following bradykinin superfusion, the histamine-induced discharge activity was enhanced and previously unresponsive C-fibers were excited by histamine [247,248]. Consequently, following bradykinin pretreatment of human skin, histamine induced burning pain, whereas in normal skin it causes a pure itch sensation when applied iontophoretically [247]. Bradykinin also facilitated the histamine-induced release of CGRP from afferent neurons in the guinea pig lung [249]. This effect was diminished by indomethacin pretreatment indicating a partial mediation by COX products. Bradykinin augmented the spike-generating action of histamine in ischemically sensitive cardiac afferents in the cat, and this effect was abolished by COX blockade [250]. Conversely, histamine reduced the response of the afferents to bradykinin. In CFA-induced inflammation in Lewis rats, bradykinin augmented the noradrenaline-evoked spike generation in cutaneous nociceptors as studied in the skin–saphenous nerve preparation *in vitro* [251]. In naïve Sprague–Dawley rats, bradykinin pretreatment enabled noradrenaline to evoke action potentials in cutaneous C-fibers [252]. Bradykinin had an enhancing effect on serotonin-evoked inward current in trigeminal sensory neurons through activation of PKC [47], and vice versa, serotonin pretreatment facilitated the subsequent excitatory response to bradykinin of rat polymodal C-fibers *in vitro* [150]. In adrenal chromaffine and PC-12 cells, B_2 receptor activation can phosphorylate src and the receptor for epidermal growth factor that leads to the facilitation of the Ca^{2+} signal caused by this factor providing evidence that a crosstalk between G-protein-coupled bradykinin receptors and tyrosine kinase-linked growth factor receptors may occur [253].

5.5 The role of NO in the excitatory and sensitizing actions of bradykinin

NO appears to be involved not only in B_2 receptor desensitization but also in some excitatory and sensitizing effects of bradykinin. Bradykinin was shown to evoke pain in humans on application into a vascularly isolated vein segment and this effect was reduced by inhibition of NO synthase or guanylyl cyclase, pointing to an involvement of NO and cGMP in the algogenic action of bradykinin in this model [186,254,255]. Consistent with this, NO solutions also evoked pain in this model [256]. Furthermore, intracutaneous administration of NO evoked pain in humans [257]. The increase in renal sympathetic nerve activity induced by epicardial bradykinin was reduced by NO synthase inhibition in dogs [191]. Excitation of perivascular afferents by bradykinin evoked a complex reflex response in the dog that was reduced by inhibitors of either NO synthase or guanylyl cyclase [190]. The mechanical hyperalgesia induced by intradermal injection of bradykinin in the rat paw was diminished by inhibition of either the NO synthase or guanylyl cyclase or the cGMP-dependent protein kinase G (PKG) indicating that activation of the NO–cGMP–PKG axis was involved in this sensitizing effect of bradykinin [258]. The hyperalgesic effect of bradykinin was also reduced by inhibition of PKA in this model. Although these studies clearly show that NO can be an important mediator of both the algogenic and sensitizing effects of bradykinin, at least in certain models, they do not provide information about the source of NO, which might be the bradykinin-responsive nociceptor or other, adjacent cells. Also suggested by these studies is that NO acts as a pronociceptive agent in the periphery.

In contrast, the mechanical hyperalgesia induced by intraplantar (subcutaneous) injection of bradykinin was potentiated by local pretreatment with an inhibitor of guanylyl cyclase suggesting that the NO–cGMP axis has an inhibitory effect on the hyperalgesic effect of bradykinin [259]. Similarly, in arthritic but not in normal rats, bradykinin-induced discharge activity of articular nociceptors was enhanced by NO synthase inhibition suggesting that under inflammatory conditions NO can reduce the excitatory effect of bradykinin [260]. These data point to a peripheral antinociceptive role of the NO–cGMP pathway. There are numerous studies supporting either a pro- or an antinociceptive role for the NO–cGMP axis in the periphery, independent of the bradykinin action. These data, however, are not covered in this chapter.

5.6 The role of endogenous bradykinin in inflammatory pain states

Findings from animal and, to a much lesser degree, human studies support the view that endogenous bradykinin is involved in the induction and maintenance of inflammatory pain. The concentration of bradykinin is elevated, and its elimination is delayed, in inflammatory exudates. Plenty of neuronal excitatory and sensitizing effects to heat, mechanical, and chemical stimuli of exogenously applied bradykinin have been described. The heat-sensitizing action of the peptide does not appear to exhibit any notable tachyphylaxis. Not only bradykinin can sensitize nociceptors to other inflammatory mediators but the reverse interaction can augment the actions of the peptide, thereby establishing a self-reinforcing cycle. It should be emphasized that experiments employing bradykinin receptor antagonists as well as bradykinin receptor knock-out animals can provide the most reliable information with regard to the pathophysiological functions of endogenous bradykinin. The nondesensitizing B_1 receptors are upregulated under inflammatory conditions and in experimental animal models of inflammatory pain B_1 and/or B_2 bradykinin receptor antagonists exert antihyperalgesic actions and animals lacking B_1 and/or B_2 bradykinin receptor genes exhibit diminished nociceptive responses. Furthermore, activation of cardiac afferents by ischemia was shown to involve endogenous bradykinin raising the possibility that the peptide contributes to pain associated with angina pectoris [54]. The writhing reaction in mice induced by i.p. injection of kaolin was inhibited by pretreatment with bromelain, a depletor of prekallikrein and high molecular weight kininogen [261]. Similarly, the acid-induced writhing was suppressed by B_2 receptor antagonists [262]. During sustained isometric trapezius muscle contraction in man, a correlation was found between tissue levels of kinins (bradykinin plus kallidin) and pain ratings [263]. Surprisingly, antagonism of either the B_1 or B_2 bradykinin receptor failed to alter the mechanical and heat hyperalgesia evoked by surgical incision [264]. In PMA-induced overt nociception in mice, activation of B_2 receptors was revealed [74]. In colonic hypersensitivity to intraluminally applied capsaicin induced by proteinase-activated receptor type 2 stimulation in mice, activation of the B_2 receptor was revealed suggesting that formation of bradykinin is involved in the response [265].

To what extent, if at all, endogenous bradykinin contributes to inflammatory or ischemic pain in humans remains to be shown by clinical studies. According to clinical trials, icatibant (HOE 140) does not appear to be efficacious in acute postoperative pain. Considering the induction of B_1 receptors under inflammatory conditions, it also seems reasonable to examine the clinical efficacy of B_1 receptor antagonists alone and in combination with B_2 receptor blockers (for review see Refs [266–268]).

ACKNOWLEDGEMENTS

This work was supported by the Alexander von Humboldt Foundation.

REFERENCES

1. Edery H, Lewis GP. Inhibition of plasma kininase activity at slightly acidic pH. Br J Pharmacol 1962;19:299–305.
2. Hargreaves KM, Troullos ES, Dionne RA, Schmidt EA, Schafer SC, Joris JL. Bradykinin is increased during acute and chronic inflammation: therapeutic implications. Clin Pharmacol Ther 1988;44:613–621.
3. Jeske NA, Berg KA, Cousins JC, Ferro ES, Clarke WP, Glucksman MJ et al. Modulation of bradykinin signaling by EP24.15 and EP24.16 in cultured trigeminal ganglia. J Neurochem 2006;97:13–21.
4. McEachern AE, Shelton ER, Bhakta S, Obernolte R, Bach C, Zuppan P et al. Expression cloning of a rat B_2 bradykinin receptor. Proc Natl Acad Sci U S A 1991;88:7724–7728.
5. Hess JF, Borkowski JA, Young GS, Strader CD, Ransom RW. Cloning and pharmacological characterization of a human bradykinin [BK_2] receptor. Biochem Biophys Res Commun 1992;184:260–268.
6. Menke JG, Borkowski JA, Bierilo KK, MacNeil T, Derrick AW, Schneck KA et al. Expression cloning of a human B_1 bradykinin receptor. J Biol Chem 1994;269:21583–21586.
7. Hall JM. Bradykinin receptors. Gen Pharmacol 1997;28:1–6.
8. Marceau F, Bachvarov DR. Kinin receptors. Clin Rev Allergy Immunol 1998;16:385–401.
9. Regoli DC, Marceau F, Lavigne J. Induction of beta 1-receptors for kinins in the rabbit by a bacterial lipopolysaccharide. Eur J Pharmacol 1981;71:105–115.

10. Marceau F, Hess JF, Bachvarov DR. The B_1 receptors for kinins. Pharm Rev 1998;50:357–386.
11. Ahluwalia A, Perretti M. B_1 receptors as a new inflammatory target. Could this B the 1? Trends Pharmacol Sci 1999;20:101–104.
12. DeBlois D, Bouthillier J, Marceau F. Pulse exposure to protein synthesis inhibitors enhances vascular responses to des-Arg^9-bradykinin: possible role of interleukin-1. Br J Pharmacol 1991;103:1057–1066.
13. Galizzi JP, Bodinier MC, Chapelain B, Ly SM, Coussy L, Giraud S et al. Up-regulation of [^{3}H]-des-Arg^{10}-kallidin binding to the bradykinin B_1 receptor by interlukin-1β in isolated smooth muscle cells: correlation with B_1 agonist-induced PGI_2 production. Br J Pharmacol 1994;113:389–394.
14. Perkins MN, Kelly D. Interleukin-1 beta induced-des-Arg^9-bradykinin-mediated thermal hyperalgesia in the rat. Neuropharmacology 1994;33:657–660.
15. Davis AJ, Perkins MN. The involvement of bradykinin B_1 and B_2 receptor mechanisms in cytokine-induced mechanical hyperalgesia in the rat. Br J Pharmacol 1994;113:63–68.
16. Campos MM, Souza GE, Calixto JB. In vivo B_1 kinin-receptor upregulation. Evidence for involvement of protein kinases and nuclear factor kappaB pathways. Br J Pharmacol 1999;127:1851–1859.
17. Campos MM, Mata LV, Calixto JB. Expression of B_1 kinin receptors mediating paw edema and formalin-induced nociception. Modulation by glucocorticoids. Can J Physiol Pharmacol 1995;73:812–819.
18. Campos MM, Souza GE, Calixto JB. Upregulation of B_1 receptor mediating des-Arg^9-BK-induced rat paw oedema by systemic treatment with bacterial endotoxin. Br J Pharmacol 1996;117:793–798.
19. Cabrini DA, Campos MM, Tratsk KS, Merino VF, Silva JA, Jr., Souza GE et al. Molecular and pharmacological evidence for modulation of kinin B_1 receptor expression by endogenous glucocorticoid hormones in rats. Br J Pharmacol 2001;132:567–577.
20. Mathis SA, Criscimagna NL, Leeb-Lundberg LM. B_1 and B_2 kinin receptors mediate distinct patterns of intracellular Ca^{2+} signaling in single cultured vascular smooth muscle cells. Mol Pharmacol 1996;50:128–139.
21. Austin CE, Faussner A, Robinson HE, Chakravarty S, Kyle DJ, Bathon JM et al. Stable expression of the human kinin B_1 receptor in Chinese hamster ovary cells. Characterization of ligand binding and effector pathways. J Biol Chem 1997;272:11420–11425.
22. Whalley ET, Clegg S, Stewart JM, Vavrek RJ. The effect of kinin agonists and antagonists on the pain response of the human blister base. Naunyn-Schmiedebergs Arch Pharmacol 1987;336:652–655.
23. Haley JE, Dickenson AH, Schachter M. Electrophysiological evidence for a role of bradykinin in chemical nociception in the rat. Neurosci Lett 1989;97:198–202.
24. Mizumura K, Minagawa M, Tsujii Y, Kumazawa T. The effects of bradykinin agonists and antagonists on visceral polymodal receptor activities. Pain 1990;40:221–227.
25. Holzer-Petsche U. Blood pressure and gastric motor responses to bradykinin and hydrochloric acid injected into somatic or visceral tissues. Naunyn-Schmiedeberg's Arch Pharmacol 1992;346:219–225.
26. Dray A, Patel IA, Perkins MN, Rueff A. Bradykinin-induced activation of nociceptors: receptor and mechanistic studies on the neonatal rat spinal cord-tail preparation in vitro. Br J Pharmacol 1992;107:1129–1134.

27. Ferreira SH, Lorenzetti BB, Cunha FQ, Poole S. Bradykinin release of TNF-alpha plays a key role in the development of inflammatory hyperalgesia. Agents Actions 1993;38:Spec No:C7–C9.
28. Ferreira SH, Lorenzetti BB, Poole S. Bradykinin initiates cytokine-mediated inflammatory hyperalgesia. Br J Pharmacol 1993;110:1227–1231.
29. Heapy CG, Shaw JS, Farmer SC. Differential sensitivity of antinociceptive assays to the bradykinin antagonist Hoe 140. Br J Pharmacol 1993;108:209–213.
30. Schuligoi R, Donnerer J, Amann R. Bradykinin-induced sensitization of afferent neurons in the rat paw. Neuroscience 1994;59:211–215.
31. Pan HL, Stahl GL, Rendig SV, Carretero OA, Longhurst JC. Endogenous BK stimulates ischemically sensitive abdominal visceral C fiber afferents through kinin B_2 receptors. Am J Physiol 1994;267:H2398–H2406.
32. Khasar SG, Miao FJP, Levine JD. Inflammation modulates the contribution of receptor-subtypes to bradykinin-induced hyperalgesia in the rat. Neuroscience 1995;69:685–690.
33. De Campos RO, Alves RV, Kyle DJ, Chakravarty S, Mavunkel BJ, Calixto JB. Antioedematogenic and antinociceptive actions of NPC 18521, a novel bradykinin B_2 receptor antagonist. Eur J Pharmacol 1996;316:277–286.
34. Rupniak NMJ, Boyce S, Webb JK, Williams AR, Carlson EJ, Hill RG et al. Effects of the bradykinin B_1 receptor antagonist des-Arg^9[Leu^8] bradykinin and genetic disruption of the B_2 receptor on nociception in rats and mice. Pain 1997;71:89–97.
35. Seabrook GR, Bowery BJ, Heavens R, Brown N, Ford H, Sirinathsinghi DJS et al. Expression of B_1 and B_2 bradykinin receptor mRNA and their functional roles in sympathetic ganglia and sensory dorsal root ganglia neurones from wild-type and B_2 receptor knockout mice. Neuropharmacology 1997;36:1009–1017.
36. Griesbacher T, Amann R, Sametz W, Diethart S, Juan H. The nonpeptide B_2 receptor antagonist FR173657: inhibition of effects of bradykinin related to its role in nociception. Br J Pharmacol 1998;124:1328–1334.
37. De Campos RO, Alves RV, Ferreira J, Kyle DJ, Chakravarty S, Mavunkel BJ et al. Oral antinociception and oedema inhibition produced by NPC 18884, a non-peptidic bradykinin B_2 receptor antagonist. Naunyn-Schmiedeberg's Arch Pharmacol 1999;360:278–286.
38. Poole S, Lorenzetti BB, Cunha JM, Cunha FQ, Ferreira SH. Bradykinin B1 and B2 receptors, tumour necrosis factor alpha and inflammatory hyperalgesia. Br J Pharmacol 1999;126:649–656.
39. Maubach KA, Grundy D. The role of prostaglandins in the bradykinin-induced activation of serosal afferents of the rat jejunum in vitro. J Physiol 1999;515:277–285.
40. Burgess GM, Perkins MN, Rang HP, Campbell EA, Brown MC, McIntyre P et al. Bradyzide, a potent non-peptide B_2 bradykinin receptor antagonist with long-lasting oral activity in animal models of inflammatory hyperalgesia. Br J Pharmacol 2000;129:77–86.
41. Kollarik M, Dinh QT, Fischer A, Undem BJ. Capsaicin-sensitive and -insensitive vagal bronchopulmonary C-fibres in the mouse. J Physiol 2003;551:869–879.
42. Jenkins DW, Sellers LA, Feniuk W, Humphrey PP. Characterization of bradykinin-induced prostaglandin E2 release from cultured rat trigeminal ganglion neurones. Eur J Pharmacol 2003;469:29–36.
43. Ferreira J, da Silva GL, Calixto JB. Contribution of vanilloid receptors to the overt nociception induced by B2 kinin receptor activation in mice. Br J Pharmacol 2004;141:787–794.

44. Oh EJ, Weinreich D. Bradykinin decreases K(+) and increases Cl(−) conductances in vagal afferent neurones of the guinea pig. J Physiol 2004;558:513–526.
45. Tang HB, Inoue A, Oshita K, Nakata Y. Sensitization of vanilloid receptor 1 induced by bradykinin via the activation of second messenger signaling cascades in rat primary afferent neurons. Eur J Pharmacol 2004;498:37–43.
46. Brierley SM, Jones RC 3rd, Xu L, Gebhart GF, Blackshaw LA. Activation of splanchnic and pelvic colonic afferents by bradykinin in mice. Neurogastroenterol Motil 2005;17:854–862.
47. Hu WP, Li XM, Wu JL, Zheng M, Li ZW. Bradykinin potentiates 5-HT3 receptor-mediated current in rat trigeminal ganglion neurons. Acta Pharmacol Sin 2005;26:428–434.
48. Soukhova-O'Hare GK, Zhang JW, Gozal D, Yu J. Bradykinin B2 receptors mediate pulmonary sympathetic afferents induced reflexes in rabbits. Life Sci 2006;78:1990–1997.
49. Cunha TM, Verri WA Jr., Fukada SY, Guerrero AT, Santodomingo-Garzón T, Poole S et al. TNF-alpha and IL-1beta mediate inflammatory hypernociception in mice triggered by B1 but not B2 kinin receptor. Eur J Pharmacol 2007;573:221–229.
50. Burgess GM, Mullaney I, McNeill M, Dunn PM, Rang HP. Second messengers involved in the mechanism of action of bradykinin in sensory neurons in culture. J Neurosci 1989;9:3314–3325.
51. Fox AJ, Barnes PJ, Urban L, Dray A. An in vitro study of the properties of single vagal afferents innervating guinea-pig airways. J Physiol 1993;469:21–35.
52. Hua XY, Yaksh TL. Pharmacology of the effects of bradykinin, serotonin and histamine on the release of calcitonin gene-related peptide from C-fiber terminals in the rat trachea. J Neurosci 1993;13:1947–1953.
53. Davis CL, Naeem S, Phagoo SB, Campbell EA, Urban L, Burgess GM. B_1 bradykinin receptors and sensory neurones. Br J Pharmacol 1996;118:1469–1476.
54. Tjen-A-Looi SC, Pan HL, Longhurst JC. Endogenous bradykinin activates ischaemically sensitive cardiac visceral afferents through kinin B_2 receptors in cats. J Physiol 1998;510:633–641.
55. Fox A, Wotherspoon G, McNair K, Hudson L, Patel S, Gentry C et al. Regulation and function of spinal and peripheral neuronal B_1 bradykinin receptors in inflammatory mechanical hyperalgesia. Pain 2003;104:683–691.
56. Khan AA, Raja SN, Manning DC, Campbell JN, Meyer RA. The effects of bradykinin and sequence-related analogs on the response properties of cutaneous nociceptors in monkeys. Somatosens Mot Res 1992;9:97–106.
57. Kollarik M, Undem BJ. Activation of bronchopulmonary vagal afferent nerves with bradykinin, acid and vanilloid receptor agonists in wild-type and TRPV1-/- mice. J Physiol 2004;555:115–123.
58. Kajekar R, Proud D, Myers AC, Meeker SN, Undem BJ. Characterization of vagal afferent subtypes stimulated by bradykinin in guinea pig trachea. J Pharmacol Exp Ther 1999;289:682–687.
59. Eisenbarth H, Rukwied R, Petersen M, Schmelz M. Sensitization to bradykinin B_1 and B_2 receptor activation in UV-B irradiated human skin. Pain 2004;110:197–204.
60. Shibata M, Ohkubo T, Takahashi H, Inoki R. Modified formalin test: characteristic biphasic pain response. Pain 1989;38:347–352.
61. Correa CR, Calixto JB. Evidence for participation of B_1 and B_2 kinin receptors in formalin-induced nociceptive response in the mouse. Br J Pharmacol 1993;110:193–198.

62. Sufka KJ, Roach JT. Stimulus properties and antinociceptive effects of selective bradykinin B_1 and B_2 receptor antagonists in rats. Pain 1996;66:99–103.
63. Pesquero JB, Araujo RC, Heppenstall PA, Stucky CL, Silva JA Jr., Walther T et al. Hypoalgesia and altered inflammatory responses in mice lacking kinin B_1 receptors. Proc Natl Acad Sci U S A 2000;97:8140–8145.
64. Gougat J, Ferrari B, Sarran L, Planchenault C, Poncelet M, Maruani J et al. SSR240612 [(2R)-2-[((3R)-3-(1,3-benzodioxol-5-yl)-3-[[(6-methoxy-2-naphthyl)sulfonyl]amino]propanoyl)amino]-3-(4-[[2R,6S)-2,6-dimethylpiperidinyl]methyl]phenyl)-N-isopropyl-N-methylpropanamide hydrochloride], a new nonpeptide antagonist of the bradykinin B1 receptor: biochemical and pharmacological characterization. J Pharmacol Exp Ther 2004;309:661–669.
65. Porreca F, Vanderah TW, Guo W, Barth M, Dodey P, Peyrou V et al. Antinociceptive pharmacology of N-[[4-(4,5-dihydro-1H-imidazol-2-yl)phenyl]methyl]-2-[2-[[(4-methoxy-2,6-dimethylphenyl) sulfonyl]methylamino]ethoxy]-N-methylacetamide, fumarate (LF22-0542), a novel nonpeptidic bradykinin B1 receptor antagonist. J Pharmacol Exp Ther 2006;318:195–205.
66. Chichorro JG, Lorenzetti BB, Zampronio AR. Involvement of bradykinin, cytokines, sympathetic amines and prostaglandins in formalin-induced orofacial nociception in rats. Br J Pharmacol 2004;141:1175–1184.
67. Perkins MN. Interactions between kinins and the inflammatory pain process. In: Brain SD, Moore PK, eds. Pain and neurogenic inflammation. Basel: Birkhäuser Verlag, 1999;103–114.
68. Petersen M, Eckert AS, Segond von Banchet G, Heppelmann B, Klusch A, Kniffki KD. Plasticity in the expression of bradykinin binding sites in sensory neurons after mechanical nerve injury. Neuroscience 1998;83:949–959.
69. Brand M, Klusch A, Kurzai O, Valdeolmillos M, Schmidt RF, Petersen M. No evidence for bradykinin B_1 receptors in rat dorsal root ganglion neurons. NeuroReport 2001;12:3165–3168.
70. Rashid MH, Inoue M, Matsumoto M, Ueda H. Switching of bradykinin-mediated nociception following partial sciatic nerve injury in mice. J Pharmacol Exp Ther. 2004;308:1158–1164.
71. Levy D, Zochodne DW. Increased mRNA expression of the B_1 and B_2 bradykinin receptors and antinociceptive effects of their antagonists in an animal model of neuropathic pain. Pain 2000;86:265–271.
72. von Banchet G, Petersen M, Heppelmann B. Bradykinin receptors in cultured rat dorsal root ganglion cells. Influence of length of time in culture. Neuroscience 1996;75:1211–1218.
73. Shughrue PJ, Ky B, Austin CP. Localization of B1 bradykinin receptor mRNA in the primate brain and spinal cord: an in situ hybridization study. J Comp Neurol 2003;465:372–384.
74. Ferreira J, Trichês KM, Medeiros R, Calixto JB. Mechanisms involved in the nociception produced by peripheral protein kinase c activation in mice. Pain 2005;117:171–181.
75. Wotherspoon G, Winter J. Bradykinin B_1 receptor is constitutively expressed in the rat sensory nervous system. Neurosci Lett 2000;294:175–178.
76. Ma QP, Hill R, Sirinathsinghji D. Basal expression of bradykinin B_1 receptor in the peripheral sensory ganglia in the rat. NeuroReport 2000;11:4003–4005.

77. Ma QP. The expression of bradykinin B(1) receptors on primary sensory neurones that give rise to small caliber sciatic nerve fibres in rats. Neuroscience 2001;107:665–673.
78. Vellani V, Zachrisson O, McNaughton PA. Functional bradykinin B_1 receptors are expressed in nociceptive neurones and are upregulated by the neurotrophin GDNF. J Physiol 2004;560:391–401.
79. McGuirk SM, Vallis Y, Pasternak CA, Dolphin AC. Bradykinin enhances excitability in cultured rat sensory neurones by a GTP-dependent mechanism. Neurosci Lett 1989;99:85–89.
80. McGehee DS, Oxford GS. Bradykinin modulates the electrophysiology of cultured rat sensory neurons through a pertussis toxin-insensitive G protein. Mol Cell Neurosci 1991;2:21–30.
81. McGuirk SM, Dolphin AC. G-protein mediation in nociceptive signal transduction: an investigation into the excitatory action of bradykinin in a subpopulation of cultured rat sensory neurons. Neuroscience 1992;49:117–128.
82. Thayer SA, Perney TM, Miller RJ. Regulation of calcium homeostasis in sensory neurons by bradykinin. J Neurosci 1988;8:4089–4097.
83. Dray A, Bettaney J, Forster P, Perkins MN. Bradykinin-induced stimulation of afferent fibres is mediated through protein kinase C. Neurosci Lett 1988;91:301–307.
84. Mizumura K, Sato J, Kumazawa T. Effects of prostaglandins and other putative chemical intermediaries on the activity of canine testicular polymodal receptors studied in vitro. Pflügers Arch 1987;408:565–572.
85. Barber LA, Vasko MR. Activation of protein kinase C augments peptide release from rat sensory neurons. J Neurochem 1996;67:72–80.
86. Kano M, Kawakami T, Hikawa N, Hori H, Takenaka T, Gotoh H. Bradykinin-responsive cells of dorsal root ganglia in culture: Cell size, firing, cytosolic calcium, and substance P. Cell Mol Neurobiol 1994;14:49–57.
87. Jackson JG, Usachev YM, Thayer SA. Bradykinin-induced nuclear factor of activated T-cells-dependent transcription in rat dorsal root ganglion neurons. Mol Pharmacol 2007;72:303–310.
88. Wu ZZ, Pan HL. Role of TRPV1 and intracellular Ca2+ in excitation of cardiac sensory neurons by bradykinin. Am J Physiol Regul Integr Comp Physiol 2007;293:R276–R283.
89. Burgess GM, Mullaney I, McNeill M, Coote PR, Minhas A, Wood JN. Activation of guanylate cyclase by bradykinin in rat sensory neurones is mediated by calcium influx: possible role of the increase in cyclic GMP. J Neurochem 1989;53:1212–1218.
90. Linhart O, Obreja O, Kress M. The inflammatory mediators serotonin, prostaglandin E2 and bradykinin evoke calcium influx in rat sensory neurons. Neuroscience 2003;118:69–74.
91. Gammon CM, Allen AC, Morell P. Bradykinin stimulates phosphoinositide hydrolysis and mobilization of arachidonic acid in dorsal root ganglion neurons. J Neurochem 1989;53:95–101.
92. Allen AC, Gammon CM, Ousley AH, McCarthy KD, Morell P. Bradykinin stimulates arachidonic acid release through the sequential actions of an sn-1 diacylglycerol lipase and a monoacylglycerol lipase. J Neurochem 1992;58:1130–1139.
93. Shin J, Cho H, Hwang SW, Jung J, Shin CY, Lee SY et al. Bradykinin-12-lipoxygenase-VR1 signaling pathway for inflammatory hyperalgesia. Proc Natl Acad Sci U S A 2002;99:10150–10155.

94. Burch RM, Axelrod J. Dissociation of bradykinin-induced prostaglandin formation from phosphatidylinositol turnover in Swiss 3T3 fibroblasts: evidence for G protein regulation of phospholipase A2. Proc Natl Acad Sci U S A 1987;84:6374–6378.
95. Bevan S. Intracellular messengers and signal transduction in nociceptors. In: Belmonte C, Cervero F, eds. Neurobiology of nociceptors. Oxford: Oxford University Press, 1996; 298–324.
96. Caterina MJ, Schumacher MA, Tominaga M, Rosen TA, Levine JD, Julius D. The capsaicin receptor: a heat-activated ion channel in the pain pathway. Nature 1997;389:816–824.
97. Gunthorpe MJ, Benham CD, Randall A, Davis JB. The diversity in the vanilloid [TRPV] receptor family of ion channels. Trends Pharmacol Sci 2002;23:183–191.
98. Montell C, Birnbaumer L, Flockerzi V, Bindels RJ, Bruford EA, Caterina MJ et al. A unified nomenclature for the superfamily of TRP cation channels. Mol Cell 2002;9:229–231.
99. Dhaka A, Viswanath V, Patapoutian A. Trp ion channels and temperature sensation. Annu Rev Neurosci 2006;29:135–161.
100. Caterina MJ, Julius D. The vanilloid receptor: a molecular gateway to the pain pathway. Annu Rev Neurosci 2001;24:487–517.
101. Bevan S, Szolcsányi J. Sensory neuron-specific actions of capsaicin: mechanisms and applications. Trends Pharmacol Sci 1990;11:330–333.
102. Kwak JY, Jung JY, Hwang SW, Lee WT, Oh U. A capsaicin-receptor antagonist, capsazepine, reduces inflammation-induced hyperalgesic responses in the rat: evidence for an endogenous capsaicin-like substance. Neuroscience 1998;86:619–626.
103. Jung J, Hwang SW, Kwak J, Lee SY, Kang CJ, Kim WB et al. Capsaicin binds to the intracellular domain of the capsaicin-activated ion channel. J Neurosci 1999;19:529–538.
104. Hwang SW, Cho H, Kwak J, Lee SY, Kang CJ, Jung J et al. Direct activation of capsaicin receptors by products of lipoxygenases: Endogenous capsaicin-like substances. Proc Natl Acad Sci U S A 2000;97:6155–6160.
105. Carr MJ, Kollarik M, Meeker SN, Undem BJ. A role for TRPV1 in bradykinin-induced excitation of vagal airway afferent nerve terminals. J Pharmacol Exp Ther 2003;304:1275–1279.
106. Wang MM, Reynaud D, Pace-Asciak CR. In vivo stimulation of 12[S]-lipoxygenase in the rat skin by bradykinin and platelet activating factor: formation of 12[S]-HETE and hepoxilins, and actions on vascular permeability. Biochim Biophys Acta 1999;1436:354–362.
107. Ringkamp M, Schmelz M, Kress M, Allwang M, Ogilvie A, Reeh PW. Activated human platelets in plasma excite nociceptors in rat skin, in vitro. Neurosci Lett 1994;170:103–106.
108. Schmelz M, Osiander G, Blunk J, Ringkamp M, Reeh PW, Handwerker HO. Intracutaneous injections of platelets cause acute pain and protracted hyperalgesia. Neurosci Lett 1997;226:171–174.
109. Premkumar LS, Ahern GP. Induction of vanilloid receptor channel activity by protein kinase C. Nature 2000;408:985–990.
110. Sugiura T, Tominaga M, Katsuya H, Mizumura K. Bradykinin lowers the threshold temperature for heat activation of vanilloid receptor 1. J Neurophysiol 2002;88:544–548.
111. Vellani V, Mapplebeck S, Moriondo A, Davis JB, McNaughton PA. Protein kinase C activation potentiates gating of the vanilloid receptor VR1 by capsaicin, protons, heat and anandamide. J Physiol 2001;534:813–825.

112. Crandall M, Kwash J, Yu W, White G. Activation of protein kinase C sensitizes human VR1 to capsaicin and to moderate decreases in pH at physiological temperatures in Xenopus oocytes. Pain 2002;98:109–117.
113. Tominaga M, Wada M, Masu M. Potentiation of capsaicin receptor activity by metabotropic ATP receptors as a possible mechanism for ATP-evoked pain and hyperalgesia. Proc Natl Acad Sci U S A 2001;98:6951–6956.
114. Numazaki M, Tominaga T, Toyooka H, Tominaga M. Direct phosphorylation of capsaicin receptor VR1 by protein kinase Cε and identification of two target serine residues. J Biol Chem 2002;277:13375–13378.
115. Lee SY, Lee JH, Kang KK, Hwang SY, Choi KD, Oh U. Sensitization of vanilloid receptor involves an increase in the phosphorylated form of the channel. Arch Pharm Res 2005;28:405–412.
116. Chuang HH, Prescott ED, Kong H, Shields S, Jordt S-E, Basbaum AI et al. Bradykinin and nerve growth factor release the capsaicin receptor from PtdIns[4,5]P_2-mediated inhibition. Nature 2001;411:957–962.
117. Pan HL, Chen SR. Sensing tissue ischemia: another new function for capsaicin receptors? Circulation 2004;110:1826–1831.
118. Rong W, Hillsley K, Davis JB, Hicks G, Winchester WJ, Grundy D. Jejunal afferent nerve sensitivity in wild-type and TRPV1 knockout mice. J Physiol 2004;560:867–881.
119. Lee MG, Macglashan DW, Jr., Undem BJ. Role of chloride channels in bradykinin-induced guinea pig airway vagal C-fibre activation. J Physiol 2005;566:205–212.
120. Bautista DM, Jordt SE, Nikai T, Tsuruda PR, Read AJ, Poblete J et al. TRPA1 mediates the inflammatory actions of environmental irritants and proalgesic agents. Cell 2006;124:1269–1282.
121. Rueff A, Dray A. Sensitization of peripheral afferent fibres in the in vitro neonatal rat spinal cord-tail by bradykinin and prostaglandins. Neuroscience 1993;54:527–535.
122. Fox AJ, Lalloo UG, Belvisi MG, Bernareggi M, Chung KF, Barnes PJ. Bradykinin-evoked sensitization of airway sensory nerves: A mechanism for ACE-inhibitor cough. Nat Med 1996;2:814–817.
123. Stucky CL, Abrahams LG, Seybold VS. Bradykinin increases the proportion of neonatal rat dorsal root ganglion neurons that respond to capsaicin and protons. Neuroscience 1998;84:1257–1265.
124. Oshita K, Inoue A, Tang HB, Nakata Y, Kawamoto M, Yuge O. CB(1) cannabinoid receptor stimulation modulates transient receptor potential vanilloid receptor 1 activities in calcium influx and substance P Release in cultured rat dorsal root ganglion cells. J Pharmacol Sci 2005;97:377–385.
125. Tang HB, Inoue A, Iwasa M, Hide I, Nakata Y. Substance P release evoked by capsaicin or potassium from rat cultured dorsal root ganglion neurons is conversely modulated with bradykinin. J Neurochem 2006;97:1412–1418.
126. Singh Tahim A, Sántha P, Nagy I. Inflammatory mediators convert anandamide into a potent activator of the vanilloid type 1 transient receptor potential receptor in nociceptive primary sensory neurons. Neuroscience 2005;136:539–548.
127. Amann R, Donnerer J, Lembeck F. Activation of primary afferent neurons by thermal stimulation. Influence of ruthenium red. Naunyn-Schmiedeberg's Arch Pharmacol 1990;341:108–113.
128. Geppetti P, Tramontana M, Santicioli P, Bianco ED, Giuliani S, Maggi CA. Bradykinin-induced release of calcitonin gene-related peptide from capsaicin-sensitive nerves in

guinea-pig atria: mechanism of action and calcium requirements. Neuroscience 1990;38:687–692.
129. Dray A, Forbes CA, Burgess GM. Ruthenium red blocks the capsaicin-induced increase in intracellular calcium and activation of membrane currents in sensory neurones as well as the activation of peripheral nociceptors in vitro. Neurosci Lett 1990;110:52–59.
130. Dickenson AH, Dray A. Selective antagonism of capsaicin by capsazepine: evidence for a spinal receptor site in capsaicin-induced antinociception. Br J Pharmacol 1991;104:1045–1049.
131. Zahner MR, Li DP, Chen SR, Pan HL. Cardiac vanilloid receptor 1-expressing afferent nerves and their role in the cardiogenic sympathetic reflex in rats. J Physiol 2003;551:515–523.
132. Qin C, Farber JP, Miller KE, Foreman RD. Responses of thoracic spinal neurons to activation and desensitization of cardiac TRPV1-containing afferents in rats. Am J Physiol Regul Integr Comp Physiol 2006;291:R1700–R1707.
133. Peier AM, Moqrich A, Hergarden AC, Reeve AJ, Andersson DA, Story GM et al. A TRP channel that senses cold stimuli and menthol. Cell 2002;705–715.
134. Premkumar LS, Raisinghani M, Pingle SC, Long C, Pimentel F. Downregulation of transient receptor potential melastatin 8 by protein kinase C-mediated dephosphorylation. J Neurosci 2005;25:11322–11329.
135. Liu B, Qin F. Functional control of cold- and menthol-sensitive TRPM8 ion channels by phosphatidylinositol 4,5-bisphosphate. J Neurosci 2005;25:1674–1681.
136. Linte RM, Ciobanu C, Reid G, Babes A. Desensitization of cold- and menthol-sensitive rat dorsal root ganglion neurones by inflammatory mediators. Exp Brain Res 2007;178:89–98.
137. Story GM, Peier AM, Reeve AJ, Eid SR, Mosbacher J, Hricik TR et al. ANKTM1, a TRP-like channel expressed in nociceptive neurons, is activated by cold temperatures. Cell 2003;112:819–829.
138. Bandell M, Story GM, Hwang SW, Viswanath V, Eid SR, Petrus MJ et al. Noxious cold ion channel TRPA1 is activated by pungent compounds and bradykinin. Neuron 2004;41:849–857.
139. Jordt SE, Bautista DM, Chuang HH, McKemy DD, Zygmunt PM, Högestätt ED et al. Mustard oils and cannabinoids excite sensory nerve fibres through the TRP channel ANKTM1. Nature 2004;427:260–265.
140. Amaya F, Wang H, Costigan M, Allchorne AJ, Hatcher JP, Egerton J et al. The voltage-gated sodium channel Na(v)1.9 is an effector of peripheral inflammatory pain hypersensitivity. J Neurosci 2006;26:12852–12860.
141. Weinreich D. Bradykinin inhibits a slow spike afterhyperpolarization in visceral sensory neurons. Eur J Pharmacol 1986;132:61–63.
142. Undem BJ, Weinreich D. Electrophysiological properties and chemosensitivity of guinea pig nodose ganglion neurons in vitro. J Auton Nerv Syst 1993;44:17–33.
143. Usachev YM, DeMarco SJ, Campbell C, Strehler EE, Thayer SA. Bradykinin and ATP accelerate Ca(2+) efflux from rat sensory neurons via protein kinase C and the plasma membrane Ca(2+) pump isoform 4. Neuron 2002;33:113–122.
144. Weinreich D, Wonderlin WF. Inhibition of calcium-dependent spike after-hyperpolarization increases excitability of rabbit visceral sensory neurones. J Physiol 1987;394:415–427.
145. Plotkin MD, Kaplan MR, Peterson LN, Gullans SR, Hebert SC, Delpire E. Expression of the Na(+)-K(+)-2Cl- cotransporter BSC2 in the nervous system. Am J Physiol 1997;272:C173–C183.

146. Sung KW, Kirby M, McDonald MP, Lovinger DM, Delpire E. Abnormal GABAA receptor-mediated currents in dorsal root ganglion neurons isolated from Na-K-2Cl cotransporter null mice. J Neurosci 2000;20:7531–7538.
147. Kurahashi T, Yau KW. Co-existence of cationic and chloride components in odorant-induced current of vertebrate olfactory receptor cells. Nature 1993;363:71–74.
148. Lai J, Luo MC, Chen Q, Ma S, Gardell LR, Ossipov MH et al. Dynorphin A activates bradykinin receptors to maintain neuropathic pain. Nat Neurosci 2006;9:1534–1540.
149. Kumazawa T, Mizumura K, Sato J. Response properties of polymodal receptors studied using in vitro testis superior spermatic nerve preparations of dogs. J Neurophysiol 1987;57:702–711.
150. Lang E, Novak A, Reeh PW, Handwerker HO. Chemosensitivity of fine afferents from rat skin in vitro. J Neurophysiol 1990;63:887–901.
151. Manning DC, Raja SN, Meyer RA, Campbell JN. Pain and hyperalgesia after intradermal injection of bradykinin in humans. Clin Pharmacol Ther 1991;50:721–729.
152. Liang YF, Haake B, Reeh PW. Sustained sensitization and recruitment of rat cutaneous nociceptors by bradykinin and a novel theory of its excitatory action. J Physiol 2001;532:229–239.
153. McGehee DS, Goy MF, Oxford GS. Involvement of the nitric oxide-cyclic GMP pathway in the desensitization of bradykinin responses of cultured rat sensory neurons. Neuron 1992;9:315–324.
154. Bauer MB, Simmons ML, Murphy S, Gebhart GF. Bradykinin and capsaicin stimulate cyclic GMP production in cultured rat dorsal root ganglion neurons via a nitrosyl intermediate. J Neurosci Res 1993;36:280–289.
155. Rueff A, Patel A, Urban L, Dray A. Regulation of bradykinin sensitivity in peripheral sensory fibres of the neonatal rat by nitric oxide and cyclic GMP. Neuropharmacology 1994;33:1139–1145.
156. Zhang X, Verge V, Wiesenfeld-Hallin Z, Ju G, Bredt D, Synder SH et al. Nitric oxide synthase-like immunoreactivity in lumbar dorsal root ganglia and spinal cord of rat and monkey and effect of peripheral axotomy. J Comp Neurol 1993;335:563–575.
157. Steel JH, Terenghi G, Chung JM, Na HS, Carlton SM, Polak JM. Increased nitric oxide synthase immunoreactivity in rat dorsal root ganglia in a neuropathic pain model. Neurosci Lett 1994;169:81–84.
158. Qian Y, Chao DS, Santillano DR, Cornwell TL, Nairn AC, Greengard P et al. cGMP-dependent protein kinase in dorsal root ganglion: relationship with nitric oxide synthase and nociceptive neurons. J Neurosci 1996;16:3130–3138.
159. Harvey JS, Burgess GM. Cyclic GMP regulates activation of phosphoinositidase C by bradykinin in sensory neurons. Biochem J 1996;316:539–544.
160. Pizard A, Blaukat A, Müller-Esterl W, Alhenc-Gelas F, Rajerison RM. Bradykinin-induced internalization of the human B_2 receptor requires phosphorylation of three serine and two threonine residues at its carboxyl tail. J Biol Chem 1999;274:12738–12747.
161. Simaan M, Bédard-Goulet S, Fessart D, Gratton JP, Laporte SA. Dissociation of beta-arrestin from internalized bradykinin B2 receptor is necessary for receptor recycling and resensitization. Cell Signal 2005;17:1074–1083.
162. Boyd RS, Donnelly LE, Macdermot J. Loss of responses to bradykinin, ATP or carbachol follows depletion of a shared pool of calcium ions. Eur J Pharmacol 1994;267:161–166.
163. Reeh PW. Sensory receptors in mammalian skin in an in vitro preparation. Neurosci Lett 1986;66:141–146.

164. Szolcsányi J. Selective responsiveness of polymodal nociceptors of the rabbit ear to capsaicin, bradykinin and ultra-violet irradiation. J Physiol [Lond] 1987;388:9–23.
165. Schmelz M, Schmidt R, Weidner C, Hilliges M, Torebjork HE, Handwerker HO. Chemical response pattern of different classes of C-nociceptors to pruritogens and algogens. J Neurophysiol 2003;89:2441–2448.
166. Kumazawa T, Mizumura K. Chemical responses of polymodal receptors of the scrotal contents in dogs. J Physiol 1980;299:219–231.
167. Toda K, Ishii N, Nakamura Y. Characteristics of mucosal nociceptors in the rat oral cavity: an in vitro study. Neurosci Lett 1997;228:95–98.
168. Takeuchi Y, Toda K. Subtypes of nociceptive units in the rat temporomandibular joint. Brain Res Bull 2003;61:603–608.
169. Takeuchi Y, Zeredo JL, Fujiyama R, Amagasa T, Toda K. Effects of experimentally induced inflammation on temporomandibular joint nociceptors in rats. Neurosci Lett 2004;354:172–174.
170. Toda K, Zeredo JL, Fujiyama R, Okada Y, Oi K, Hayashi Y et al. Characteristics of nociceptors in the periodontium–an in vitro study in rats. Brain Res Bull 2004;62:345–349.
171. Baker DG, Coleridge HM, Coleridge JC, Nerdrum T. Search for a cardiac nociceptor: stimulation by bradykinin of sympathetic afferent nerve endings in the heart of the cat. J Physiol 1980;306:519–536.
172. Pan HL, Chen SR. Myocardial ischemia recruits mechanically insensitive cardiac sympathetic afferents in cats. J Neurophysiol 2002;87:660–668.
173. Guo ZL, Symons JD, Longhurst JC. Activation of visceral afferents by bradykinin and ischemia: independent roles of PKC and prostaglandins. Am J Physiol 1999;276:1884–1891.
174. Kopp UC, Farley DM, Smith LA. Bradykinin-mediated activation of renal sensory neurons due to prostaglandin-dependent release of substance P. Am J Physiol 1997;272:R2009–R2016.
175. Szolcsányi J. Capsaicin-sensitive chemoceptive neural system with dual sensory-efferent function. In: Chahl LA, Szolcsányi J, Lembeck F, eds. Antidromic Vasodilatation and Neurogenic Inflammation. Budapest: Akadémiai Kiadó, 1984; pp. 27–52.
176. Hisata Y, Zeredo JL, Eishi K, Toda K. Cardiac nociceptors innervated by vagal afferents in rats. Auton Neurosci 2006;126–127:174–178.
177. Hillsley K, McCaul C, Aerssens J, Peeters PJ, Gijsen H, Moechars D et al. Activation of the cannabinoid 2 (CB2) receptor inhibits murine mesenteric afferent nerve activity. Neurogastroenterol Motil 2007;19:769–777.
178. Beck PW, Handwerker HO. Bradykinin and serotonin effects on various types of cutaneous nerve fibers. Pflügers Arch 1974;347:209–222.
179. Undem BJ, Chuaychoo B, Lee MG, Weinreich D, Myers AC, Kollarik M. Subtypes of vagal afferent C-fibres in guinea-pig lungs. J Physiol 2004;556:905–917.
180. Kanaka R, Schaible HG, Schmidt RF. Activation of fine articular afferent units by bradykinin. Brain Res 1985;327:81–90.
181. Kumazawa T, Mizumura K. Thin-fibre receptors responding to mechanical, chemical, and thermal stimulation in the skeletal muscle of the dog. J Physiol 1977;273:179–194.
182. Birrell GJ, McQueen DS, Iggo A, Grubb BD. Prostanoid-induced potentiation of the excitatory and sensitizing effects of bradykinin on articular mechanonociceptors in the rat ankle joint. Neuroscience 1993;54:537–544.

183. Longhurst JC, Kaufman MP, Ordway GA, Musch TI. Effects of bradykinin and capsaicin on endings of afferent fibers from abdominal visceral organs. Am J Physiol 1984;247:R552–R559.
184. Haupt P, Jänig W, Kohler W. Response pattern of visceral afferent fibres, supplying the colon, upon chemical and mechanical stimuli. Pflugers Arch 1983;398:41–47.
185. Keele CA, Armstrong D. Substances Producing Pain and Itch. London: Edward Arnold, 1964.
186. Kindgen-Milles D, Klement W, Arndt JO. The nociceptive systems of skin, paravascular tissue and hand veins of humans and their sensitivity to bradykinin. Neurosci Lett 1994;181:39–42.
187. Szolcsányi J. A pharmacological approach to elucidation of the role of different nerve fibres and receptor endings in mediation of pain. J Physiol [Paris] 1977;73:251–259.
188. Lembeck F, Juan H. Interaction of prostaglandins and indomethacin with algesic substances. Naunyn-Schmiedeberg's Arch Pharmacol 1974;285:301–313.
189. Lembeck F, Popper H, Juan H. Release of prostaglandins by bradykinin as an intrinsic mechanism of its algesic effect. Naunyn-Schmiedeberg's Arch Pharmacol 1976;294:69–73.
190. Corrado AP, Ballejo G. Is guanylate cyclase activation through the release of nitric oxide or a related compound involved in bradykinin-induced perivascular primary afferent excitation? Agents Actions Suppl 1992;36:238–250.
191. Wang W. Cardiac sympathetic afferent stimulation by bradykinin in heart failure: role of NO and prostaglandins. Am J Physiol 1998;275:783–788.
192. Seyedi N, Maruyama R, Levi R. Bradykinin activates a cross-signalling pathway between sensory and adrenergic nerve endings in the heart: A novel mechanism of ischemic norepinephrine release? J Pharmacol Exp Ther 1999;290:656–663.
193. Vasko MR, Campbell WB, Waite KJ. Prostaglandin E_2 enhances bradykinin-stimulated release of neuropeptides from rat sensory neurons in culture. J Neurosci 1994;14:4987–4997.
194. Weinreich D, Koschorke GM, Undem BJ, Taylor GE. Prevention of the excitatory actions of bradykinin by inhibition of PGI_2 formation in nodose neurones of the guinea-pig. J Physiol 1995;483:735–746.
195. Chopra B, Giblett S, Little JG, Donaldson LF, Tate S, Evans RJ et al. Cyclooxygenase-1 is a marker for a subpopulation of putative nociceptive neurons in rat dorsal root ganglia. Eur J Neurosci 2000;12:911–920.
196. Inoue A, Iwasa M, Nishikura Y, Ogawa S, Nakasuka A, Nakata Y. The long-term exposure of rat cultured dorsal root ganglion cells to bradykinin induced the release of prostaglandin E2 by the activation of cyclooxygenase-2. Neurosci Lett 2006;401:242–247.
197. Tang HB, Inoue A, Oshita K, Hirate K, Nakata Y. Zaltoprofen inhibits bradykinin-induced responses by blocking the activation of second messenger signaling cascades in rat dorsal root ganglion cells. Neuropharmacology 2005;48:1035–1042.
198. Franco-Cereceda A. Prostaglandins and CGRP release from cardiac sensory nerves. Naunyn-Schmiedeberg's Arch Pharmacol 1989;340:180–184.
199. Pethő G, Derow A, Reeh PW. Bradykinin-induced nociceptor sensitization to heat is mediated by cyclooxygenase products in isolated rat skin. Eur J Neurosci 2001;14:210–218.
200. Hong Y, Abbott FV. Behavioural effects of intraplantar injection of inflammatory mediators in the rat. Neuroscience 1994;63:827–836.

201. Kumazawa T, Mizumura K, Minagawa M, Tsujii Y. Sensitizing effects of bradykinin on the heat responses of the visceral nociceptor. J Neurophysiol 1991;66:1819–1824.
202. Koltzenburg M, Kress M, Reeh PW. The nociceptor sensitization by bradykinin does not depend on sympathetic neurons. Neuroscience 1992;46:465–473.
203. Meyer RA, Davis KD, Raja SN, Campbell JN. Sympathectomy does not abolish bradykinin-induced cutaneous hyperalgesia in man. Pain 1992;51:323–327.
204. Cesare P, McNaughton P. A novel heat-activated current in nociceptive neurons and its sensitization by bradykinin. Proc Natl Acad Sci U S A 1996;93:15435–15439.
205. Cesare P, Dekker LV, Sardini A, Parker PJ, McNaughton PA. Specific involvement of PKC-epsilon in sensitization of the neuronal response to painful heat. Neuron 1999;23:617–624.
206. Günther S, Reeh PW, Kress M. Rises in $[Ca^{2+}]_i$ mediate capsaicin- and proton-induced heat sensitization of rat primary nociceptive neurons. Eur J Neurosci 1999;11:3143–3150.
207. Mizumura K, Koda H, Kumazawa T. Evidence that protein kinase C activation is involved in the excitatory and facilitatory effects of bradykinin on canine visceral nociceptors in vitro. Neurosci Lett 1997;237:29–32.
208. Kessler W, Kirchhoff C, Reeh PW, Handwerker HO. Excitation of cutaneous afferent nerve endings in vitro by a combination of inflammatory mediators and conditioning effect of substance P. Exp Brain Res 1992;91:467–476.
209. Fischer MJ, Reeh PW. Sensitization to heat through G-protein-coupled receptor pathways in the isolated sciatic mouse nerve. Eur J Neurosci 2007;25:3570–3575.
210. Aley KO, Messing RO, Mochly-Rosen D, Levine JD. Chronic hypersensitivity for inflammatory nociceptor sensitization mediated by the ε isozyme of protein kinase C. J Neurosci 2000;20:4680–4685.
211. Taiwo YO, Bjerknes LK, Goetzl EJ, Levine JD. Mediation of primary afferent peripheral hyperalgesia by the cAMP second messenger system. Neuroscience 1989;32:577–580.
212. Mizumura K, Koda H, Kumazawa T. Augmenting effects of cyclic AMP on the heat response of canine testicular polymodal receptors. Neurosci Lett 1993;162:75–77.
213. Cui M, Nicol GD. Cyclic AMP mediates the prostaglandin E_2-induced potentiation of bradykinin excitation in rat sensory neurons. Neuroscience 1995;66:459–466.
214. Rathee PK, Distler C, Obreja O, Neuhuber W, Wang GK, Wang SY et al. PKA/AKAP/VR1 module: A common link of G_s-mediated signaling to thermal hyperalgesia. J Neurosci 2002;22:4740–4745.
215. Graness A, Adomeit A, Ludwig B, Müller WD, Kaufmann R, Liebmann C. Novel bradykinin signalling events in PC-12 cells: stimulation of the cAMP pathway leads to cAMP-mediated translocation of protein kinase Cε. Biochem J 1997;327:147–154.
216. Bell JD, Buxton IL, Brunton LL. Enhancement of adenylate cyclase activity in S49 lymphoma cells by phorbol esters. Putative effect of C kinase on alpha s-GTP-catalytic subunit interaction. J Biol Chem 1985;260:2625–2628.
217. Sibley DR, Jeffs RA, Daniel K, Nambi P, Lefkowitz RJ. Phorbol diester treatment promotes enhanced adenylate cyclase activity in frog erythrocytes. Arch Biochem Biophys 1986;244:373–381.
218. Jacobowitz O, Chen J, Premont RT, Iyengar R. Stimulation of specific types of G_s-stimulated adenylyl cyclases by phorbol ester treatment. J Biol Chem 1993;268:3829–3832.
219. Mayer S, Izydorczyk I, Reeh PW, Grubb BD. Bradykinin-induced nociceptor sensitisation to heat depends on cox-1 and cox-2 in isolated rat skin. Pain 2007;130:14–24.

220. Sauer SK, Averbeck B, Reeh PW. Denervation and NK_1 receptor block modulate stimulated CGRP and PGE_2 release from rat skin. NeuroReport 2000;11:283–286.
221. Caterina MJ, Rosen TA, Tominaga M, Brake AJ, Julius D. A capsaicin-receptor homologue with a high threshold for noxious heat. Nature 1999;398:436–441.
222. Smith GD, Gunthorpe MJ, Kelsell RE, Hayes PD, Reilly P, Facer P et al. TRPV3 is a temperature-sensitive vanilloid receptor-like protein. Nature 2002;418:186–190.
223. Levine JD, Lau W, Kwiat G, Goetzl EJ. Leukotriene B_4 produces hyperalgesia that is dependent on polymorphonuclear leukocytes. Science 1984;225:743–745.
224. Levine JD, Taiwo YO, Collins SD, Tam JK. Noradrenaline hyperalgesia is mediated through interaction with sympathetic postganglionic neurone terminals rather than activation of primary afferent nociceptors. Nature 1986;323:158–160.
225. Steranka LR, Manning DC, DeHaas CJ, Ferkany JW, Borosky SA, Connor JR et al. Bradykinin as a pain mediator: Receptors are localized to sensory neurons, and antagonists have analgesic actions. Proc Natl Acad Sci U S A 1988;85:3245–3249.
226. Taiwo YO, Goetzl EJ, Levine JD. Hyperalgesia onset latency suggests a hierarchy of action. Brain Res 1987;423:333–337.
227. Taiwo YO, Levine JD. Characterization of the arachidonic acid metabolites mediating bradykinin and noradrenaline hyperalgesia. Brain Res 1988;458:402–406.
228. Taiwo YO, Levine JD. Effects of cyclooxygenase products of arachidonic acid metabolism on cutaneous nociceptive threshold in the rat. Brain Res 1990;537:372–374.
229. White DM, Taiwo YO, Coderre TJ, Levine JD. Delayed activation of nociceptors: correlation with delayed pain sensations induced by sustained stimuli. J Neurophysiol 1991;66:729–734.
230. Chen X, Tanner K, Levine JD. Mechanical sensitization of cutaneous C-fiber nociceptors by prostaglandin E2 in the rat. Neurosci Lett 1999;267:105–108.
231. Gonzales R, Goldyne ME, Taiwo YO, Levine JD. Production of hyperalgesic prostaglandins by sympathetic postganglionic neurons. J Neurochem 1989;53:1595–1598.
232. Taiwo YO, Heller PH, Levine JD. Characterization of distinct phospholipases mediating bradykinin and noradrenaline hyperalgesia. Neuroscience 1990;39:523–531.
233. Seabrook GR, Bowery BJ, Hill RG. Bradykinin receptors in mouse and rat isolated superior cervical ganglia. Br J Pharmacol 1995;115:368–372.
234. Borkowski JA, Ransom RW, Seabrook GR, Trumbauer M, Chen H, Hill RG et al. Targeted disruption of a B_2 bradykinin receptor gene in mice eliminates bradykinin action in smooth muscle and neurons. J Biol Chem 1995;270:13706–13710.
235. Khasar SG, Green PG, Levine JD. Comparison of intradermal and subcutaneous hyperalgesic effects of inflammatory mediators in the rat. Neurosci Lett 1993;153:215–218.
236. Mense S, Meyer H. Bradykinin-induced modulation of the response behaviour of different types of feline group III and IV muscle receptors. J Physiol 1988;398:49–63.
237. Neugebauer V, Schaible HG, Schmidt RF. Sensitization of articular afferents to mechanical stimuli by bradykinin. Pflügers Arch 1989;415:330–335.
238. Grubb BD, Birrell GJ, McQueen DS, Iggo A. The role of PGE_2 in the sensitization of mechanoreceptors in normal and inflamed ankle joints of the rat. Exp Brain Res 1991;84:383–392.
239. Cunha FQ, Lorenzetti BB, Poole S, Ferreira SH. Interleukin-8 as a mediator of sympathetic pain. Br J Pharmacol 1991;104:765–767.
240. Cunha FQ, Poole S, Lorenzetti BB, Ferreira SH. The pivotal role of tumour necrosis factor alpha in the development of inflammatory hyperalgesia. Br J Pharmacol 1992;107:660–664.

241. Lorenzetti BB, Veiga FH, Canetti CA, Poole S, Cunha FQ, Ferreira SH. Cytokine-induced neutrophil chemoattractant 1 (CINC-1) mediates the sympathetic component of inflammatory mechanical hypersensitivitiy in rats. Eur Cytokine Netw 2002;13:456–461.
242. Jensen K, Tuxen C, Pedersen-Bjergaard U, Jansen I, Edvinsson L, Olesen J. Pain and tenderness in human temporal muscle induced by bradykinin and 5-hydroxytryptamine. Peptides 1990;11:1127–1132.
243. Babenko VV, Graven-Nielsen T, Svensson P, Drewes AM, Jensen TS, Arendt-Nielsen L. Experimental human muscle pain induced by intramuscular injections of bradykinin, serotonin, and substance P. Eur J Pain 1999;3:93–102.
244. Babenko V, Graven-Nielsen T, Svensson P, Drewes AM, Jensen TS, Arendt-Nielsen L. Experimental human muscle pain and muscular hyperalgesia induced by combinations of serotonin and bradykinin. Pain 1999;82:1–8.
245. Chahl LA, Iggo A. The effects of bradykinin and prostaglandin E_1 on rat cutaneous afferent nerve activity. Br J Pharmacol 1977;59:343–347.
246. Kress M, Reeh PW, Vyklicky L. An interaction of inflammatory mediators and protons in small diameter dorsal root ganglion neurons of the rat. Neurosci Lett 1997;224:37–40.
247. Koppert W, Reeh PW, Handwerker HO. Conditioning of histamine by bradykinin alters responses of rat nociceptors and human itch sensation. Neurosci Lett 1993;152:117–120.
248. Koppert W, Martus P, Reeh PW. Interactions of histamine and bradykinin on polymodal C-fibres in isolated rat skin. Eur J Pain 2001;5:97–106.
249. Schuligoi R, Peskar BA, Donnerer J, Amann R. Bradykinin-evoked sensitization of neuropepetide release from afferent neurons in the guinea-pig lung. Br J Pharmacol 1998;125:388–392.
250. Fu LW, Longhurst JC. Interactions between histamine and bradykinin in stimulation of ischaemically sensitive cardiac afferents in felines. J Physiol. 2005;565:1007–1017.
251. Banik RK, Sato J, Giron R, Yajima H, Mizumura K. Interactions of bradykinin and norepinephrine on rat cutaneous nociceptors in both normal and inflamed conditions in vitro. Neurosci Res 2004;49:421–425.
252. Sato J, Yajima H, Banik RK, Kumazawa T, Mizumura K. Norepinephrine reduces heat responses of cutaneous C-fiber nociceptors in Sprague-Dawley rats in vitro. Neurosci Lett 2005;378:111–116.
253. Hur EM, Park YS, Lee BD, Jang IH, Kim HS, Kim TD et al. Sensitization of epidermal growth factor-induced signaling by bradykinin is mediated by c-Src. Implications for a role of lipid microdomains. J Biol Chem 2004;279:5852–5860.
254. Kindgen-Milles D, Arndt JO. Nitric oxide as a chemical link in the generation of pain from veins in humans. Pain 1996;64:139–142.
255. Holthusen H. Involvement of the NO/cyclic GMP pathway in bradykinin–evoked pain from veins in humans. Pain 1997;69:87–92.
256. Holthusen H, Arndt JO. Nitric oxide evokes pain at nociceptors of the paravascular tissue and veins in humans. J Physiol 1995;487:253–258.
257. Holthusen H, Arndt JO. Nitric oxide evokes pain in humans on intracutaneous injection. Neurosci Lett 1994;165:71–74.
258. Nakamura A, Fujita M, Shiomi H. Involvement of endogenous nitric oxide in the mechanism of bradykinin-induced peripheral hyperalgesia. Br J Pharmacol 1996;117:407–412.
259. Cunha FQ, Teixeira MM, Ferreira SH. Pharmacological modulation of secondary mediator systems–cyclic AMP and cyclic GMP–on inflammatory hyperalgesia. Br J Pharmacol 1999;127:671–678.

260. Kelly DC, Asghar AUR, Marr CG, McQueen DS. Nitric oxide modulates articular sensory discharge and responsiveness to bradykinin in normal and arthritic rats in vivo. NeuroReport 2001;12:121–125.
261. Fujiyoshi T, Hayashi I, Oh-ishi S. Kaolin-induced writhing response in mice: activation of the plasma kallikrein-kinin system by kaolin. J Pharmacobiodyn 1989;12:483–487.
262. Ikeda Y, Ueno A, Naraba H, Oh-ishi S. Involvement of vanilloid receptor VR1 and prostanoids in the acid-induced writhing responses of mice. Life Sci 2001;69:2911–2919.
263. Boix F, Røe C, Rosenborg L, Knardahl S. Kinin peptides in human trapezius muscle during sustained isometric contraction and their relation to pain. J Appl Physiol 2005;98:534–540.
264. Leonard PA, Arunkumar R, Brennan TJ. Bradykinin antagonists have no analgesic effect on incisional pain. Anesth Analg 2004;99:1166–1172.
265. Kawabata A, Kawao N, Kitano T, Matsunami M, Satoh R, Ishiki T et al. Colonic hyperalgesia triggered by proteinase-activated receptor-2 in mice: involvement of endogenous bradykinin. Neurosci Lett 2006;402:167–172.
266. Moreau ME, Garbacki N, Molinaro G, Brown NJ, Marceau F, Adam A. The kallikrein-kinin system: current and future pharmacological targets. J Pharmacol Sci 2005;99:6–38.
267. Leeb-Lundberg LM, Marceau F, Müller-Esterl W, Pettibone DJ, Zuraw BL. International union of pharmacology. XLV. Classification of the kinin receptor family: from molecular mechanisms to pathophysiological consequences. Pharmacol Rev 2005;57:27–77.
268. Campos MM, Leal PC, Yunes RA, Calixto JB. Non-peptide antagonists for kinin B1 receptors: new insights into their therapeutic potential for the management of inflammation and pain. Trends Pharmacol Sci 2006;12:646–651.

Edited by Gábor Jancsó

The Inhibition of Neurogenic Inflammation

JOSEF DONNERER and ULRIKE HOLZER-PETSCHE

Institute of Experimental and Clinical Pharmacology, Medical University of Graz, Graz, Austria

ABSTRACT

The components of neurogenic inflammation include plasma extravasation, vasodilatation, smooth muscle contraction, release of mast cell mediators, recruitment and activation of inflammatory cells, and possibly other proinflammatory effects. All these effects can be evoked by neuropeptides released from the peripheral terminals of primary afferent neurons. The neuropeptides involved include the tachykinins substance P and neurokinin A, and calcitonin gene-related peptide. Antagonists blocking their receptors in the target tissue are able to inhibit neurogenic inflammation. Additionally, the prevention of the sensitization and excitation of afferent neurons, and modulators of the release of neuropeptides, are known to be effective in the control of neurogenic inflammation. In many inflammatory conditions of the skin, the joints, the upper respiratory system, and the bronchial tree, as well as the intestinal and the urinary tract, a neurogenic component may be involved, although in certain instances only partially. The neurogenic component might play a role in the initiation, the perpetuation, or the exacerbation of inflammatory processes. Therefore, the overall therapeutic significance of a specific blockade of the neurogenic component in chronic inflammatory conditions is not yet clear.

1. INTRODUCTION

Inflammation is a defense reaction of the organism toward various harmful stimuli, aimed at eliminating or destroying the noxious agent. In most instances, inflammation is triggered by resident, migrating, or circulating cells of the local defense and the immune system, without apparent involvement of the nervous system. However, the immune, the endocrine, and the neural systems may interact in many inflammatory processes, leading to a perpetuation, an exacerbation, or the termination of the inflammation. In contrast, neurogenic inflammation is a well-defined process in which inflammation is triggered by the nervous system. Originally, chemical stimulation of sensory nerves was recognized as a trigger of neurogenic inflammation. Jancsó et al. [1] first used the term "neurogenic inflammation" to describe the increase in vascular permeability, plasma extravasation, and tissue swelling, which occurred in the skin and conjunctiva when sensory nerves were stimulated by topical irritants. The original concept of neurogenic inflammation has been extended to include the various components of the efferent function of the sensory nerves [2–4]. This type of inflammation is caused by the local release of the neuropeptides substance P, neurokinin A (NKA), and calcitonin gene-related peptide (CGRP) known to coexist in sensory neurons. Nowadays, any process by which neuropeptides are directly released from peripheral axon terminals of sensory nerves to produce vasodilatation, edema, and other manifestations of inflammation is considered to be neurogenic inflammation, independent of the type of stimulation. The peripheral release of the transmitters of sensory neurons can be due to direct excitation of the nerve terminals, or due to "antidromic stimulation" of the axons by the so-called axon reflex [2].

The sensory neurons which can give rise to axon reflexes arise from the dorsal root ganglia and the trigeminal ganglia – primary afferent neurons, as well as vagal afferents from the nodose ganglia that innervate the viscera. Neurons with small-diameter cell bodies and unmyelinated slowly conducting C-fiber axons play the key role in neurogenic inflammation. According to their sensory qualities they are classified as C-polymodal (mechanical/chemical/heat) nociceptors; however, a special subclass of heat nociceptors may be responsible for the neurogenic vasodilatation in primates [5].

2. ACTIVATION OF SENSORY NEURONS IN NEUROGENIC INFLAMMATION

Under experimental conditions the sensory neurons can be stimulated either electrically or chemically [6]. Under pathophysiological conditions direct mechanical activation is less important than chemical activation. Many of the mediators released locally during tissue injury or during an inflammation are known to stimulate polymodal nociceptors. The chemical irritant capsaicin, which has been particularly useful in the elucidation of neurogenic inflammation, activates a specific ion channel for Na^+, Ca^{2+}, and other cations in the plasma membrane of sensory neurons, called transient receptor potential vanilloid type 1 receptor (TRPV1) [7]. Noxious heat is an endogenous activator of TRPV1 channels and they are sensitized by protons [7]. Due to the Ca^{2+} overload, afferent nerves stimulated via opening of TRPV1 channels can become desensitized to further stimuli for a prolonged period of time. Because for many years capsaicin was used as the typical stimulating agent for this class of afferents they have also been termed "capsaicin-sensitive primary afferents."

After local tissue injury, a release of chemical mediators such as K^+, H^+, ATP, bradykinin as well as prostaglandin E_2 (PGE_2) can directly activate nerve endings [6] that are endowed with nicotinic, ATP-P2X, 5-hydroxytryptamine (5-HT), and bradykinin receptors. Their agonists trigger the release of algesic mediators such as histamine from mast cells, 5-HT from platelets, as well as nerve growth factor (NGF), and prostanoids from various cell types in the vicinity. These mediators in turn sensitize surrounding nociceptors. Glutamate released from damaged cells may also act on sensory nerve endings via kainate receptors and contribute to neurogenic inflammation. Impulses generated in the stimulated branches of sensory nerves propagate to other terminal branches where they induce the release of neuropeptides such as substance P, NKA, and CGRP [2].

Short-term sensitization most likely occurs via prostaglandin-induced elevation of cAMP levels and activation of protein kinase A [6]. The augmentation by PGE_2 of the bradykinin-evoked release of substance P and CGRP from sensory neurons may account for the inflammatory and hyperalgesic actions of this eicosanoid [8]. In contrast, NGF typically induces long-term sensitization [9] by inducing the expression of receptors and channels that ultimately enhance nociceptor activation and neurogenic inflammation. Whereas under some pathophysiological conditions a direct mechanical, chemical, or neural stimulus can initiate neurogenic inflammation, in others a cascade of mediators is leading to sensory neuron stimulation.

Neurogenic bronchoconstriction or bronchial edema formation can be evoked by mechanical irritation or by chemical irritants, such as capsaicin, ether, formalin, and cigarette smoke. In the airways of asthmatics, C-fiber nerve endings have been shown to become more sensitive to irritants, which results in axon reflexes and symptoms of neurogenic inflammation [10]. Moreover, in part due to epithelial shedding, C-fiber endings are more likely to be exposed to various airway irritants during chronic inflammation of the bronchial tree. Such pathological processes lead to activation of C-fibers followed by release of neuropeptides in the vicinity of smooth muscle, glands, and blood vessels via an antidromic reflex to peripheral nerve branches [11]. Thus, inflammation of the airways has a neurogenic component that may be initiated and perpetuated by inflammatory stimuli at the epithelial surface. These stimuli may further recruit inflammatory cells like eosinophils or neutrophils. If, however, the mucosa of the respiratory tract is challenged by antigens as in allergic asthma, plasma extravasation is initiated indirectly by endogenous kinins that activate the sensory nerve endings resulting in release of tachykinins [12,13]. This response to bradykinin is mediated by B_2 receptors.

Sensitization of afferent nerves is also implicated in the so-called "functional" disorders of the gut or the urogenital tract. Although this pathophysiology can be influenced by many of the mechanisms described below, there is no evidence for the involvement of "neurogenic inflammation" *sensu stricto* in these disorders.

In several primary forms of headache such as migraine or cluster headache, the trigeminal nerves can be activated by the same compounds that also act on the wall of blood vessels, as for example 5-HT. This initiates the trigeminovascular reflex, causing antidromic release of CGRP [14–16]. CGRP is the major neuropeptide involved in this neurogenic vasodilator response within the dura mater and in the cerebral vasculature.

NGF, released from inflammatory cells, not only enhances synthesis, axonal transport, and release of substance P and CGRP in arthritis, but it also sensitizes the nerve terminals. This favors stimulation of sensory neurons by any irritant and provokes excess release of neuropeptide transmitters. Thus, neurogenic inflammation may be a component in arthritis [17].

3. EFFECTOR MECHANISMS AND RECEPTORS INVOLVED IN NEUROGENIC INFLAMMATION

In the skin, C-fibers form a complex network which extends from the subcutis to the epidermis running parallel to the cutaneous vascular supply. It is particularly dense in the papillary and subpapillary dermis. Therefore, it is plausible that each cutaneous cell or structure is connected to this neural network and thus susceptible to neural influences. CGRP is the major mediator of neurogenic vasodilatation in the skin. The vasodilator reaction is mediated by $CGRP_1$ receptors, coupled to the adenylate cyclase system, arises from a direct action on the vascular smooth muscle, and does not involve nitric oxide or prostaglandins [18,19]. Substance P elicits vasodilatation in the skin and in mucosal surfaces via activation of tachykinin NK_1 receptors, coupled to the phosphoinositide signaling pathway; yet, in many species substance P seems not to be the mediator of neurogenic vasodilatation. However, substance P and NKA are the prime mediators of the increase in vascular permeability [3,20]. Substance P-induced extravasation arises from a direct action on the endothelium of postcapillary venules. Later phases of extravasation may depend on the release of mediators from mast cells in the skin [2]. Substance P and CGRP interact insofar as CGRP enhances substance P-induced plasma extravasation in the skin and in joints [19–22].

Substance P released from sensory neurons activates both vascular endothelial cells and granulocytes by inducing the expression of adhesion molecules; this results in infiltration of granulocytes into the tissue [23]. Substance P stimulates cell trafficking, cell activation, and cytokine release from neutrophils, macrophages, and mononuclear cells [24,25]. Cell-derived mediators of neurogenic inflammation in the skin may originate from mast cells, leukocytes, basophils, or monocytes. The distribution of mast cells corresponds to that of sensory nerve fibers showing substance P and CGRP immunoreactivities, and close contacts between mast cells and neuropeptide-containing nerve fibers have been proposed to be present in the skin and mucosa [26,27].

Experiments with neurogenic inflammatory reactions in the skin of pigs have provided a valuable model for the situation in humans [28,29]. No plasma extravasation, but only a flare reaction was observed in pig skin upon neurogenic stimulation. This resembles the situation in the human skin, where only a neurogenic vasodilatation can be observed in response to electrical stimulation of sensory nerves. However, plasma extravasation is observed upon direct histamine iontophoresis. Moreover, chemical neurogenic activation of sensory C-fibers in human skin does not involve the release of histamine from mast cells. One may assume that the amount of substance P released from peptidergic nerve endings is too low to cause mast cell degranulation or plasma extravasation. The mediator profile in such reactions may be a combination of several peptides and not a single one, possibly CGRP plus substance P and other tachykinins [30,31].

Neurogenic control of the airway smooth muscle is highly variable depending on the physiological and pathological conditions of the lung. In general, parasympathetic vagal nerves narrow the airway caliber, whereas there is little evidence for a direct smooth muscle relaxing effect by sympathetic innervation in human bronchi. In many cases, sensory nerve fibers are excited via activation of TRPV1 channels [32] and subsequently release neuropeptides antidromically. The ensuing response of the bronchial smooth muscle and plasma extravasation is, however, species-dependent (see below). Also in the human respiratory tract substance P- and NKA-immunoreactive sensory nerves densely innervate the epithelium, submucosal glands, blood vessels, and the bronchial smooth muscle [33]. Substance P and NKA contract human bronchi and bronchioli, NKA was found to be more potent than substance P [34].

The NK_2 receptor has been reported to be responsible for contracting the airways, whereas the NK_1 receptor is responsible for most of the other airway effects. Axon reflex bronchoconstriction and plasma extravasation is most prominent in guinea pigs and less pronounced in rats. In other species, bronchoconstriction seems to be brought about mainly by a cholinergic mechanism [10]. Stimulation of sensory nerves in the airway mucosa of the rat evokes the release of inflammatory peptides such as substance P, which can increase microvascular permeability via NK_1 receptors in the wall of postcapillary venules, resulting in neurogenic plasma extravasation [35]. In view of the distribution of substance P-positive nerve endings, the most likely source for substance P are afferent nerve terminals in the venular wall or the arterioles from where the peptide is transported with the blood to the venular endothelium [10,36]. This neurogenic inflammation has however so far been described only in airways of rodents [10].

In rat airways substance P is the main agent of neurogenic vasodilatation [37]. NK_1 receptors are present on mucous cells, vessels, epithelium, and inflammatory cells. They are predominant in the creation of bronchial inflammation, vasodilatation and edema, mucosecretion, chemotaxis, and activation of inflammatory cells. Via NK_1 receptors substance P has the capacity to increase neutrophil adhesion to bronchial epithelial cells and it is suggested that substance P plays an important role in modulating airway inflammation [38]. Tachykinins can regulate neutrophil recruitment into the lower respiratory tract, in part by inducing the release of neutrophil chemotactic activity from airway epithelial cells [39]. Overall, in the airways the sensory neuropeptides have important proinflammatory effects such as stimulation of T lymphocyte proliferation, stimulation of macrophages, mast cells, and eosinophils, and the chemoattraction of eosinophils and neutrophils; neuropeptides may also cause mucosecretion [40].

The local release of tachykinins is considered important in the pathophysiology of asthma [41]. Neurogenic inflammation in the airways might contribute to the development of bronchial hyperresponsiveness and to the late asthmatic response in the patients with bronchial asthma [11]. Furthermore, there is evidence that in allergic inflammation sensory nerve endings become more sensitive to irritants and contain higher levels of proinflammatory neuropeptides [42]. That sensory neuropeptides can indeed be released from sensory nerves into the human airways during an allergic reaction has been measured in nasal and bronchoalveolar lavage fluid [43]. In the upper airways of pigs, antidromic sensory nerve stimulation of the nasal innervation causes an increase in nasal blood flow and a release of CGRP. An axon reflex underlies fluid secretion in the upper airways in both animals and humans [44].

Neurogenic inflammation can also be evoked in the gastrointestinal tract; however, regional differences seem to exist as to the effects of antidromic activity of afferent nerves. In rats, neurogenic plasma extravasation in response to capsaicin has only been observed in the small intestine [45]. In the foregut, it is rather an increase in blood flow and initiation of additional protective mechanisms that prevail [46]. In rats as well as in humans, efferent activity of capsaicin-sensitive peptidergic afferents is involved in the pathophysiology of inflammatory and infectious diseases of the gastrointestinal tract [47–50]. In the urogenital tract, release of tachykinins from capsaicin-sensitive afferent endings is made responsible for symptoms induced by inflammatory or irritating stimuli [51,52].

The dura mater and its accompanying blood vessels are innervated by peptidergic sensory nerves mainly from ophthalmic branches of the trigeminal nerves. Antidromic stimulation of small caliber trigeminal axons causes neurogenic inflammation in the dura mater as evidenced by marked increases in mast cell activation, protein extravasation, and platelet aggregates within postcapillary venules [16]. Activation of the trigeminovascular system also leads to neurogenic vasodilatation within the dura mater and to cerebral vasodilatation. These processes have been implicated in the pathogenesis of migraine [16]. CGRP is the major neuropeptide involved in this neurogenic vasodilator response [53] and it is elevated in the cranial venous blood of patients suffering acute attacks of migraine. In other forms of primary headache, the whole trigeminovascular reflex is also initiated, causing, on the one hand, antidromic nerve activity leading to release of CGRP, and, on the other hand, orthodromic central stimulation, pain and autonomic reflexes [14,15].

4. MODULATION OF NEUROGENIC INFLAMMATION BY ENDOGENOUS MECHANISMS

Once an inflammation, or specifically a neurogenic inflammation, is initiated, interactions between resident cells, inflammatory cells, immune cells, and the nervous system may perpetuate or terminate the inflammatory response. Several feedback and feedforward loops have been elucidated that can play a role, their anatomical and chemical details, however, depend very much on the site of neurogenic inflammation. Recent findings about the connections between the neural mediators, the neuroendocrine, and the immune system have greatly expanded the understanding of neurogenic inflammation in various diseases.

A negative feedback pathway has been described by which an ongoing inflammatory response at one site produces an inhibition of an inflammatory response at a second site. This pathway may involve activation of C-fibers in ascending tracts in the spinal cord which then stimulate the hypothalamus/pituitary to release adrenocorticotropic hormone (ACTH). ACTH in turn stimulates the cortex of the adrenal gland to release glucocorticoids which act at a secondary site [54]. In view of the peripheral anti-inflammatory actions of Corticotrophin-Releasing Factor (CRF) and of the corticosteroids, any activation of the hypothalamus/pituitary/adrenal axis may exert an anti-inflammatory effect. However, after stimulation of the stress axis for a longer period, the number of substance P-positive nerve fibers increases in mouse skin concomitantly with an increase in mast cell degranulation [55].

Activation of the peripheral sympathoadrenal system and increases in plasma adrenaline and noradrenaline levels can either inhibit or facilitate peripheral neurogenic inflammatory responses depending on species and/or location. The mechanisms range from influencing blood vessel diameter to inhibition of neuropeptide release and inhibition or release of secondary mediators from a variety of cells [56].

Somatostatin has been shown to exert a tonic inhibition at TRPV1-positive nerve endings [57]. Antidromic activation of primary afferent nerve fibers induces not only local but also systemic anti-inflammatory effects: somatostatin released from such nerves has been proposed to be the mediator of this neurohumoral regulatory mechanism of sensory fibers and provides an explanation for the anti-inflammatory effect of counterirritation or acupuncture [58]. In contrast, in another study somatostatin itself induced vasodilatation and plasma protein extravasation in the rat knee joint [59]. Endogenous galanin released from sensory nerves can also exert a tonic inhibitory effect on inflammatory processes [60].

Endogenous opioids, released into the systemic circulation from the pituitary gland or locally from immunocompetent cells, can suppress neurogenic inflammation. In very elegant studies, Stein [61] has demonstrated how the immune system can interact with peripheral sensory nerve endings to inhibit inflammatory pain. Opioid receptors are expressed and transported to the peripheral terminals of sensory neurons. Endogenous opioid peptides released from macrophages or lymphocytes in chronically inflamed tissue can stimulate these opioid receptors located on peripheral sensory nerves, to inhibit pain and the release of excitatory proinflammatory neuropeptides and the further perpetuation of neurogenic inflammation [61]. Opioids released from nerve fibers supplying the airways can influence airway functions via modification of neural transmission [62].

The opioid-like neuropeptide nociceptin/orphanin FQ is also a candidate modulator of neurogenic inflammation. In guinea pigs nociceptin occurs in afferent neurons but in a different set than the tachykinins [63]. The peptide has been shown to reduce vasodilatation and plasma extravasation in response to electrical nerve stimulation or capsaicin by inhibiting the release of tachykinins from capsaicin-sensitive afferents in the skin of rats and the airways of both rats and guinea pigs. Most likely, this occurs via stimulation of prejunctional Opioid Receptor Like (ORL1) receptors [63–66]. However, other studies in skin or inflamed joints, also performed in rats, reveal a proinflammatory action of nociceptin via degranulation of mast cells. The subsequent release of histamine and 5-HT, in turn, leads to stimulation of capsaicin-sensitive afferent nerve endings with concomitant release of substance P and CGRP [67–69]. Vasodilatation in response to nociceptin due to a direct action on vascular smooth muscle has also been observed [70]. It remains to be shown which type of mechanism prevails in models resembling human pathology more closely.

Recently, cannabinoids have been shown to inhibit capsaicin-induced neurogenic inflammation in rat skin and neurogenic vasodilatation in human skin and rat mesenteric arteries [71,72]. This is brought about by activation of cannabinoid receptors on primary afferent nerve endings, leading to inhibition of neuropeptide release. In rat skin, this effects is mediated via CB_1 receptors [73]; however, in the guinea pig bronchi, CB_2 receptors seem to be responsible [74]. There is also evidence for cannabinoids activating CB_2 receptors on keratinocytes, thereby inducing release of β-endorphin and inhibiting primary afferent nerve endings indirectly [75].

Neurogenic inflammation can also play a role in the perpetuation of a disease process, as, for example, in chronic inflammation of the airways. In the intact epithelium neutral endopeptidase (NEP), a cell surface enzyme, degrades substance P and NKA released from sensory nerves thereby suppressing neurogenic inflammation. In asthmatic airways, when the epithelium is shed or NEP is downregulated, tachykinins have an exaggerated effect [11]. In the lung this enzyme is inhibited by cigarette smoke, by viral infections, or by volatile organic chemicals. Typically, when NEP is downregulated or inhibited in the airways, the contribution of the tachykinins and bradykinin to the anaphylactic response following antigen challenge is more prominent. A deficit of NEP could play a role in favoring an exaggerated neurogenic response in upper airway disease [76]. The importance of NEP in curbing the symptoms of neurogenic inflammation has also been demonstrated in models as diverse as electrically stimulated human skin [77] and intestinal inflammation in rats [78].

Enzymatic breakdown of substance P is also affected by angiotensin converting enzyme and another membrane-bound endopeptidase specific for substance P. The resulting N-terminal metabolite, substance P1-7, has been shown to inhibit substance P-induced vasodilatation in a blister model of inflammation in the rat paw [79]. This might represent another endogenous mechanism to curb the local inflammatory process.

Peripheral neural mechanisms in arthritis include the release of substance P and CGRP, which synergistically exert proinflammatory paracrine effects in the synovium. Synthesis, axonal transport, and release of substance P and CGRP are enhanced in arthritis, probably due to the stimulation of NGF synthesis in, and release from, local inflammatory cells [17]. The important role of NGF in the sensitization and perpetuation of neurogenic inflammation has been shown for a wide variety of tissues and disease states, ranging from irritation of the nasal mucosa by toxic vapors [80] over asthma [81,82] to colonic inflammation [83] and overactivity of the urinary bladder [84]. In rat skin NGF leads to edema and, after subchronic administration, increases skin content of CGRP resulting in greater amounts of the peptide being released after stimulation with capsaicin [9,85].

On the whole it is not easy to predict which pathway predominates in a certain inflammatory response. The use of selective inhibitors toward the various transmitters and components of neurogenic inflammation has helped to clarify many pathophysiological mechanisms and prepared the way for future therapeutic interventions.

5. INHIBITION OF NEUROGENIC INFLAMMATION BY EXOGENOUS AGENTS

Though not always initiated by the afferent nervous system, a neurogenic component may be involved in several chronic disorders such as asthma, allergic rhinitis, and other allergic upper airway diseases, rheumatoid arthritis, migraine headache, cystitis, skin diseases like contact

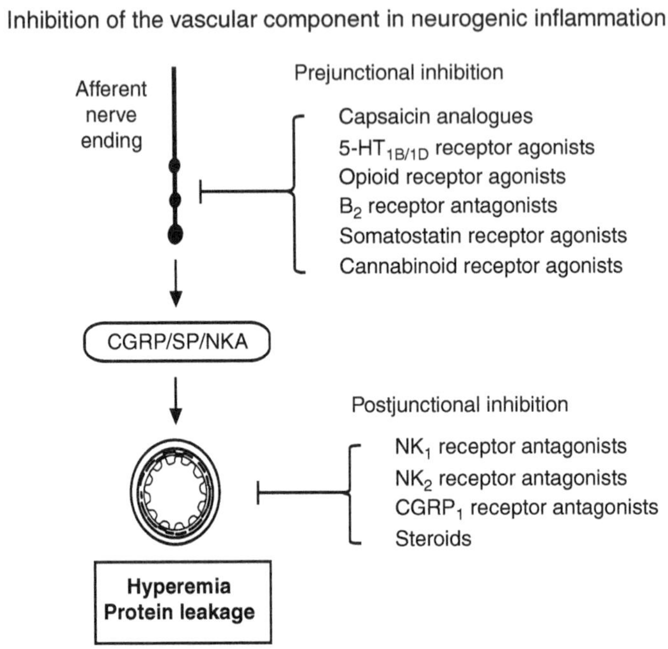

Figure 1. Presentation of several factors that inhibit the activity of afferent nerve fibres to elicit vasodilatation and plasma protein extravasation (modified from Ref. [27] with permission).

dermatitis, or psoriasis [4]. A distinction between neurogenic and immunologic pathomechanisms (e.g., in asthma) is limited because the two systems influence each other at multiple stages after the release of mediators. Since the sensory neuropeptides substance P, NKA, and CGRP appear to be the principal transmitters in these processes, antagonists directed toward their postjunctional receptors, or prejunctional modulators of their release, are the prime targets for developing inhibitors of neurogenic inflammation [27,34,86].

An overview of possible inhibitory influences on the vascular component of neurogenic inflammation is given in Fig. 1.

5.1 Direct inhibition of the effect of substance P and CGRP

As soon as the importance of tachykinins and CGRP for neurogenic inflammation was recognized, research aimed not only at further characterizing the pathophysiology of the relevant diseases but also at developing tachykinin or CGRP receptor antagonists to be used as therapeutics. Potent and metabolically stable antagonists to postjunctional NK_1 receptors can inhibit neurogenic plasma extravasation in the rat skin, but also in visceral organs (CP-96,345 [87,88], SR140333 [89]). With the help of RP67580 it has also been shown that endogenous tachykinins play a role in edema formation in acute allergic responses [90]. Thus, NK_1 receptor antagonists might be useful drugs under these conditions. The intricate connections between peptidergic neuronal mechanisms and the immune system offer numerous opportunities for therapeutic measures [91].

In the airways, bronchoconstriction is mediated primarily by NK_2 receptors, whereas the NK_1 receptor is responsible for other effects such as bronchial inflammation, edema formation, mucosecretion, chemotaxis, and activation of inflammatory cells. Therefore, both NK_1 and NK_2 antagonists might be useful anti-inflammatory and bronchospasmolytic drugs [33]. Tachykinins are likely to be the final mediators and, at least in the guinea pig, the NK_1 antagonist CP96,345 inhibits plasma extravasation in the trachea [92] and nasal mucosa [12], and a NK_2 antagonist, MEN10,627, blocks the contraction of airway smooth muscle [93]. Neurogenic plasma leakage might occur after inhibition of enzymes that normally degrade inflammatory peptides, and neurogenic plasma exudation occurs in the nasal mucosa. These effects are mediated by NK_1 receptors and can be blocked by NK_1 antagonists, but also by a B_2 antagonist [94]. In the nose of rats, the NK_1 receptor antagonist, RP67580, inhibits neurogenic fluid secretion [95].

Edema is a major contributor to airflow obstruction and leads to increased airway fluid. Cellular infiltration and activation participate in this process and increase airway reactivity. In asthma, therapy aimed at preventing inflammation is important to reduce not only airflow obstruction, but also airway reactivity. Neurogenic inflammation in the airways contributes to the development of bronchial hyperresponsiveness and the late asthmatic response in patients with bronchial asthma [11,96]. Although human airways contain sensory nerves, it is not clear whether neurogenic inflammation really occurs in human airways as it does in rats or guinea pigs, but it seems likely. Research with tachykinin antagonists described above have led to the proposal that NK_1 as well as NK_2 receptor antagonists might be useful in treating asthmatics [34].

This hypothesis is further supported by the observation that MDL105,212, a nonselective NK_1/NK_2 receptor antagonist, inhibits respiratory effects produced in guinea pigs exposed to chemical irritants as well as delayed responses to antigen challenge in sensitized animals. These data strongly suggest a role for tachykinins in allergic diseases and prove the utility of tachykinin antagonists [97]. Cyclic peptide antagonists blocking both the NK_1 and the NK_2 receptor can inhibit bradykinin-induced bronchoconstriction and cough in asthmatics when given by aerosol [43].

In earlier studies in humans, emphasis was laid on studying NK_2 receptors in airway smooth muscle, because the expression of NK_2 receptors was found to be increased in chronic obstructive pulmonary disease [98]. Indeed, under these conditions NK_2 receptors play an important role in mediating bronchoconstriction, plasma protein extravasation, and activation of immune cells [99]. Thus, selective NK_2 receptor antagonists have been proposed as good candidates for a treatment of allergic responses and of airway hyperresponsiveness in asthma [34,93]. However, recently the importance of the other types of tachykinin receptors was recognized and the tendency is now to look for dual NK_1/NK_2 or even triple $NK_1/NK_2/NK_3$ antagonists [100].

In the urinary tract, tachykinins released from sensory nerves located in the muscle layer, mucosa, and submucosa can trigger plasma protein extravasation and increased urinary bladder motor activity via NK_1 and NK_2 receptors, while they do not seem to play a role in normal bladder function [34]. Increasing evidence points to an involvement of neurogenic inflammation in painful pelvic diseases [101] where tachykinin antagonists may represent a treatment option. An early study demonstrated the effectivity of the NK_1 antagonist CP96,345 against capsaicin-induced plasma extravasation in the guinea pig ureter [88]. Both NK_1 and NK_2 receptors are involved in neurogenically mediated bladder dysfunctions in rats and guinea pigs. However, in most other species including humans only NK_2 receptor antagonists seem to be useful against such symptoms [34,102].

Although the largest source of tachykinins in the gut is represented by intrinsic neurons, the neurogenic release of tachykinins in inflammatory responses in the gastrointestinal tract is likely [103–106]. There is a potential for tachykinin antagonists in inflammatory bowel disease, specifically to prevent pain and reflex hypermotility [107]. The NK_1 antagonists CP96,345 and RP67580 have been found beneficial in rats and mice with colitis of various origin [108,109]. In humans, tachykinins released from peripheral endings of sensory nerves can cause exaggerated smooth muscle contraction in irritable bowel syndrome or cystitis for the most part via NK_2 receptors. Studies with nepadutant, a new and potent NK_2 receptor antagonist, in human intestinal tissue *in vitro* as well as in healthy volunteers yielded promising results to justify the development of NK_2 antagonists as drugs against gastrointestinal diseases that involve sensory nerves [107].

Contrary to the tachykinins, CGRP is the prime mediator of neurogenic vasodilatation but not of the extravasation. Therefore, research concentrated on migraine and other forms of headache, where CGRP-mediated vasodilatation has been shown to play a major role. Activation of a trigeminovascular reflex causing neurogenic inflammation in the meninges and cerebral vasodilatation due to the release of CGRP are major components in the pathogenesis of migraine headache [16,110]. Of all possible mediators, CGRP is the most important neuropeptide involved in this pathomechanism and antagonists targeting the trigeminovascular system may be useful drugs against migraine and other forms of headache [111,112].

In some animal models of meningeal neurogenic inflammation, release of both substance P and CGRP is relevant [113]; however, vasodilatation in rat dura after topical application of capsaicin can be abolished by the CGRP antagonist hCGRP (8-37) [114]. Also in humans only CGRP seems to play a role. This neuropeptide is elevated in the cranial venous blood of patients suffering acute attacks of migraine and other forms of primary headache [15,112].

Several preclinical studies proved the effectivity of BIBN4096BS (olcegepant) as a $CGRP_1$ receptor antagonist that inhibited CGRP-induced vasodilatation as well as vasodilatation induced by antidromic stimulation of a sensory nerve [115–118]. In patients, this non-peptide CGRP antagonist showed a significant effect against acute migraine attacks, without causing any cardiovascular side effects [119]. The latter was also confirmed in healthy volunteers, where infusion of BIBN4096BS did not induce any changes in hemodynamics [120].

Any systemic application of CGRP antagonists has, however, to be seen in the light that CGRP is considered an endogenous myocardial protective substance. In fact, the nonpeptide CGRP antagonist WO98/11128 was shown to inhibit CGRP-induced relaxation of human coronary arteries [121].

Furthermore, BIBN4096BS has been shown to potently block CGRP-induced dilatation of human pial blood vessels *in vitro* [122] as well as the capsaicin-induced dilatation of the carotid artery in pigs, which was mediated by the release of CGRP [123].

5.2 Inhibition of neurogenic inflammation by indirect means: inhibition of mediator release and functional antagonism

Another very efficient method to inhibit neurogenic inflammation is to block the release of mediators from the sensory nerve endings. This is shown best in the therapy of migraine, where patients have to rely on the triptans until CGRP antagonists are introduced into the medical practice. Administration of the $5\text{-}HT_{1B/1D}$ receptor agonist sumatriptan returned the elevated CGRP during migraine attack to control levels [15,111]. Triptans and also dihydroergotamine inhibit CGRP release by hyperpolarizing prejunctional sensory nerve terminals following $5\text{-}HT_{1B/1D}$ receptor activation [124]. Thereby they limit overdistension of cerebral arterioles. $5\text{-}HT_{1B/1D}$ receptors are also located on central parts of the trigeminovascular reflex and their activation interrupts central aspects of the headache process [15]. Other consequences of neurogenic inflammation in the dura mater, such as increases in mast cell activation, protein extravasation, and platelet aggregation within postcapillary venules, which most likely are induced by substance P release, are also inhibited by sumatriptan [125]. A recent study showed that in rats also the cyclooxygenase (COX)-2 inhibitor parecoxib was able to inhibit plasma protein extravasation from dura in response to antidromic trigeminal activation. [126]. However, neither the inhibition of COX-1 nor of COX-2 diminished plasma protein extravasation in response to mustard oil in rat skin [127].

A more general approach to block neurogenic inflammation – and the most important one under experimental conditions – is the desensitization or long-lasting impairment of the neuropeptide-containing primary afferents by pretreatment with capsaicin [1,128,129]. Two types of desensitization play a role: a functional desensitization to short-term local application to nerve endings leads to reduction or loss of responsiveness of sensory neurons to many stimuli. This may underlie the anti-inflammatory effect of capsaicin and related vanilloids. The second type is a pharmacological desensitization, where repeated administration of capsaicin leads to a decline in the response that is normally evoked by capsaicin. Both desensitization processes depend on calcium uptake by the sensory neuron through the TRPV1 channel. Pretreatment with high doses of capsaicin decreases the number of substance P fibers in peripheral tissues and reduces symptoms of neurogenic inflammation [20]. This procedure, however, has mainly experimental significance for the inhibition of neurogenic inflammation, since the local treatment in humans with capsaicin turned out to be too pungent. As an example, the pathophysiological significance of neurogenic inflammation in the perpetuation of arthritis has been recognized [130]. Capsaicin, owing to its peculiar capacity to markedly reduce the neurogenic component of inflammation through the depletion of neuropeptides from nerve endings, has been shown to be of certain benefit for the treatment of pain and inflammation in rheumatoid joint diseases. However, only a nonpungent, desensitizing and neuropeptide-depleting analog would be of value [131]. In the meantime such nonpungent capsaicin analogs have been developed and may turn out to be useful as analgesic/antihyperalgesic agents in inflammatory conditions [132]. Until such compounds are readily available, there still remains the possibility to administer local anesthetics before capsaicin during the first treatment sessions. With such a regimen a beneficial effect of repeated capsaicin application has been demonstrated in patients with idiopathic rhinitis [133].

Since the capsaicin receptor was identified as the TRPV1 channel, another line of research aims at using TRPV1 antagonists to prevent the response of afferent nerve endings to irritating stimuli. Among these antagonists are iodo-resiniferatoxin and JYL1421 (SC0030), which have been shown to block several neurogenic responses to capsacin or acetic acid [134,135]. Therapeutically useful TRPV1 antagonists are also under development (see Ref. [136]).

Somatostatin inhibits presynaptically the release of substance P and CGRP from primary afferent nerve terminals thereby preventing plasma extravasation [22]. In rats, the somatostatin analog octreotide was ineffective against the effects of trigeminal stimulation with capsaicin [137]. In contrast, a clinical study in patients with cluster headache showed a beneficial effect of subcutaneously injected octreotide, although this peptide does not directly affect blood vessels [138]. Newly developed somatostatin analogs without any endocrine function have also been shown to inhibit neurogenic inflammation by blocking the release of the neuropeptides from sensory nerve terminals [58].

Opioid agonists inhibit primary afferent-dependent plasma extravasation [139,140]. A μ-agonist has been shown to inhibit edema formation prejunctionally only when the sensory nerve was stimulated at lower frequencies, but not at high-frequency stimulation [89]. Agonists at the μ- and κ-opioid receptor can exert an inhibitory action on neurogenic extravasation in rat skin which appears independent of their central nervous system and systemic effects [141]. However, other studies suggest a central rather than peripheral site of action for opioids [142] and complex interactions of opioids with sympathetic and histaminergic mechanisms [143]. Also in the mouse intestine inflammation-induced plasma extravasation could be ameliorated by administration of opioids, κ agonists being most potent in this model [144]. The distinction between peripheral and central sites of action is important since the systemic administration of peripherally selective μ-agonists may be considered undesirable because of their constipating effects; therefore, a local application might be advantageous, e.g., in a joint [61,145,146].

Several lines of evidence suggest that the sensory nerves are not directly stimulated after antigen challenge, but indirectly by endogenous kinins, to initiate neurogenic inflammation [147]. This response to bradykinin is mediated by B_2 receptors. The B_2 receptor antagonist HOE140 (icatibant) can inhibit the increase in plasma extravasation in the trachea and nasal mucosa, and it can inhibit the contraction of airway smooth muscle induced by antigen challenge in guinea pigs [12]. Also in immediate hypersensitivity/inflammatory reactions in other organs such as the urinary bladder or the airways, bradykinin, acting via B_2 receptors, and prostaglandin formation, is involved in C-fiber excitation. B_2 antagonists and COX inhibitors can be used to suppress these reactions [148,149]. The B_2 antagonists FR173657 and icatibant have been shown to reduce plasma protein extravasation in models of visceral and cutaneous inflammation in rats and guinea pigs [150]. However, icatibant was inactive when plasma protein extravasation was elicited by saphenous nerve stimulation in rat skin 2705 [89]. Nonpeptide bradykinin antagonists are currently being developed as therapeutics [151].

Glucocorticoids, applied systemically or locally, inhibit all types of inflammation and therefore also reduce neurogenic plasma extravasation in nasal mucosa, trachea, and urinary bladder postjunctionally [152]. However, this is not a specific effect.

In conclusion, it can be noted that the newly developed selective tachykinin, CGRP and bradykinin antagonists have been a clear step forward in the understanding of neurogenic inflammation. They also might have beneficial effects in certain diseases processes. However, in the light of the widespread distribution of these peptide receptors in different tissues and organs also their possible side effects have to be considered.

REFERENCES

1. Jancsó N, Jancsó-Gábor A, Szolcsányi J. Direct evidence for neurogenic inflammation and its prevention by denervation and by pretreatment with capsaicin. Br J Pharmacol 1967;31:138–151.
2. Holzer P. Local effector functions of capsaicin-sensitive sensory nerve endings: involvement of tachykinins, calcitonin gene-related peptide and other neuropeptides. Neuroscience 1988;24:739–768.
3. Otsuka M, Yoshioka K. Neurotransmitter functions of mammalian tachykinins. Physiol Rev 1993;73:229–308.
4. Meggs W. Neurogenic inflammation and sensitivity to environmental chemicals. Environ Health Perspect 1993;101:234–238.
5. Lynn B, Schütterle S, Pierau F. The vasodilator component of neurogenic inflammation is caused by a special subclass of heat-sensitive nociceptors in the skin of the pig. J Physiol 1996;494:587–593.

6. Wood J, Docherty R. Chemical activators of sensory neurons. Annu Rev Physiol 1997;59:457–482.
7. Caterina MJ, Schumacher MA, Tominaga M, Rosen TA, Levine JD, Julius D. The capsaicin receptor: a heat-activated ion channel in the pain pathway. Nature 1997;389:816–824.
8. Vasko M, Campbell W, Waite K. Prostaglandin E_2 enhances bradykinin-stimulated release of neuropeptides from rat sensory neurons in culture. J Neurosci 1994;14:4987–4997.
9. Amann R, Schuligoi R, Herzeg G, Donnerer J. Intraplantar injection of nerve growth factor into the rat hind paw: local edema and effects on thermal nociceptive threshold. Pain 1995;64:323–329.
10. Coleridge H, Coleridge J. Airway axon reflexes – where now? News Physiol Sci 1995;10:91–96.
11. Barnes P, Baraniuk J, Belvisi M. Neuropeptides in the respiratory tract. Am Rev Respir Dis 1991;144:1187–1198.
12. Ricciardolo F, Nadel J, Bertrand C, Yamawaki I, Chan B, Geppetti P. Tachykinins and kinins in antigen-evoked plasma extravasation in guinea-pig nasal mucosa. Eur J Pharmacol 1994;261:127–132.
13. Bertrand C, Geppetti P. Tachykinin and kinin receptor antagonists: therapeutic perspectives in allergic airway disease. Trends Pharmacol Sci 1996;17:255–259.
14. Vass Z, Steyger PS, Hordichok AJ, Trune DR, Jancsó G, Nuttall AL. Capsaicin stimulation of the cochlea and electric stimulation of the trigeminal ganglion mediate vascular permeability in cochlear and vertebro-basilar arteries: a potential cause of inner ear dysfunction in headache. Neuroscience 2001;103:189–201.
15. Edvinsson L. Aspects of the pathophysiology of migraine and cluster headache. Pharmacol Toxicol 2001;89:65–73.
16. Goadsby PJ. Recent advances in understanding migraine mechanisms, molecules and therapeutics. Trends Mol Med 2007;13:39–44.
17. Donnerer J, Schuligoi R, Stein C. Increased content and transport of substance P and calcitonin gene-related peptide in sensory nerves innervating inflamed tissue: evidence for a regulatory function of nerve growth factor in vivo. Neuroscience 1992;49:693–698.
18. Wimalawansa S. Calcitonin gene-related peptide and its receptors: molecular genetics, physiology, pathophysiology, and therapeutic potentials. Endocrine Rev 1996;17:533–585.
19. Brain S. Sensory neuropeptides: their role in inflammation and wound healing. Immunopharmacology 1997;37:133–152.
20. Györfi A, Fazekas A, Irmes F, Jakab G, Sütö T, Rosivall L. Role of substance P (SP) in development of symptoms of neurogenic inflammation in the oral mucosa of the rat. J Periodont Res 1993;28:191–196.
21. Cruwys S, Kidd B, Mapp P, Walsh D, Blake D. The effects of calcitonin gene-related peptide on formation of intra-articular oedema by inflammatory mediators. Br J Pharmacol 1992;107:116–119.
22. Green P, Basbaum A, Levine J. Sensory neuropeptide interactions in the production of plasma extravasation in the rat. Neuroscience 1992;50:745–749.
23. Nakagawa N, Sano H, Iwamoto I. Substance P induces the expression of intercellular adhesion molecule-1 on vascular endothelial cells and enhances neutrophil transendothelial migration. Peptides 1995;16:721–725.

24. Wiedermann C, Niedermühlbichler M, Zilian U, Geissler D, Lindley I, Braunsteiner H. Priming of normal human neutrophils by tachykinins: tuftsin-like inhibition of invitro chemotaxis stimulated by formylpeptide or interleukin-8. Regul Pept 1991;36:359–368.
25. Fantini F, Magnoni C, Pincelli C, Gianetti A. Neurogenic inflammation and the skin: neural modulation of cutaneous inflammatory reactions. Eur J Dermatol 1995;5:349–357.
26. Domeij S, Carlsöö B, Dahlqvist A, Forsgren S. Occurrence of mast cells in relation to the distribution of nerve fibres in the rat larynx. Acta Otolaryngol 1991;111:981–989.
27. Holzer P. Neurogenic vasodilatation and plasma leakage in the skin. Gen Pharmacol 1998;30:5–11.
28. Jancsó G, Pierau F, Sann H. Mustard oil-induced cutaneous inflammation in the pig. Agents Actions 1993;39:31–34.
29. Pierau F, Ernst R, Faulstroh K, Sann H, Donnerer J. Neurogenic inflammation in the pig's skin: flare reaction and plasma extravasation. In: Hökfelt T, Schaible H, Schmidt R, eds. Neuropeptides, Nociception, and Pain. London: Chapman & Hall, 1994; pp. 85–102.
30. Petersen L, Church M, Skov P. Histamine is released in the wheal but not the flare following challenge of human skin in vivo: a microdialysis study. Clin Exp Allergy 1996;27:284–295.
31. Petersen L, Winge K, Brodin E, Skov P. No release of histamine and substance P in capsaicin-induced neurogenic inflammation in intact human skin in vivo: a microdialysis study. Clin Exp Allergy 1997;27:957–965.
32. Geppetti P, Materazzi S, Nicoletti P. The transient receptor potential vanilloid 1: role in airway inflammation and disease. Eur J Pharmacol 2006;533:207–214.
33. Joos G, Germonpre P, Kips J, Peleman R, Pauwels R. Sensory neuropeptides and the human lower airways: present state and future directions. Eur Respir J 1994;7:1161–1171.
34. Lecci A, Giuliani S, Tramontana M, Carini F, Maggi C. Peripheral actions of tachykinins. Neuropeptides 2000;34:303–313.
35. McDonald D. Infections intensify neurogenic plasma extravasation in the airway mucosa. Am Rev Respir Dis 1992;146:S40–S44.
36. Baluk P, Nadel J, McDonald D. Substance P-immunoreactive sensory axons in the rat respiratory tract: a quantitative study of their distribution and role in neurogenic inflammation. J Comp Neurol 1992;319:586–598.
37. Piedimonte G, Hoffman J, Husseini W, Hiser W, Nadel J. Effect of neuropeptides released from sensory nerves on blood flow in the rat airway microcirculation. J Appl Physiol 1992;72:1563–1570.
38. DeRose V, Robbins R, Snider R, Spurzem J, Thiele G, Rennard S et al. Substance P increases neutrophil adhesion to bronchial epithelial cells. J Immunol 1994;152:1339–1346.
39. Von Essen S, Rennard S, O'Neill D, Ertl R, Robbins R, Koyama S et al. Bronchial epithelial cells release neutrophil chemotactic activity in response to tachykinins. Am J Physiol 1992;263:L226–L231.
40. Maggi C. The effects of tachykinins on inflammatory and immune cells. Regul Pept 1997;70:75–90.
41. Joos GF, De Swert KO, Schelfhout V, Pauwels RA. The role of neural inflammation in asthma and chronic obstructive pulmonary disease. Ann N Y Acad Sci 2003;992:218–230.

42. Undem B, Hunter D, Liu M, Haak-Frendscho M, Oakragly A, Fischer A. Allergen-induced sensory neuroplasticity in airways. Int Arch Allergy Immunol 1999;118:150–153.
43. Joos G, Kips R, Peleman R, Pauwels R. Tachykinin antagonists and the airways. Arch Int Pharmakodyn Ther 1995;329:205–219.
44. Joos G. Neurogenic inflammation in human airways: is it important? Thorax 1995;50:217–219.
45. Sann H, Dux M, Schemann M, Jancsó G. Neurogenic inflammation in the gastrointestinal tract of the rat. Neurosci Lett 1996;219:147–150.
46. Holzer P. Efferent-like roles of afferent neurons in the gut: blood flow regulation and tissue protection. Auton Neurosci 2006;125:70–75.
47. Prystowsky JB, Rege RV. Neurogenic inflammation in cholecystitis. Dig Dis Sci 1997;42:1489–1494.
48. McVey DC, Vigna SR. The capsaicin VR1 receptor mediates substance P release in toxin A-induced enteritis in rats. Peptides 2001;22:1439–1446.
49. Yiangou Y, Facer P, Dyer NH, Chan CL, Knowles C, Williams NS et al. Vanilloid receptor 1 immunoreactivity in inflamed human bowel. Lancet 2001; 357:1338–1339.
50. Matthews PJ, Aziz Q, Facer P, Davis JB, Thompson DG, Anand P. Increased capsaicin receptor TRPV1 nerve fibres in the inflamed human oesophagus. Eur J Gastroenterol Hepatol 2004;16:897–902.
51. Ahluwalia A, Giuliani S, Scotland R, Maggi CA. Ovalbumin-induced neurogenic inflammation in the bladder of sensitized rats. Br J Pharmacol 1998;124:190–196.
52. Candenas L, Lecci A, Pinto FM, Patak E, Maggi CA, Pennefather JN. Tachykinins and tachykinin receptors: effects in the genitourinary tract. Life Sci 2005;76:835–862.
53. Edvinsson L, Goadsby PJ. Neuropeptides in headache. Eur J Neurol 1998;5:329–341.
54. Green P, Miao F, Jänig W, Levine J. Negative feedback neuroendocrine control of the inflammatory response in rats. J Neurosci 1995;15:4678–4686.
55. Peters EMJ, Kuhlmei A, Tobin DJ, Müller-Röver S, Klapp BF, Arck PC. Stress exposure modulates peptidergic innervation and degranulates mast cells in murine skin. Brain Behav Immun 2005;19:252–262.
56. Häbler H-J. Neurogenic inflammation and sympathetic nervous system. In: Schmidt RF, Willis WD, eds. Encyclopedia of Pain. Berlin-Heidelberg-. New York: Springer, 2007; pp. 1276–1278.
57. Carlton SA, Zhou ST, Du JH, Hargett GL, Ji GC, Coggeshall RE. Somatostatin modulates the transient receptor potential vanilloid 1 (TRPV1) ion channel. Pain 2004;110:616–627.
58. Pintér E, Helyes Z, Szolcsányi J. Inhibitory effect of somatostatin on inflammation and nociception. Pharmacol Ther 2006;112:440–456.
59. Lam FFY, Ng ESK. Characterisation of somatostatin actions on knee joint blood vessels of the rat. Eur J Pharmacol 2003;474:295–301.
60. Jancsó G, Sántha P, Horváth V, Pierau F. Inhibitory neurogenic modulation of histamine-induced cutaneous plasma extravasation in the pigeon. Regul Pept 2000;95:75–80.
61. Stein C. The control of pain in peripheral tissue by opioids. N Engl J Med 1995;332:1685–1690.
62. Groneberg DA, Quarcoo D, Frossard N, Fischer A. Neurogenic mechanisms in bronchial inflammatory diseases. Allergy 2004;59:1139–1152.

63. Fischer A, Forssmann W-G, Undem BJ. Nociceptin-induced inhibition of tachykinergic neurotransmission in guinea pig bronchus. J Pharmacol Exp Ther 1998;285:902–907.
64. Helyes Z, Németh J, Pintér E, Szolcsányi J. Inhibition by nociceptin of neurogenic inflammation and the release of SP and CGRP from sensory nerve terminals. Br J Pharmacol 1997;121:613–615.
65. Häbler H-J, Timmermann L, Stegmann J-U, Jänig W. Effects of nociceptin and nocistatin on antidromic vasodilatation in hairless skin of the rat hindlimb in vivo. Br J Pharmacol 1999;127:1719–1727.
66. Corboz MR, Rivelli MA, Egan RW, Tulshian D, Matasi J, Fawzi AB et al. Nociceptin inhibits capsaicin-induced bronchoconstriction in isolated guinea pig lung. Eur J Pharmacol 2000;402:171–179.
67. Kimura T, Kitaichi K, Hiramatsu K, Yoshida M, Ito Y, Kume H et al. Intradermal application of nociceptin increases vascular permeability in rats: the possible involvement of histamine release from mast cells. Eur J Pharmacol 2000;407:327–332.
68. Grond S, Gabriel A, Pietruck C, Yu LC, Xie GX, Palmer PP. Bi-directional modulation of 5-hydroxytryptamine-induced plasma extravasation in the rat knee joint by nociceptin. Neuroscience 2001;103:1085–1092.
69. Zhang CF, Mcdougall JJ. Stimulation of sensory neuropeptide release by nociceptin/orphanin FQ leads to hyperaemia in acutely inflamed rat knees. Br J Pharmacol 2006;148:938–946.
70. McDougall JJ. Involvement of sympathetic efferents but not capsaicin-sensitive afferents in nociceptin-mediated dual control of rat synovial blood flow. Am J Physiol 2003;284:R1477–R1485.
71. Dvorak M, Watkinson A, McGlone F, Rukwied R. Histamine induced responses are attenuated by a cannabinoid receptor agonist in human skin. Inflamm Res 2003;52:238–245.
72. Ralevic V. Cannabinoid modulation of peripheral autonomic and sensory neurotransmission. Eur J Pharmacol 2003;472:1–21.
73. Richardson JD, Kilo S, Hargreaves KM. Cannabinoids reduce hyperalgesia and inflammation via interaction with peripheral CB1 receptors. Pain 1998;75:111–119.
74. Yoshihara S, Morimoto H, Yamada Y, Abe T, Arisaka O. Cannabinoid receptor agonists inhibit sensory nerve activation in guinea pig airways. Am J Respir Crit Care Med 2004;170:941–946.
75. Ibrahim MM, Porreca F, Lai J, Albrecht PJ, Rice FL, Khodorova A et al. CB2 cannabinoid receptor activation produces antinociception by stimulating peripheral release of endogenous opioids. Proc Natl Acad Sci U S A 2005;102:3093–3098.
76. Evangelista S, Paoli S, Giachetti A, Manzini S. Involvement of tachykinin NK_1 receptors in plasma protein extravasation induced by tachykinins in the guinea pig upper airways. Neuropeptides 1997;31:65–70.
77. Kramer HH, Schmidt K, Leis S, Schmelz M, Sommer C, Birklein F. Inhibition of neutral endopeptidase (NEP) facilitates neurogenic inflammation. Exp Neurol 2005;195:179–184.
78. Kirkwood KS, Bunnett NW, Maa J, Castagliolo I, Liu B, Gerard N et al. Deletion of neutral endopeptidase exacerbates intestinal inflammation induced by Clostridium difficile toxin A. Am J Physiol 2001;281:G544–G551.
79. Wiktelius D, Khalil Z, Nyberg F. Modulation of peripheral inflammation by the substance P N-terminal metabolite substance P1-7. Peptides 2006;27:1490–1497.

80. Wilfong ER, Dey RD. Nerve growth factor and substance P regulation in nasal sensory neurons after toluene diisocyanate exposure. Am J Respir Cell Mol Biol 2004;30:793–800.
81. Frossard N, Freund W, Advenier C. Nerve growth factor and its receptors in asthma and inflammation. Eur J Pharmacol 2004;500:453–465.
82. Nassenstein C, Kutschker J, Tumes D, Braun A. Neuro-immune interaction in allergic asthma: role of neurotrophins. Biochem Soc Trans 2006;34:591–593.
83. Delafoy L, Gelot A, Ardid D, Eschalier A, Bertrand C, Doherty AM et al. Interactive involvement of brain derived neurotrophic factor, nerve growth factor, and calcitonin gene related peptide in colonic hypersensitivity in the rat. Gut 2006;55:940–945.
84. Lamb K, Gebhart GF, Bielefeldt K. Increased nerve growth factor expression triggers bladder overactivity. J Pain 2004;5:150–156.
85. Bowles WR, Sabino M, Harding-Rose C, Hargreaves KM. Chronic nerve growth factor administration increases the peripheral exocytotic activity of capsaicin-sensitive cutaneous neurons. Neurosci Lett 2006;403:305–308.
86. Donnerer J, Amann R. The inhibition of neurogenic inflammation. Gen Pharmacol 1993;24:519–529.
87. Lembeck F, Donnerer J, Tsuchiya M, Nagahisa A. The non-peptide tachykinin antagonist, CP-96,345, is a potent inhibitor of neurogenic inflammation. Br J Pharmacol 1992;105:527–530.
88. Nagahisa A, Kanai Y, Suga O, Taniguchi K, Tsuchiya M, Lowe J III. et al. Antiinflammatory and analgesic activity of a non-peptide substance P receptor antagonist. Eur J Pharmacol 1992;217:191–195.
89. Towler P, Brain S. Activity of tachykinin NK1 and bradykinin B2 receptor in neurogenic inflammation in the rat. Neurosci Lett 1998;257:5–8.
90. Wilsoncroft P, Euzger H, Brain S. Effect of a neurokinin-1 (NK1) receptor antagonist on oedema formation induced by tachykinins, carrageenin and an allergic response in guinea-pig. Neuropeptides 1994;26:405–411.
91. Peters EMJ, Ericson ME, Hosoi J, Seiffert K, Hordinsky MK, Ansel JC et al. Neuropeptide control mechanisms in cutaneous biology: physiological and clinical significance. J Invest Dermatol 2006;126:1937–1947.
92. Bertrand C, Geppetti P, Baker J, Yamawaki I, Nadel J. Role of neurogenic inflammation in antigen-induced vascular extravasation in guinea pig trachea. J Immunol 1993;150:1479–1485.
93. Perretti F, Ballati L, Manzini S, Maggi C, Evangelista S. Antibronchospastic activity of MEN10,627, a novel tachykinin NK_2 receptor antagonist, in guinea-pig airways. Eur J Pharmacol 1995;273:129–135.
94. Baluk P, Thurston G, Murphy T, Bunnett N, Mcdonald D. Neurogenic plasma leakage in mouse airways. Br J Pharmacol 1999;126:522–528.
95. Moussaoui S, Montier F, Carruette A, Blanchard J, Laduron P, Garret C. A non-peptide NK1-receptor antagonist, RP 67580, inhibits neurogenic inflammation postsynaptically. Br J Pharmacol 1993;109:259–264.
96. Kamikawa Y. Neurogenic control of airway smooth muscle function. J Pharmacol Toxicol Methods 1994;31:207–213.
97. Kudlacz E, Knippenberg R, Logan D, Burkholder T. Effect of MDL 105,212, a nonpeptide NK-1/ NK-2 receptor antagonist in an allergic guinea pig model. J Pharmacol Exp Ther 1996;279:732–739.

98. Bai TR, Zhou DY, Weir T, Walker B, Hegele R, Hayashi S et al. Substance-P (NK_1)-receptor and neurokinin-A (NK_2)-receptor gene-expression in inflammatory airway diseases. Am J Physiol 1995;13:L309–L317.
99. Van Schoor J, Joos GF, Chasson BL, Brouard RJ, Pauwels RA. The effect of the NK_2 tachykinin receptor antagonist SR 48968 (saredutant) on neurokinin A-induced bronchoconstriction in asthmatics. Eur Respir J 1998;12:17–23.
100. De Swert KO, Joos GF. Extending the understanding of sensory neuropeptides. Eur J Pharmacol 2006;533:171–181.
101. Wesselmann U. Neurogenic inflammation and chronic pelvic pain. World J Urol 2001;19:180–185.
102. Patacchini R, Maggi CA. Peripheral tachykinin receptors as targets for new drugs. Eur J Pharmacol 2001;429:13–21.
103. Mantyh CR, Gates TS, Zimmerman RP, Welton ML, Passaro EP, Jr., Vigna SR et al. Receptor binding sites for substance P, but not substance K or neuromedin K, are expressed in high concentrations by arterioles, venules, and lymph nodules in surgical specimens obtained from patients with ulcerative colitis and Crohn disease. Proc Natl Acad Sci U S A 1988;85:3235–3239.
104. Buéno L, Fioramonti J, Delvaux M, Frexinos J. Mediators and pharmacology of visceral sensitivity: from basic to clinical investigations. Gastroenterology 1997;112:1714–1743.
105. Holzer P, Holzer-Petsche U. Tachykinins in the gut. Part II. Roles in neural excitation, secretion and inflammation. Pharmacol Ther 1997;73:219–263.
106. Holzer P. Implication of tachykinins and calcitonin gene-related peptide in inflammatory bowel disease. Digestion 1998;59:269–283.
107. Lecci A, Capriati A, Altamura M, Maggi CA. Tachykinins and tachykinin receptors in the gut, with special reference to NK_2 receptors in human. Auton Neurosci 2006;126:232–249.
108. Stucchi AF, Shofer S, Leeman S, Materne O, Beer E, McClung J et al. NK-1 antagonist reduces colonic inflammation and oxidative stress in dextran sulfate-induced colitis in rats. Am J Physiol 2000;279:G1298–1306.
109. Rijnierse A, Van Zijl KMF, Koster AS, Nijkamp FP, Kraneveld AD. Beneficial effect of tachykinin NK_1 receptor antagonism in the development of hapten-induced colitis in mice. Eur J Pharmacol 2006;548:150–157.
110. Durham PL. Calcitonin gene-related peptide (CGRP) and migraine. Headache 2006;46:S3–S8.
111. Edvinsson L. New therapeutic target in primary headaches – blocking the CGRP receptor. Ext Opin Ther Targets 2003;7:377–383.
112. Edvinsson L. Blockade of CGRP receptors in the intracranial vasculature: a new target in the treatment of headache. Cephalalgia 2004;24:611–622.
113. Ebersberger A, Averbeck B, Messlinger K, Reeh P. Release of substance P, calcitonin gene-related peptide and prostaglandin E_2 from rat dura mater encephali following electrical and chemical stimulation in vitro. Neuroscience 1999;89:901–907.
114. Dux M, Sántha P, Jancsó G. Capsaicin-sensitive neurogenic sensory vasodilatation in the dura mater of the rat. J Physiol 2003;552:859–867.
115. Doods H, Hallermayer G, Wu D, Entzeroth M, Rudolf K, Engel W et al. Pharmacological profile of BIBN4096BS, the first selective small molecule CGRP antagonist. Br J Pharmacol 2000;129:420–423.

116. Grant AD, Tam CW, Lazar Z, Shih MK, Brain SD. The calcitonin gene-related peptide (CGRP) receptor antagonist BIBN4096BS Blocks CGRP and adrenomedullin vasoactive responses in the microvasculature. Br J Pharmacol 2004;142:1091–1098.
117. Sheykhzade M, Lind H, Edvinsson L. Noncompetitive antagonism of BIBN4096BS on CGRP-induced responses in human subcutaneous arteries. Br J Pharmacol 2004;143:1066–1073.
118. Hay DL, Poyner D. The preclinical pharmacology of BIBN4096BS, a CGRP antagonist. Cardiovasc Drug Rev 2005;23:31–42.
119. Olesen J, Diener HC, Husstedt PJ, Goadsby PJ, Hall D, Meier U et al., for the BIBN 4096 BS Clinical Proof of Concept Study Group. Calcitonin gene-related peptide receptor antagonist BIBN4096BS for the acute treatment of migraine. N Engl J Med 2004;350:1104–1110.
120. Petersen KA, Birk S, Lassen LH, Kruuse C, Jonassen O, Olesen J. The novel CGRP-antagonist, BIBN4096BS does not affect cerebral hemodynamics in healthy volunteers. Cephalalgia 2003;23:729.
121. Hasbak P, Opgaard OS, Eskesen K, Schifter S, Arendrup H, Longmore J et al. Investigation of CGRP receptors and peptide pharmacology in human coronary arteries. Characterization with a nonpeptide antagonist. J Pharmacol Exp Ther 2003;304:326–333.
122. Moreno MJ, Abounader R, Hebert E, Doods H, Hamel E. Efficacy of the non-peptide CGRP receptor antagonist BIBN4096BS in blocking CGRP-induced dilations in human and bovine cerebral arteries: potential implications in acute migraine treatment. Neuropharmacology 2002;42:568–576.
123. Kapoor K, Arulmani U, Heiligers JPC, Garrelds IM, Willems EW, Doods H et al. Effects of the CGRP receptor antagonist BIBN4096BS on capsaicin-induced carotid haemodynamic changes in anaesthetised pigs. Br J Pharmacol 2003;140:329–338.
124. Limmroth V, Katsarava Z, Liedert B, Guehring H, Schmitz K, Diener H et al. An in vivo rat model to study calcitonin gene-related peptide release following activation of the trigeminal vascular system. Pain 2001;92:101–106.
125. Buzzi M, Dimitriadou V, Theoharides T, Moskowitz M. 5-Hydroxytryptamine receptor agonists for the abortive treatment of vascular headaches block mast cell, endothelial and platelet activation within the rat dura mater after trigeminal stimulation. Brain Res 1992;583:137–149.
126. Schuh-Hofer S, Tayefeh M, Reuter U, Dirnagl U, Arnold G. Effects of parecoxib on plasma protein extravasation and C-fos expression in the rat. Headache 2006;46:276–285.
127. Helyes Z, Pintér E, Németh J, Keri G, Than M, Oroszi G et al. Anti-inflammatory effect of synthetic somatostatin analogues in the rat. Br J Pharmacol 2001;134:1571–1579.
128. Holzer P. Capsaicin: cellular targets, mechanisms of action, and selectivity for thin sensory neurons. Pharmacol Rev 1991;43:143–201.
129. Dux M, Sann H, Schemann M, Jancsó G. Changes in fibre population of the rat hairy skin following selective chemodenervation by capsaicin. Cell Tissue Res 1999;296:471–477.
130. Konttinen Y, Kemppinen P, Segersberg M, Hukkanen M, Rees R, Santavirta S et al. Peripheral and spinal neural mechanisms in arthritis, with particular reference to treatment of inflammation and pain. Arthritis Rheum 1994;37:965–982.
131. Matucci Cerinic M, McCarthy G, Lombardi A, Pignone A, Partsch G. Neurogenic influences in arthritis: potential modification by capsaicin. J Rheumatol 1995;22:1447–1449.

132. Urban L, Campbell E, Panesar M, Patel S, Chaudhry N, Kane S et al. In vivo pharmacology of SDZ 249-665, a novel, non-pungent capsaicin analogue. Pain 2000;89:65–74.
133. Van Rijswijk JB, Boeke EL, Keizer JM, Mulder PGH, Blom HM, Fokkens WJ. Intranasal capsaicin reduces nasal hyperreactivity in idiopathic rhinitis: a double-blind randomized application regimen study. Allergy 2003;58:754–761.
134. Rigoni M, Trevisani M, Gazzieri D, Nadaletto R, Tognetto M, Creminon C et al. Neurogenic Responses mediated by vanilloid receptor-1 (TRPV1) are blocked by the high affinity antagonist, iodo-resiniferatoxin. Br J Pharmacol 2003;138:977–985.
135. Jakab B, Helyes Z, Varga A, Bölcskei K, Szabó A, Sándor K et al. Pharmacological characterization of the TRPV1 receptor antagonist JYL1421 (SC0030) in vitro and in vivo in the rat. Eur J Pharmacol 2005;517:35–44.
136. Szallasi A, Cruz F, Geppetti P. TRPV1: a therapeutic target for novel analgesic drugs? Trends Mol Med 2006;12:545–554.
137. Kemper RHA, Jeuring M, Meijler WJ, Korf J, Ter Horst GJ. Intracisternal octreotide does not ameliorate orthodromic trigeminovascular nociception. Cephalalgia 2000;20:114–121.
138. Matharu MS, Levy MJ, Meeran K, Goadsby PJ. Subcutaneous octreotide in cluster headache: randomized placebo-controlled double-blind crossover study. Ann Neurol 2004;56:488–494.
139. Lembeck F, Donnerer J. Opioid control of the function of primary afferent substance P fibres. Eur J Pharmacol 1985;114:241–246.
140. Williamson DJ, Shepheard SL, Cook DA, Hargreaves RJ, Hill RG, Cumberbatch MJ. Role of opioid receptors in neurogenic dural vasodilation and sensitization of trigeminal neurones in anaesthetized rats. Br J Pharmacol 2001;133:807–814.
141. Barber A. μ- and κ-opioid receptor agonists produce peripheral inhibition of neurogenic plasma extravasation in rat skin. Eur J Pharmacol 1993;236:113–120.
142. Whiteside GT, Boulet JM, Walker K. The role of central and peripheral μ opioid receptors in inflammatory pain and edema: a study using morphine and DiPOA ([8-(3,3-diphenyl-propyl)-4-oxo-1-phenyl-1,3,8-triaza-spiro[4.5]dec-3-Yl]-acetic acid). J Pharmacol Exp Ther 2005;314:1234–1240.
143. Romero A, Planas E, Poveda R, Sanchez S, Pol O, Puig MM. Anti-exudative effects of opioid receptor agonists in a rat model of carrageenan-induced acute inflammation of the paw. Eur J Pharmacol 2005;511:207–217.
144. Pol O, Puig MM. Expression of opioid receptors during peripheral inflammation. Curr Top Med Chem 2004;4:51–61.
145. Barber A, Gottschlich R. Opioid agonists and antagonists: an evaluation of their peripheral actions in inflammation. Med Res Rev 1992;12:525–562.
146. McDougall JJ, Baker CL, Hermann PM. Attenuation of knee joint inflammation by peripherally administered endomorphin-1. J Mol Neurosci 2003;22:125–137.
147. Bertrand C, Nadel J, Yamawaki I, Geppetti P. Role of kinins in the vascular extravasation evoked by antigen and mediated by tachykinins in guinea pig trachea. J Immunol 1993;151:4902–4907.
148. Ahluwalia A, Giuliani S, Scotland R, Maggi C. Ovalbumin-induced neurogenic inflammation in the bladder of sensitized rats. Br J Pharmacol 1998;124:190–196.
149. Hirayama Y, Miyayasu K, Yamagami K, Imai T, Ohkubo Y, Mutoh S. Effect of FK3657, a non-peptide bradykinin B-2 receptor antagonist, on allergic airway disease models. Eur J Pharmacol 2003;467:197–203.

150. Griesbacher T, Legat FJ. Effects of the non-peptide B2 receptor antagonist FR173657 in models of visceral and cutaneous inflammation. Inflamm Res 2000;49:535–540.
151. Marceau F, Regoli D. Bradykinin receptor ligands: therapeutic perspectives. Nature Rev Drug Disc 2004;3:845–852.
152. Bacci E, Bertrand C, Geppetti P, Baker J, Nadel J, Laitinen L et al. Budesonide inhibits plasma extravasation induced by capsaicin and by substance P in the nasal mucosa. Regul Pept 1993;49:159–166.

IV. THE PATHOLOGICAL SIGNIFICANCE OF NEUROGENIC INFLAMMATION

Neurogenic Inflammation in Health and Disease
Edited by Gábor Jancsó

Neurogenic Vascular Responses in the Dura Mater and their Relevance for the Pathophysiology of Headaches

MÁRIA DUX[1] and KARL MESSLINGER[2]

[1] *Department of Physiology, University of Szeged, Szeged, Hungary*
[2] *Institute for Physiology and Pathophysiology, University of Erlangen-Nürnberg, Erlangen, Germany*

ABSTRACT

Neurogenic inflammation of the dura mater encephali is considered to be involved in the generation of primary vascular headaches. Different components of neurogenic inflammation – increased blood flow and vascular permeability, activation of endothelial and mast cells – seem to have differential significance in the pathophysiology of primary headaches. While the release of vasoactive substances from primary sensory neurons, endothelium, or mast cells has clearly been shown to occur during headache attacks, vascular permeability changes seem not to play any role. Clinical and experimental observations indicate that the sensory neuropeptide calcitonin gene-related peptide (CGRP) has a key position in meningeal vasodilatation. Beside its direct vasorelaxant effect, CGRP releases the inflammatory mediator histamine from mast cells and interacts with another potent vasodilator, nitric oxide, originating mainly from dural endothelial cells. Parasympathetic nerve fibers of the dura mater are also activated in some types of primary headaches, possibly by a trigeminal-parasympathetic reflex involving brain stem mechanisms. Interactions between different vasoactive regulators of meningeal blood flow may contribute to the pathogenesis of primary headaches.

1. INTRODUCTION

The peripheral and central mechanisms responsible for the generation of primary headaches are not clear, but some components of the underlying pathophysiological processes have been extensively studied. The neurogenic hypothesis of headache deals with the anatomical and functional relationship between trigeminal sensory nerve fibers and blood vessels of the intracranial tissues. Although the innervation of extracranial blood vessels may partly be involved in nociceptive processes [1], intracranial perivascular sensory fibers seem to play the most important role. Arteries of the dura mater encephali, the sagittal sinus, and the basal intracerebral arteries are sensitive to noxious stimuli and may therefore be involved in the generation of headaches [2,3]. Stimuli activating trigeminal sensory nerve fibers of these structures transmit nociceptive information to the brainstem, particularly to the trigeminal nucleus caudalis, and also promote a sterile inflammation in the tissue by releasing vasoactive peptides from perivascular nerve endings [4]. An increase in blood flow, plasma extravasation, as well as activation and degranulation of mast cells are the consequences of neuropeptide release from the stimulated sensory nerve fibers in the dura mater [5]. Calcitonin gene-related peptide (CGRP) seems to be responsible for vasodilatation, whereas substance P (SP) and neurokinin A (NKA) increase vascular permeability in response to trigeminal nerve stimulation [6–8]. Neuropeptides released by sensory nerve fibers may also influence vascular reactions indirectly by releasing vasoactive substances from mast cells [9,10]. Thus, the early phase of neurogenic inflammation in the dura mater results from a direct action of neuropeptides on arterioles and venules, whereas the later phase seems to depend, at least in part, on the release of other vasoactive substances from mast cells (e.g., histamine, 5-hydroxytryptamine, and prostaglandins) [11].

As one of the few pain-sensitive intracranial structures, the dura mater encephali has become the preferential target for studying mechanisms of meningeal nociception and headache [12–14]. A model of neurogenic inflammation in the dura mater has been developed by Moskowitz [15]. Although neurogenic inflammation seems not to be causal for headaches, the results obtained with this model indicate that blood flow changes occurring in parallel with pain sensation in headache patients may be explained by the development of neurogenic inflammation in meningeal tissues. Mechanical hyperalgesia (e.g., vessel pulsation) experienced as a painful stimulus may also occur following neurogenic inflammatory processes [16]. Clinical observations support the significance of neuropeptide release and associated mast cell reactions in the pathophysiology of primary headaches. Blood of the ipsilateral jugular vein showed increased levels of the vasodilator CGRP during migraine and cluster headache attacks [17,18]. Moreover, serum levels of histamine in patients with migraine and cluster headache were also increased indicating activation of mast cells [19–21]. Drugs reducing CGRP release from sensory nerve fibers or blocking its vasodilator effect have also been used for the therapy of cluster headache [18].

Several animal models have been developed to study pathophysiological mechanisms involved in the generation of headaches. Anatomical and physiological similarities of the trigeminovascular system in rats and humans provide a solid basis to study nociceptive components likely to be involved in the generation and control of headaches [22].

2. PEPTIDERGIC INNERVATION OF THE DURA MATER

In rats, a rich trigeminal and a less pronounced upper cervical sensory innervation supplies the dura mater encephali. Autonomic nerve fibers present in meningeal tissues are mainly of sympathetic origin, whereas a sparse parasympathetic innervation has also been described [23–28]. The neuropeptide content of nerve fibers in the dura has been determined by immunohistochemical methods. CGRP, SP, and NKA immunoreactivities are localized in sensory neurons, while nerve fibers immunoreactive for neuropeptide Y (NPY) and vasoactive intestinal polypeptide (VIP) are most likely of sympathetic and parasympathetic origin, respectively [29,30]. CGRP-immunoreactive nerve fibers are more abundant than SP-immunoreactive fibers, accounting for approximately 20 and 10% of the total number of dural nerve fibers, respectively [31]. Coexistence of CGRP- and SP-like immunoreactivity has been shown in the trigeminal ganglia of different species [32,33]. Immunoelectron microscopic preparations have revealed that CGRP- and SP-like immunoreactivity is present generally in thin unmyelinated nerve fibers forming bundles around the blood vessels and running as single fibers in the dural connective tissue [31].

In various tissues, a population of chemosensitive neurons containing CGRP, SP, and NKA has been described giving rise to polymodal C-fiber and Aδ mechano-heat nociceptors. Capsaicin (8-methyl-*N*-vanillyl-6-nonenamide), the pungent agent of red pepper, selectively stimulates the chemosensitive neuron population and, after repeated application, functionally inactivates these capsaicin-sensitive neurons [34–39]. Selective denervation with loss of capsaicin-sensitive neurons reduced the SP-like immunoreactivity in the rat trigeminal ganglion [40]. Capsaicin acts on the transient receptor potential vanilloid 1 (TRPV1) receptor. TRPV1 receptor immunoreactivity could be demonstrated in a distinct population of CGRP containing human trigeminal ganglion cells [41]. Recent observations provided evidence for the presence of capsaicin-sensitive sensory nerves in the rat dura mater. Activation of capsaicin-sensitive afferent nerves produces a TRPV1 receptor-mediated vasodilatory response through the release of CGRP. Histological studies also confirmed the existence of functional capsaicin-sensitive nerves in the rat dura mater. Capsaicin-sensitive sensory ganglion neurons may participate in the transmission of stimuli causing chemogenic and inflammatory pain. These nociceptive neurons may also play a fundamental role in local regulatory functions, including vascular reactions, through the release of vasoactive peptides from their activated peripheral endings [42,43].

3. RELEASE OF NEUROPEPTIDES FROM MENINGEAL NERVE FIBERS

Release of peptides from sensory terminals inducing neurogenic inflammation in the dura mater has been proposed as an important pathophysiological component in the generation of migraine pain and other severe headaches [5,44]. Electrical or chemical stimulation of trigeminal fibers results in neuropeptide release from sensory terminals in both *in vivo* and *in vitro* experimental models. In rats and cats, electrical stimulation of the trigeminal ganglion increased CGRP content in venous blood samples of the superior sagittal sinus [45,46]. Long lasting (20–30 min) local electrical stimulation of the parietal dura mater with stimulus parameters inducing blood flow elevation in meningeal blood vessels depleted CGRP immunoreactivity in dural nerve fibers within a distance of 2–3 mm from the stimulating electrodes [47]. Similar observations following electrical stimulation of trigeminal ganglion in the rat were described by Knyihár-Csillik et al. [48].

An *in vitro* preparation of rat dura mater has been developed to measure the neuropeptide release upon stimulation using an immunoenzyme assay. Electrical stimulation of the trigeminal ganglion and superfusion of the exposed arachnoid surface of the dura mater with a mixture of inflammatory mediators [5-hydroxytriptamine (5-HT), histamine, and bradykinin, each at 10^{-5} M, pH 6.1] induced a significant increase in CGRP concentration in the superfusate; however, SP concentration was not altered [49].

4. INCREASE IN VASCULAR PERMEABILITY IN THE DURA MATER ENCEPHALI

Neurogenic plasma protein extravasation in the dura mater has been extensively studied as one of the characteristic manifestations of neurogenic inflammation induced by SP and NKA release from trigeminal nerve fibers. Electrical stimulation of the trigeminal ganglion or intravenous administration of capsaicin resulted in increased permeability in postcapillary venules of the rat dura mater [50,51]. In those experiments, increased vascular permeability was detected by injecting different indicator substances into the bloodstream of animals prior to stimulating the sensory nerve fibers of interest. Increased permeability of blood vessels in dural tissue has been indicated by increases in extravasation of Evans blue dye or radiolabeled albumin [52,53]. Extravasation techniques are of considerable value in functional studies; however, because of their limited spatial and temporal resolution, they are less suited for exact histological demonstration of leaky blood vessels and studies on the cellular mechanisms of plasma extravasation.

Light microscopical techniques have been used to demonstrate the localization of leaky blood vessels. Injection of horseradish peroxidase and detection of its product in a histological preparation indicate the localization of leaky blood vessels in the dura mater [4]. Administration of a colloidal substance (e.g., colloidal silver solution) into the bloodstream of an experimental animal prior to the induction of neurogenic inflammation results in an accumulation of the colloidal substance at the basal lamina of leaky blood vessels [54–56]. Using this vascular labeling method, histological signs of increased vascular permeability after topical application of the irritant mustard oil (5%) and acidic phosphate buffer (pH 5.5), but not after local electrical stimulation (15 V, 10 Hz, for 10 min) of dural nerve fibers, could be detected [57]. To explain the lack of silver-labeled blood vessels after electrical stimulation, different mechanisms responsible for increased vascular permeability in dural blood vessels should be considered. Electronmicroscopical observations indicate that electrical stimulation of the trigeminal ganglion increases the number of endothelial pinocytotic vesicles and endoluminal pits. In addition, after intravenous administration of SP, the formation of interendothelial gaps has also been observed [58]. Opening of interendothelial junctions is visualized by the vascular labeling technique, whereas other mechanisms such as vesicular transport of plasma components, which can be the major pathway of plasma transport in this case, are probably not indicated by this method.

Intravenous injection of SP as well as release of SP evoked by systemic administration of capsaicin increased vascular permeability in the dura mater, and this leakage was inhibited by pretreatment with a NK_1 receptor antagonist [59,60]. Destruction of thin unmyelinated nerve fibers with neonatal capsaicin treatment abolished the increase in vascular permeability normally resulting from electrical stimulation of the trigeminal ganglion [4].

5. VASODILATATION IN THE DURA MATER ENCEPHALI

Neuropeptides released by trigeminal nerve fibers are potent vasodilators of intracranial blood vessels [33,61–63]. Two *in vivo* preparations have recently been developed to study vascular reactions controlling meningeal perfusion in the rat. One of these preparations uses laser Doppler flowmetry [64], the other uses intravital microscopy to measure blood vessel diameter [65]. In both experimental models, local electrical stimulation of the exposed dura mater caused transient vasodilatation and increases in dural arterial flow in the vicinity of the stimulating electrodes. Electrically evoked increases in meningeal blood flow have been shown to be neurogenically mediated as they could be blocked by the topical application of local anesthetics to the exposed dura mater [47]. Moreover, these increases in flow could be inhibited by topical preapplication of the CGRP receptor antagonist $CGRP_{8-37}$ to the dura mater or by systemic administration of the CGRP receptor antagonist BIBN4096BS, indicating that CGRP released from stimulated nerve fibers caused the flow increase [64,65]. In the experiments that used videomicroscopy, increases in arterial diameters were observed by stimulating the dura mater electrically through a thin intact bone layer. These increases in blood vessel diameter could also be inhibited by systemically administered $CGRP_{8-37}$ [66,67]. Local or intravenous administration of CGRP was reported to increase meningeal blood flow and vessel diameter [64,67].

In the dura mater, the vasodilator effect of the tachykinins SP and NKA is controversial. Systemic administration of SP and NKA induced significant increases in blood vessel diameter when visualized by videomicroscopy [67]. Pretreatment with the NK_1 receptor antagonist RP67580 abolished both SP- and NKA-induced vasodilatation in the same experimental model, suggesting that tachykinins may play a role in the regulation of meningeal perfusion. However, results obtained using laser Doppler flowmetry do not support these observations, because local application of the NK_1 receptor antagonist had no significant effect on the basal flow and the electrically evoked increases in flow. Topical application of SP was also without effect on dural blood flow in this preparation [68]. These results indicate that different mechanisms may be responsible for neuropeptide-induced vasodilatation. CGRP released from perivascular nerve fibers acts directly on smooth muscle cells of arterioles in which there is a subsequent accumulation of the second messenger cAMP [61,69,70], whereas the vasodilator effect of SP is mediated by the endothelium [71]. SP-mediated vasodilatation seems to activate endothelial nitric oxide (NO) production [61,71]. SP released by SP-immunoreactive perivascular dural nerve fibers may not be able to penetrate the arterial wall to the endothelium at functionally sufficient concentrations, whereas SP present in the circulation may easily reach endothelial NK_1 receptors. It is concluded that CGRP, one of the most potent vasodilators, is probably the essential neuropeptide of trigeminal afferents that induces vasodilatation upon its stimulated release, thereby increasing blood flow in dural blood vessels. SP, if released from nerve terminals, does not significantly contribute to this effect.

6. INVOLVEMENT OF MAST CELLS IN THE NEUROGENIC INFLAMMATION OF MENINGEAL TISSUES

It has been known for a long time that infusion of histamine induces headaches particularly in migraineurs [21,72,73]. Plasma levels of histamine have been reported to be increased in migraine and cluster headache patients [19–21]. The source of histamine in these patients is not known; however, the dura mater with its rich mast cell population is a likely candidate; histaminergic nerve fibers have not been identified in the dura [74]. Mast cells are closely associated with sensory nerve fibers and blood vessels, supporting the hypothesis that all these structures can functionally interact to produce vasodilatation and pain in headache patients [9]. Histamine is a potent vasodilator and is also able to increase vascular permeability in different tissues [75]. Results from *in vivo* and *in vitro* experiments alike suggest that there is a role for different histamine receptors in intracranial blood vessels. Relaxation of dural arteries is mediated by H_2 receptors, most likely located on vascular smooth muscle cells, and also by endothelial H_1 receptors. In addition, H_1 receptors on smooth muscle cells may mediate vasoconstriction [76,77]. Apart from the direct vascular effects of histamine, a bidirectional communication between sensory nerve fibers and dural mast cells may be involved in neurogenic inflammatory processes modulating the vasodilator and permeability increasing effects of sensory neuropeptides [78,79]. Activation and degranulation of mast cells by sensory neuropeptides may play an important role in the course of migraine and cluster headache attacks [19,80]. Mast cells in meningeal tissue may be degranulated by both sensory neuropeptides, CGRP and SP [10], but CGRP could be confirmed as a histamine releasing substance in an *in vitro* preparation of the rat cranial dura mater [81]. Histamine released by mast cells may in turn influence transmitter release from both sensory and autonomic nerve fibers. Histamine directly stimulates CGRP- and SP-containing cultured trigeminal ganglion neurons by the activation of H_1 receptors [82]. Stimulation of H_3 receptors, present on both autonomic and sensory fibers, has been reported to reduce transmitter release from nerve terminals [83,84].

7. MODULATION OF CGRP-INDUCED VASODILATATION BY NITRIC OXIDE

NO is an important mediator of vasodilatation in different tissues including the meningeal circulation [85–87]. NO is suggested to be a key substance in the generation of nociceptive processes in migraine pain and other vascular headaches [88]. In migraineurs, the infusion of nitrovasodilators leads to headaches that are reported to be very similar to spontaneous migraine or cluster headache attacks [89–92]. NO-producing structures in the rat dura mater have been localized using different methods. NO-synthase activity in endothelial cells and some dural nerve fibers has been verified by immunohistochemical methods [93,94]. Nicotinamide adenine dinucleotide phosphate diaphorase (NADPH-diaphorase) activity also indicates NO-synthase activity in brain and peripheral tissues [95]. Enzymatic activity of NADPH-diaphorase was found to be localized exclusively in the wall of arteries and arterioles up to their transition into capillaries. Nerve fibers and other dural cells showed no significant NADPH-diaphorase activity [96]. Experiments using laser Doppler flowmetry indicate the role of NO, released by endothelial or neuronal structures, in the regulation of dural arterial flow [97]. Systemic administration or local application of the NO-synthase inhibitor *N*ω-nitro-L-arginine methyl ester lowered the basal blood flow in the medial meningeal artery and reduced flow increases evoked by local electrical stimulation. Because electrically induced flow increases depend on CGRP release from afferent nerve fibers, a functional connection between these vasodilatory mediators was suggested. NO may facilitate the release of CGRP from perivascular afferents or interact with its vasorelaxant effect on the vascular smooth muscle cells. Wei et al. [98] have found that the vasodilatory effect of the NO donors, nitroglycerin and nitroprusside, on cerebral arterioles of the cat was reduced after trigeminal ganglionectomy and after preapplication of the CGRP receptor antagonist $CGRP_{8-37}$. In a recent series of experiments, our group has shown a similar inhibitory effect of $CGRP_{8-37}$ on blood flow increases induced by NO donors in the rat dura mater. The same NO donors caused concentration-dependent increases in CGRP release in an *in vitro* preparation of rat cranial dura mater [96]. Endothelial cells seem to produce NO constitutively in the dura mater as in other tissues. However, nerve fibers, mast cells, and immune cells are also potential sources of NO, especially under inflammatory conditions, when NO-synthase activity is upregulated. Interactions of NO with CGRP release and the dilatory effect of CGRP on vascular smooth muscle cells may contribute to the sensitization of perivascular afferents and could contribute to the peripheral mechanism of nociception in headaches.

8. MODIFICATION OF THE CGRP RECEPTOR SENSITIVITY AS A POTENTIAL REASON FOR ENHANCED INFLAMMATORY AND NOCICEPTIVE REACTIONS

The CGRP receptor can activate multiple signal transduction pathways although its actions are most commonly mediated by a G-protein-coupled receptor increasing the intracellular cAMP levels. The CGRP receptor consists of three different parts: the seven-transmembrane calcitonin-like receptor (CLR) and two further subunits, receptor activity-modifying protein-1 (RAMP1) and receptor component protein (RCP) [99]. RAMP1 is a small single-transmembrane protein that is required for CGRP binding to CLR [100]. RCP is an intracellular protein that interacts with CLR and facilitates cAMP production [101].

CGRP receptors are located on the meningeal blood vessels, dural mast cells, trigeminal ganglion cells, and sensory neuronal structures within the trigeminal nuclei. All these localizations of CGRP receptors may be relevant for the pathophysiology of headaches. RAMP1 seems to be functionally limiting for CGRP effects in the trigeminovascular system. Factors elevating neuronal RAMP1 level could potentially sensitize the individuals to CGRP. Sensitization may also include increased CGRP synthesis and enhanced intensity of neurogenic inflammation, which could potentially prolong and intensify the nociceptive reactions in migraine [102].

9. INVOLVEMENT OF THE PARASYMPATHETIC NERVOUS SYSTEM IN THE NEUROGENIC INFLAMMATION OF THE DURA MATER

Based on clinical and experimental observations, a role of parasympathetic nerve fibers in neurogenic vasomotor reactions of intracranial blood vessels has been suggested. An increased plasma level of the vasodilatory transmitter VIP, which is normally of parasympathetic origin, was measured in external jugular vein blood samples taken from patients during cluster headache [18]. Experimental studies in the cat indicate that electrical stimulation of the trigeminal ganglion and the superior sagittal sinus activates a reflex circuit involving trigeminal afferents, the sphenopalatine ganglion, and parasympathetic efferents, which induces an increase in cortical blood flow [103,104]. Using the laser speckle-contrast imaging technique, the role of parasympathetic nerve fibers in the cortical spreading depression-induced elevation of meningeal blood flow was studied in rats. Cortical spreading depression is probably the phenomenon underlying visual aura and other aura phenomena that are typically experienced by migraineurs who suffer from migraine with aura. In these experiments, blood flow elevation was found to be neurogenically mediated, in part through the brainstem connections of parasympathetic efferents. Perivascular trigeminal afferents with cell bodies in the trigeminal ganglion synapse predominantly in the trigeminal nucleus caudalis. Neurons with cell bodies in the trigeminal nucleus caudalis project to central structures that are involved in nociception. Activation of this pathway also activates the superior salivatory nucleus and parasympathetic efferents via the sphenopalatine ganglia. Release of vasoactive substances such as VIP, acetylcholine, and NO from postganglionic parasympathetic nerve fibers promotes vasodilatation and augments blood flow increases generated by sensory neuropeptides in the dura mater. In this experimental model, transection of parasympathetic nerve fibers innervating the dura mater significantly suppressed blood flow elevation in the medial meningeal artery supporting the significance of this central trigeminal-parasympathetic reflex [14].

The role of parasympathetic nerve fibers in the generation of plasma extravasation is not yet clear. Direct electrical stimulation of the sphenopalatine ganglion of rats induces a significant increase in plasma extravasation of bovine serum albumin in the dura mater [105]. However, another study has shown that experimental activation of the central trigeminal-parasympathetic reflex did not influence plasma leakage in dural venules [14]. The release of parasympathetic vasodilator substances involved in meningeal blood flow regulation may in turn be regulated by stimuli acting at parasympathetic nerve terminals. [106].

10. INVOLVEMENT OF 5-HYDROXYTRYPTAMINE RECEPTORS IN NEUROGENIC INFLAMMATORY REACTIONS

Clinical and experimental observations have shown that 5-HT, for which several types of receptors are located on blood vessels, can exert opposite effects on the processes that presumably underlie the generation of headaches. It is therefore not surprising that 5-HT receptor agonists can either initiate or abort migraine attacks. Cerebrovascular endothelial 5-HT_{2B} receptors have been suggested to be activated at the beginning of a migraine attack [107,108]. 5-HT_2 receptor antagonists are therefore used in the prophylaxis of migraine. Activation of other 5-HT receptor types, 5-$HT_{1B/1D}$ receptors, cause vasoconstriction and inhibition of trigeminovascular reactions. When used clinically, 5-$HT_{1B/1D}$ agonists can alleviate headache and other symptoms of migraine. Sumatriptan, a 5-HT_1 receptor agonist, is effectively used in the treatment of acute pain in migraine and other forms of primary headache. The beneficial effects of 5-HT_1 receptor agonists on headache seem to be associated with their direct vasoconstrictor effect on dilated arterioles, the so-called postjunctional effect. The prejunctional effect of 5-HT_1 receptor agonists is the inhibition of CGRP and SP release from trigeminal afferents. The postjunctional vasoconstrictor effect of 5-HT_1 receptor agonists is mediated by 5-HT_{1B} receptors on vascular smooth muscle cells. Prejunctional suppression of transmitter release is mediated by 5-HT_{1D} and possibly 5-HT_{1F} receptors localized on trigeminal nerve fibers [109–112]. A wide variety of experimental results have shown that 5-HT_1 receptor agonists inhibit plasma extravasation and other elements of neurogenic inflammation in the dura mater encephali of rats and guinea pigs [107–110]. Increased levels of CGRP measured in venous blood after electrical stimulation of the trigeminal ganglion or the superior sagittal sinus in rats and cats could be reduced by 5-HT_1 receptor agonists such as sumatriptan and dihydroergotamine [44,46]. Pretreatment with the 5-HT_1 receptor agonist sumatriptan and the 5-HT_{1B} receptor agonist CP93,129 has been shown to be effective in inhibiting neurogenic plasma extravasation in the rat dura mater [113–116]. Blood flow increases in meningeal blood vessels, induced by electrical stimulation of dural nerve fibers, seem to be much less sensitive to 5-HT_1 receptor agonist pretreatment. Doses of sumatriptan and CP93,129, which have been shown to be effective in inhibiting plasma extravasation caused by electrical and chemical stimulation of trigeminal afferents, were much less effective in inhibiting neurogenic vasodilatation, measured as increases in dural blood flow induced by electrical stimulation [117]. Moderate inhibition of neurogenic vasodilatation was achieved only with the local application or systemic (i.v.) administration of higher doses of 5-HT_1 receptor agonists. The different threshold doses required to inhibit neurogenic plasma extravasation and blood flow increases in the dura mater may be explained by different sensitivities of trigeminal neuron populations to 5-HT_1 receptor agonists: the small SP-immunoreactive population of trigeminal neurons seems to be more sensitive to 5-HT_1 receptor agonists than the much larger CGRP-immunoreactive population, which is thought to be mainly responsible for vasodilatation in the meninges.

11. RELEVANCE FOR THE PATHOPHYSIOLOGY OF HEADACHES

The intensity of pain experienced during headache depends on the inflow of sensory information to the pain processing system via the trigeminal nerve. Although it is not yet clear what factors are responsible for the activation or sensitization of nociceptive afferents under these conditions, disturbances of peripheral and/or central mechanisms seem to be present in different types of headaches. Neurogenic inflammation induced by neuropeptide release and mast cell degranulation has been proposed to be an important pronociceptive peripheral mechanism in pain-sensitive intracranial structures. Moskowitz and his group focused their attention on the plasma extravasation component of neurogenic inflammation in meningeal tissues. Although drugs used in migraine therapy (e.g., the 5-HT_1 receptor agonist sumatriptan) can effectively block plasma extravasation from dural blood vessels under experimental conditions, peripherally acting NK_1 receptor antagonists, inhibiting plasma extravasation in experimental models [118,119], failed to show any antimigraine effects in humans [120]. The endothelin antagonist bosentan also blocks plasma extravasation in rat dura mater, but this antagonist was also ineffective in the treatment of migraine pain [121]. It is very likely that the plasma-extravasation component of neurogenic inflammation is not an important factor contributing to headache generation, because neither release of the neuropeptide SP nor signs of plasma extravasation could clearly be shown to occur in any type of headache [18].

The vasodilatation component of neurogenic inflammation seems to be a more important mechanism in the pathophysiology of headaches. Clinical data show increased levels of vasodilator neuropeptides in the venous outflow from the head in headache patients. During attacks of migraine, increased blood CGRP levels have been reported, whereas during cluster headache attacks, both CGRP and VIP content in venous blood was significantly increased. Treatments reducing CGRP levels during migraine attacks were effective in producing pain relief in most patients and, in parallel, normalizing blood flow in intracerebral arteries [18,44]. A few years ago, the introduction of a highly specific and potent CGRP-receptor antagonist BIBN4096BS was a breakthrough in CGRP receptor pharmacology. Preclinical and clinical investigations have shown that BIBN4096BS has a significant antimigraine potential [122–124]. Recent electrophysiological recordings from the rat spinal trigeminal nucleus have shown that BIBN4096BS reduces the neuronal activity of neurons with afferent input from the cranial dura, confirming the antinociceptive principle of CGRP receptor inhibition in the trigeminal system [125]. Thus, animal experiments that examine changes in neurogenic vasodilatation and trigeminal neuronal activation have a high predictive value in testing the efficacy of potential antimigraine drugs.

Although there is an obvious correlation between CGRP release, vasodilatation, and pain, the role of vasodilator peptides in headache attacks is not yet clear. Migraine patients often report throbbing pain, the origin of which is most likely a pulsatile mechanical activation of perivascular afferents innervating meningeal or intracerebral arteries [126]. Techniques measuring regional cerebral blood flow during migraine attacks indicate a significant dilatation of intracerebral blood vessels during the painful phase in most of the patients [127]. Infusion of vasodilator substances (e.g., the NO-donor nitroglycerin or histamine) also produces pain [128]. A likely common pathway for migraine pain, therefore, is the increased level of trigeminal afferent activation in response to vasodilatation although vasodilatation alone is probably not sufficient to produce pain. The temporal relationship between vasodilatation and headache is complex. Experimental headache is produced following the infusion of vasodilators although in some cases the latency to headache onset is long. In these cases, it is not clear whether the pain sensation results from the vasodilatation [129,130]. Delayed headache after infusion of NO donors or histamine is probably independent of vasodilatory processes. Instead, they seem to result from the sensitization of meningeal and/or central trigeminal processes [131]. CGRP-induced vasodilatation might even have a secondary protective function, because it may accelerate washout of nociceptive metabolites in the meninges.

ACKNOWLEDGMENTS

We thank Dr. R. Carr for valuable comments on the manuscript. This study was supported by the Deutsche Forschungsgemeinschaft (SFB 353, B3), the Sander-Stiftung, and OTKA T 46469, K63663, ETT 193/2006, and RET-08/2004.

REFERENCES

1. Drummond PD, Lance JW. Extracranial vascular changes and the source of pain in migraine headache. Ann Neurol 1983;13:32–37.
2. Ray BS, Wolff HG. Experimental studies on headache: pain sensitive structures of the head and their significance in headache. Arch Surg 1940;1:813–856.
3. Penfield W, McNaughton FL. Dural headache and innervation of the dura mater. Arch Neurol Psychiatry 1940;44:43–75.
4. Dimitriadou V, Buzzi MG, Theoharides TC, Moskowitz MA. Ultrastructural evidence for neurogenically mediated changes in blood vessels of the rat dura mater and tongue following antidromic trigeminal stimulation. Neuroscience 1992;48:187–203.
5. Moskowitz MA. Neurogenic inflammation in the pathophysiology and treatment of migraine. Neurology 1993;43:S16–S20.
6. Moskowitz MA, Cutrer FM. Possible importance of neurogenic inflammation within the meninges to migraine headache. Fields HL, Liebeskind JC, eds. Progress in Pain Research and Management, Vol. 1. Seattle, Washington, DC: IASP Press, 1994; pp. 43–49.
7. Brain SD, Williams TJ, Tippins JR, Morris HR, MacIntyre I. Calcitonin gene-related peptide is a potent vasodilator. Nature 1985;313:54–56.
8. Goadsby PJ, Edvinsson L, Ekman R. Release of vasoactive peptides in the extracerebral circulation of humans and the cat during activation of the trigeminovascular system. Ann Neurol 1988;23:193–196.
9. Dimlich RVW, Keller JT, Strauss TA, Fritts MJ. Linear arrays of homogenous mast cells in the dura mater of the rat. J Neurocytol 1991;20:485–503.
10. Ottosson A, Edvinsson L. Release of histamine from dural mast cells by substance P and calcitonin gene-related peptide. Cephalalgia 1997;17:166–174.
11. Waeber C, Moskowitz MA. Migraine as an inflammatory disorder. Neurology 2005;64:S9–S15.
12. Strassman AM, Mason P, Moskowitz MA, Maciewicz RJ. Response of brainstem trigeminal neurons to electrical stimulation of the dura. Brain Res 1986;379:242–250.
13. Davis KD, Dostrovsky JO. Responses of feline trigeminal spinal tract nucleus neurons to stimulation of the middle meningeal artery and sagittal sinus. J Neurophysiol 1988;59:648–665.
14. Bolay H, Reuter U, Dunn AK, Huang Z, Boas DA, Moskowitz MA. Intrinsic brain activity triggers trigeminal meningeal afferents in a migraine model. Nat Med 2002;8:136–142.
15. Moskowitz MA. The neurobiology of vascular head pain. Ann Neurol 1984;16:157–168.
16. Sanchez del Rio M, Moskowitz MA. The trigeminal system. In: Olesen J, Tfelt-Hansen P, Welch KMA, eds. The Headaches. Philadelphia, PA: Lippincott Williams & Wilkins, 2000; pp. 141–149.

17. Goadsby PJ, Edvinsson L, Ekman R. Vasoactive peptide release in the extracerebral circulation of humans during migraine headache. Ann Neurol 1990;28:183–187.
18. Goadsby PJ, Edvinsson L. Human in vivo evidence for trigeminovascular activation in cluster headache. Neuropeptide changes and effects of acute attacks therapies. Brain 1994;117:427–434.
19. Heatley RV, Denburg JA, Bayer N, Bienenstock J. Increased plasma histamine levels in migraine patients. Clin Allergy 1982;12:145–149.
20. Anthony M, Lance JW. Histamine and serotonin in cluster headache. Arch Neurol 1971;25:225–231.
21. Krabbe A, Olesen J. Headache provocation by continuous intravenous infusion of histamine. Clinical results and receptor mechanisms. Pain 1980;8:253–259.
22. Suzuki N, Hardebo JE, Owman C. Origins and pathways of cerebrovascular nerves storing substance P and calcitonin gene-related peptide in rat. Neuroscience 1989;31:427–438.
23. Edvinsson L, Uddman R. Adrenergic, cholinergic and peptidergic nerve fibres in dura mater – involvement in headache? Cephalalgia 1981;1:175–179.
24. Steiger HJ, Tew JM, Keller JT. The sensory representation of the dura mater in the trigeminal ganglion of the cat. Neurosci Lett 1982;31:231–236.
25. Keller JT, Marfurt CF, Dimlich RVW, Tierney BE. Sympathetic innervation of the supratentorial dura mater of the rat. J Comp Neurol 1989;290:310–321.
26. Uddman R, Hara H, Edvinsson L. Neuronal pathways to the rat middle meningeal artery revealed by retrograde tracing and immunocytochemistry. J Auton Nerv Syst 1989;26:69–75.
27. Amenta F, Sancesario G, Ferrante F, Cavallotti C. Acetylcholinesterase-containing nerve fibers in the dura mater of guinea pig, mouse, and rat. J Neural Transm 1980;47:237–242.
28. Edvinsson L, Goadsby PJ. Neuropeptides in headache. Eur J Neurol 1998;5:329–341.
29. Von Düring M, Bauersachs M, Böhmer B, Veh RW, Andres KH. Neuropeptide Y- and substance P-like immunoreactive nerve fibers in the rat dura mater encephali. Anat Embryol 1990;182:363–373.
30. Edvinsson L, Jansen I, Uddman R, Gulbenkian S. Innervation of the human cerebral circulation. J Auton Nerv Syst 1994;49:S91–S96.
31. Messlinger K, Hanesch U, Baumgärtel M, Trost B, Schmidt RF. Innervation of the dura mater encephali of cat and rat: ultrastructure and calcitonin gene-related peptide-like and substance P-like immunoreactivity. Anat Embryol 1993;188:219–237.
32. Lee Y, Kawai Y, Shiosaka S, Takami K, Kiyama H, Hillyard CJ et al. Co-existence of calcitonin gene-related peptide and substance P-like peptide in single cells of the trigeminal ganglion of the rat: immunohistochemical analysis. Brain Res 1985;330:194–196.
33. McCulloch J, Uddman R, Kingman TA, Edvinsson L. Calcitonin gene-related peptide: functional role in cerebrovascular regulation. Proc Natl Acad Sci U S A 1986;83:5731–5735.
34. Jancsó N. Role of the nerve terminals in the mechanism of inflammatory reactions. Bull Millard Fillmore Hosp Buffalo NY 1960;7:53–77.
35. Jancsó G, Király E, Jancsó-Gábor A. Pharmacologically induced selective degeneration of chemosensitive primary sensory neurons. Nature 1977;270:741–743.
36. Jancsó G, Király E, Jancsó-Gábor A. Chemosensitive pain fibers and inflammation. Int J Tissue React 1980;2:57–66.
37. Jancsó G, Király E. Sensory neurotoxins: chemically induced selective destruction of primary sensory neurons. Brain Res 1981;210:83–89.

38. Jancsó G. Pathobiological reactions of C-fibre primary sensory neurons to peripheral nerve injury. Exp Physiol 1992;77:405–431.
39. Caterina MJ, Schumacher MA, Tominaga M, Rosen TA, Levine JD, Julius D. The capsaicin receptor: a heat-activated ion channel in the pain pathway. Nature 1997;389:816–824.
40. Mione MC, Cavanagh JF, Kirkpatrick KA, Burnstock G. Plasticity in expression of calcitonin gene-related peptide and substance P immunoreactivity in ganglia and fibres following guanethidine and/or capsaicin denervation. Cell Tissue Res 1992;268:491–504.
41. Hou M, Uddman R, Tajti J, Kanje M, Edvinsson L. Capsaicin receptor immunoreactivity in the human trigeminal ganglion. Neurosci Lett 2002;330:223–226.
42. Dux M, Sántha P, Jancsó G. Capsaicin-sensitive neurogenic sensory vasodilatation in the dura mater of the rat. J Physiol 2003;552:859–867.
43. Akerman S, Kaube H, Goadsby PJ. Vanilloid type 1 receptors (VR1) on trigeminal sensory nerve fibres play a minor role in neurogenic dural vasodilatation, and are involved in capsaicin-induced dural dilation. Br J Pharmacol 2003;140:718–724.
44. Goadsby PJ, Edvinsson L. The trigeminovascular system and migraine: studies characterizing cerebrovascular and neuropeptide changes seen in humans and cats. Ann Neurol 1993;33:48–56.
45. Buzzi MG, Carter WB, Shimizu T, Heath HG, Moskowitz MA. Dihydroergotamine and sumatriptan attenuate levels of CGRP in plasma in rat superior sagittal sinus during electrical stimulation of the trigeminal ganglion. Neuropharmacology 1991;30:1193–1200.
46. Zagami AS, Goadsby PJ, Edvinsson L. Extracranial peptide release following stimulation of the superior sagittal sinus in the cat. Cephalalgia 1989;9(Suppl. 10):294–295.
47. Messlinger K, Pawlak M. Regulation of meningeal blood flow by neuropeptides: relevance to migraine. In: Brain SD, Moore PK, eds. Pain and Neurogenic Inflammation. Basel: Birkhäuser Verlag, 1999; pp. 245–274.
48. Knyihár-Csillik E, Tajti J, Samsam M, Sáry G, Vécsei L. Electrical stimulation of the Gasserian ganglion induces structural alterations of calcitonin gene-related peptide-immunoreactive perivascular sensory nerve terminals in the rat cerebral dura mater: a possible model of migraine headache. Neurosci Lett 1995;184:189–192.
49. Ebersberger A, Averbeck B, Messlinger K, Reeh PW. Release of substance P, calcitonin gene-related peptide and prostaglandin E2 from rat dura mater encephali following electrical and chemical stimulation in vitro. Neuroscience 1999;89:901–907.
50. Markowitz S, Saito K, Moskowitz MA. Neurogenically mediated leakage of plasma protein occurs from blood vessels in dura mater but not brain. J Neurosci 1987;7:4129–4136.
51. Buzzi MG, Bonamini M, Moskowitz MA. Neurogenic model of migraine. Cephalalgia 1995;15:277–280.
52. Saito K, Markowitz S, Moskowitz MA. Ergot alkaloids block neurogenic extravasation in dura mater: proposed action in vascular headaches. Ann Neurol 1988;24:732–737.
53. Buzzi MG, Moskowitz MA, Peroutka SJ, Byun BJ. Further characterization of the putative 5-HT receptor which mediates blockade of neurogenic plasma extravasation in rat dura mater. Br J Pharmacol 1991;103:1421–1428.
54. Jancsó N, Jancsó-Gábor A, Szolcsányi J. The role of the sensory nerve endings in neurogenic inflammation induced in human skin and in the eye and paw of the rat. Br J Pharmacol Chemother 1968;33:32–41.

55. Majno G, Palade GE, Schoefl GI. Studies on Inflammation. II. The site of action of histamine and serotonin along the vascular tree: a topographic study. J Biophys Biochem Cytol 1961;11:607–625.
56. Dux M, Jancsó G. A new technique for the direct demonstration of overlapping cutaneous innervation territories of peptidergic C-fibre afferents of rat hindlimb nerves. J Neurosci Methods 1994;55:47–52.
57. Dux M, Messlinger K. Histological demonstration of increased vascular permeability in the dura mater of the rat. Microsc Res Tech 2001;53:229–231.
58. Ghabriel MN, Lu MX, Leigh C, Cheung WC, Allt G. Substance P-induced enhanced permeability of dura mater microvessels is accompanied by pronounced ultrastructural changes, but is not dependent on the density of endothelial cell anionic sites. Acta Neuropathol 1999;97:297–305.
59. Moussaoui SM, Philippe L, Le Prado N, Garret C. Inhibition of neurogenic inflammation in the meninges by a non-peptide NK1 receptor antagonist, RP 67580. Eur J Pharmacol 1993;238:421–424.
60. Shepheard SL, Williamson DJ, Hill RG, Hargreaves RJ. The non-peptide neurokinin1 receptor antagonist, RP 67580, blocks neurogenic plasma extravasation in the dura mater of rats. Br J Pharmacol 1993;108:11–12.
61. Edwards RM, Stack EJ, Trizna W. Calcitonin gene-related peptide stimulates adenylate cyclase and relaxes intracerebral arterioles. J Pharmacol Exp Ther 1991;257:1020–1024.
62. Jansen I, Alafaci C, McCulloch J, Uddman R, Edvinsson L. Tachykinins (substance P, neurokinin A, neuropeptide K, and neurokinin B) in the cerebral circulation: vasomotor responses in vitro and in situ. J Cereb Blood Flow Metab 1991;11:567–575.
63. Williamson DJ, Hargreaves RJ. Neurogenic inflammation in the context of migraine. Microsc Res Tech 2001;53:167–178.
64. Messlinger K, Hanesch U, Kurosawa M, pawlak M, Schmidt RF. Calcitonin gene related peptide released from dural nerve fibres mediates increase of meningeal blood flow in the rat. Can J Physiol Pharmacol 1995;73:1020–1024.
65. Tröltzsch M, Denekas T, Messlinger K. The CGRP receptor antagonist BIBN4096BS reduces neurogenic increases in meningeal blood flow. Eur J Pharmacol 2007;562:103–110.
66. Williamson DJ, Hargreaves RJ, Hill RG, Shepheard SL. Sumatriptan inhibits neurogenic vasodilation of dural blood vessels in the anaesthetized rat – intravital microscope studies. Cephalalgia 1997;17:525–531.
67. Williamson DJ, Hargreaves RJ, Hill RG, Shepheard SL. Intravital microscope studies on the effects of neurokinin agonists and calcitonin gene-related peptide on dural vessels diameter in the anaesthetized rat. Cephalalgia 1997;17:518–524.
68. Carmody J, Pawlak M, Messlinger K. Lack of a role for substance P in the control of dural arterial flow. Exp Brain Res 1996;111:424–428.
69. Edvinsson L, Fredholm BB, Hamel E, Jansen I, Verrecchia C. Perivascular peptides relax cerebral arteries concomitant with stimulation of cyclic adenosine monophosphate accumulation or release of an endothelium-derived relaxing factor in the cat. Neurosci Lett 1985;58:213–217.
70. Jansen-Olesen I, Mortensen A, Edvinsson L. Calcitonin gene-related peptide is released from capsaicin-sensitive nerve fibres and induces vasodilatation of human cerebral arteries concomitant with activation of adenyl cyclase. Cephalalgia 1996;16:310–316.
71. Enokibori M, Okamura T, Toda N. Mechanism underlying substance P-induced relaxation in dog isolated superficial temporal arteries. Br J Pharmacol 1994;11:77–82.

72. Clark D, Hough H, Wolff HG. Experimental studies on headache: observations on histamine headache. Arch Neurol Psychiatry (Chicago) 1936;35:1054–1070.
73. Pickering GW. Observations on the mechanisms of headache produced by histamine. Clin Sci 1933;1:77–101.
74. Rozniecki JJ, Dimitriadou V, Lambracht-Hall M, Pang X, Theoharis CT. Morphological and functional demonstration of rat dura mater mast cell – neuron interactions in vitro and in vivo. Brain Res 1999;849:1–15.
75. Church MK, Lowman MA, Rees PH, Benyon RC. Mast cells, neuropeptides and inflammation. Agents Actions 1989;27:8–16.
76. Jansen-Olesen I, Ottosson A, Cantera L, Strunk S, Lassen LH, Olesen J et al. Role of endothelium and nitric oxide in histamine-induced responses in human cranial arteries and detection of mRNA encoding H_1- and H_2-receptors by RT-PCR. Br J Pharmacol 1997;121:41–48.
77. Dux M, Schwenger N, Messlinger K. H1 and H2 receptors are differentially involved in meningeal blood flow elevation. Regul Pept 2000;89:59.
78. Dimitriadou V, Lambracht-Hall M, Reichler J, Theoharides TC. Histochemical and ultrastructural characteristics of rat brain perivascular mast cells stimulated with compound 48/80 and carbachol. Neuroscience 1990;39:209–224.
79. Dimitriadou V, Buzzi MG, Moskowitz MA, Theoharides TC. Trigeminal sensory fiber stimulation induces morphological changes reflecting secretion in rat dura mater mast cells. Neuroscience 1991;44:97–112.
80. Liberski PP, Prusinski A. Further observations on the mast cells over the painful region in cluster headache patients. Headache 1982;22:115–117.
81. Schwenger N, Dux M, de Col R, Carr R, Messlinger K. Interaction of calcitonin gene-related peptide, nitric oxide and histamine release in neurogenic blood flow and afferent activation in the rat cranial dura mater. Cephalalgia 2007;27:481–491.
82. Tani E, Ishikawa T. Histamine acts directly on calcitonin gene-related peptide- and substance P-containing trigeminal ganglion neurons as assessed by calcium influx and immunocytochemistry. Auris Nasus Larynx 1990;17:267–274.
83. Matsubara T, Moskowitz MA, Huang Z. UK-14,304, R(-)-alpha-methyl-histamine and SMS 201-995 block plasma protein leakage within dura mater by prejunctional mechanisms. Eur J Pharmacol 1992;224:145–150.
84. Malinowska B, Godlewski G, Schlicker E. Histamine H_3 receptors – general characterization and their function in the cardiovascular system. J Physiol Pharmacol 1998;49:191–211.
85. Buchanan JE, Phillis JW. The role of nitric oxide in the regulation of cerebral blood flow. Brain Res 1993;610:248–255.
86. Korytko PJ, Boje KM. Pharmacological characterization of nitric oxide production in a rat model of meningitis. Neuropharmacology 1996;35:231–237.
87. Buzzi MG, Moskowitz MA. The pathophysiology of migraine: year 2005. J Headache Pain 2005;6:105–111.
88. Olesen J, Thomsen LL, Iversen HK. Nitric oxide is a key molecule in migraine and other vascular headaches. Trends Pharmacol Sci 1994;15:149–153.
89. Olesen J, Thomsen LL, Lassen LH, Olesen IJ. The nitric oxide hypothesis of migraine and other vascular headaches. Cephalalgia 1995;15:94–100.
90. Thomsen LL, Iversen HK, Brinck TA, Olesen J. Arterial supersensitivity to nitric oxide (nitroglycerin) in migraine sufferers. Cephalalgia 1993;13:395–399.

91. Iversen HK, Olesen J, Tfelt-Hansen P. Intravenous nitroglycerin as an experimental model of vascular headache. Basic characteristics. Pain 1989;38:17–24.
92. Olesen J, Iversen HK, Thomsen LL. Nitric oxide supersensitivity: a possible molecular mechanism of migraine pain. NeuroReport 1993;4:1027–1030.
93. Nemade RV, Lewis AI, Zuccarello M, Keller JT. Immunohistochemical localization of endothelial nitric oxide synthase in vessels of the dura mater of the Sprague-Dawley rat. Neurosci Lett 1995;197:78–80.
94. Berger RJ, Zuccarello M, Keller JT. Nitric oxide synthase immunoreactivity in the rat dura mater. NeuroReport 1994;5:519–521.
95. Dawson TM, Brendt DS, Fotuhi M, Hwang PM, Snyder SH. Nitric oxide synthase and neuronal NADPH diaphorase are identical in brain and peripheral tissues. Proc Natl Acad Sci U S A 1991;88:7797–7801.
96. Strecker T, Dux M, Messlinger K. Increase in meningeal blood flow by nitric oxide – interaction with calcitonin gene-related peptide receptor and prostaglandin synthesis inhibition. Cephalalgia 2002;22:233–241.
97. Messlinger K, Suzuki A, Pawlak M, Zehnter A, Schmidt RF. Involvement of nitric oxide in the modulation of dural arterial blood flow in the rat. Br J Pharmacol 2000;129:1397–1404.
98. Wei EP, Moskowitz MA, Boccalini P, Kontos HA. Calcitonin gene-related peptide mediates nitroglycerin and sodium nitroprusside-induced vasodilation in feline cerebral arterioles. Circ Res 1992;70:1313–1319.
99. Poyner DR, Sexton PM, Marshall I, Smith DM, Quirion R, Born W et al. International Union of Pharmacology. XXXII. The mammalian calcitonin gene-related peptides, adrenomedullin, amylin, and calcitonin receptors. Pharmacol Rev 2002;54:233–246.
100. McLatchie LM, Fraser NJ, Main MJ, Wise A, Brown J, Thompson N et al. RAMPs regulate the transport and ligand specificity of the calcitonin-receptor-like receptor. Nature 1998;393:333–339.
101. Brain SD, Grant AD. Vascular actions of calcitonin gene-related peptide and adrenomedullin. Physiol Rev 2004;84:903–934.
102. Zhang Z, Winborn CS, Marquez de Prado B, Russo AF. Sensitization of Calcitonin Gene-Related Peptide Receptors by Receptor Activity-Modifying Protein-1 in the Trigeminal Ganglion. J Neurosci 2007;27:2693–2703.
103. Lambert GA, Goadsby PJ, Zagami AS, Duckworth JW. Comparative effects of stimulation of the trigeminal ganglion and the superior sagittal sinus on cerebral blood flow and evoked potentials in the cat. Brain Res 1988;453:143–149.
104. Goadsby PJ. Characteristics of facial nerve-elicited cerebral vasodilatation determined using laser Doppler flowmetry. Am J Physiol 1991;260:R250–R262.
105. Délépine L, Aubineau P. Plasma protein extravasation induced in the rat dura mater by stimulation of the parasympathetic sphenopalatine ganglion. Exp Neurol 1997;147:389–400.
106. Drummond PD. Sweating and vascular responses in the face: normal regulation and dysfunction in migraine, cluster headache and harlequin syndrome. Clin Auton Res 1994;4:273–285.
107. Fozard JR, Kalkman HO. 5-Hydroxytryptamine (5-HT) and the initiation of migraine: new perspectives. Naunyn Schmiedebergs Arch Pharmacol 1994;350:225–229.
108. Kalkman HO. Is migraine prophylactic activity caused by 5-HT_{2B} or 5-HT_{2C} receptor blockade? Life Sci 1994;54:641–644.

109. Rebeck GW, Maynard KI, hyman BT, moskowitz MA. Selective 5-$HT_{1D\alpha}$ serotonin receptor gene expression in trigeminal ganglia: implications for antimigraine drug development. Proc Natl Acad Sci U S A 1994;91:3666–3669.
110. Bruinvels AT, Landwehrmeyer B, Gustafson EL, Durkin MM, Mengod G, Branchek TA et al. Localization of 5-HT1B, 5-HT1D alpha, 5-HT1E and 5-HT1F receptor messenger RNA in rodent and primate brain. Neuropharmacology 1994;33:367–386.
111. Bouchelet I, Cohen Z, Case B, Seguela P, Hamel E. Differential expression of sumatriptan-sensitive 5-hydroxytryptamine receptors in human trigeminal ganglia and cerebral blood vessels. Mol Pharmacol 1996;60:219–223.
112. Goadsby PJ, Lipton RB, Ferrari MD. Migraine-current understanding and treatment. N Engl J Med 2002;346:257–270.
113. Buzzi MG, Moskowitz MA. The antimigaine drug, sumatriptan (GR 43175), selectively blocks neurogenic plasma extravasation from blood vessels in dura mater. Br J Pharmacol 1990;99:202–206.
114. Matsubara T, Moskowitz MA, Byun B. CP-93,129, a potent and selective 5-HT1B receptor agonist blocks neurogenic plasma extravasation within rat but not guinea-pig dura mater. Br J Pharmacol 1991;104:3–4.
115. Buzzi MG, Dimitriadou V, Theoharides TC, Moskowitz MA. 5-Hydroxytryptamine receptor agonists for the abortive treatment of vascular headaches block mast cell, endothelial and platelet activation within the rat dura mater after trigeminal stimulation. Brain Res 1992;583:137–149.
116. Williamson DJ, Shepheard SL, Hill RG, Hargreaves RJ. The novel anti-migraine agent rizatriptan inhibits neurogenic dural vasodilation and extravasation. Eur J Pharmacol 1997;328:61–64.
117. Messlinger K, Hotta H, Pawlak M, Schmidt RF. Effects of the 5-HT_1 receptor agonists, sumatriptan and CP 93,129, on dural arterial flow in the rat. Eur J Pharmacol 1997;332:173–181.
118. Lee WS, Moussaoui SM, Moskowitz MA. Blockade by oral or parenteral RPR 100893 (a non-peptide NK1 receptor antagonist) of neurogenic plasma protein extravasation within guinea-pig dura mater and conjunctiva. Br J Pharmacol 1994;112:920–924.
119. Phebus LA, Johnson KW, Stengel PW, Lobb KL, Nixon JA, Hipskind PA. The non-peptide NK-1 receptor antagonist LY303870 inhibits neurogenic dural inflammation in guinea pigs. Life Sci 1997;60:1553–1561.
120. Goldstein DJ, Wang O, Saper JR, Stoltz R, Silberstein SD, Mathew NT. Ineffectiveness of neurokinin-1 antagonist in acute migraine: a crossover study. Cephalalgia 1997;17:785–790.
121. May A, Gijsman HJ, Wallnöfer A, Jones R, Diener HC, Ferrari MD. Endothelin antagonist blocks neurogenic inflammation, but is not effective in aborting migraine attacks. Pain 1996;67:375–378.
122. Olesen J, Diener HC, Husstedt IW, Goadsby PJ, Hall D, Meier U et al. BIBN 4096 BS Clinical Proof of Concept Study Group. Calcitonin gene-related peptide receptor antagonist BIBN 4096 BS for the acute treatment of migraine. N Engl J Med 2004;350:1104–1110.
123. Arulmani U, Maassenvandenbrink A, Villalon CM, Saxena PR. Calcitonin gene-related peptide and its role in migraine pathophysiology. Eur J Pharmacol 2004;500:315–330.
124. Geppetti P, Capone JG, Trevisani M, Nicoletti P, Zagli G, Tola MR. CGRP and migraine: neurogenic inflammation revisited. J Headache Pain 2005;6:61–70.

125. Fischer MJC, Koulchitsky S, Messlinger K. The nonpeptide calcitonin gene-related peptide receptor antagonist BIBN4096BS lowers the activity of neurons with meningeal input in the rat spinal trigeminal nucleus. J Neurosci 2005;25:5877–5883.
126. Humphrey PP, Goadsby PJ. The mode of action of sumatriptan is vascular? A debate. Cephalalgia 1994;14:401–410.
127. Cutrer FM, O'Donnell A, Sanchez del Rio M. Functional neuroimaging: enhanced understanding of migraine pathophysiology. Neurology 2000;55:S36–45.
128. Iversen HK. Experimental headache in humans. Cephalalgia 1995;15:281–287.
129. Thomsen LL, Kruuse C, Iversen HK, Olesen J. A nitric oxide donor (nitroglycerin) triggers genuine migraine attacks. Eur J Neurol 1994;1:73–80.
130. Fanciullacci M, Alessandri M, Figini M, Geppetti P, Michelacci S. Increase in plasma calcitonin gene-related peptide from the extracerebral circulation during nitroglycerin-induced cluster headache attack. Pain 1995;60:119–123.
131. Reuter U, Bolay H, Jansen-Olesen I, Chiarugi A, Sanchez del Rio M, Letourneau R et al. Delayed inflammation in rat meninges: implications for migraine pathophysiology. Brain 2001;124:2490–2502.

Neurogenic Inflammation in Health and Disease
Edited by Gábor Jancsó

Neurogenic Mechanisms in Arthritis

LUCY F. DONALDSON

Department of Physiology and Pharmacology, School of Medical Sciences, University of Bristol, Bristol, UK

ABSTRACT

Neurogenic inflammation occurs in many tissues, but has been principally studied in skin. Evidence accumulated over the last two decades supports the hypothesis that neurogenic inflammation is of prime importance in arthritis. Joints are richly innervated by neuropeptide-containing primary afferent and autonomic fibers, peptides are released into the synovial cavity of arthritic joints, and exogenous and endogenous neuropeptides exert proinflammatory actions on articular tissues. Acute and chronic interruption of the articular nerve supply renders joints less susceptible to arthritis.

Evidence also indicates that neuronal activation and neurogenic inflammation may underlie the symmetrical nature of arthritis, in both humans and experimental animals. The evidence for neurogenic inflammation in the maintenance of arthritis is reviewed, and a hypothesis is suggested for mechanisms by which the spread of arthritis may occur.

1. INTRODUCTION

Neurogenic inflammation, whereby inflammatory changes are caused by activation of sensory afferents, was described in detail by Bayliss (1900), who showed that electrical stimulation of sensory neurons and activation of antidromic action potentials could result in peripheral vasodilatation that was not due to activation of either spinal or sympathetic neurons [1]. This work was extended by many others, notably Thomas Lewis [2] who proposed a physiological mechanism whereby sensory nerves could contribute to inflammatory processes in the periphery. He observed that stimulation of sensory terminals evoked a flare and wheal, the so-called "triple response" in skin. His hypothesis was that invasion of branched sensory afferent terminals by antidromic action potentials resulted in the release of substances that acted to cause hyperemia and edema in local tissues (Fig. 1). Although Lewis thought it likely that histamine or acetylcholine were mediating these vascular effects [2], work over the last century has demonstrated that more likely candidates are the sensory neuropeptides; primarily the tachykinins substance P (SP) and neurokinin A (NKA), and calcitonin gene-related peptide (CGRP) (for review see Ref. [3]). It is now well accepted that SP and NKA act to increase vascular permeability in the capillary beds and postcapillary venules [4], and that CGRP acts on precapillary arterioles to cause vasodilatation and increased local blood flow [5]. The actions of these peptides on the vascular system therefore results in hyperemia and local

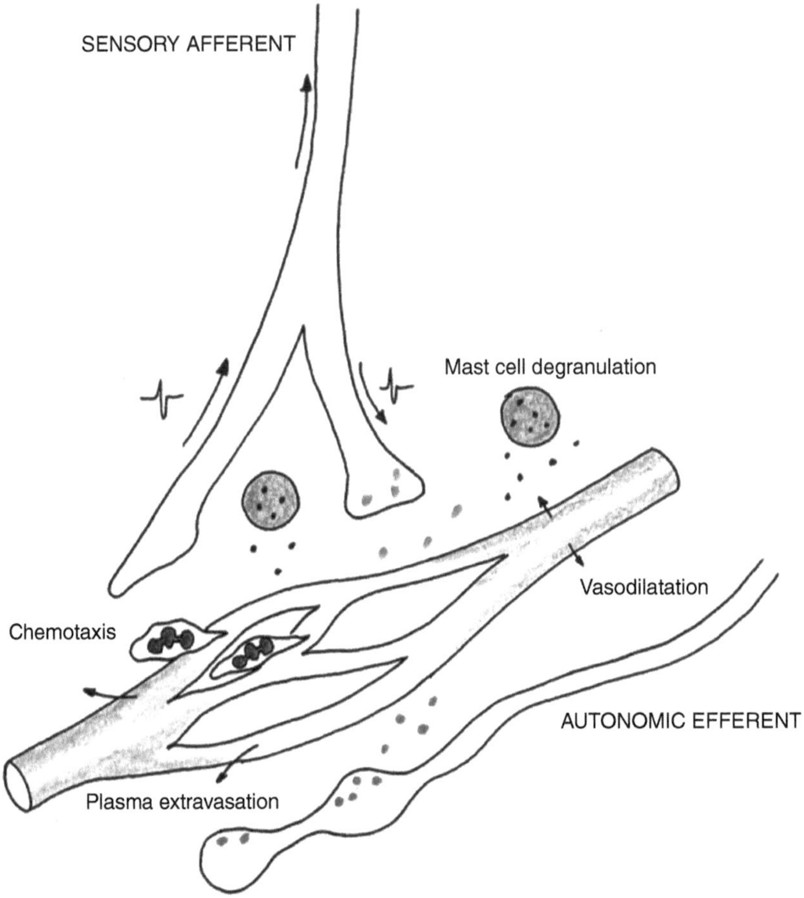

Figure 1. The axon reflex as first proposed by Lewis. Stimulation of primary afferent terminals results in orthodromic action potentials conducted to the CNS. Invasion of primary afferent terminal branches by antidromic action potentials (arrows) results in the release of proinflammatory neuropeptides such as substance P and calcitonin gene-related peptide (CGRP). These act on arterioles to cause vasodilatation (CGRP), on capillaries and postcapillary venules to cause plasma extravasation, on mast cells resulting in degranulation, and attract leukocytes out of the bloodstream (chemotaxis). Autonomic fibers may also release neuropeptide Y and vasoactive intestinal polypeptide, which also cause vasodilatation. (See color Plate 3.)

edema that affect both the site of injury or stimulation and surrounding areas innervated by terminal branches of the activated primary afferents (Fig. 1). Other vasoactive neuropeptides such as vasoactive intestinal polypeptide (VIP) and neuropeptide Y (NPY), derived principally from autonomic nerve terminals also exert similar effects on local vasculature, at least in the skin [6].

Although neurogenic inflammation was initially demonstrated in skin there is no clear reason why inflammation in joints should not also have a neurogenic component. It was, however, only in the 1980s that the possibility was first mooted that neuronal peptides and neurogenic inflammation could affect the degree of inflammation in arthritis [7]. The contribution of neurogenic mechanisms in arthritis has been debated over the last two decades but it is now clear that the majority of evidence supports a role for neurogenic inflammation in arthritis.

In this review I will consider the evidence for neurogenic inflammation in arthritis through discussion of (i) articular innervation by neuropeptide-containing nerve fibers, and the maintenance of innervation during arthritis; (ii) release of neuropeptides into articular tissues or the synovial space during arthritis allowing these peptides to exert their effects locally; (iii) proinflammatory effects of neuropeptides on the articular tissues; and (iv) the effect of interruption of the articular innervation on both experimental and clinical arthritis. It seems clear that in order for neurogenic inflammation to play a major role in the pathogenesis of arthritis, neuropeptides must be found in the locale, must be released from sensory fibers and must have proinflammatory actions on joints, and finally that abolition of innervation inhibits arthritis. I will then discuss the evidence, and proposed mechanisms for a central component to neurogenic inflammation in arthritis and its contribution to the spread of arthritis to distant joints.

2. PEPTIDERGIC JOINT INNERVATION

2.1 Innervation of normal joint tissues

Many studies have investigated innervation of synovial joints either directly to visualize the sensory and autonomic fibers innervating specific articular tissues, or by using retrograde labeling to verify the origin of the articular nerve terminals. In the immunochemical study of joint tissue innervation, it is assumed that the neuropeptides SP and CGRP are markers of sensory fibers, and that VIP and NPY are markers of parasympathetic and sympathetic fibers, respectively. In general VIP and NPY are found only in small numbers of uninjured dorsal root ganglion (DRG) neurons [8–10]. Many rigorous studies using retrograde labeling have shown that synovial joints in different species are innervated by sensory, sympathetic, and parasympathetic fibers containing a range of neuropeptides [11–14]. In determining whether specific articular tissues have a substantial innervation by neuropeptide-containing afferents that is maintained during arthritis, I will consider evidence garnered from studies directly localizing peptidergic fibers to articular tissues, rather than studies on peptide-containing neurons in innervating ganglia, as these investigations give direct evidence of the structure/tissue innervated within the joint.

Peptidergic fibers have been found to innervate most if not all articular tissues in many different species (Table 1). There is clear evidence of sensory innervation of the joint capsule [15] and connective tissue [16], synovium (both superficial and deep layers) [16–19], bone marrow [20], bone and periosteum [17,20], and ligaments and tendons [21] (see Table 1). There has been some controversy over the sensory innervation of cartilage, as this tissue has long been considered to be devoid of innervation. In joints, however there is evidence that peptidergic fibers innervate the epiphyses of growing long bones in young animals [22], knee joint menisci (fibrocartilage) show peptidergic nerve terminals innervating the superficial layers [23,24] and sensory fibers may also directly innervate hyaline cartilage [25].

In agreement with the findings that sensory neuropeptides act to regulate blood flow and vascular permeability in joints (see Section 3.1), many peptidergic fibers have been found in close apposition to blood vessels, in particular CGRP-positive fibers are found in close apposition to arterioles in the stroma or synovium [16,19,21,26]. Many of the sensory terminals found in articular tissue present as free nerve endings rather than being associated with specialized end

Table 1 Peptidergic innervation of normal articular tissues

Peptides	Structures innervated	Species	Ref.
Sensory neuropeptides			
Substance P and/or CGRP	Connective tissue and synovium of knee joint Synovial blood vessels	Human	[16]
	Bone marrow and subchondral regions of the proximal tibia	Rat	[20]
	Cruciate and collateral ligaments, meniscus, fat pad, synovium, and capsule of knee joint Associated with blood vessels in all tissues plus free nerve endings	Cat	[21]
	Joint capsule, ligaments, synovium, periosteum, subchondral bone, and bone marrow (mostly around blood vessels) of metacarpal joint	Horse	[19]
	Joint capsule, disc, periosteum, and synovium of Temporomandibular Joint. Some fibers very close to the synovial lining surface.	Rat	[18]
	Joint capsule, periosteum, synovium, fat pad, subchondral bone, and patella of knee joint	Mouse	[17]
	Arteries in the medial anterior capsule of the knee joint	Cat	[26]
Autonomic neuropeptides			
VIP and NPY	Posterior attachment of Temporomandibular Joint Nerve endings associated with vascular structures	Human	[29]
	Synovium of Temporomandibular Joint (VIP, Pituitary Adenylate Cyclase-Activating Polypeptide)	Rat	[15]
	NPY-positive nerve endings in blood vessels in sublining layer of synovium and bone marrow	Rat	[27]
	Perivascular fibers (NPY) and free endings (VIP) in lumbar facet capsule	Human	[28]

organs or terminal boutons. Autonomic fibers containing VIP or NPY are most often found in close apposition to blood vessels in synovium [15,27] and connective tissue or joint capsule [28,29].

Table 1 summarizes the data from selected studies on peptidergic joint innervation and illustrates the types of joints, the species, and the articular tissues in which peptidergic nerve fibers have been localized using immunochemical techniques. It is clear from many investigations that most, if not all articular tissues in many different species receive both sensory and autonomic innervation. These fibers contain neuropeptides known to exert proinflammatory effects, and therefore if these peptides are released into inflamed joints, they may contribute the degree of inflammation.

2.2 Innervation of joint tissues in arthritis

Evidence is accumulating that peptidergic innervation of joints may be altered in many experimental articular disease states. It is now clear that there are increases not only in the number of DRG neurons expressing SP and CGRP in arthritis [30,31], but also in the absolute amount of peptide expressed per neuron [9]. In addition, the expression of peptides considered to be analgesic or anti-inflammatory, such as somatostatin (SS) [32] or galanin [8] are also increased in primary afferents in experimental arthritis, suggesting that complex controls may exist on both central peptidergic neurotransmission and peripheral neurogenic inflammation. The dual regulation of neuropeptide expression in primary afferent neurons in terms of neuronal expression level and neuronal number could result in increased neuropeptide synthesis and release at both central terminals in the spinal cord and peripheral terminals in articular tissues. It is estimated that 80% of SP synthesized in the afferent cell body is transported to peripheral, rather than central terminals [33]. Evidence for the transport and release of neuropeptides in joints has come from studies of synovial neuropeptide content in clinical and experimental studies (see Section 2.1). There are few studies of the effect of arthritis on neuropeptides in peripheral nerve terminals, but in studies in which quantitation has been attempted it appears that articular peptidergic innervation may be decreased in chronic arthritis in animals [17,34–36] and humans [37–39]. This is possibly a consequence of terminal degeneration [17,35]; however, clinical data have demonstrated an increase in SP labeling, particularly in synovium, in joint disease [40,41]. Close inspection of studies suggests that apparent reductions in the innervation of synovium and other joint tissues may be explained by the release of SP and CGRP from terminals, resulting in less intense immunostaining. In addition proliferation of synovium with no concurrent increase in innervation density, could also be incorrectly interpreted as an overall loss of nerve fibers [42]. This combination of effects would be interpreted as a net loss of innervation. There is however ultrastructural evidence of damage to afferent terminals in arthritis [35], so a net loss of afferent terminals cannot be definitively ruled out.

3. LOCAL RELEASE OF NEUROPEPTIDES INTO JOINTS

3.1 Release of sensory neuropeptides into arthritic joints

Numerous studies have investigated the neuropeptide content of synovial fluid in various arthropathies in humans to determine the relationship (if any) between joint disease and articular neuropeptide release/synovial fluid content. Many other cell types, such as fibroblasts may express and release SP [43], although neuropeptides found in synovial fluid in experimental arthritis were not derived from synoviocytes [44], excluding synovium as a potential local source of neuropeptides.

Only a limited number of studies have investigated alterations in synovial neuropeptide content in experimental arthritis, but all have shown clear increases in the concentrations of sensory and autonomic neuropeptides in arthritis in synovial fluid and articular tissues. Injection of Freund's complete adjuvant (FCA) injection into the temporomandibular (TMJ) or knee joint resulted in significant increases in synovial fluid content of SP, neurokinin (NKA), CGRP, and NPY [45,46]. In neither study were significant changes in plasma or cerebrospinal fluid neuropeptides observed, indicating a local rather than a systemic mechanism for the alteration in synovial neuropeptide content.

In two studies in the horse, osteoarthritis and osteochondrosis resulted in increased intra-articular SP concentrations [47,48]. These latter findings are of particular interest as neurogenic mechanisms have been assumed to be predominant in inflammatory arthritis; these studies in the horse suggest that traumatic, noninflammatory arthritis may also have a neurogenic component.

Unlike experimental studies where disease duration and progression can be accurately assessed, in many clinical studies the interpretation of the data is confounded by a lack of information on whether patients suffer from acute, chronic, or acute exacerbation of a chronic arthritis. In addition, in some studies adequate disease-free controls are lacking for direct comparison with arthritic patients. Table 2 summarizes the clinical studies on synovial neuropeptide content reported to date. The majority of studies have compared patients with osteoarthritis and rheumatoid arthritis (RA), and found that SP levels are generally increased in patients

Table 2 Neuropeptide content of synovial fluid in joint disease

Conditions studied	Intra-articular peptides studied	Findings – intra-articular peptide concentrations	Ref.
RA vs. osteoarthritis (no disease-free controls)	SP, CGRP, VIP	RA > osteoarthritis	[49]
	SP	No significant difference	[38]
	SP	RA > osteoarthritis	[52]
	SP, CGRP, VIP	SP: RA > osteoarthritis. No significant difference in other peptides	[50] [180]
	SP, NGF	SP: RA > osteoarthritis	[51]
	SP, Somatostatin	SP, Somatostatin: RA > osteoarthritis	[58]
	VIP, Somatostatin	VIP: RA > osteoarthritis	
Inflammatory vs. degenerative joint disease (no disease-free controls)	SP, NKA, CGRP, NPY	No significant difference between groups	[53]
RA, osteoarthritis, traumatic arthritis (no disease-free controls)	SP	SP: Posttrauma Other conditions – no significant difference	[55]
RA (no disease-free controls)	SP, CGRP, NKA, NPY	High peptide concentrations in RA associated with pain and joint destruction	[56]
RA, ankylosing spondylitis, psoriatic arthritis (no disease-free controls)	CGRP	Peptide concentration correlated with intra-articular	
	NPY	temperature CGRP +ve correlation NPY –ve correlation	[57, 190]
Temporomandibular Joint disc derangement (some with RA, some with osteoarthritis) (no disease-free controls)	SP, NKA, CGRP, VIP, NPY	No significant difference between groups Temporomandibular Joint concentrations > knee joint	[54]
RA vs. ligament injury (no disease-free controls)	SP, NKA, CGRP, NPY, VIP	No SP found in any group NKA: ligament injury > RA CGRP and NPY: RA > ligament injury	[59]
Arthritis	SP and metabolites	Only very low levels detected	[61]

with RA [38,44,49–52]. Generally, changes in other neuropeptide levels have not been reported [38,53–55] although one study has shown increased somatostatin levels in RA [51]. It is particularly surprising that no significant changes in synovial CGRP levels have been found, although high intra-articular CGRP content has been associated with pain, joint destruction, and intra-articular temperature [56,57], potential symptoms of active disease. Neuropeptides postulated to be released from autonomic fibers, NPY, and VIP have an even less clear profile than sensory neuropeptides when studied in synovial fluid. One study has found significantly higher VIP levels in RA compared to osteoarthritis [58], and one has identified higher NPY levels in a similar comparison [59].

If osteoarthritis (and indeed other arthritides) does have a neurogenic component, as is suggested by both studies in the horse [48] and the human [60], then alteration in synovial neuropeptide concentration would also be predicted in this condition. A lack of significant alteration in neuropeptide concentration in RA when compared to osteoarthritis may therefore mask any changes, resulting in a nonsignificant difference, due to the use of a diseased control group. Alternatively, lack of significant findings may be due to differences between studies in the disease duration or diagnostic factors. Significantly, one group has reported that they were unable to find significant quantities of SP in synovial fluid from patients with arthritis and that the levels they identified were up to 10,000-fold lower than those reported in other studies [61]. It is therefore also a possibility that synovial neuropeptide concentrations could be affected by the way in which clinical samples are treated before assay.

4. ACTIONS OF NEUROPEPTIDES ON ARTICULAR TISSUES

4.1 Proinflammatory effects of exogenous neuropeptides

Exogenous neuropeptides injected into normal joints are known to have proinflammatory effects on many parameters of joint physiology that are similar to those seen in other tissues. Perfusion of normal joints with SP in anesthetized animals results in a dose-dependent synovial plasma extravasation [62], as does perfusion with CGRP [63]. CGRP also enhances the plasma extravasation stimulated by SP, histamine, or bradykinin [5,64], probably because of the concurrent CGRP-dependent vasodilatation and subsequent increase in articular blood flow [65]. The effect of the interaction of SP and CGRP on articular blood flow is complex; both peptides induce vasodilatation with that produced by CGRP being more prolonged than SP whereas coadministration of the peptides results in a more transient vasodilatation than that seen with CGRP alone [66].

SP and CGRP can both exert direct effects on the immune system. SP can activate, induce proliferation of, and induce chemotaxis in peripheral leukocytes [67] through both NK1 and NK2 receptors [68] and through receptor-independent mechanisms [67]. CGRP is also a chemoattractant for peripheral blood leukocytes [69]. Neuropeptides also induce endothelial cell adhesion molecule expression [70,71], aiding extravasation of leukocytes, and may recruit immature dendritic cells to sites of inflammation [72].

With specific reference to articular tissues, CGRP in particular has profound effects on osteoclastic activity and therefore bone resorption. CGRP-positive fibers have been shown to contact osteoclasts and osteoblasts [73], and sensory denervation with capsaicin reduces osteoclast number and active bone resorption [74]. SP can also stimulate proliferation of synoviocytes, which is a particular feature of RA [75,76].

4.2 Proinflammatory effects of endogenous neuropeptides

Activation of articular sensory afferents by electrical stimulation or by local capsaicin application results in the release of tachykinins into the synovial fluid [77], and so provides a model for the study of endogenous neuropeptide action in joints. Electrical stimulation of mixed sensory and autonomic articular nerves results in sympathetically mediated vasoconstrictor and sensory nerve mediated prolonged vasodilator responses. The latter is inhibited by the NK1 receptor antagonist D-Pro4D-Trp7,9,10-SP(4-11) [78] and is attributable to SP release. The related tachykinins, NKA and NKB also cause transient vasodilatation in the knee [66]. The sympathetic vasoconstrictor response seen in normal joints is reversed to give a vasodilator response in acute arthritis and when exogenous sensory neuropeptides are administered concurrently with sympathetic stimulation, suggesting that under inflammatory conditions normal sympathetic vasomotor control is modified by neuropeptides released from sensory afferents [79]. In chronic inflammation, normal vasoconstrictor and vasodilator responses disappear [80]. T cell infiltrates, swelling (edema), and joint destruction are all reduced following sensory denervation of joints, suggesting that the source of peptides mediating these effects is indeed the articular afferent innervation [81].

Most studies have concentrated on the proinflammatory neuropeptides SP and CGRP, but evidence is accumulating that other sensory neuropeptides such as somatostatin may also modulate the inflammatory response in arthritis. Although the numbers of somatostatin containing afferents is not altered in arthritis [30], the expression level per DRG neuron is increased in the chronic stages of an experimental model [32]. Somatostatin can also be released from sensory nerve terminals by antidromic electrical stimulation [82] where it exerts an anti-inflammatory effect, probably through inhibition of SP and CGRP release from afferent terminals [83]. This peptide may therefore act as an endogenous anti-inflammatory agent; exogenous somatostatin can reduce synovial hyperplasia in RA [84] and inflammation in experimental arthritis [84].

In addition, somatostatin may also act as an endogenous analgesic peptide. Clinical studies show that intra-articular injection of somatostatin is an effective analgesic in both RA and osteoarthritis [85,86], possibly acting to directly inhibit nociceptors innervating the inflamed joint [87].

In general therefore, the evidence appears to support the hypothesis that neuropeptides are released intra-articularly in inflammatory arthritis in humans and experimental animals, peptidergic fibers innervate most if not all articular tissue types and neuropeptides exert both proinflammatory and anti-inflammatory effects in articular tissues. However, to verify that neuropeptides are derived from sensory afferents, and to substantiate the claim that neurogenic mechanisms are of significance in the pathogenesis of arthritis then the hypothesis that denervation protects joints from the development of arthritis, or attenuates preexisting arthritis in both clinical and experimental situations needed to be tested.

5. PERIPHERAL NEUROGENIC MECHANISMS IN ARTHRITIS

Levine and colleagues were the first to suggest that SP may contribute to the severity of experimental joint inflammation, and that the source of intra-articular SP may be neuronal [88]. In 1985 Levine, Moskowitz, and Basbaum first hypothesized that there may be a neurogenic component in inflammatory arthritis [7] based on experimental and clinical data.

This hypothesis was later reiterated by others [89,90] and the use of many denervation strategies in experimental arthritis over many years has shown that in most animal models, arthritis either does not develop or is significantly attenuated in denervated joints [91–98], supporting the Levine hypothesis.

5.1 Effect of articular denervation on arthritis

Axotomy is the most profound denervation injury. Initial reports suggested that this injury did protect joints from arthritis in the FCA-induced polyarthritis model [99] although this has been difficult to reproduce [91]. A less profound denervation injury, either systemic capsaicin treatment in the neonate or adult, local application of capsaicin to nerve trunks, or local capsaicin injection in the joint results in degeneration of only a subset of primary afferents expressing the capsaicin receptor. This injury attenuates experimental arthritis [81,93,94,96,100], articular neuropeptide concentrations [92] and the density of peptidergic innervation in joints [98]. According to Lewis' hypothesis of the triple response, where neurogenic inflammation results from the activation of a peripheral sensory terminal and the antidromic invasion of terminal branches (Fig. 1), local capsaicin application would be predicted to yield these results, as terminal invasion by antidromic action potentials would be prevented. From these experimental data, it is now widely accepted that capsaicin-sensitive transient receptor potential vanilloid receptor-1 (TRPV1) (also known as VR1) expressing primary afferents are the fibers upon which neurogenic arthritis is dependent.

Interestingly, blockade of the sciatic nerve with local anesthetic also inhibits the development of paw edema and mechanical allodynia in rats [101], an intervention that also prevents the increase in neuropeptide mRNA usually seen in acute inflammation [102]. Capsaicin lesion also inhibits alterations in primary afferent neuropeptide expression and release [92,94], suggesting that the mechanism of attenuation of inflammation by both capsaicin and local anesthetic is through blockade of neuropeptide upregulation and consequent increased peripheral release.

5.2 Effect of denervation injury on remote joint involvement in arthritis – neurogenic mechanisms in the spread of disease

The evidence outlined above for the involvement of the peripheral nervous system, particularly capsaicin-sensitive primary afferents, in arthritis is strong. In their original hypothesis, Levine et al. [103] suggested that not only was the peripheral nervous system intimately involved in the maintenance of inflammation in an arthritic joint, but that the peripheral and central nervous systems may also be involved in remote effects in inflammatory arthritis. Consequently, arthritis may initially present in a single joint but then spreads to other, specifically symmetrical contralateral joints through a neurogenic action.

This hypothesis is supported by clinical observation in patients with RA, although increasingly osteoarthritis may be considered to be subject to similar neurogenic inflammatory effects. Clinically, RA is defined as a symmetrical disease [104]. When patients with RA are followed over several years, their disease shows a high tendency to become more symmetrical over time, particularly in patients who are rheumatoid factor positive [105]. Although osteoarthritis is often not thought to be an inflammatory arthritis, recent epidemiological study shows that osteoarthritis also appears to have a higher incidence of symmetrical joint involvement than nonsymmetrical joint involvement [60]. Many reports have documented that patients who develop inflammatory joint disease after nerve injury, hemiplegia, or poliomyelitis show sparing of the denervated joints [106–109], closely agreeing with the experimental data that an intact nerve supply seems to be necessary for the development of arthritis at a site distant to the initial lesion. Although clinical and experimental studies have usually reported on the effect of prior denervation on the subsequent development of inflammation, it was recently reported that a patient with established RA showed resolution of joint inflammation in the paralyzed limbs on subsequent development of hemiplegia, even though paresis occurred some time after the onset of inflammatory joint disease [110]. Experimental studies have also shown that in experimental arthritis, capsaicin inhibits the spread of disease to distant joints in addition to ameliorating disease at the initial inflammatory focus [94,111].

These findings are not limited to arthritis; dermatitis does not affect denervated skin in which sensory nerve terminal numbers are significantly reduced [6] and psoriasis is also reported to resolve following denervation [112].

It is difficult to logically attribute symmetry in arthritis to immunological or systemic mechanisms as an explanation as to why specific joints, particularly symmetrical homologous joints, are more prone to inflammation than others. Circulating factors or increased contralateral joint usage could be suggested as explanations for the symmetry in RA. Experimentally, however, immobilization does not protect joints from arthritic involvement as would be predicted if increased usage or altered weight bearing were to act as a predisposing factor. Indeed, immobilization can both exacerbate arthritis [113,114] and neuronal activity in articular nerves [115]. The involvement of circulating systemic factors has also been addressed in experimental models, where interruption of the blood supply to a joint does not prevent the development of disease [97].

Therefore, the experimental evidence to date shows that an intact innervation by capsaicin-sensitive primary afferents is key to both neurogenic maintenance of an existing arthritis, and the spread of arthritis to remote joints. Alterations in neuropeptide expression levels and/or peripheral release are important. Clinical evidence also suggests that the central nervous system (CNS) may have a role to play (as hemiplegia or poliomyelitis also attenuate arthritis), and it is logical to assume that the capsaicin-sensitive primary afferents must have some connection with their mirror image counterparts through the CNS.

6. NEURONAL MECHANISMS IN THE SPREAD OF ARTHRITIS

The hypothesis for a neurogenic component in the spread of arthritis therefore proposes that an initial arthritic focus activates ipsilateral afferents that can act to maintain that inflammatory process. In addition, activation of these innervating afferents must somehow activate the homologous contralateral afferents, initiating neurogenic inflammation in the contralateral homologous joint. There are few direct connections between primary afferent neurons on each side of the spinal cord; therefore, this contralateral activation must occur either through the spinal cord, or through higher CNS centers.

In consideration of this hypothesis, I will discuss evidence for the ipsilateral and contralateral activation of spinal neurons, ipsilateral and contralateral primary afferent activation, and putative mechanisms through which these events may occur.

6.1 Contralateral neuronal activation – alteration of contralateral neuronal phenotype

In experimental models of unilateral arthritis, there have been frequent reports of bilateral contralateral alterations in neuronal phenotype similar to those seen on the inflamed side, but often with slightly different kinetics or of a smaller magnitude. Alteration in neuronal phenotype in any pathological condition is often considered as a reactive process – which is reflective of some activity or perturbation in the neuron because of the pathological process. It has therefore been assumed that changes in phenotype (mRNA or protein expression) may reflect an activation of the cell as a consequence of the pathology. As with most studies of neurogenic inflammation, the majority of studies have concentrated on the proinflammatory neuropeptides, SP, and CGRP. SP and its receptors have been shown to be upregulated contralaterally in unilateral arthritis [111,116,117] although this is not a universal finding [9,118]. Many other molecules expressed in the spinal cord show bilaterally altered expression levels or activation. These molecules have varied functions but include second messenger molecules such as NADPH-diaphorase [119] (thought to equate with nitric oxide synthase expression), cyclooxygenase-2 [120], protein kinase C [121], and CREB (cAMP response element binding protein) [122], proinflammatory molecules such as tumor necrosis factor α and interleukin-1 [123], and neurotransmitter receptors N-methyl-D aspartic acid (NMDA receptors) [124]. Alterations in expression of such molecules could be interpreted as being indicative of contralateral spinal neuronal activation. Other transmitters and receptors such as gamma-aminobutyric acid (GABA) are found to show only unilateral alteration in monoarthritis [125,126] suggesting that bilateral spinal alteration of phenotype is not a universal, nonspecific consequence of peripheral inflammation but rather a specific, appropriate, possibly protective response to a peripheral inflammatory focus [127].

In addition to alterations in these proinflammatory molecules in spinal neurons, contralateral DRG neurons also show altered expression levels of different molecules including many receptors; e.g., tachykinin, prostaglandin, and bradykinin receptors in monoarthritis [117,128], 5-HT (5-hydroxytryptamine) receptors following bee venom injection [129], and markers of neuronal activation such as phosphorylated p38 [123,130]. Altered primary afferent receptor expression could underpin the altered primary afferent neuronal function also seen in monoarthritis (Section 6.2).

6.2 Contralateral neuronal activation – alteration of ipsilateral and contralateral primary afferent neuronal function

Do alterations in cellular phenotype necessarily reflect cellular activation? No study to date has yet investigated spinal neuronal function, in terms of electrophysiological properties, in neurons contralateral to an inflamed joint. It has been known for some time that deep dorsal horn neurons receiving afferent input from an inflamed joint develop expanded receptive fields (RFs) that often include the mirror image areas on the contralateral uninflamed limb [131]. It is therefore not possible to determine how alteration in spinal neuronal phenotype relates to cellular function in the contralateral spread of arthritis. It would seem feasible, however, that the molecules that show bilateral regulation in unilateral arthritis may have some role in the process, and are therefore potential "target genes" for initial studies investigating this process.

The majority of evidence suggesting that contralateral neurons do indeed show alterations in function, in addition to phenotype during unilateral arthritis has been gathered by the groups headed by Westlund and Willis in Galveston, Texas, in a series of studies on both ipsilateral and contralateral primary afferent neuronal activity in monoarthritis. Initially Rees and colleagues demonstrated that mechanically evoked antidromic activity could be recorded in ipsilateral articular primary afferent neurons innervating an inflamed joint [132]. These antidromic action potentials (also known as dorsal root reflexes (DRRs) as they were first identified in dorsal roots) have been shown in sensory afferents prior to the studies of Willis and Westlund, but generally in hypothermic anesthetized animals (for an in depth review of DRRs, see Ref. [133]). This DRR activity was inhibited by local anesthetic application but not by sympathectomy, showing that the activity was truly antidromic in sensory neurons. This was the first study in which DRRs, either occurring spontaneously or evoked by somatosensory stimulation have been demonstrated to occur in normothermic animals. Subsequent studies from the same groups have demonstrated that blockade of ipsilateral DRRs reduces parameters of inflammation such as joint swelling, further supporting a neurogenic contribution to experimental arthritis and secondly that DRR generation may be a process that is fundamental to the generation of neurogenic inflammation in joints [132,134]. DRRs are not unique to arthritic lesions; however, such activity can also be recorded in sensory afferents following a local injection of capsaicin into the footpad where dermal neurogenic inflammation caused by capsaicin can be blocked by sciatic nerve section [135]. The occurrence of spontaneous and evoked action potentials recorded in sensory afferents at their central terminals (in dorsal roots), and the demonstration that inhibition of this activity reduces inflammation in that joint suggests that antidromic activation of the afferents does not only occur in the afferent terminals as proposed by Lewis, but may (also) occur through a central mechanism. Supporting a role for the CNS in contralateral activation of primary afferents are the significant findings that mechanically evoked and spontaneous DRRs can also be recorded in the contralateral homologous articular sensory afferents in monoarthritis [136,137]. In unilateral inflammation, blockade of peripheral neurotransmission with local anesthetic applied either ipsilaterally or contralaterally reduces swelling [138]. These findings indicate that bilateral alteration in spinal and primary afferent neuronal phenotype is accompanied by bilateral activation of primary afferent neurons and the generation of antidromic activity that may initiate or contribute to the remote inflammatory effects seen in clinical and experimental arthritis and inflammation. Significant spontaneous antidromic activity is seen in homologous contralateral afferents at a frequency known to result in neuropeptide release [139] and it is therefore likely that this neuronal activity contributes to the sustained neurogenic inflammation seen in experimental models. Indeed, very early in the process of arthritis when a single joint is inflamed but there are no obvious signs of remote joint disease, but when spontaneous and evoked DRRs can be recorded, the vasculature in the contralateral joint is more permeable to albumin-bound Evans blue [137].

In a search for potential spinal mechanisms through which arthritis spreads, understanding of how DRRs are generated centrally could give ideas as to how homologous contralateral afferents may also be activated. Current understanding is that primary afferents do alter their electrophysiological properties in a manner that could predispose a joint to inflammation both ipsilateral and contralateral to a monoarthritis. How then could these alterations be effected centrally?

6.3 Potential spinal mechanisms through which ipsilateral neurogenic inflammation may occur in arthritis

A potential spinal mechanism through which primary afferent activation can occur has also been studied and described by the Galveston groups. In their studies on monoarthritis, the spinal pathways through which ipsilateral DRRs can be generated have been partially described. DRRs are generated by the action of GABA on the central terminals of primary afferent neurons; this has also been termed primary afferent depolarization and is one mechanism through which neurotransmitter release from primary afferent central terminals can be inhibited. When GABA activates $GABA_A$ receptors on the central terminals of these neurons, as a result of a high intracellular chloride concentration in these neurons, Cl^- movement is outward and hence neurons are depolarized. The high intracellular chloride concentration in primary afferent neurons is maintained at a high level by the action of the sodium–potassium–chloride cotransporter, NKCC1. Antagonists at $GABA_A$ but not $GABA_B$ receptors applied to spinal cord during arthritis block DRRs [140]. In addition, spinal administration of glutamate antagonists reduces DRRs [140] and both joint inflammation and thermal hyperalgesia [141] in the same model, showing that spinal generation of DRRs by the central actions of both GABA and glutamate can contribute to the degree of inflammation found in the joint. Modulation of GABAergic effects can also come about through changes in chloride gradients in DRG or spinal neurons [142,143]. Inhibition of the NKCC1 and KCC chloride cotransporters in the spinal cord inhibits DRRs and neurogenic edema caused by intradermal capsaicin injection [144], suggesting that neuronal chloride gradients are key to the generation of DRRs.

Other neurotransmitters implicated in ipsilateral DRR generation in monoarthritis are SP [145] and 5-HT [146]; the latter is probably released by descending neurons projecting from supraspinal sites. The reason that SP and glutamate antagonists also block DRRs is that primary afferent nociceptors release these two neurotransmitters at their central terminals. SP and glutamate then activate GABAergic interneurons, which then depolarize the central terminal of the primary afferent. When depolarization is sufficient, DRRs may be generated, a hypothesis suggested by Willis [133]. Many other neurochemicals are also involved in the central sensitization of primary afferent terminals, including centrally released prostaglandins and nitric oxide [147–152].

Studies have clearly demonstrated a central component in unilateral neurogenic arthritis, and have begun to elucidate a potential mechanism through which spinal neurotransmitters could act to cause this. The challenge facing us now is to determine how arthritis on one side of the body activates the central and peripheral nervous systems on the other side of the body to initiate development of arthritis in remote joints.

Figure 2 shows that neurogenic inflammation involves interactions between the nervous and immune systems. Although the nervous system appears to have a role in the initiation, development, and maintenance of inflammatory joint disease, I would not wish to underplay the major role of the immune system. It is clear however that the spread of arthritis to involve remote joints cannot be solely dependent on immune mechanisms. The often symmetrical nature of the disease cannot be completely explained by immune system activation, as why should one symmetrical homologous joint be targeted above others? The hypothesis presented in Fig. 2 suggests an intimate relationship between the nervous and immune systems, involving priming and targeting of specific joints by the nervous system that then become the site at which destructive immune system activation subsequently occurs. The hypothesis extended in Fig. 2 for the generation of contralateral inflammation is that ipsilateral arthritis activates ipsilateral afferents resulting in centrally generated DRRs that add to the neurogenic component of the existing monoarthritis. In addition, the contralateral primary afferents are also activated through an as yet unknown spinal mechanism. Contralateral DRRs are generated and the contralateral homologous joint is "primed" for inflammation by the local release of proinflammatory neuropeptides. This is supported by data on increased vascular perfusion/permeability in the contralateral joint [137]. This then predisposes the contralateral joint to the development of arthritis due to the combined effects of the vascular and chemotaxic effects of, in particular, SP and CGRP and an activated immune system as is found in RA. Working on the hypothesis that centrally generated DRRs reflect ipsilateral and contralateral primary afferent activation, how then could contralateral DRRs be generated?

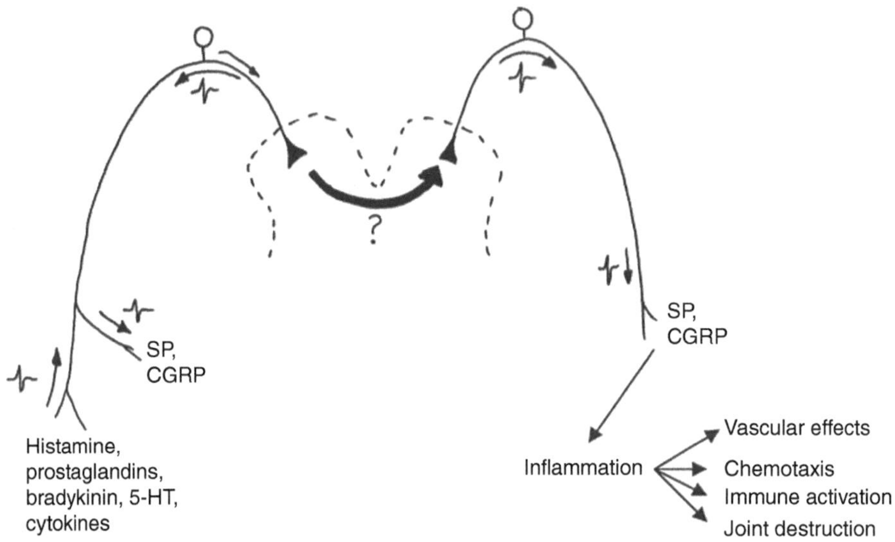

Figure 2. Hypothesis for the neurogenic spread of arthritis. Activation of capsaicin-sensitive primary afferents by an initial inflammatory focus (left) through the actions of inflammatory mediators causes increased afferent input to the spinal cord. The initial left-sided arthritis is maintained in part by neurogenic mechanisms; both the peripheral axon reflex and centrally generated activity through generation of DRRs. Through a transneuronal mechanism that is still unclear, the homologous contralateral capsaicin-sensitive primary afferents are activated centrally to produce antidromic action potentials (arrows). This results in the peripheral release of SP, CGRP, and other neuropeptides in the mirror image joint. These peptides initiate an inflammatory reaction that, together with actions of the immune system, results in joint destruction. The central anatomical and neurochemical pathways through which contralateral capsaicin-sensitive primary afferents are activated are still unclear.

6.4 Hypothesized spinal pathways through which contralateral DRRs could be generated

If a peripheral sensory nerve is damaged, resulting in changes in phenotype or function, similar changes, but of lower magnitude or slower kinetics, have often been reported in the contralateral homologous neurons. This area was thoroughly reviewed by Koltzenburg and colleagues [153]. This review suggested potential anatomical pathways through which these contralateral changes could be mediated. It is probable that contralateral inflammatory changes are mediated through similar or even identical pathways.

It is clear from electrophysiological evidence that functional connections between the two sides of the spinal cord exist but that these connections may be under tonic inhibitory control by glycinergic and glutamatergic systems [154]. Many dorsal horn neurons have contralateral inhibitory RFs [155], which expand, and in some cases become excitatory in arthritis. In addition, the numbers of neurons with a bilateral RF also increases in chronic inflammation [131]. In the trigeminal system, monoarthritis of the temporomandibular joint causes long-term bilateral CNS activation, evidenced by bilateral cFos expression [156].

Figure 3 summarizes different hypothesized anatomical pathways that could explain contralateral DRR generation. Figure 3A proposes an activation of a GABAergic interneuron (crosshatched) by primary afferent activity and subsequent DRR generation (arrow). The same primary afferent has a contralateral projection that also activates contralateral neurons, either interneurons or the homologous primary afferent. This generates contralateral antidromic

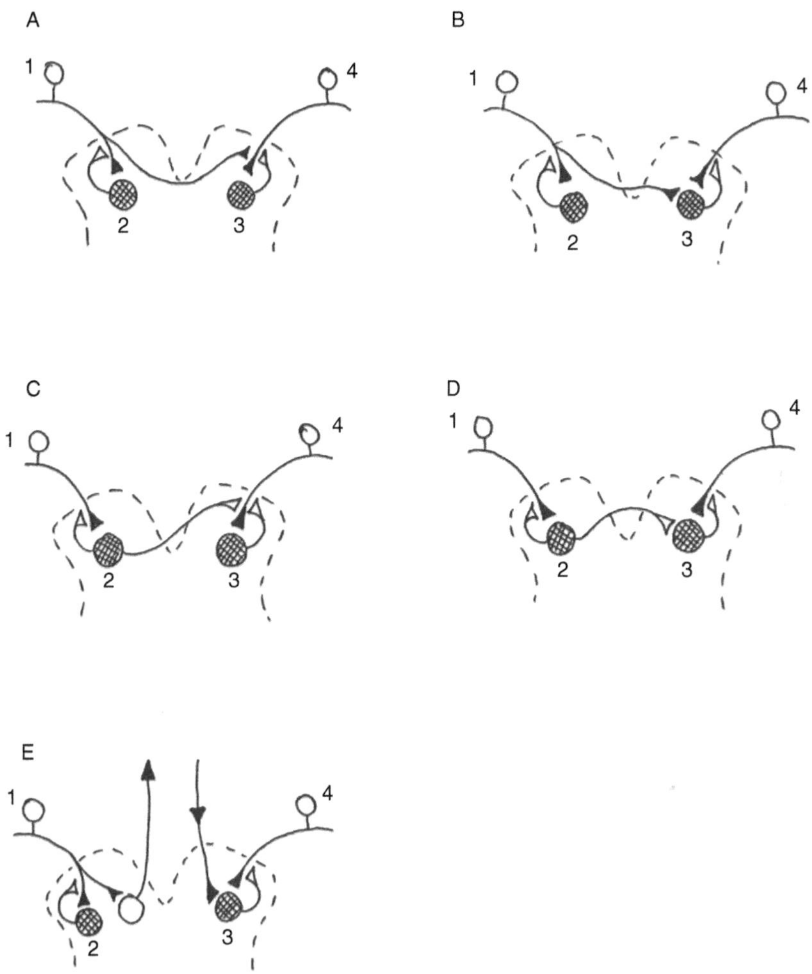

Figure 3. Proposed anatomical pathways through which the spread of arthritis to contralateral joints may be mediated. See main text for full explanations. GABAergic interneurons are crosshatched. (A) Direct contralateral projection by capsaicin-sensitive primary afferents onto the central terminals of homologous primary afferent neurons. (B) Direct contralateral primary afferent projection onto contralateral GABAergic interneuron that activates the contralateral primary afferent through primary afferent depolarization. (C) Contralateral projection at the segmental level through a single GABAergic interneuron that activates the contralateral primary afferent through primary afferent depolarization. (D) Contralateral projection at the segmental level through multiple GABAergic interneurons. Activation of the contralateral interneuron may be through nitric oxide release (see Fig. 4). (E) Connection between spinal dorsal horns through a purely supraspinal, rather than spinal segmental route. Projection neurons (unshaded) receive excitatory input from primary afferent neurons. Descending neurons activate GABAergic interneurons resulting in contralateral primary afferent depolarization.

action potentials, possibly through GABA in a similar manner to ipsilateral DRRs, or by direct activation by other primary afferent transmitters.

Direct projections to the contralateral dorsal (laminae I–IV) and ventral horns have been described in many mammalian species; however, most of these studies have been carried out in the sacral cord, where the dorsal horns fuse in the midline [157,158]. These contralateral projections are seen in the superficial (lamina I [159]) and deeper dorsal horn [160,161] and mostly represent the central terminals of visceral afferents. Direct contralateral projections in the lumbar region, the segment of the spinal cord receiving hind limb joint input, are sparse in the rat, as noted by Koltzenburg and colleagues [153], and the numbers seen in this region may also be species dependent [162]. It is therefore unlikely that direct primary afferent projections to the contralateral dorsal horn are of importance in contralateral DRR generation, partly due to the relatively small numbers of such projections and partly through the possible neurotransmitters involved. It is possible that contralateral DRRs are not generated by the central action of GABA, as ipsilateral DRRs are, but there is no evidence on which to speculate which other transmitters may be involved, as the neurochemical basis of contralateral DRRs has not yet been fully tested.

Figure 3C and D proposes a contralateral projection by GABAergic interneurons. There is evidence that there are interneuronal connections at the segmental level in the spinal cord through the commisure [163], although their neurotransmitter content has not been determined. Very few experimental studies have been able to investigate the effect of spinal cord lesion on the spread of arthritis, due to obvious ethical considerations in working with animals with both hindquarter paralysis and arthritis. In one study where bilateral cartilage degradation was found in monoarthritis [164], spinal cord compression injury inhibited contralateral cartilage degradation. This could indicate that a direct segmental spinal connection exists between homologous primary afferents as shown in Fig. 3B and C. This could either be through interneurons that project directly from one dorsal horn to the contralateral homologous primary afferent (Fig. 3C), or through a multisynaptic pathway involving two or more interneurons (Fig. 3D). Supporting the hypothesis that the contralateral projection is segmental is the observation that joint inflammation is unaffected by thoracic spinal cord transection [165], and that contralateral DRG neuronal activation is restricted to the segments receiving inputs from the inflamed joint [123]. If one accepts the premise that contralateral changes in neuronal phenotype are indicative of neuronal activation in this pathway, activation of the neurons as depicted in Fig. 3D would be in the order $1 \rightarrow 2 \rightarrow 3 \rightarrow 4$, as spinal phenotypic changes (minutes to hours) occur prior to contralateral primary afferent activation (days). If model 3C is considered, activation of primary afferent 4 would occur prior to activation of interneuron 3 (i.e., $1 \rightarrow 2 \rightarrow 4 \rightarrow 3$), which does not fit with experimental observations of phenotypic and electrophysiological changes.

Figure 3E depicts circuitry in which there is no direct connection between dorsal horns at the spinal level, i.e., that contralateral spread of arthritis is solely dependent on supraspinal activation. However, as spinal transection did not inhibit contralateral DRRs but spinal compression injury at the thoracic level did inhibit contralateral cartilage degradation [165], it is possible that both spinal and supraspinal sites play a role. Direct electrical activation of the periaqueductal gray, a brainstem area of pivotal importance in descending control of spinal cord function, enhances DRRs in normal animals [146], but there is as yet no evidence that supraspinal sites modulate DRRs in inflamed animals. The potential involvement of supraspinal sites in neurogenic arthritis is therefore still not resolved. It is much more likely that none of these relatively simple models accurately depicts the correct wiring of this pathway and that it is a combination of Fig. 3C–E, involving both direct spinal projections, and ascending and descending pathways that results in the contralateral activation of primary afferents and antidromic action potential generation.

6.5 Spinal neuronal phenotype and putative neurotransmitters involved in the generation of contralateral DRRs

The anatomical pathways involved in the remote spread of arthritis are one area in which there is still much work to be done before we clearly understand the neuronal processes involved. There is a little more information on the potential involvement of specific neurotransmitters on which speculation can be based. Our knowledge on the GABAergic dependence of ipsilateral DRR generation suggests that it is unlikely that GABA is not involved, and pharmacological study on DRRs in monoarthritis gives insight into other potential neurotransmitter candidates, such as glutamate and SP. It is also feasible that regulatory molecules such as CREB, nitric oxide, or cyclooxygenases that show bilateral alteration in expression in monoarthritis may be involved in contralateral neuronal activation, either upstream or downstream of the GABA, glutamate, or SP. As yet, no studies have fully investigated the neurochemical basis for contralateral DRRs in monoarthritis, although our work shows that contralateral spontaneous DRRs can be inhibited by the centrally acting analgesic dipyrone (metamizol, related to acetaminophen) [137]. The exact mechanism of action of this drug is under debate, but its effect on contralateral DRRs suggests a role for either the nitric oxide [166] or cyclooxygenase enzymes [167,168] in generation of this neuronal activity.

Willis, in an excellent review on DRRs [133] proposed that in ipsilateral DRR generation, primary afferent neurons release both SP and glutamate onto spinal GABAergic interneurons resulting in activation. Subsequent release of GABA on the central terminal of the primary afferent initiates antidromic activity (Fig. 4, left side). This model explains in a simple and logical fashion the findings of Willis and Westlund, and fits well with data from other groups [145].

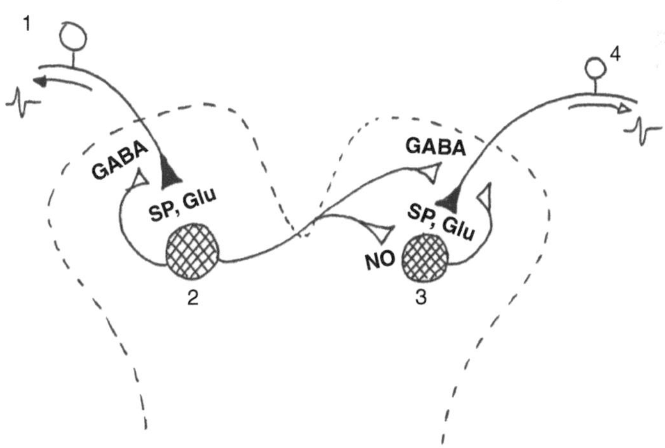

Figure 4. Proposed multisynaptic pathway for the contralateral activation of primary afferent neurons (composite of Fig. 3B–D for simplicity) with proposed neurotransmitters at each synapse. Primary afferent stimulation in unilateral inflammation results in SP and glutamate release onto GABAergic interneurons (crosshatched). GABA depolarizes the ipsilateral primary afferent, causing DRRs (arrow). Contralateral projection of the GABAergic interneuron either directly causes contralateral DRRs by release of GABA onto primary afferent terminals, or activates a second GABAergic interneuron through nitric oxide (NO) release. Contralateral DRRs result in the initiation of contralateral inflammation (see Fig. 2).

There is as yet no experimental evidence on which to base a hypothesis for the neurotransmitters released at contralateral synapses in Fig. 4. If there is a direct interneuronal synapse onto the contralateral primary afferent, DRRs could be generated by direct GABA release onto primary afferent 4. The interneuron 2 (crosshatched) is proposed to be GABAergic and therefore inhibitory, but some GABAergic interneurons are known to express nitric oxide synthase [152]. It is therefore possible that contralateral GABAergic interneuron three may be excited through the release of nitric oxide or some other excitatory transmitter; however, this hypothesis is highly speculative and requires experimental evidence.

6.6 Does the sympathetic nervous system contribute to neurogenic inflammation in arthritis?

The original hypothesis proposed for a neurogenic contribution arthritis included actions of both the sensory and sympathetic nervous systems in contributing to arthritic inflammation [90,100]. Joints have a rich sympathetic innervation [169] and sympathetic regulation of vascular tone is clearly seen in joints as in other tissues, where sympathetic stimulation results in vasoconstriction [170]. Whether sympathetic efferents contribute to arthritis in ways other than direct effects on the vasculature remains equivocal.

Kidd et al. [171], proposed that sympathetic activity could persistently stimulate articular afferents thereby contributing to the progress of the disease and the pain experienced. Normal blood flow in joints depends upon endogenous tachykinins as shown in a series of studies by Ferrell and colleagues. Endogenous tachykinins maintain basal blood flow, (as NK1 antagonists reduce basal blood flow), opposing the sympathetically mediated vasoconstriction seen in these structures [172]. Therefore, in the normal joint the sympathetic and sensory nervous systems seem to have opposing effects on blood flow. Plasma extravasation and alterations in blood flow in acute arthritis also appear to be mediated though the sensory rather than the sympathetic nervous system [79,96,173]. Indeed the peptidergic vasodilatory effects seen on sensory stimulation increase while the sympathetic vasoconstriction decreases in acute arthritis [79]. Exogenous SP or CGRP attenuate sympathetically mediated articular vasoconstriction in acute arthritis, suggesting that in inflammation, the sensory nervous system may modulate sympathetic function in joints rather than the converse [79].

Bradykinin-induced plasma extravasation in the knee joint has been hypothesized to be dependent on postganglionic sympathetic neurons (PGSN) [174]. Evidence supports the hypothesis that nociceptive afferents innervating inflamed sites can be activated by adrenergic agonists, particularly noradrenaline [175], but this effect is not dependent on the PGSN [176]. Intra-articular perfusion with 6-hydroxydopamine (6-OHDA), thought to be a stimulator of PGSN causes plasma extravasation, a finding that was interpreted as sympathetic contribution to arthritis [177], although again this has not been shown by all groups [178]. 6-OHDA has been shown to also exert effects through sensory rather than sympathetic neurons [179], which could suggest that there is little sympathetic involvement.

The evidence from studies of joint perfusion and plasma extravasation would seem to suggest that rather than the PGSN further activating sensory afferents, and thereby exacerbating inflammation, sympathetic and sensory afferents may have opposing actions with most proinflammatory effects being largely attributable to the sensory nervous system.

If PGSNs do play a role in neurogenic inflammation in arthritis, and there is some evidence that this is the case, lesion of the sympathetic nervous system should reduce inflammation and hyperalgesia in arthritis, as the proposed "feedforward" stimulation of the primary afferents through the PGSN would be lost. Sympathectomy has been shown to reduce inflammation in experimental arthritis [100,180], and pain and stiffness in patients [181]. This finding has not been universally observed however – combined surgical and chemical sympathetic lesioning did not alter joint swelling or hyperalgesia in kaolin/carrageenan arthritis in the rat [165], other models of arthritis show little or no sympathetic dependence [95], but capsaicin-enhanced DRRs are blocked by sympathectomy [182]. In contrast, arthritis even be exacerbated following sympathectomy [183]. Surgical sympathectomy does exert profound effects on immune system stimulation in the early stages of adjuvant arthritis [184] and may therefore affect disease progression through this action. Chemical sympathectomy using 6-OHDA, in addition to the potential effects on sensory neurons noted above, seems to be dependent on the route of administration – systemic 6-OHDA administration exacerbated arthritis while local lymph node administration attenuated the disease [183]. The potential role of the PGSN in neurogenic arthritis is therefore not clear. One study suggests that activation of the sympathetic system is of importance and that adrenaline and noradrenaline can have profound effects on sensory afferents [175], but the interaction may be at the level of the hypothalamic–pituitary–adrenal (HPA) axis rather than the PGSN [176,185–187]. Certainly HPA activation is seen as a stress response in animals with relatively mild arthritis [188]; therefore, endogenous adrenergic agonists, in addition to glucocorticoids may modify the inflammatory response. Concurrent HPA and sympathetic activation has recently been proposed to contribute to a generalized sensitization of sensory neurons [189].

With respect to the potential involvement of the sympathetic nervous system in the spread of arthritis, Kidd and colleagues suggested that joint damage activates ipsilateral PGSN. Ipsilateral PGSN project to the contralateral PGSN and activate contralateral sympathetic output, which then activates contralateral nociceptors resulting in contralateral inflammation [90]. In this scenario, lesion of either the sensory or sympathetic nervous systems should inhibit the spread of arthritis; therefore this hypothesis is difficult to test experimentally. We did note increased sympathetic efferent activity contralateral to arthritis, but the possibility that this contributes to the spread of arthritis requires further investigation [137]. Chemical sympathectomy has been reported to inhibit the spread of arthritis [100] but as mentioned above chemical sympathectomy may have nonspecific effects on other, and importantly, sensory systems. Significantly, the only study to investigate surgical sympathectomy and arthritis found no effect [165], but this study did not look at the spread of arthritis to remote sites.

The major challenge for future study in this area is the mechanism of the spread of arthritis. It is clear from the experimental and clinical evidence that neurogenic inflammation, involving capsaicin-sensitive primary afferents adds to the severity of the disease process. Synovial joints receive a rich sensory peptidergic innervation to all joint tissues and in clinical and experimental arthritis, neuropeptides are released from the sensory terminals into the joint space, where it is likely they induce proinflammatory changes. These peptidergic neurons increase their activity, firing spontaneous and evoked antidromic action potentials, giving a functional correlate for the behavioral observation that interruption of the sensory supply to joints in both the experimental and clinical settings results in an attenuation of arthritis. Sympathetic activation may also exert proinflammatory effects in arthritis.

The neurogenic spread of arthritis requires a great deal of further study. Our understanding of the anatomical pathways, sympathetic versus sensory, the spinal versus higher center controls is currently rudimentary. Acceptance of the hypothesis that arthritis can affect remote joints through a neurogenic mechanism is now widespread. Rigorous scientific investigation of the mechanisms involved may aid in the amelioration of a disease that affects significant numbers of people worldwide.

ACKNOWLEDGMENTS

I would like to thank Dr. D.O. Bates for constructive criticism of the manuscript and Dr. J.C. Hancox for helpful assistance with the figures.

REFERENCES

1. Bayliss W. On the origin from the spinal cord of the vaso-dilator fibres of the hind-limb, and on the nature of these fibres. J Physiol 1900;26:173–209.
2. Lewis T. The nocifensor system of nerves and its reactions. Br Med J 1937;1(3974):491–494.
3. Holzer P. Peptidergic sensory neurons in the control of vascular functions: mechanisms and significance in the cutaneous and splanchnic vascular beds. Rev Physiol Biochem Pharmacol 1992;121:49–146.
4. Holzer P. Neurogenic vasodilatation and plasma leakage in the skin. Gen Pharmacol 1998;30(1):5–11.
5. Brain SD, Williams TJ. Inflammatory oedema induced by synergism between calcitonin gene-related peptide (CGRP) and mediators of increased vascular permeability. Br J Pharmacol 1985;86(4):855–860.
6. Wallengren J, Tegner E, Sundler F. Cutaneous sensory nerve fibers are decreased in number after peripheral and central nerve damage. J Am Acad Dermatol 2002;46(2):215–217.
7. Levine J, Collier D, Basbaum A, Moskowitz M, Helms C. Hypothesis: the nervous system may contribute to the pathophysioogy of rheumatoid arthritis. J Rheumatol 1985;12(3):406–411.
8. Calza L, Pozza M, Arletti R, Manzini E, Hokfelt T. Long-lasting regulation of galanin, opioid, and other peptides in dorsal root ganglia and spinal cord during experimental polyarthritis. Exp Neurol 2000;164(2):333–343.
9. Donaldson LF, Harmar AJ, McQueen DS, Seckl JR. Increased expression of preprotachykinin, calcitonin gene-related peptide, but not vasoactive intestinal peptide messenger RNA in dorsal root ganglia during the development of adjuvant monoarthritis in the rat. Brain Res Mol Brain Res 1992;16(1–2):143–149.
10. Noguchi K, De Leon M, Nahin RL, Senba E, Ruda MA. Quantification of axotomy-induced alteration of neuropeptide mRNAs in dorsal root ganglion neurons with special reference to neuropeptide Y mRNA and the effects of neonatal capsaicin treatment. J Neurosci Res 1993;35(1):54–66.
11. Hanesch U, Heppelmann B, Schmidt RF. Somatostatin-like immunoreactivity in primary afferents of the medial articular nerve and colocalization with substance P in the cat. J Comp Neurol 1995;354(3):345–352.
12. Hanesch U, Heppelmann B, Schmidt RF. Substance P- and calcitonin gene-related peptide immunoreactivity in primary afferent neurons of the cat's knee joint. Neuroscience 1991;45(1):185–193.
13. Salo PT, Theriault E. Number, distribution and neuropeptide content of rat knee joint afferents. J Anat 1997;190(Pt 4):515–522.
14. Tamura R, Hanesch U, Schmidt RF, Kumazawa T, Mizumura K. Examination of colocalization of calcitonin gene-related peptide- and substance P-like immunoreactivity in the knee joint of the dog. Neurosci Lett 1998;254(1):53–56.

15. Uddman R, Grunditz T, Kato J, Sundler F. Distribution and origin of nerve fibers in the rat temporomandibular joint capsule. Anat Embryol (Berl) 1998;197(4):273–282.
16. Hirasawa Y, Okajima S, Ohta M, Tokioka T. Nerve distribution to the human knee joint: anatomical and immunohistochemical study. Int Orthop 2000;24(1):1–4.
17. Buma P, Verschuren C, Versleyen D, Van der Kraan P, Oestreicher AB. Calcitonin gene-related peptide, substance P and GAP-43/B-50 immunoreactivity in the normal and arthrotic knee joint of the mouse. Histochemistry 1992;98(5):327–339.
18. Kido MA, Kiyoshima T, Kondo T, Ayasaka N, Moroi R, Terada Y et al. Distribution of substance P and calcitonin gene-related peptide-like immunoreactive nerve fibers in the rat temporomandibular joint. J Dent Res 1993;72(3):592–598.
19. Nixon AJ, Cummings JF. Substance P immunohistochemical study of the sensory innervation of normal subchondral bone in the equine metacarpophalangeal joint. Am J Vet Res 1994;55(1):28–33.
20. Imai S, Tokunaga Y, Maeda T, Kikkawa M, Hukuda S. Calcitonin gene-related peptide, substance P, and tyrosine hydroxylase-immunoreactive innervation of rat bone marrows: an immunohistochemical and ultrastructural investigation on possible efferent and afferent mechanisms. J Orthop Res 1997;15(1):133–140.
21. Marshall KW, Theriault E, Homonko DA. Distribution of substance P and calcitonin gene related peptide immunoreactivity in the normal feline knee. J Rheumatol 1994;21(5):883–889.
22. Edoff K, Grenegard M, Hildebrand C. Retrograde tracing and neuropeptide immunohistochemistry of sensory neurones projecting to the cartilaginous distal femoral epiphysis of young rats. Cell Tissue Res 2000;299(2):193–200.
23. Gronblad M, Korkala O, Liesi P, Karaharju E. Innervation of synovial membrane and meniscus. Acta Orthop Scand 1985;56(6):484–486.
24. Mine T, Kimura M, Sakka A, Kawai S. Innervation of nociceptors in the menisci of the knee joint: an immunohistochemical study. Arch Orthop Trauma Surg 2000;120(3–4):201–204.
25. Schwab W, Funk RH. Innervation pattern of different cartilaginous tissues in the rat. Acta Anat (Basel) 1998;163(4):184–190.
26. Heppelmann B, Shahbazian Z, Hanesch U. Quantitative examination of calcitonin gene-related peptide immunoreactive nerve fibres in the cat knee joint capsule. Anat Embryol (Berl) 1997;195(6):525–530.
27. Iwasaki A, Inoue K, Hukuda S. Distribution of neuropeptide-containing nerve fibers in the synovium and adjacent bone of the rat knee joint. Clin Exp Rheumatol 1995;13(2):173–178.
28. Ashton IK, Ashton BA, Gibson SJ, Polak JM, Jaffray DC, Eisenstein SM. Morphological basis for back pain: the demonstration of nerve fibers and neuropeptides in the lumbar facet joint capsule but not in ligamentum flavum. J Orthop Res 1992;10(1):72–78.
29. Haeuchi Y, Matsumoto K, Ichikawa H, Maeda S. Immunohistochemical demonstration of neuropeptides in the articular disk of the human temporomandibular joint. Cells Tissues Organs 1999;164(4):205–211.
30. Hanesch U, Blecher F, Stiller RU, Emson PC, Schaible HG, Heppelmann B. The effect of a unilateral inflammation at the rat's ankle joint on the expression of preprotachykinin-A mRNA and preprosomatostatin mRNA in dorsal root ganglion cells – a study using non-radioactive in situ hybridization. Brain Res 1995;700(1–2):279–284.

31. Hanesch U, Heppelmann B, Schmidt RF. Quantification of cat's articular afferents containing calcitonin gene-related peptide or substance P innervating normal and acutely inflamed knee joints. Neurosci Lett 1997;233(2–3):105–108.
32. Abd El-Aleem SA, Morales-Aza BM, McQueen DS, Donaldson LF. Inflammation alters somatostatin mRNA expression in sensory neurons in the rat. Eur J Neurosci 2005;21(1):135–141.
33. Keen P, Harmar AJ, Spears F, Winter E. Biosynthesis, axonal transport and turnover of neuronal substance P. Ciba Found Symp 1982;91:145–164.
34. da Silva JA, Fonseca JE, Graca L, Moita L, Carmo-Fonseca M. Reinnervation of post-arthritic joints in the rat. Clin Exp Rheumatol 1996;14(1):43–51.
35. Imai S, Tokunaga Y, Konttinen YT, Maeda T, Hukuda S, Santavirta S. Ultrastructure of the synovial sensory peptidergic fibers is distinctively altered in different phases of adjuvant induced arthritis in rats: ultramorphological characterization combined with morphometric and immunohistochemical study for substance P, calcitonin gene related peptide, and protein gene product 9.5. J Rheumatol 1997;24(11):2177–2187.
36. Tahmasebi-Sarvestani A, Tedman R, Goss AN. The influence of experimentally induced osteoarthrosis on articular nerve fibers of the sheep temporomandibular joint. J Orofac Pain 2001;15(3):206–217.
37. Mapp PI, Kidd BL, Gibson SJ, Terry JM, Revell PA, Ibrahim NB et al. Substance P-, calcitonin gene-related peptide- and C-flanking peptide of neuropeptide Y-immunoreactive fibres are present in normal synovium but depleted in patients with rheumatoid arthritis. Neuroscience 1990;37(1):143–153.
38. Menkes CJ, Renoux M, Laoussadi S, Mauborgne A, Bruxelle J, Cesselin F. Substance P levels in the synovium and synovial fluid from patients with rheumatoid arthritis and osteoarthritis. J Rheumatol 1993;20(4):714–717.
39. Pereira da Silva JA, Carmo-Fonseca M. Peptide containing nerves in human synovium: immunohistochemical evidence for decreased innervation in rheumatoid arthritis. J Rheumatol 1990;17(12):1592–1599.
40. Witonski D, Wagrowska-Danielewicz M. Distribution of substance-P nerve fibers in the knee joint in patients with anterior knee pain syndrome. A preliminary report. Knee Surg Sports Traumatol Arthrosc 1999;7(3):177–183.
41. Yoshida H, Fujita S, Nishida M, Iizuka T. The expression of substance P in human temporomandibular joint samples: an immunohistochemical study. J Oral Rehabil 1999;26(4):338–344.
42. Buma P, Elmans L, Van Den Berg WB, Schrama LH. Neurovascular plasticity in the knee joint of an arthritic mouse model. Anat Rec 2000;260(1):51–61.
43. Inoue H, Shimoyama Y, Hirabayashi K, Kajigaya H, Yamamoto S, Oda H et al. Production of neuropeptide substance P by synovial fibroblasts from patients with rheumatoid arthritis and osteoarthritis. Neurosci Lett 2001;303(3):149–152.
44. Matucci-Cerinic M, Partsch G, Marabini S, Cagnoni M. High levels of substance P in rheumatoid arthritis synovial fluid. Lack of substance P production by synoviocytes in vitro. Clin Exp Rheumatol 1991;9(4):440–441.
45. Bileviciute I, Lundeberg T, Ekblom A, Theodorsson E. Substance P-, neurokinin A-, calcitonin gene-related peptide- and neuropeptide Y-like immunoreactivity (-LI) in rat knee joint synovial fluid during acute monoarthritis is not correlated with concentrations of neuropeptide-LI in cerebrospinal fluid and plasma. Neurosci Lett 1994;167(1–2):145–148.

46. Carleson J, Alstergren P, Appelgren A, Appelgren B, Kopp S, Srinivasan GR et al. Effects of adjuvant on neuropeptide-like immunoreactivity in experimentally induced temporomandibular arthritis in rats. Arch Oral Biol 1996;41(7):705–712.
47. Caron JP, Bowker RM, Abhold RH, Toppin DS, Sonea IM, Vex KB. Substance P in the synovial membrane and fluid of the equine middle carpal joint. Equine Vet J 1992;24(5):364–366.
48. Kirker-Head CA, Chandna VK, Agarwal RK, Morris EA, Tidwell A, O'Callaghan MW et al. Concentrations of substance P and prostaglandin E2 in synovial fluid of normal and abnormal joints of horses. Am J Vet Res 2000;61(6):714–718.
49. Arnalich F, de Miguel E, Perez-Ayala C, Martinez M, Vazquez JJ, Gijon-Banos J et al. Neuropeptides and interleukin-6 in human joint inflammation relationship between intraarticular substance P and interleukin-6 concentrations. Neurosci Lett 1994;170(2):251–254.
50. Hernanz A, De Miguel E, Romera N, Perez-Ayala C, Gijon J Arnalich F. Calcitonin gene-related peptide II, substance P and vasoactive intestinal peptide in plasma and synovial fluid from patients with inflammatory joint disease. Br J Rheumatol 1993;32(1):31–35.
51. Marabini S, Matucci-Cerinic M, Geppetti P, Del Bianco E, Marchesoni A, Tosi S et al. Substance P and somatostatin levels in rheumatoid arthritis, osteoarthritis, and psoriatic arthritis synovial fluid. Ann N Y Acad Sci 1991;632:435–436.
52. Matucci-Cerinic M, Lombardi A, Leoncini G, Pignone A, Sacerdoti L, Spillantini MG et al. Neutral endopeptidase (3.4.24.11) in plasma and synovial fluid of patients with rheumatoid arthritis. A marker of disease activity or a regulator of pain and inflammation? Rheumatol Int 1993;13(1):1–4.
53. Alstergren P, Appelgren A, Appelgren B, Kopp S, Lundeberg T, Theodorsson E. Co-variation of neuropeptide Y, calcitonin gene-related peptide, substance P and neurokinin A in joint fluid from patients with temporomandibular joint arthritis. Arch Oral Biol 1995;40(2):127–135.
54. Holmlund A, Ekblom A, Hansson P, Lind J, Lundeberg T, Theodorsson E. Concentrations of neuropeptides substance P, neurokinin A, calcitonin gene-related peptide, neuropeptide Y and vasoactive intestinal polypeptide in synovial fluid of the human temporomandibular joint. A correlation with symptoms, signs and arthroscopic findings. Int J Oral Maxillofac Surg 1991;20(4):228–231.
55. Marshall KW, Chiu B, Inman RD. Substance P and arthritis: analysis of plasma and synovial fluid levels. Arthritis Rheum 1990;33(1):87–90.
56. Appelgren A, Appelgren B, Kopp S, Lundeberg T, Theodorsson E. Neuropeptides in the arthritic TMJ and symptoms and signs from the stomatognathic system with special consideration to rheumatoid arthritis. J Orofac Pain 1995;9(3):215–225.
57. Appelgren A, Appelgren B, Kopp S, Lundeberg T, Theodorsson E. Relation between the intra-articular temperature of the temporomandibular joint and the presence of neuropeptide Y-like immunoreactivity in the joint fluid. A clinical study. Acta Odontol Scand 1993;51(1):1–8.
58. Lygren I, Ostensen M, Burhol PG, Husby G. Gastrointestinal peptides in serum and synovial fluid from patients with inflammatory joint disease. Ann Rheum Dis 1986;45(8):637–640.
59. Larsson J, Ekblom A, Henriksson K, Lundeberg T, Theodorsson E. Concentration of substance P, neurokinin A, calcitonin gene-related peptide, neuropeptide Y and vasoactive intestinal polypeptide in synovial fluid from knee joints in patients suffering from rheumatoid arthritis. Scand J Rheumatol 1991;20(5):326–335.

60. Cooper C, Egger P, Coggon D, Hart DJ, Masud T, Cicuttini F et al. Generalized osteoarthritis in women: pattern of joint involvement and approaches to definition for epidemiological studies. J Rheumatol 1996;23(11):1938–1942.
61. Joyce TJ, Yood RA, Carraway RE. Quantitation of substance-P and its metabolites in plasma and synovial fluid from patients with arthritis. J Clin Endocrinol Metab 1993;77(3):632–637.
62. Scott DT, Lam FY, Ferrell WR. Time course of substance P-induced protein extravasation in the rat knee joint measured by micro-turbidimetry. Neurosci Lett 1991;129(1):74–76.
63. Karimian M, Ferrell WR. Plasma protein extravasation into the rat knee joint induced by calcitonin gene-related peptide. Neurosci Lett 1994;166(1):39–42.
64. Cruwys SC, Kidd BL, Mapp PI, Walsh DA, Blake DR. The effects of calcitonin gene-related peptide on formation of intra-articular oedema by inflammatory mediators. Br J Pharmacol 1992;107(1):116–119.
65. Brain SD, Williams TJ, Tippins JR, Morris HR, MacIntyre I. Calcitonin gene-related peptide is a potent vasodilator. Nature 1985;313(5997):54–56.
66. Lam FY, Ferrell WR. Effects of interactions of naturally-occurring neuropeptides on blood flow in the rat knee joint. Br J Pharmacol 1993;108(3):694–699.
67. Kavelaars A, Jeurissen F, Heijnen CJ. Substance P receptors and signal transduction in leukocytes. Immunomethods 1994;5(1):41–48.
68. Hood VC, Cruwys SC, Urban L, Kidd BL. Differential role of neurokinin receptors in human lymphocyte and monocyte chemotaxis. Regul Pept 2000;96(1–2):17–21.
69. Schratzberger P, Reinisch N, Prodinger WM, Kahler CM, Sitte BA, Bellmann R et al. Differential chemotactic activities of sensory neuropeptides for human peripheral blood mononuclear cells. J Immunol 1997;158(8):3895–3901.
70. Lambert N, Lescoulie PL, Yassine-Diab B, Enault G, Mazieres B, De Preval C et al. Substance P enhances cytokine-induced vascular cell adhesion molecule-1 (VCAM-1) expression on cultured rheumatoid fibroblast-like synoviocytes. Clin Exp Immunol 1998;113(2):269–275.
71. Smith CH, Barker JN, Morris RW, MacDonald DM, Lee TH. Neuropeptides induce rapid expression of endothelial cell adhesion molecules and elicit granulocytic infiltration in human skin. J Immunol 1993;151(6):3274–3282.
72. Dunzendorfer S, Kaser A, Meierhofer C, Tilg H, Wiedermann CJ. Cutting edge: peripheral neuropeptides attract immature and arrest mature blood-derived dendritic cells. J Immunol 2001;166(4):2167–2172.
73. Imai S, Rauvala H, Konttinen YT, Tokunaga T, Maeda T, Hukuda S et al. Efferent targets of osseous CGRP-immunoreactive nerve fiber before and after bone destruction in adjuvant arthritic rat: an ultramorphological study on their terminal-target relations. J Bone Miner Res 1997;12(7):1018–1027.
74. Adam C, Llorens A, Baroukh B, Cherruau M, Saffar JL. Effects of capsaicin-induced sensory denervation on osteoclastic resorption in adult rats. Exp Physiol 2000;85(1):62–66.
75. Lotz M, Carson DA, Vaughan JH. Substance P activation of rheumatoid synoviocytes: neural pathway in pathogenesis of arthritis. Science 1987;235(4791):893–895.
76. Partsch G, Matucci-Cerinic M, Marabini S, Jantsch S, Pignone A, Cagnoni M. Collagenase synthesis of rheumatoid arthritis synoviocytes: dose-dependent stimulation by substance P and capsaicin. Scand J Rheumatol 1991;20(2):98–103.

77. Yaksh TL. Substance P release from knee joint afferent terminals: modulation by opioids. Brain Res 1988;458(2):319–324.
78. Khoshbaten A, Ferrell WR. Nerve-mediated responses of blood vessels in the rabbit knee joint. J Vasc Res 1993;30(2):102–107.
79. Lam FY, Ferrell WR. Acute inflammation in the rat knee joint attenuates sympathetic vasoconstriction but enhances neuropeptide-mediated vasodilatation assessed by laser Doppler perfusion imaging. Neuroscience 1993;52(2):443–449.
80. McDougall JJ, Karimian SM, Ferrell WR. Prolonged alteration of vasoconstrictor and vasodilator responses in rat knee joints by adjuvant monoarthritis. Exp Physiol 1995;80(3):349–357.
81. Hood VC, Cruwys SC, Urban L, Kidd BL. The neurogenic contribution to synovial leucocyte infiltration and other outcome measures in a guinea pig model of arthritis. Neurosci Lett 2001;299(3):201–204.
82. Szolcsanyi J, Helyes Z, Oroszi G, Nemeth J, Pinter E. Release of somatostatin and its role in the mediation of the anti-inflammatory effect induced by antidromic stimulation of sensory fibres of rat sciatic nerve. Br J Pharmacol 1998;123(5):936–942.
83. Green PG, Basbaum AI, Levine JD. Sensory neuropeptide interactions in the production of plasma extravasation in the rat. Neuroscience 1992;50(3):745–749.
84. Matucci-Cerinic M, Borrelli F, Generini S, Cantelmo A, Marcucci I, Martelli F. Somatostatin-induced modulation of inflammation in experimental arthritis. Arthritis Rheum 1995;38(11):1687–1693.
85. Fioravanti A, Govoni M, La Montagna G, Perpignano G, Tirri G, Trotta F et al. Somatostatin 14 and joint inflammation: evidence for intraarticular efficacy of prolonged administration in rheumatoid arthritis. Drugs Exp Clin Res 1995;21(3):97–103.
86. Silveri F, Morosini P, Brecciaroli D, Cervini C. Intra-articular injection of somatostatin in knee osteoarthritis: clinical results and IGF-1 serum levels. Int J Clin Pharmacol Res 1994;14(2):79–85.
87. Heppelmann B, Pawlak M. Inhibitory effect of somatostatin on the mechanosensitivity of articular afferents in normal and inflamed knee joints of the rat. Pain 1997;73(3):377–382.
88. Levine JD, Clark R, Devor M, Helms C, Moskowitz MA, Basbaum AI. Intraneuronal substance P contributes to the severity of experimental arthritis. Science 1984;226(4674):547–549.
89. Fitzgerald M. Arthritis and the nervous system. Trends Neurosci 1989;12(3):86–87.
90. Kidd BL, Mapp PI, Gibson SJ, Polak JM, O'Higgins F, Buckland-Wright JC et al. A neurogenic mechanism for symmetrical arthritis. Lancet 1989;2(8672):1128–1130.
91. Ahmed M, Srinivasan GR, Theodorsson E, Schultzberg M, Kreicbergs A. Effects of surgical denervation on substance P and calcitonin gene-related peptide in adjuvant arthritis. Peptides 1995;16(4):569–579.
92. Carleson J, Kogner P, Bileviciute I, Theodorsson E, Appelgren A, Appelgren B et al. Effects of capsaicin in temporomandibular joint arthritis in rats. Arch Oral Biol 1997;42(12):869–876.
93. Cruwys SC, Garrett NE, Kidd BL. Sensory denervation with capsaicin attenuates inflammation and nociception in arthritic rats. Neurosci Lett 1995;193(3):205–207.
94. Donaldson LF, McQueen DS, Seckl JR. Neuropeptide gene expression and capsaicin-sensitive primary afferents: maintenance and spread of adjuvant arthritis in the rat. J Physiol 1995;486(Pt 2):473–482.
95. Lam FY, Ferrell WR. Neurogenic component of different models of acute inflammation in the rat knee joint. Ann Rheum Dis 1991;50(11):747–751.

96. Lam FY, Ferrell WR. Capsaicin suppresses substance P-induced joint inflammation in the rat. Neurosci Lett 1989;105(1–2):155–158.
97. Levine JD, Dardick SJ, Basbaum AI, Scipio E. Reflex neurogenic inflammation. I. Contribution of the peripheral nervous system to spatially remote inflammatory responses that follow injury. J Neurosci 1985;5(5):1380–1386.
98. Mapp PI, Kerslake S, Brain SD, Blake DR, Cambridge H. The effect of intra-articular capsaicin on nerve fibres within the synovium of the rat knee joint. J Chem Neuroanat 1996;10(1):11–18.
99. Courtright LJ, Kuzell WC. Sparing effect of neurological deficit and trauma on the course of adjuvant arthritis in the rat. Ann Rheum Dis 1965;24(4):360–368.
100. Levine JD, Dardick SJ, Roizen MF, Helms C, Basbaum AI. Contribution of sensory afferents and sympathetic efferents to joint injury in experimental arthritis. J Neurosci 1986;6(12):3423–3429.
101. Gentili ME, Mazoit JX, Samii KK, Fletcher D. The effect of a sciatic nerve block on the development of inflammation in carrageenan injected rats. Anesth Analg 1999;89(4):979–984.
102. Donaldson LF, McQueen DS, Seckl JR. Local anaesthesia prevents acute inflammatory changes in neuropeptide messenger RNA expression in rat dorsal root ganglia neurons. Neurosci Lett 1994;175(1–2):111–113.
103. Levine JD, Moskowitz MA, Basbaum AI. The contribution of neurogenic inflammation in experimental arthritis. J Immunol 1985;135(Suppl 2):843s–847s.
104. Arnett FC, Edworthy SM, Bloch DA, McShane DJ, Fries JF, Cooper NS et al. The American Rheumatism Association 1987 revised criteria for the classification of rheumatoid arthritis. Arthritis Rheum 1988;31(3):315–324.
105. Zangger P, Keystone EC, Bogoch ER. Asymmetry of small joint involvement in rheumatoid arthritis: prevalence and tendency towards symmetry over time. Joint Bone Spine 2005;72(3):241–247.
106. Glick EN. Asymmetrical rheumatoid arthritis after poliomyelitis. Br Med J 1967;3(556):26–28.
107. Glynn JJ, Clayton ML. Sparing effect of hemiplegia on tophaceous gout. Ann Rheum Dis 1976;35(6):534–535.
108. Thompson M, Bywaters EG. Unilateral rheumatoid arthritis following hemiplegia. Ann Rheum Dis 1962;21:370–377.
109. Veale D, Farrell M, Fitzgerald O. Mechanism of joint sparing in a patient with unilateral psoriatic arthritis and a longstanding hemiplegia. Br J Rheumatol 1993;32(5):413–416.
110. Lapadula G, Iannone F, Zuccaro C, Covelli M, Grattagliano V, Pipitone V. Recovery of erosive rheumatoid arthritis after human immunodeficiency virus-1 infection and hemiplegia. J Rheumatol 1997;24(4):747–751.
111. Kidd BL, Cruwys SC, Garrett NE, Mapp PI, Jolliffe VA, Blake DR. Neurogenic influences on contralateral responses during experimental rat monoarthritis. Brain Res 1995;688(1–2):72–76.
112. Raychaudhuri SP, Farber EM. Are sensory nerves essential for the development of psoriatic lesions? J Am Acad Dermatol 1993;28(3):488–489.
113. van Lent PL, van den Bersselaar L, van de Putte LB, van den Berg WB. Immobilization aggravates cartilage damage during antigen-induced arthritis in mice. Attachment of polymorphonuclear leukocytes to articular cartilage. Am J Pathol 1990;136(6):1407–1416.

114. Kim HK, Kerr RG, Cruz TF, Salter RB. Effects of continuous passive motion and immobilization on synovitis and cartilage degradation in antigen induced arthritis. J Rheumatol 1995;22(9):1714–1721.
115. Okamoto T, Atsuta Y, Shimazaki S. Sensory afferent properties of immobilised or inflamed rat knees during continuous passive movement. J Bone Joint Surg Br 1999;81(1):171–177.
116. Mapp PI, Terenghi G, Walsh DA, Chen ST, Cruwys SC, Garrett N et al. Monoarthritis in the rat knee induces bilateral and time-dependent changes in substance P and calcitonin gene-related peptide immunoreactivity in the spinal cord. Neuroscience 1993;57(4):1091–1096.
117. Segond von Banchet GG, Petrow PK, Brauer R, Schaible HG. Monoarticular antigen-induced arthritis leads to pronounced bilateral upregulation of the expression of neurokinin 1 and bradykinin 2 receptors in dorsal root ganglion neurons of rats. Arthritis Res 2000;2(5):424–427.
118. Smith GD, Harmar AJ, McQueen DS, Seckl JR. Increase in substance P and CGRP, but not somatostatin content of innervating dorsal root ganglia in adjuvant monoarthritis in the rat. Neurosci Lett 1992;137(2):257–260.
119. Traub RJ, Solodkin A, Gebhart GF. NADPH-diaphorase histochemistry provides evidence for a bilateral, somatotopically inappropriate response to unilateral hindpaw inflammation in the rat. Brain Res 1994;647(1):113–123.
120. Ichitani Y, Shi T, Haeggstrom JZ, Samuelsson B, Hokfelt T. Increased levels of cyclooxygenase-2 mRNA in the rat spinal cord after peripheral inflammation: an in situ hybridization study. NeuroReport 1997;8(13):2949–2952.
121. Yashpal K, Pitcher GM, Parent A, Quirion R, Coderre TJ. Noxious thermal and chemical stimulation induce increases in 3H-phorbol 12,13-dibutyrate binding in spinal cord dorsal horn as well as persistent pain and hyperalgesia, which is reduced by inhibition of protein kinase C. J Neurosci 1995;15(5 Pt 1):3263–3272.
122. Messersmith DJ, Kim DJ, Iadarola MJ. Transcription factor regulation of prodynorphin gene expression following rat hindpaw inflammation. Brain Res Mol Brain Res 1998;53(1–2):260–269.
123. Beloeil H, Ji RR, Berde CB. Effects of bupivacaine and tetrodotoxin on carrageenan-induced hind paw inflammation in rats (Part 2): cytokines and p38 mitogen-activated protein kinases in dorsal root ganglia and spinal cord. Anesthesiology 2006;105(1):139–145.
124. Kus L, Sanderson JJ, Beitz AJ. N-methyl-d-aspartate R1 messenger RNA and [125I]MK-801 binding decrease in rat spinal cord after unilateral hind paw inflammation. Neuroscience 1995;68(1):159–165.
125. Ji RR, Zhang X, Wiesenfeld-Hallin Z, Hokfelt T. Expression of neuropeptide Y and neuropeptide Y (Y1) receptor mRNA in rat spinal cord and dorsal root ganglia following peripheral tissue inflammation. J Neurosci 1994;14(11 Pt 1):6423–6434.
126. Zhang Q, Schaffer M, Elde R, Stein C. Effects of neurotoxins and hindpaw inflammation on opioid receptor immunoreactivities in dorsal root ganglia. Neuroscience 1998;85(1):281–291.
127. Donaldson LF. Unilateral arthritis: contralateral effects. Trends Neurosci 1999;22(11):495–496.
128. Donaldson LF, Humphrey PS, Oldfield S, Giblett S, Grubb BD. Expression and regulation of prostaglandin E receptor subtype mRNAs in rat sensory ganglia and spinal cord in

response to peripheral inflammation. Prostaglandins Other Lipid Mediat 2001;63(3):109–122.
129. Liu XY, Wu SX, Wang YY, Wang W, Zhou L, Li YQ. Changes of 5-HT receptor subtype mRNAs in rat dorsal root ganglion by bee venom-induced inflammatory pain. Neurosci Lett 2005;375(1):42–46.
130. Ji RR, Samad TA, Jin SX, Schmoll R, Woolf C.J., p38 MAPK activation by NGF in primary sensory neurons after inflammation increases TRPV1 levels and maintains heat hyperalgesia. Neuron 2002;36(1):57–68.
131. Grubb BD, Stiller RU, Schaible HG. Dynamic changes in the receptive field properties of spinal cord neurons with ankle input in rats with chronic unilateral inflammation in the ankle region. Exp Brain Res 1993;92(3):441–452.
132. Rees H, Sluka KA, Westlund KN, Willis WD. Do dorsal root reflexes augment peripheral inflammation? NeuroReport 1994;5(7):821–824.
133. Willis WD Jr. Dorsal root potentials and dorsal root reflexes: a double-edged sword. Exp Brain Res 1999?124(4):395–421.
134. Willis WD, Rees H, Sluka KA, Westlund KN. Dorsal-root reflexes augment peripheral inflammation in the anesthetized rat. J Physiol (Lond) 1994; 476P:48-P 48.
135. Lin Q, Wu J, Willis WD. Dorsal root reflexes and cutaneous neurogenic inflammation after intradermal injection of capsaicin in rats. J Neurophysiol 1999;82(5):2602–2611.
136. Rees H, Sluka KA, Lu Y, Westlund KN, Willis WD. Dorsal root reflexes in articular afferents occur bilaterally in a chronic model of arthritis in rats. J Neurophysiol 1996;76(6):4190–4193.
137. Kelly S, Dunham JP, Donaldson LF. Sensory nerves have altered function contralateral to a monoarthritis and may contribute to the symmetrical spread of inflammation. Eur J Neurosci 2007;26(4):935–942.
138. Beloeil H, Ababneh Z, Chung R, Zurakowski D, Mulkern RV, Berde CB. Effects of bupivacaine and tetrodotoxin on carrageenan-induced hind paw inflammation in rats (Part 1): hyperalgesia, edema, and systemic cytokines. Anesthesiology 2006;105(1):128–138.
139. White DM, Helme RD. Release of substance P from peripheral nerve terminals following electrical stimulation of the sciatic nerve. Brain Res 1985;336(1):27–31.
140. Rees H, Sluka KA, Westlund KN, Willis WD. The role of glutamate and GABA receptors in the generation of dorsal root reflexes by acute arthritis in the anaesthetized rat. J Physiol (Lond) 1995;484(Pt 2):437–445.
141. Sluka KA, Jordan HH, Westlund KN. Reduction in joint swelling and hyperalgesia following post-treatment with a non-NMDA glutamate receptor antagonist. Pain 1994;59(1):95–100.
142. Morales-Aza BM, Chillingworth NL, Payne JA, Donaldson LF. Inflammation alters cation chloride cotransporter expression in sensory neurons. Neurobiol Dis 2004;17(1):62–69.
143. Price TJ, Cervero F, de Koninck Y. Role of cation-chloride-cotransporters (CCC) in pain and hyperalgesia. Curr Top Med Chem 2005;5(6):547–555.
144. Valencia-de Ita S, Lawand NB, Lin Q, Castaneda-Hernandez G, Willis WD. Role of the Na^+-K^+-$2Cl^-$ cotransporter in the development of capsaicin-induced neurogenic inflammation. J Neurophysiol 2006;95(6):3553–3561.
145. Gouze-Decaris E, Philippe L, Minn A, Haouzi P, Gillet P, Netter P et al. Neurophysiological basis for neurogenic-mediated articular cartilage anabolism alteration. Am J Physiol Regul Integr Comp Physiol 2001;280(1):R115–R122.

146. Peng YB, Wu J, Willis WD, Kenshalo DR. GABA(A) and 5-HT(3) receptors are involved in dorsal root reflexes: possible role in periaqueductal gray descending inhibition. J Neurophysiol 2001;86(1):49–58.
147. Beiche F, Klein T, Nusing R, Neuhuber W, Goppelt-Struebe M. Localization of cyclooxygenase-2 and prostaglandin E2 receptor EP3 in the rat lumbar spinal cord. J Neuroimmunol 1998;89(1–2):26–34.
148. Dirig DM, Konin GP, Isakson PC, Yaksh TL. Effect of spinal cyclooxygenase inhibitors in rat using the formalin test and in vitro prostaglandin E2 release. Eur J Pharmacol 1997;331(2–3):155–160.
149. Nakayama Y, Omote K, Kawamata T, Namiki A. Role of prostaglandin receptor subtype EP1 in prostaglandin E2-induced nociceptive transmission in the rat spinal dorsal horn. Brain Res 2004;1010(1–2):62–68.
150. Tegeder I, Niederberger E, Vetter G, Brautigam L, Geisslinger G. Effects of selective COX-1 and -2 inhibition on formalin-evoked nociceptive behaviour and prostaglandin E(2) release in the spinal cord. J Neurochem 2001;79(4):777–786.
151. Vanegas H, Schaible HG. Prostaglandins and cyclooxygenases [correction of cycloxygenases] in the spinal cord. Prog Neurobiol 2001;64(4):327–363.
152. Bernardi PS, Valtschanoff JG, Weinberg RJ, Schmidt HH, Rustioni A. Synaptic interactions between primary afferent terminals and GABA and nitric oxide-synthesizing neurons in superficial laminae of the rat spinal cord. J Neurosci 1995;15(2):1363–1371.
153. Koltzenburg M, Wall PD, McMahon SB. Does the right side know what the left is doing? Trends Neurosci 1999;22(3):122–127.
154. Sotgiu ML, Brambilla M, Valente M, Biella GE. Contralateral input modulates the excitability of dorsal horn neurons involved in noxious signal processes. Potential role in neuronal sensitization. Somatosens Mot Res 2004;21(3–4):211–215.
155. Fitzgerald M. The contralateral input to the dorsal horn of the spinal cord in the decerebrate spinal rat. Brain Res 1982;236(2):275–287.
156. Bereiter DA, Okamoto K, Bereiter DF. Effect of persistent monoarthritis of the temporomandibular joint region on acute mustard oil-induced excitation of trigeminal subnucleus caudalis neurons in male and female rats. Pain 2005;117(1–2):58–67.
157. Matsushita M, Tanami T. Contralateral termination of primary afferent axons in the sacral and caudal segments of the cat, as studied by anterograde transport of horseradish peroxidase. J Comp Neurol 1983;220(2):206–218.
158. Rethelyi M, Trevino DL, Perl ER. Distribution of primary afferent fibers within the sacrococcygeal dorsal horn: an autoradiographic study. J Comp Neurol 1979;185(4):603–621.
159. Light AR, Perl ER. Spinal termination of functionally identified primary afferent neurons with slowly conducting myelinated fibers. J Comp Neurol 1979;186(2):133–150.
160. Sprague JM. The distribution of dorsal root fibres on motor cells in the lumbosacral spinal cord of the cat, and the site of excitatory and inhibitory terminals in monosynaptic pathways. Proc R Soc Lond B Biol Sci 1958;149(937):534–556.
161. Sprague JM, Hongchien HA. The terminal fields of dorsal root fibers in the lumbosacral spinal cord of the cat, and the dendritic organization of the motor nuclei. Prog Brain Res 1964;11:120–154.
162. Rivero-Melian C., Grant G. Distribution of lumbar dorsal root fibers in the lower thoracic and lumbosacral spinal cord of the rat studied with choleragenoid horseradish peroxidase conjugate. J Comp Neurol 1990;299(4):470–481.

163. Willis WD, Coggeshall RE. Sensory Mechanisms of the Spinal Cord, 2 ed., New York: Plenum Press, 1991.
164. Decaris E, Guingamp C, Chat M, Philippe L, Grillasca JP, Abid A et al. Evidence for neurogenic transmission inducing degenerative cartilage damage distant from local inflammation. Arthritis Rheum 1999;42(9):1951–1960.
165. Sluka KA, Lawand NB, Westlund KN. Joint inflammation is reduced by dorsal rhizotomy and not by sympathectomy or spinal cord transection. Ann Rheum Dis 1994;53(5):309–314.
166. Sachs D, Cunha FQ, Ferreira SH. Peripheral analgesic blockade of hypernociception: activation of arginine/NO/cGMP/protein kinase G/ATP-sensitive K+ channel pathway. Proc Natl Acad Sci U S A 2004;101(10):3680–3685.
167. Abbate R, Gori AM, Pinto S, Attanasio M, Paniccia R, Coppo M et al. Cyclooxygenase and lipoxygenase metabolite synthesis by polymorphonuclear neutrophils: in vitro effect of dipyrone. Prostaglandins Leukot Essent Fatty Acids 1990;41(2):89–93.
168. Abbate R, Pinto S, Gori AM, Paniccia R, Coppo M, Neri Serneri GG. Activity of dipyrone on intraplatelet arachidonic acid metabolism: an in vitro study. Pharmacol Res 1989;21(1):43–50.
169. Catre MG, Salo PT. Quantitative analysis of the sympathetic innervation of the rat knee joint. J Anat 1999;194(Pt 2):233–239.
170. Khoshbaten A., Ferrell WR. Alterations in cat knee joint blood flow induced by electrical stimulation of articular afferents and efferents. J Physiol 1990;430:77–86.
171. Kidd BL, Cruwys S, Mapp PI, Blake DR. Role of the sympathetic nervous system in chronic joint pain and inflammation. Ann Rheum Dis 1992;51(11):1188–1191.
172. Ferrell WR, Lockhart JC, Karimian SM. Tachykinin regulation of basal synovial blood flow. Br J Pharmacol 1997;121(1):29–34.
173. Ferrell WR, Russell NJ. Extravasation in the knee induced by antidromic stimulation of articular C fibre afferents of the anaesthetized cat. J Physiol 1986;379:407–416.
174. Green PG, Luo J, Heller P, Levine JD. Modulation of bradykinin-induced plasma extravasation in the rat knee joint by sympathetic co-transmitters. Neuroscience 1993;52(2):451–458.
175. Sato J, Suzuki S, Iseki T, Kumazawa T. Adrenergic excitation of cutaneous nociceptors in chronically inflamed rats. Neurosci Lett 1993;164(1–2):225–258.
176. Sato J, Suzuki S, Tamura R, Kumazawa T. Norepinephrine excitation of cutaneous nociceptors in adjuvant-induced inflamed rats does not depend on sympathetic neurons. Neurosci Lett 1994;177(1–2):135–138.
177. Green PG, Luo J, Heller PH, Levine JD. Further substantiation of a significant role for the sympathetic nervous system in inflammation. Neuroscience 1993;55(4):1037–1043.
178. Cambridge H, Brain SD. Mechanism of bradykinin-induced plasma extravasation in the rat knee joint. Br J Pharmacol 1995;115(4):641–647.
179. Sulakvelidze I, Baluk P, McDonald DM. Plasma extravasation induced in rat trachea by 6-OHDA is mediated by sensory nerves, not by sympathetic nerves. J Appl Physiol 1994;76(2):701–707.
180. Aloe L, Tuveri MA, Levi-Montalcini R. Studies on carrageenan-induced arthritis in adult rats: presence of nerve growth factor and role of sympathetic innervation. Rheumatol Int 1992;12(5):213–216.
181. Levine JD, Fye K, Heller P, Basbaum AI, Whiting-O'Keefe Q. Clinical response to regional intravenous guanethidine in patients with rheumatoid arthritis. J Rheumatol 1986;13(6):1040–1043.

182. Wang J, Ren Y, Zou X, Fang L, Willis WD, Lin Q. Sympathetic influence on capsaicin-evoked enhancement of dorsal root reflexes in rats. J Neurophysiol 2004;92(4):2017–2026.
183. Lorton D, Lubahn C, Klein N, Schaller J, Bellinger DL. Dual role for noradrenergic innervation of lymphoid tissue and arthritic joints in adjuvant-induced arthritis. Brain Behav Immun 1999;13(4):315–334.
184. Esquifino AI, Castrillon PO, Chacon F, Cutrera R, Cardinali DP. Effect of local sympathectomy on 24-h changes in mitogenic responses and lymphocyte subset populations in rat submaxillary lymph nodes during the preclinical phase of Freund's adjuvant arthritis. Brain Res 2001;888(2):227–234.
185. Coderre TJ, Basbaum AI, Dallman MF, Helms C, Levine JD. Epinephrine exacerbates arthritis by an action at presynaptic B2-adrenoceptors. Neuroscience 1990;34(2):521–523.
186. Green PG, Miao FJ, Janig W, Levine JD. Negative feedback neuroendocrine control of the inflammatory response in rats. J Neurosci 1995;15(6):4678–4686.
187. Green PG, Miao FJ, Strausbaugh H, Heller P, Janig W, Levine JD. Endocrine and vagal controls of sympathetically dependent neurogenic inflammation. Ann N Y Acad Sci 1998;840:282–288.
188. Donaldson LF, McQueen DS, Seckl JR. Endogenous glucocorticoids and the induction and spread of monoarthritis in the rat. J Neuroendocrinol 1994;6(6):649–654.
189. Khasar SG, Burkham J, Dina OA, Brown AS, Bogen O, Alessandri-Haber N et al. Stress induces a switch of intracellular signaling in sensory neurons in a model of generalized pain. J Neurosci 2008;28(22):5721–5730.
190. Appelgren A, Appelgren B, Kopp S, Lundeberg T, Theodorsson E. Substance P-associated increase of intra-articular temperature and pain threshold in the arthritic TMJ. J Orofac Pain 1998;12(2):101–107.

Edited by Gábor Jancsó

Neurogenic Regulation of Bradykinin-Induced Synovitis

PAUL G. GREEN

Department of Oral and Maxillofacial Surgery, University of California San Francisco, San Francisco, CA, USA

ABSTRACT

Substantial evidence is described supporting the hypothesis that bradykinin-induced synovial inflammation is dependent on the sympathetic postganglionic neuron. In addition, experimental data indicate that a sympathetic–C-fiber afferent neuron integration plays a role in acute and chronic inflammation in the synovium and other tissues. For example, sympathetic–C-fiber coupling takes place during the skin inflammatory response [146,212,213]. Specifically, sympathetic efferent activity acts on peripheral adrenergic receptors to enhance C-fiber sensitization, thereby augmenting the generation of dorsal root reflexes that produce vasodilatation. However, to firmly establish such integration in synovial tissue, further studies are needed that employ controlled selective activation of the afferents supplying the synovium, e.g., by electrical stimulation of the corresponding dorsal roots [211].

Furthermore, the evidence suggests that there exists a functionally distinct sympathetic efferent pathway that enables the central nervous system to modulate the inflammatory response by acting on components of the immune system [103,116]. Specifically, at a systemic level activity in the sympathetic neurons modulates immune cell function via innervation of lymphoid organs and by direct contact with lymphocytes and macrophages [214], and at the local level the sympathetic modulates the sensitivity of nociceptors, directly or via immune cells and their signaling molecules (e.g., cytokines). A schematic representation of the interaction of some key systems in the control of synovial inflammation is shown in Fig. 4.

1. INTRODUCTION

Inflammation is normally a protective response to injury, irritation, or infection and serves to remove invading microorganisms and aid in tissue repair through multiple mechanisms. It is characterized by increased blood flow, vascular permeability, attraction of leukocytes, and the sensitization of primary afferent neurons, and constitutes a positive feedback inflammatory cascade. While inflammation is a protective mechanism, if allowed to continue unimpeded, it results in tissue destruction and chronic disease. For example, chronic synovitis (inflammation of synovial membranes) leads to joint damage and eventually rheumatoid arthritis [1]. Arthritis and other chronic inflammatory diseases are among the most prevalent chronic medical conditions, with rheumatoid arthritis affecting an estimated 46 million adults in the United States alone [2]. Synovitis is even more common, but in up to half of patients the disease resolves spontaneously over a few months [3–5]. This suggests that mechanisms exist to regulate the magnitude and duration of synovitis and similar inflammatory responses, so that they do not develop into chronic disabling inflammatory diseases. There is evidence that the control of inflammation is a multicomponent system, involving both the innate immune system and the nervous system. Understanding the complexities of this regulation is therefore of great importance.

It has been known since the early nineteenth century that stimulation of peripheral nervous system produces inflammation and changes in blood flow and capillary diameter [6–10]. More recently, it has been shown that both sympathetic neurons as well as sensory (C-fiber) neurons contribute to inflammatory processes. In particular, synovial joints are abundantly innervated by primary afferent and sympathetic neurons and the inflammatory response in the synovium appears to be under extensive modulation by these neurons. Classically, synovial C-fiber primary afferent neurons have a sensory function, while sympathetic neurons serve to control blood flow. However, it is now appreciated that efferent primary afferent C-fibers and sympathetic neurons regulate multiple aspects of the early inflammatory responses including blood flow, plasma extravasation, and immune cell recruitment. In addition to affecting the acute inflammatory response, there is clinical and experimental evidence that suggest that sympathetic and C-fiber neurons affect the course of chronic inflammatory diseases. For example, it has long been known that surgical or chemical sympathectomy reduces the severity of rheumatoid arthritis in humans [11–13] and experimental arthritis in animals [14–16], and destruction of C-fibers by treatment with capsaicin also attenuates the severity of experimental arthritis [14,17]. The peripheral nervous system exerts both proinflammatory and anti-inflammatory actions at multiple levels, including C-fiber afferent, sympathetic postganglionic, and vagal neurons, and integrates immune and vascular responses. Furthermore, there is compelling evidence that sympathetic efferent activity plays a key role in mediating the central nervous system control of the inflammatory response. I shall therefore review details of both C-fiber and sympathetic neuron roles and their interactions and mediators in inflammation.

2. C-FIBER NEURONS

Several lines of evidence have established a role of small-diameter sensory neuron (C-fiber) afferents in neurogenic inflammation [18–21]. For example, activation of C-fiber afferents by noxious chemical or mechanical stimuli [22–25] or C-fiber strength electrical stimulation [26–31] locally increases signs of inflammation, such as plasma extravasation and blood flow. Such noxious stimulus-induced inflammation is attenuated by physical ablation of primary afferents or chemical ablation with high-dose capsaicin treatment [32–35]. This noxious stimulus-induced inflammation is believed to be due to release of C-fiber sensory neuron neuropeptides such as substance P, neurokinin A, and neurokinin B, which have been shown to markedly increase plasma extravasation and blood flow [21,36–38], and calcitonin gene-related peptide (CGRP) enhances plasma extravasation produced by other mediators [38–40]. In *in vitro* studies, inflammatory mediators, e.g., bradykinin [41,42], prostaglandin E_2 [43], and interleukin 1β [44] release substance P and other proinflammatory neuropeptides from primary afferent neurons.

In experimental models of chronic inflammation, there is evidence that substance P and other C-fiber-derived neuropeptides affect the magnitude of the inflammatory response. For example, substance P exacerbated experimental arthritis when infused into the knee joint of a rat [45], and a selective substance P receptor (NK1) antagonist reduced joint damage [46] reduced inflammation in mouse models of inflammatory bowel disease [47,48]. While there is evidence for a contribution of C-fiber neurons to neurogenic inflammation in many tissues, including skin [18,49,50], airways [51–53], gastrointestinal tract [54–56], the trigeminovascular system [57,58] as well as synovium [59–61], the significance of C-fiber-induced neurogenic inflammation in chronic inflammatory disease is unclear, and, to date, clinical trials involving tachykinin receptor antagonists appear not to be effective in asthma [62], arthritis [63,64], or migraine [65]. Importantly, plasma extravasation observed in inflammatory disease is quantitatively different from that produced by activation of unmyelinated afferents. Specifically, in experimental inflammatory models C-fiber afferent-mediated plasma extravasation is smaller in magnitude than that seen in inflammatory disease states, and unlike the plasma extravasation of chronic inflammation, C-fiber afferent plasma extravasation evoked by capsaicin demonstrates rapid tachyphylaxis [38,66] (Fig. 1). However, there are data to suggest (see below) that C-fiber afferent-mediated neurogenic inflammation has a key role as one component of a constellation of coordinated responses that constitute a full inflammatory response.

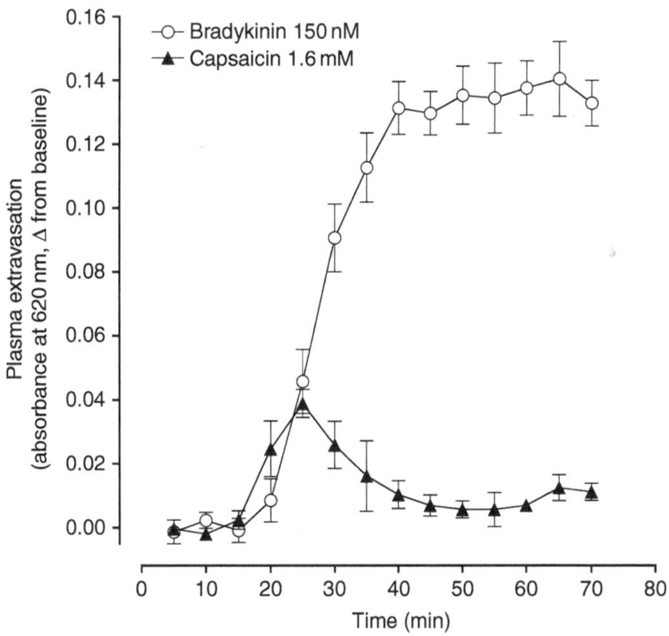

Figure 1. Magnitude and duration of synovial plasma extravasation produced by capsaicin (C-fiber-dependent) is less than that produced by bradykinin (sympathetic neuron-dependent). Pentobarbital-anesthetized rats were pretreated with Evans blue dye given i.v. (Evans blue binds to albumin and does not normally leave the vascular space), and then the knee joint was perfused with vehicle (250 μl/min). Plasma extravasation of the synovia into the knee joint cavity was determined by collecting perfusate every 5 min and measuring the concentration of Evans blue spectrophotometrically (at a wavelength of 620 nm), which is proportional to the degree of extravasation (ordinate scale) shown as an increase over baseline. Bradykinin (BK, 150 nM, $n = 8$ open circles) or capsaicin (1.6 mM, $n = 8$ filled triangles) was added to the perfusate 15 min after the beginning of the perfusion and remained in the perfusion fluid for the remainder of the experiment (55 min after baseline). Both the bradykinin and capsaicin increased plasma extravasation, but capsaicin produced a smaller increase in plasma extravasation than did bradykinin and unlike bradykinin exhibited complete tachyphylaxis within 40 min (adapted from Refs [38,99]).

3. POSTGANGLIONIC SYMPATHETIC NEURON

Sympathetic postganglionic neurons are the final motor neurons of the autonomic nervous system, and there is substantial evidence suggesting that the sympathetic postganglionic neuron terminal plays an important role in the regulation of inflammation [67–69] and integrates with C-fiber-dependent inflammation (see below). The role of the sympathetic postganglionic neuron in inflammation has been most extensively studied in the synovium (see Ref. [16] for review). For example, selective activation of the sympathetic postganglionic neuron terminal with 6-hydroxydopamine, to displace sympathetic neurotransmitters from vesicular stores [70,71], increases plasma extravasation in the rat knee joint [72], and plasma extravasation is decreased by surgical [73–79] or chemical (with guanethidine or 6-hydroxydopamine) [72,80–89] sympathectomy. In particular, surgical or chemical sympathectomy attenuates the magnitude of both basal plasma extravasation and that induced by bradykinin, a principal inflammatory mediator, in the rat knee joint by 60–70% [88,90,91] and producing a large leftward shift in the dose–response curve for bradykinin (Fig. 2).

Importantly, at least within the synovium, sympathetic postganglionic neuron-dependent plasma extravasation can be maintained for hours [91], demonstrating considerably less

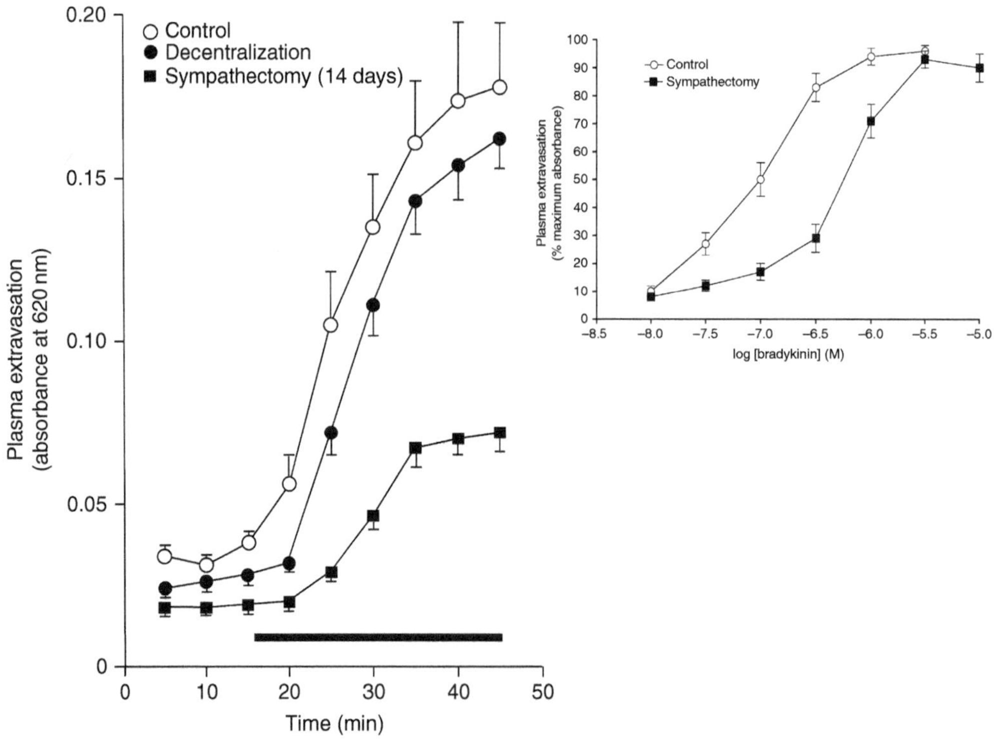

Figure 2. Bradykinin-induced plasma extravasation in the synovium is dependent on the postganglionic sympathetic neuron. Increase of plasma extravasation induced by bradykinin in control rats (open circles, $N=12$ knees) is significantly attenuated 14 days after sympathectomy (closed squares, $N=12$ knees) but was not significantly different from control after decentralization (closed circles, $N=12$ knees). Horizontal black bar indicates presence of bradykinin in perfusion fluid. Inset: Shift in bradykinin dose–response curve (presented as percent of maximum effect produced). Surgical sympathectomy shifted the dose–response curve for bradykinin to the right (data modified from Refs [90,91]).

tachyphylaxis (as well as greater magnitude of response) than that produced by C-fiber afferent-mediated plasma extravasation (which lasts <40 min) [38] (see Fig. 2).

Furthermore, destruction of C-fibers with neurotoxic doses of capsaicin in neonatal rats does not affect the magnitude of sympathetic postganglionic neuron-dependent plasma extravasation produced by bradykinin [73]. While the synovium may have unique characteristics regarding the dependence on the sympathetic postganglionic neuron in neurogenic inflammation, possibly due to the density of sympathetic postganglionic neuron innervation which accounts for between half and two-thirds of the nerve fibers in the synovium [92,93], sympathetic postganglionic neurons also contribute to neurogenic inflammation in other tissues, such as skin, adipose tissue, colon, and lungs [67,69,74,80,86,89,94].

It is not clear what mediators contribute to sympathetic-dependent plasma extravasation in the synovium. As mentioned above, this effect is independent of impulse propagation in the sympathetic neuron. Importantly, in this regard, there is a stimulus dependence in the generation and/or release specific sympathetic postganglionic neuron neurotransmitters [95], with high and low frequency of electrical stimulation differentially releasing mediators such as norepinephrine and neuropeptide Y (NPY) [96], that *inhibit* plasma extravasation [86,97,98]. Furthermore, activation of the sympathetic postganglionic neurons by electrical stimulation of the lumbar sympathetic chain reduces both resting and bradykinin-induced plasma extravasation, possibly because of a reduction of blood flow through the synovium [91]. However, catecholamines, acting via α- and β-adrenergic receptors, can also sensitize nociceptors and recruit leukocytes to the site of inflammation, thereby releasing proinflammatory neuropeptides and cytokines to increase plasma extravasation and vasodilatation. It has also been suggested that the sympathetic postganglionic neuron releases *proinflammatory* mediators such as prostaglandin E_2 [91] and nitric oxide [99] that act, either directly or via cells closely associated with the sympathetic neuron terminal, to enhance the inflammatory process. This contribution of the sympathetic nervous system, at least at initial stages of inflammatory response, is key to the establishment of severe, long-lasting inflammatory diseases. For example, administering α or β-adrenergic (but not the mixed α/β-adrenergic receptor agonist SH1293) given during administration of Freund's adjuvant, exacerbated experimental arthritis in rats, but reduced it when given later at disease onset [100]. Furthermore, the sympathetic nervous system may play a regulatory role in secondary lymphoid organs as it has been shown that selective sympathectomy in secondary lymphoid organs exacerbates experimental arthritis [16,101].

4. ROLE OF BRADYKININ IN SYMPATHETIC POSTGANGLIONIC NEURON SYNOVIAL INFLAMMATION

Perfusion of bradykinin through the synovium produces a marked increase in the magnitude of plasma extravasation [73,84,90,91,97,102–104]. In patients with joint pain, there is a positive correlation between concentration of bradykinin in the synovium and magnitude of synovitis [105], with an increase in expression of the constitutive G-protein-coupled bradykinin B_2 receptors [106]. Bradykinin acts directly on sympathetic postganglionic neuron terminals [107–110] via bradykinin B_2 receptors present on sympathetic neuron terminals [111] to enhance sympathetic neurotransmission [112]. Importantly, while bradykinin-induced plasma extravasation is markedly attenuated by sympathectomy (see above), in the rat synovium it is unchanged following ablation of *preganglionic axons* that innervate the postganglionic neurons to the hind limb (i.e., decentralization of the lumbar sympathetic chain), acute interruption of the lumbar sympathetic chain, or perfusion of tetrodotoxin through the knee joint synovial cavity [90,91]. This provides strong evidence that sympathetic postganglionic neuron-dependent synovitis produced by bradykinin is not dependent on the excitability of the sympathetic postganglionic neuron, but is due to proinflammatory mediator synthesis and release from the sympathetic postganglionic neuron terminal itself. It is not known whether the other inflammatory mediators, such as 5-HT (5-hydroxy tryptamine) [113,114], that produce sympathetic neuron-dependent synovial inflammation also do so in an action potential-independent manner.

5. INTEGRATIVE CONTROLS IN SYNOVITIS

There is a growing understanding that the development and maintenance of acute and chronic inflammation is affected by cross talk between the sympathetic nervous system and peptidergic C-fiber afferents, and further modulated by mediators released from the hypothalamo–pituitary–adrenal (HPA) and the sympathoadrenal axes, as well as by vagal afferents [16,103,115,116].

5.1 Sympathetic–C-fiber afferent coupling in the inflammatory response

Sympathetic–C-fiber afferent coupling has been most comprehensively studied in the context of nociception (see Ref. [117] for review), but is also highly relevant in the inflammatory response (see Ref. [16] for review). Coupling is believed to occur following either excitation-dependent or independent release of norepinephrine from sympathetic postganglionic neuron to act directly or indirectly on primary afferent neurons in the periphery. Importantly, sympathetic–C-fiber coupling only becomes apparent following injury, such that C-fiber nociceptor excitatory responses may increase with sympathetic efferent activity or exogenously applied adrenergic agonists [118–121], at least in part due to the increase of α_2-adrenergic receptors in C-fiber nociceptors after nerve injury [122]. In addition to nerve injury, inflammation is also believed to promote sympathetic–C-fiber afferent coupling to further enhance the inflammatory response (i.e., a local positive feedback loop) as well as the sensitization of nociceptive afferents [123]. In contrast to sympathetic mediated neurogenic inflammation, C-fiber afferent neurogenic inflammation is dependent on dorsal root reflexes conducted antidromically, since the inflammatory response after capsaicin administration is prevented by interruption of peripheral nerves, dorsal rhizotomy, or pharmacological blockade of spinal glutamate or gamma-amino butyric acid (GABA)-A interneurons [124,125].

In the synovium, sympathetic neuron-dependent plasma extravasation produced by bradykinin is independent of impulse activity in sympathetic neurons (see above), and acting through B_2 receptors, bradykinin depolarizes mouse and rat sympathetic ganglia [109,126–129]. This Ca^{2+}-independent protein kinase C sympathoexcitation [109] results in the release of neurotransmitters [130] and other mediators such as prostaglandin E_2 from sympathetic neurons [131] or from nonneuronal cells in the synovium such as fibroblasts [132]. These mediators sensitize C-fiber nociceptors that, in turn, release substance P and CGRP; the sensitization of C-fibers by bradykinin is further enhanced in the presence of these sympathetic-derived mediators such as norepinephrine and prostaglandin E_2 [133]. Importantly, norepinephrine excites and sensitizes C-fibers only after bradykinin [134], other inflammatory stimulus and after nerve injury there is a well-described phenomenon of sympathetic–afferent coupling wherein norepinephrine contributes to the generation of pain and underlies the pathophysiological mechanisms of sympathetically maintained pain (see Ref. [117] for review). Sympathetic neurons also release NPY, and it has been shown that activation of its Y1 receptor activation is necessary and sufficient for substance P release and the subsequent development of neurogenic inflammation and plasma extravasation [135]. Despite the inferential evidence for sympathetic–C-fiber sensory coupling in synovial inflammation, it has not yet been fully established how the sympathetic neuron and C-fiber neuropeptide mediators co-operate to control the inflammatory response.

It is likely that during chronic inflammation the pattern of mediator release and sympathetic–C-fiber coupling changes with time, since chronic exposure to inflammatory cytokines upregulates the bradykinin B_1 receptor component of bradykinin's effect in a prostaglandin-dependent process [136] and while bradykinin acts on a small ($\sim$14%) subpopulation of C-fibers [136], chronic inflammation upregulates both B_2 and B_1 receptors on sensory neurons [137,138]. Since B_1 receptors are only expressed in nonpeptidergic neurons [138], this suggests that bradykinin acts through B_2 receptors to mediate the C-fiber-dependent inflammatory response. In addition to these changes in bradykinin receptor expression, there are important anatomical changes. There is a great loss of sympathetic neurons in the synovial tissue of patients with rheumatoid arthritis, but there is a concomitant gain in C-fiber neurons, with an increase in abundance of substance P [139]. These changes have several implications: loss of sympathetic neurons would be expected to decrease sympathetic-dependent inflammation, while an enhance C-fiber population would be expected to both increase C-fiber-dependent inflammation (through antidromic neuropeptide release) as well as enhance (orthodromic) nociceptive transmission thereby activating the negative feedback inhibition of sympathetic-dependent inflammation (see below).

In addition to a change in receptor expression with chronic inflammation, there is also a change in the immune cell population (see below) as well as sympathetic and C-fiber innervation. Specifically, it has been shown that sympathetic innervation is reduced while C-fiber innervation is increased in the synovium from patients with chronic rheumatoid arthritis compared to osteoarthritis patients, and that there is a differential patterns of innervation that is dependent on the severity of the inflammation [140].

5.2 Stress axes

The inflammatory response in the synovium is affected not only by sympathetic–C-fiber afferent neuronal activity, but also by release of mediators from stress axes and immune cells (see below) that act to affect synovial micromilieu. Acute inflammatory stimuli activate nociceptive C-fiber afferents that, in turn, activate negative feedback circuits to regulate the magnitude and duration of the inflammatory response. For example, bradykinin-induced synovial inflammation is modulated by two neuroendocrine circuits, the HPA and the sympathoadrenal stress axes, both of which are activated by noxious stimuli.

In rats, activation of the HPA axis with cutaneous electrical stimulation at C-fiber strengths inhibits synovial bradykinin-induced plasma extravasation. This effect is sympathetic neuron-specific, since following surgical sympathectomy, residual (i.e., sympathetic neuron-independent) plasma extravasation produced by bradykinin is not inhibited by the noxious stimulation-induced feedback inhibition, and neither is the sympathetic neuron-independent plasma extravasation produced by platelet activating factor. Systemically administered corticosterone also inhibits bradykinin-induced plasma extravasation at a time course that was similar during reflex activation of the hypothalamo–pituitary axis, an effect that is not seen in sympathectomized joints [104,141]. However, corticosterone does not appear to act directly on the sympathetic neuron terminal, but rather via peptide annexin I (lipocortin I), which is released from many cells, e.g., polymorphonuclear leukocytes, upon corticosterone exposure [142].

Activation of the sympathoadrenal axis with capsaicin activating somatic or visceral nociceptors inhibits bradykinin-induced plasma extravasation, and is abolished following denervation of the adrenal medulla or adrenal enucleation [143]. This negative feedback sympathoadrenal axis is likely to be mediated by epinephrine acting on β-adrenergic receptors Jänig et al. 2000 [143,144,145]. Of note, activation of α_1-adrenergic receptors is believed to enhance C-fiber-dependent neurogenic inflammation [146]. While it is not known on which cells epinephrine may be acting, C-fiber neurons (cultured dorsal root ganglion neurons) respond to β_2-adrenergic agonists [147] and also express α_{2c}-adrenergic receptors [148], or it could act on mast cells immune cells (eosinophils, macrophages, T-lymphocytes, neutrophils, or on receptors on postcapillary venule cell [149]). While noxious input activates circuits that inhibit a sympathetic dependent inflammatory response, it is not yet known whether this feedback system also affect C-fiber-dependent inflammation.

Thus, central nervous system activity, which controls sympathoadrenal and HPA function, may markedly affect the inflammatory response. Activation of these two neuroendocrine circuits appear to be differentially activated depending on pattern and intensity of noxious input. Thus, continuous cutaneous noxious electrical stimulation, which *synchronously activates* all C-, Aδ-, and Aβ-fibers, primarily activates the ***HPA axis*** to release corticosterone. By contrast, somatic or visceral stimulation with capsaicin, which selectively activates primary afferent C-fiber and a subpopulation of Aδ-afferents, primarily activates the ***sympathoadrenal axis*** to release catecholamines and possibly other adrenal medulla mediators. This differential pattern of activation could be responsible for specific encoding of nociceptive signals to selectively activate specific neuroendocrine circuits, but it is yet to be determined how specific natural stimuli (e.g., chronic inflammatory pain or traumatic injury pain) are differentially encoded to produce divergent effects on the stress axes.

As mentioned above, there is a loss of sympathetic nerve fibers with a concomitant increase in substance P-positive C-fibers in synovia from rheumatoid arthritis patients [140,150,151]. These changes are likely to markedly decrease the contribution of sympathetic-dependent inflammation, due to loss of sympathetic efferent activity as well as an enhancement of noxious input mediated negative feedback and a significant alteration in sympathetic–C-fiber coupling. It is reasonable to hypothesize that such changes have significant consequences regarding the maintenance of rheumatoid arthritis, and, in fact, a disturbance in the sympathetic–C-fiber equilibrium may signal the progression from short-term synovitis to the development of full-blown chronic inflammatory disease.

5.3 Vagal visceral afferents

Integral to sympathoadrenal and HPA axis modulation of bradykinin-induced synovitis is the activity of vagal visceral afferents. Complete subdiaphragmatic vagotomy markedly inhibited noxious stimulation-induced plasma extravasation, an effect which increases over time with a maximal effect occurring 7–14 days postvagotomy [152] (Fig. 3). Furthermore, celiac branch vagotomy or surgical removal of the duodenum mimics this effect of total subdiaphragmatic vagotomy [153].

This suggests that celiac branch vagal afferent activity tonically inhibits ascending impulse transmission in the neuraxis projecting to both the hypothalamus and the adrenal medulla, and that vagotomy removes central inhibition acting on these neuroendocrine pathways, thus leading to the release of mediators from the adrenal medulla (e.g., catecholamines) and cortex

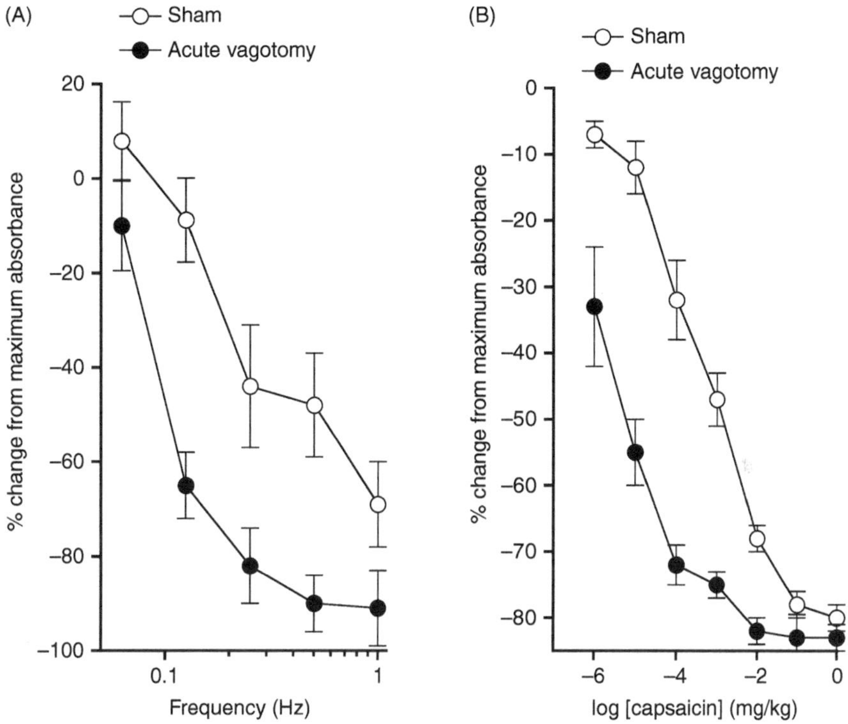

Figure 3. Vagotomy enhances noxious stimulation-induced inhibition of plasma extravasation. Noxious stimulation of the hind paw by either electrical stimulation (25 mA) (A) or capsaicin (B) depresses bradykinin-induced plasma extravasation. Compared to sham surgery control rats (open circles, $n=7$ (A); $n=9$ (B)), acute subdiaphragmatic vagotomy (filled circles) markedly potentiated the effect of noxious electrical ($n=6$) or capsaicin ($n=9$) stimulation, shifting dose–response curve significantly to the left (adapted from Refs [152,153]).

(e.g., corticosterone). A similar enhancement of the inhibitory effect of capsaicin was seen in rats that were fasted, which removes the stimulation of mechanically sensitive polymodal afferents in the proximal gastrointestinal tract. This effect was reversed in fasted rats by mechanically activating duodenal celiac branch vagal afferents using a balloon, while mechanical stimulation in postvagotomy had no effect. Mechano-sensitive duodenal vagal afferents appear to play an integral role in the physiological response to infection, perhaps since it is the first segment of the gastrointestinal tract that has a predominantly absorptive function. Inflammation, or infectious or toxic agents in the duodenum can rapidly signal central circuits via polymodal duodenal vagal afferents to coordinate specific neural and endocrine signals that can differentially regulate vascular and immune function in different parts of the body.

5.4 Immune cell component

Leukocyte infiltration is a key pathological feature of many inflammatory diseases [154–157]. Leukocytes are found in arthritic joints at the site of joint-destroying erosions [158], and neutrophils within the joint participate directly in both acute and chronic inflammation [159]. Within the synovium, bradykinin induces leukocytes migration [160,161] and accumulation

[162,163] by direct action on its receptors on leukocytes [164], or indirectly by acting on sympathetic neurons [162], to release sympathetic neurotransmitters, such as norepinephrine [107] that, in turn, releases other chemoattractant mediators such as prostaglandin [165], or on C-fibers to release mediators such as substance P and CGRP [42], which attract leukocytes [166].

Peripheral sensitization of C-fibers by bradykinin and other inflammatory mediators, such as prostaglandin, during synovial inflammation may activate both stress axes described above that powerfully affect this leukocyte component of the inflammatory response, in part by the effect of stress hormones on leukocyte function [167–172]. In particular, release of catecholamines from the adrenal medulla or from sympathetic nerve terminals that innervate the lymphoid organs [173] mediate interactions between the sympathetic and the immune systems, to alter immune cell activity [174–178]. Norepinephrine, released from sympathetic nerve terminals, and epinephrine, released from the adrenal medulla [175], act primarily on α_1-, α_2-, and β_2-adrenergic receptors expressed on most resting and activated immune cells [149,179–183] to play a role in regulating immune cell proliferation, trafficking, and functional activity [175,184–186]. For example, epinephrine suppresses reactive oxygen species generation [187,188] but enhance phagocytic activity [189] in neutrophils. The sympathetic cotransmitter NPY increases or decreases neutrophil migration, phagocytosis, and reactive oxygen species production, depending on the pathogen or specific leukocyte studied [190,191]. These adrenergic or NPY effects on leukocyte function may be due to a direct effect on cell or due to release of proinflammatory factors such as tumor necrosis factor alpha that can then act on other leukocytes to affect their function [192]. It is well established that C-fiber-derived neuropeptides, such as substance P, CGRP, and gastrin-releasing peptide, have proinflammatory and anti-inflammatory actions on leukocyte function, including a role in modulating synovial inflammation [193–200]. For example, substance P, released locally from C-fiber terminals, firstly acts as a chemoattractant for monocytes and neutrophils [201–204] and then as a stimulus for cytokine release [205,206]. Neurogenic recruitment of immune cells also facilitates wound healing with norepinephrine [207] and stimulation of sympathetic nerves [208], as well as substance P [209] and CGRP [210] having beneficial effects on tissue repair.

6. SUMMARY

The foregoing sections have described substantial evidence supporting the hypothesis that bradykinin-induced synovial inflammation is dependent on the sympathetic postganglionic neuron. While there is good evidence that a sympathetic–C-fiber afferent neuron integration plays a role in acute and chronic inflammation in the synovium and other tissues, to convincingly substantiate such an integration requires further studies employing controlled selective activation of the afferents supplying the synovia, e.g., by electrical stimulation of the corresponding dorsal roots [211], similar to that demonstrated for sympathetic–C-fiber coupling in skin inflammatory response [146,212,213] wherein sympathetic efferent activity acts on peripheral adrenergic receptors to enhance C-fiber sensitization, thereby augmenting the generation of dorsal root reflexes that produce vasodilatation.

Furthermore, the evidence suggests that there exists a functionally distinct sympathetic efferent pathway that enables the central nervous system to modulate the inflammatory response by acting on components of the immune system [103,116]. Specifically, at a systemic level activity in the sympathetic neurons modulates immune cell function via innervation lymphoid organs and direct contact with lymphocytes and macrophages [214], and at the local level the sympathetic modulates the sensitivity of nociceptors, directly or via immune cells and their signaling molecules (e.g., cytokines, see above). A schematic representation of the interaction of some key systems in the control of synovial inflammation is shown in Fig. 4.

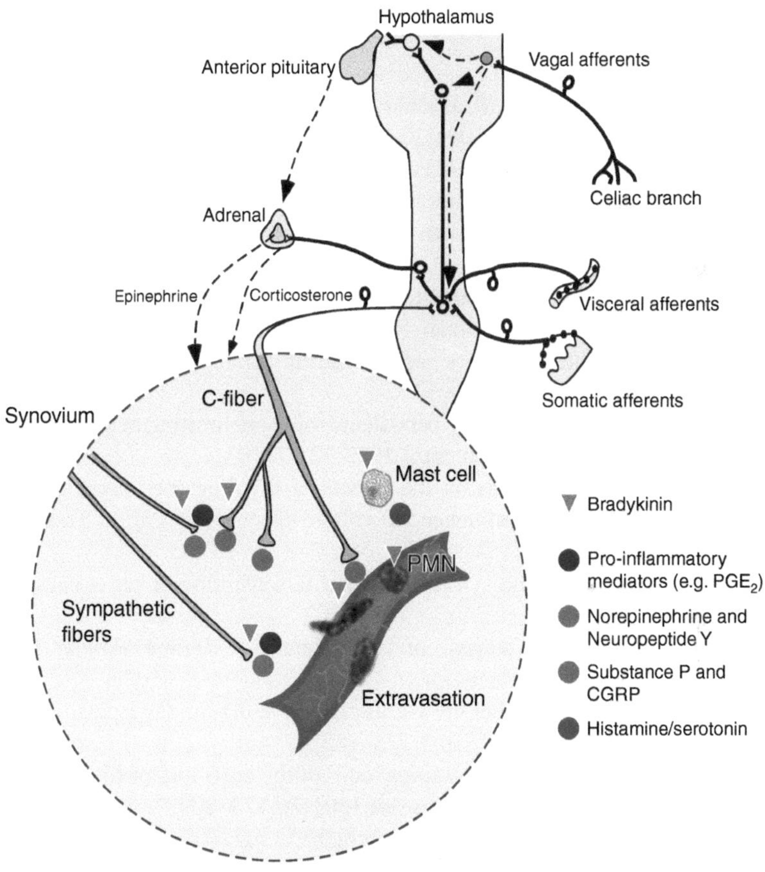

Figure 4. Production of synovitis by bradykinin. Bradykinin acts on B_2 or B_1 bradykinin receptors on the sympathetic postganglionic neuron to release pro-inflammatory mediators. These inflammatory mediators (e.g., prostaglandin E_2) are released independent of electrical activity of the sympathetic neuron, and act directly on the postcapillary venule and/or on other cell types associated with the sympathetic neuron. Bradykinin can also act directly on its receptors on C-fiber neurons to release proinflammatory mediators (e.g., substance P). Sympathetic C-fiber peptidergic coupling is bidirectional, with norepinephrine/neuropeptide Y acting on C-fibers and substance P/CGRP acting on sympathetic neuron terminals. Two neuroendocrine pathways, HPA and sympathoadrenal, that are activated by ongoing noxious input from inflammatory sites, mediate a negative feedback control of inflammation by releasing mediators (e.g., glucocorticoids and epinephrine); these circuits are tonically inhibited by activity in duodenal vagal afferents, exerted at the level of the spinal cord [144]. Immune cells, which play an essential role in the inflammatory response, are attracted to the synovium and activated by mediators released from sympathetic and C-fiber afferent neurons as well as by bradykinin and other tissue-generated inflammatory mediators. Other mechanisms, including the proinflammatory and anti-inflammatory cytokine network, sympathetic innervation of lymphoid organs [214], and the anti-inflammatory cholinergic efferent vagal pathway [215], contribute to the pathophysiology of synovial inflammation, but for the sake of clarity are not included in this schematic figure. (See color Plate 4.)

ACKNOWLEDGMENTS

Supported by a grant (15RT-0032) from the Tobacco-Related Disease Program and in part from an NIH grant R01AR05210.

REFERENCES

1. Raza K, Buckley CE, Salmon M, Buckley CD. Treating very early rheumatoid arthritis. Best Pract Res Clin Rheumatol 2006;20:849–863.
2. Yelin E, Murphy L, Cisternas MG, Foreman AJ, Pasta DJ, Helmick CG. Medical care expenditures and earnings losses among persons with arthritis and other rheumatic conditions in 2003, and comparisons with 1997. Arthritis Rheum 2007;56:1397–1407.
3. Pincus T, Kavanaugh A, Aletaha D, Smolen J. Complexities in defining remission in rheumatic diseases. Clin Exp Rheumatol 2006;24:S-1-6.
4. Harrison BJ, Symmons DP, Brennan P, Barrett EM, Silman AJ. Natural remission in inflammatory polyarthritis: issues of definition and prediction. Br J Rheumatol 1996;35:1096–1100.
5. Tunn EJ. Bacon PA. Differentiating persistent from self-limiting symmetrical synovitis in an early arthritis clinic. Br J Rheumatol 1993;32:97–103.
6. Philip APW. Some observations on the effects of dividing the nerves of the lungs, and subjecting the latter to the influence of voltaic electricity. Philos Trans R Soc Lond 1827;117:297–300.
7. Bernard C. Influence du grand sympathique sur la sensibilité et sur la calorification. C R Soc Biol 1851;3:163–164.
8. Lister J. On the early stages of inflammation. Philos Trans R Soc Lond 1858;148:645–702.
9. Stricker M. Untersuchungen über die Gefässnerven Wurzeln des Ischiadicus. Sitzungsber Kaiserl Akad Wissensch Wien 1876;3:173–185.
10. Bayliss WM. On the origin from the spinal cord of the vaso-dilator fibres of the hind-limb, and on the nature of these fibres. J Physiol 1901;26:173–209.
11. Rountree L, Adson A. Bilateral lumbar sympathetci ganlionectomy and ramisectomy for polyarthritis of the lower extremeties. JAMA 1927;88:694.
12. Young A. Ganglionectomy in the treatment of severe polyarthritis and osteoarthritis. Br Med J 1936;2:375.
13. Levine JD, Fye K, Heller P, Basbaum AI, Whiting-O'Keefe Q. Clinical response to regional intravenous guanethidine in patients with rheumatoid arthritis. J Rheumatol 1986;13:1040–1043.
14. Levine JD, Dardick SJ, Roizen MF, Helms C, Basbaum AI. Contribution of sensory afferents and sympathetic efferents to joint injury in experimental arthritis. J Neurosci 1986;6:3423–3429.
15. Aloe L, Tuveri MA, Levi-Montalcini R. Studies on carrageenan-induced arthritis in adult rats: presence of nerve growth factor and role of sympathetic innervation. Rheumatol Int 1992;12:213–216.
16. Straub RH, Harle P. Sympathetic neurotransmitters in joint inflammation. Rheum Dis Clin North Am 2005;31:43–59, viii.

17. Colpaert FC, Donnerer J, Lembeck F. Effects of capsaicin on inflammation and on the substance P content of nervous tissues in rats with adjuvant arthritis. Life Sci 1983;32:1827–1834.
18. Holzer P. Neurogenic vasodilatation and plasma leakage in the skin. Gen Pharmacol 1998;30:5–11.
19. Jancso N, Jancso-Gabor A, Szolcsanyi J. Direct evidence for neurogenic inflammation and its prevention by denervation and by pretreatment with capsaicin. Br J Pharmacol Chemother 1967;31:138–151.
20. McDonald DM, Bowden JJ, Baluk P, Bunnett NW. Neurogenic inflammation. A model for studying efferent actions of sensory nerves. Adv Exp Med Biol 1996;410:453–462.
21. Pernow B. Role of tachykinins in neurogenic inflammation. J Immunol 1985;135:812s–815s.
22. Andrews PV, Helme RD. Neurogenic plasma extravasation in response to mechanical, chemical and thermal stimuli. Clin Exp Neurol 1987;23:95–100.
23. Imai Y. A pharmacological study of the participation of the peripheral endings of primary afferent neurons in the inflammatory response evoked by heat and mechanical noxious stimulation. Osaka Daigaku Shigaku Zasshi 1990;35:307–320.
24. Koltzenburg M, McMahon SB. Plasma extravasation in the rat urinary bladder following mechanical, electrical and chemical stimuli: evidence for a new population of chemosensitive primary sensory afferents. Neurosci Lett 1986;72:352–356.
25. Siney L, Brain SD. Involvement of sensory neuropeptides in the development of plasma extravasation in rat dorsal skin following thermal injury. Br J Pharmacol 1996;117:1065–1070.
26. Baraniuk JN, Kowalski ML, Kaliner MA. Relationships between permeable vessels, nerves, and mast cells in rat cutaneous neurogenic inflammation. J Appl Physiol 1990;68:2305–2311.
27. Ferrell WR, Russell NJ. Plasma extravasation in the cat knee-joint induced by antidromic articular nerve stimulation. Pflugers Arch 1985;404:91–93.
28. Maggi CA, Abelli L, Giuliani S, Somma V, Furio M, Patacchini R et al. Motor and inflammatory effect of hyperosmolar solutions on the rat urinary bladder in relation to capsaicin-sensitive sensory nerves. Gen Pharmacol 1990;21:97–103.
29. Sann H, Pierau FK. Efferent functions of C-fiber nociceptors. Z Rheumatol 1998;57(Suppl 2):8–13.
30. Umeno E, Nadel JA, Huang HT, McDonald DM. Inhibition of neutral endopeptidase potentiates neurogenic inflammation in the rat trachea. J Appl Physiol 1989;66:2647–2652.
31. Wiesenfeld-Hallin Z, Xu XJ, Hakanson R, Feng DM, Folkers K. The specific antagonistic effect of intrathecal spantide II on substance P- and C-fiber conditioning stimulation-induced facilitation of the nociceptive flexor reflex in rat. Brain Res 1990;526:284–290.
32. Bester H, Allchorne AJ, Woolf CJ. Recovery of C-fiber-induced extravasation following peripheral nerve injury in the rat. Exp Neurol 1998;154:628–636.
33. Brenan A, Jones L, Owain NR. The demonstration of the cutaneous distribution of saphenous nerve C-fibres using a plasma extravasation technique in the normal rat and following nerve injury. J Anat 1988;157:57–66.
34. Nathan JD, Peng RY, Wang Y, McVey DC, Vigna SR, Liddle RA. Primary sensory neurons: a common final pathway for inflammation in experimental pancreatitis in rats. Am J Physiol Gastrointest Liver Physiol 2002;283:G938–G946.

35. Saria A, Lundberg JM. Capsaicin pretreatment inhibits heat-induced oedema in the rat skin. Naunyn Schmiedebergs Arch Pharmacol 1983;323:341–342.
36. Gamse R, Posch M, Saria A, Jancso G. Several mediators appear to interact in neurogenic inflammation. Acta Physiol Hung 1987;69:343–354.
37. Habler HJ, Timmermann L, Stegmann JU, Janig W. Involvement of neurokinins in antidromic vasodilatation in hairy and hairless skin of the rat hindlimb. Neuroscience 1999;89:1259–1268.
38. Green PG, Basbaum AI, Levine JD. Sensory neuropeptide interactions in the production of plasma extravasation in the rat. Neuroscience 1992;50:745–749.
39. Cruwys SC, Kidd BL, Mapp PI, Walsh DA, Blake DR. The effects of calcitonin gene-related peptide on formation of intra-articular oedema by inflammatory mediators. Br J Pharmacol 1992;107:116–119.
40. Brain SD. Grant AD. Vascular actions of calcitonin gene-related peptide and adrenomedullin. Physiol Rev 2004;84:903–934.
41. Geppetti P. Sensory neuropeptide release by bradykinin: mechanisms and pathophysiological implications. Regul Pept 1993;47:1–23.
42. Vasko MR, Campbell WB, Waite KJ. Prostaglandin E2 enhances bradykinin-stimulated release of neuropeptides from rat sensory neurons in culture. J Neurosci 1994;14:4987–4997.
43. Smith JA, Davis CL, Burgess GM. Prostaglandin E2-induced sensitization of bradykinin-evoked responses in rat dorsal root ganglion neurons is mediated by cAMP-dependent protein kinase A. Eur J Neurosci 2000;12:3250–3258.
44. Morioka N, Takeda K, Kumagai K, Hanada T, Ikoma K, Hide I et al. Interleukin-1beta-induced substance P release from rat cultured primary afferent neurons driven by two phospholipase A2 enzymes: secretory type IIA and cytosolic type IV. J Neurochem 2002;80:989–997.
45. Levine JD, Clark R, Devor M, Helms C, Moskowitz MA, Basbaum AI. Intraneuronal substance P contributes to the severity of experimental arthritis. Science 1984;226:547–549.
46. Binder W, Scott C, Walker JS. Involvement of substance P in the anti-inflammatory effects of the peripherally selective kappa-opioid asimadoline and the NK1 antagonist GR205171. Eur J Neurosci 1999;11:2065–2072.
47. Sonea IM, Palmer MV, Akili D, Harp JA. Treatment with neurokinin-1 receptor antagonist reduces severity of inflammatory bowel disease induced by Cryptosporidium parvum. Clin Diagn Lab Immunol 2002;9:333–340.
48. Rijnierse A, van Zijl KM, Koster AS, Nijkamp FP, Kraneveld AD. Beneficial effect of tachykinin NK1 receptor antagonism in the development of hapten-induced colitis in mice. Eur J Pharmacol 2006;548:150–157.
49. Rossi R, Johansson O. Cutaneous innervation and the role of neuronal peptides in cutaneous inflammation: a minireview. Eur J Dermatol 1998;8:299–306.
50. Steinhoff M, Stander S, Seeliger S, Ansel JC, Schmelz M, Luger T. Modern aspects of cutaneous neurogenic inflammation. Arch Dermatol 2003;139:1479–1488.
51. Barnes PJ. Neurogenic inflammation in the airways. Respir Physiol 2001;125:145–154.
52. Belvisi MG. Sensory nerves and airway inflammation: role of a delta and C-fibres. Pulm Pharmacol Ther 2003;16:1–7.
53. Geppetti P, Materazzi S, Nicoletti P. The transient receptor potential vanilloid 1: role in airway inflammation and disease. Eur J Pharmacol 2006;533:207–214.

54. Sann H, Dux M, Schemann M, Jancso G. Neurogenic inflammation in the gastrointestinal tract of the rat. Neurosci Lett 1996;219:147–150.
55. Figini M, Emanueli C, Grady EF, Kirkwood K, Payan DG, Ansel J et al. Substance P and bradykinin stimulate plasma extravasation in the mouse gastrointestinal tract and pancreas. Am J Physiol 1997;272:G785–G793.
56. Sipos G, Altdorfer K, Pongor E, Chen LP, Feher E. Neuroimmune link in the mucosa of chronic gastritis with Helicobacter pylori infection. Dig Dis Sci 2006;51:1810–1817.
57. Dalkara T, Zervas NT, Moskowitz MA. From spreading depression to the trigeminovascular system. Neurol Sci 2006;27(Suppl 2):S86–S90.
58. Durham PL. Calcitonin gene-related peptide (CGRP) and migraine. Headache 2006;46(Suppl 1):S3–S8.
59. Niissalo S, Hukkanen M, Imai S, Tornwall J, Konttinen YT. Neuropeptides in experimental and degenerative arthritis. Ann N Y Acad Sci 2002;966:384–399.
60. Schaible HG, Del Rosso A, Matucci-Cerinic M. Neurogenic aspects of inflammation. Rheum Dis Clin North Am 2005;31:77–101, ix.
61. Levine JD, Khasar SG, Green PG. Neurogenic inflammation and arthritis. Ann N Y Acad Sci 2006;1069:155–167.
62. Boot JD, de Haas S, Tarasevych S, Roy C, Wang L, Amin D et al. Effect of an NK1/NK2 receptor antagonist on airway responses and inflammation to allergen in asthma. Am J Respir Crit Care Med 2007;175:450–457.
63. Keeble JE, Brain SD. A role for substance P in arthritis? Neurosci Lett 2004;361:176–179.
64. Brain SD, Cox HM. Neuropeptides and their receptors: innovative science providing novel therapeutic targets. Br J Pharmacol 2006;147(Suppl 1):S202–S211.
65. Goldstein DJ, Offen WW, Klein EG, Phebus LA, Hipskind P, Johnson KW et al. Lanepitant, an NK-1 antagonist, in migraine prevention. Cephalalgia 2001;21:102–106.
66. Brokaw JJ, Hillenbrand CM, White GW, McDonald DM. Mechanism of tachyphylaxis associated with neurogenic plasma extravasation in the rat trachea. Am Rev Respir Dis 1990;141:1434–1440.
67. Rosell S. Neuronal control of microvessels. Annu Rev Physiol 1980;42:359–371.
68. Pietruck C, Grond S, Xie GX, Palmer PP. Local anesthetics differentially inhibit sympathetic neuron-mediated and C fiber-mediated synovial neurogenic plasma extravasation. Anesth Analg 2003;96:1397–1402, table of contents.
69. Romero A, Planas E, Poveda R, Sanchez S, Pol O, Puig MM. Anti-exudative effects of opioid receptor agonists in a rat model of carrageenan-induced acute inflammation of the paw. Eur J Pharmacol 2005;511:207–217.
70. Head GA, Korner PI. Mechanisms of acute hypertension and bradycardia following intracisternal 6-hydroxydopamine in conscious rabbits. Eur J Pharmacol, 1980;66:111–115.
71. Korner PI, Badoer E, Head GA. Cardiovascular role of the major noradrenergic cell groups in the rabbit: analysis based on 6-hydroxydopamine-induced transmitter release. Brain Res 1987;435:258–272.
72. Coderre TJ, Chan AK, Helms C, Basbaum AI, Levine JD. Increasing sympathetic nerve terminal-dependent plasma extravasation correlates with decreased arthritic joint injury in rats. Neuroscience 1991;40:185–189.
73. Coderre TJ, Basbaum AI, Levine JD. Neural control of vascular permeability: interactions between primary afferents, mast cells, and sympathetic efferents. J Neurophysiol 1989;62:48–58.

74. Linde B, Chisolm G, Rosell S. The influence of sympathetic activity and histamine on the blood-tissue exchange of solutes in canine adipose tissue. Acta Physiol Scand 1974;92:145–155.
75. Engel D. The influence of the sympathetic nervous system on capillary permeability. J Physiol 1941;99:161–181.
76. Engel D. The influence of the sympathetic nervous system on capillary permeability. Res Exp Med (Berl) 1978;173:1–8.
77. Fredholm BB, Oberg B, Rosell S. Effects of vasoactive drugs on circulation in canine subcutaneous adipose tissue. Acta Physiol Scand 1970;79:564–574.
78. Grubb RL Jr., Raichle ME, Eichling JO. Peripheral sympathetic regulation of brain water permeability. Brain Res 1978;144:204–207.
79. Oberg B, Rosell S. Sympathetic control of consecutive vascular sections in canine subcutaneous adipose tissue. Acta Physiol Scand 1967;71:47–56.
80. Donnerer J, Amann R, Lembeck F. Neurogenic and non-neurogenic inflammation in the rat paw following chemical sympathectomy. Neuroscience 1991;45:761–765.
81. Gonzales R, Coderre TJ, Sherbourne CD, Levine JD. Postnatal development of neurogenic inflammation in the rat. Neurosci Lett 1991;127:25–27.
82. Gozsy B, Kato L. Role of norepinephrine and 5-hydroxytryptamine in the delayed phase of the inflammatory reaction in rats. Int Arch Allergy Appl Immunol 1966;30:553–560.
83. Green KL. Role of endogenous catecholamines in the anti-inflammatory activity of alpha-adrenoceptor blocking agents. Br J Pharmacol 1974;51:45–53.
84. Green PG, Luo J, Heller PH, Levine JD. Further substantiation of a significant role for the sympathetic nervous system in inflammation. Neuroscience 1993;55:1037–1043.
85. Helme RD, Andrews PV. The effect of nerve lesions on the inflammatory response to injury. J Neurosci Res 1985;13:453–459.
86. Khalil Z, Helme RD. Sympathetic neurons modulate plasma extravasation in the rat through a non-adrenergic mechanism. Clin Exp Neurol 1989;26:45–50.
87. Lam FY, Ferrell WR. Neurogenic component of different models of acute inflammation in the rat knee joint. Ann Rheum Dis 1991;50:747–751.
88. Lee A, Coderre TJ, Basbaum AI, Levine JD. Sympathetic neuron factors involved in bradykinin-induced plasma extravasation in the rat. Brain Res 1991;557:146–148.
89. Mathison R, Davison JS. The involvement of sympathetic nerves in plasma extravasation induced by prostaglandin E2 and substance P. Brain Res 1994;644:251–256.
90. Miao FJ, Green PG, Coderre TJ, Janig W, Levine JD. Sympathetic-dependence in bradykinin-induced synovial plasma extravasation is dose-related. Neurosci Lett 1996;205:165–168.
91. Miao FJ, Janig W, Levine J. Role of sympathetic postganglionic neurons in synovial plasma extravasation induced by bradykinin. J Neurophysiol 1996;75:715–724.
92. Langford LA, Schmidt RF. Afferent and efferent axons in the medial and posterior articular nerves of the cat. Anat Rec 1983;206:71–78.
93. Hildebrand C, Oqvist G, Brax L, Tuisku F. Anatomy of the rat knee joint and fibre composition of a major articular nerve. Anat Rec 1991;229:545–555.
94. McCafferty DM, Wallace JL, Sharkey KA. Effects of chemical sympathectomy and sensory nerve ablation on experimental colitis in the rat. Am J Physiol 1997;272:G272–G280.
95. Pernow J. Co-release and functional interactions of neuropeptide Y and noradrenaline in peripheral sympathetic vascular control. Acta Physiol Scand Suppl 1988;568:1–56.

96. Lundberg JM, Rudehill A, Sollevi A, Theodorsson-Norheim E, Hamberger B. Frequency- and reserpine-dependent chemical coding of sympathetic transmission: differential release of noradrenaline and neuropeptide Y from pig spleen. Neurosci Lett 1986;63:96–100.
97. Green PG, Luo J, Heller P, Levine JD. Modulation of bradykinin-induced plasma extravasation in the rat knee joint by sympathetic co-transmitters. Neuroscience 1993;52:451–458.
98. Takahashi T, Ichinose M, Yamauchi H, Miura M, Nakajima N, Ishikawa J et al. Neuropeptide Y inhibits neurogenic inflammation in guinea pig airways. J Appl Physiol 1993;75:103–107.
99. Green PG, Luo J, Heller PH, Levine JD. Neurogenic and non-neurogenic mechanisms of plasma extravasation in the rat. Neuroscience 1993;52:735–743.
100. Lubahn CL, Schaller JA, Bellinger DL, Sweeney S, Lorton D. The importance of timing of adrenergic drug delivery in relation to the induction and onset of adjuvant-induced arthritis. Brain Behav Immun 2004;18:563–571.
101. Lorton D, Lubahn C, Klein N, Schaller J, Bellinger DL. Dual role for noradrenergic innervation of lymphoid tissue and arthritic joints in adjuvant-induced arthritis. Brain Behav Immun 1999;13:315–334.
102. Cambridge H, Brain SD. Mechanism of bradykinin-induced plasma extravasation in the rat knee joint. Br J Pharmacol 1995;115:641–647.
103. Green PG, Jänig W. Sympathetic postganglionic neurons in neurogenic inflammation of the synovia. In: Schmidt R, Willis W, eds. Encyclopedia of Pain. Heidleberg, Berlin: Springer-Verlag, 2007; pp. 2369–2373.
104. Green PG, Miao FJ, Janig W, Levine JD. Negative feedback neuroendocrine control of the inflammatory response in rats. J Neurosci 1995;15:4678–4686.
105. Nishimura M, Segami N, Kaneyama K, Suzuki T, Miyamaru M. Relationships between pain-related mediators and both synovitis and joint pain in patients with internal derangements and osteoarthritis of the temporomandibular joint. Oral Surg Oral Med Oral Pathol Oral Radiol Endod 2002;94:328–332.
106. Cassim B, Naidoo S, Ramsaroop R, Bhoola KD. Immunolocalization of bradykinin receptors on human synovial tissue. Immunopharmacology 1997;36:121–125.
107. Boehm S, Kubista H. Fine tuning of sympathetic transmitter release via ionotropic and metabotropic presynaptic receptors. Pharmacol Rev 2002;54:43–99.
108. Fuder H, Muscholl E. Heteroreceptor-mediated modulation of noradrenaline and acetylcholine release from peripheral nerves. Rev Physiol Biochem Pharmacol 1995;126:265–412.
109. Scholze T, Moskvina E, Mayer M, Just H, Kubista H, Boehm S. Sympathoexcitation by bradykinin involves Ca^{2+}-independent protein kinase C. J Neurosci 2002;22:5823–5832.
110. Starke K, Peskar BA, Schumacher KA, Taube HD. Bradykinin and postganglionic sympathetic transmission. Naunyn Schmiedebergs Arch Pharmacol 1977;299:23–32.
111. Jones S, Brown DA, Milligan G, Willer E, Buckley NJ, Caulfield MP. Bradykinin excites rat sympathetic neurons by inhibition of M current through a mechanism involving B2 receptors and G alpha q/11. Neuron 1995;14:399–405.
112. Kansui Y, Fujii K, Goto K, Abe I. Bradykinin enhances sympathetic neurotransmission in rat blood vessels. Hypertension 2002;39:29–34.
113. Xie G, Wang Y, Sharma M, Gabriel A, Mitchell J, Xing Y et al. 5-Hydroxytryptamine-induced plasma extravasation in the rat knee joint is mediated by multiple prostaglandins. Inflamm Res 2003;52:32–38.

114. Pierce PA, Xie GX, Peroutka SJ, Green PG, Levine JD. 5-Hydroxytryptamine-induced synovial plasma extravasation is mediated via 5-hydroxytryptamine2A receptors on sympathetic efferent terminals. J Pharmacol Exp Ther 1995;275:502–508.
115. Straub RH, Baerwald CG, Wahle M, Janig W. Autonomic dysfunction in rheumatic diseases. Rheum Dis Clin North Am 2005;31:61–75, viii.
116. Jänig W, Chapman CR, Green PG. Pain and body protection: sensory, autonomic, neuroendocrine, and behavioral mechanisms in the control of inflammation and hyperalgesia. In: Flor H, Kalso E, Dostrovsky JO, eds. Proceedings of the 11th World Congress on Pain. Seattle. IASP Press, 2006; pp. 331–347.
117. Jänig W, Baron R. Sympathetic nervous system and pain, In: Schmidt RF, Willis WD, eds. Encyclopedia of Pain. Berlin, Heidelberg: Springer-Verlag, 2006; pp. 2356–2366.
118. Sato J, Perl ER. Adrenergic excitation of cutaneous pain receptors induced by peripheral nerve injury. Science 1991;251:1608–1610.
119. Bossut DF, Perl ER. Effects of nerve injury on sympathetic excitation of A delta mechanical nociceptors. J Neurophysiol 1995;73:1721–1723.
120. Abdulla FA, Smith PA. Ectopic alpha2-adrenoceptors couple to N-type Ca^{2+} channels in axotomized rat sensory neurons. J Neurosci 1997;17:1633–1641.
121. Pertovaara A. Noradrenergic pain modulation. Prog Neurobiol 2006;80:53–83.
122. Birder LA, Perl ER. Expression of alpha2-adrenergic receptors in rat primary afferent neurones after peripheral nerve injury or inflammation. J Physiol 1999;515 (Pt 2):533–542.
123. Janig W, Levine JD, Michaelis M. Interactions of sympathetic and primary afferent neurons following nerve injury and tissue trauma. Prog Brain Res 1996;113:161–184.
124. Lin Q, Wu J, Willis WD. Dorsal root reflexes and cutaneous neurogenic inflammation after intradermal injection of capsaicin in rats. J Neurophysiol 1999;82:2602–2611.
125. Willis WD, Jr. Dorsal root potentials and dorsal root reflexes: a double-edged sword. Exp Brain Res 1999;124:395–421.
126. Seabrook GR, Bowery BJ, Hill RG. Bradykinin receptors in mouse and rat isolated superior cervical ganglia. Br J Pharmacol 1995;115:368–372.
127. Babbedge R, Dray A, Urban L. Bradykinin depolarises the rat isolated superior cervical ganglion via B2 receptor activation. Neurosci Lett 1995;193:161–164.
128. Lewis GP, Reit E. The action of angiotensin and bradykinin on the superior cervical ganglion of the cat. J Physiol 1965;179:538–553.
129. Trendelenburg U. Observations on the ganglion-stimulating action of angiotensin and bradykinin. J Pharmacol Exp Ther 1966;154:418–425.
130. Boehm S, Huck S. Noradrenaline release from rat sympathetic neurones triggered by activation of B2 bradykinin receptors. Br J Pharmacol 1997;122:455–462.
131. Gonzales R, Goldyne ME, Taiwo YO, Levine JD. Production of hyperalgesic prostaglandins by sympathetic postganglionic neurons. J Neurochem 1989;53:1595–1598.
132. Bathon JM, Chilton FH, Hubbard WC, Towns MC, Solan NJ, Proud D. Mechanisms of prostanoid synthesis in human synovial cells: cytokine-peptide synergism. Inflammation 1996;20:537–554.
133. Neugebauer V, Schaible HG, Schmidt RF. Sensitization of articular afferents to mechanical stimuli by bradykinin. Pflugers Arch 1989;415:330–335.
134. Banik RK, Sato J, Giron R, Yajima H, Mizumura K. Interactions of bradykinin and norepinephrine on rat cutaneous nociceptors in both normal and inflamed conditions in vitro. Neurosci Res 2004;49:421–425.

135. Naveilhan P, Hassani H, Lucas G, Blakeman KH, Hao JX, Xu XJ et al. Reduced antinociception and plasma extravasation in mice lacking a neuropeptide Y receptor. Nature 2001;409:513–517.
136. Seabrook GR, Bowery BJ, Heavens R, Brown N, Ford H, Sirinathsinghi DJ et al. Expression of B1 and B2 bradykinin receptor mRNA and their functional roles in sympathetic ganglia and sensory dorsal root ganglia neurones from wild-type and B2 receptor knockout mice. Neuropharmacology 1997;36:1009–1017.
137. von Banchet GS, Petrow PK, Brauer R, Schaible HG. Monoarticular antigen-induced arthritis leads to pronounced bilateral upregulation of the expression of neurokinin 1 and bradykinin 2 receptors in dorsal root ganglion neurons of rats. Arthritis Res 2000;2:424–427.
138. Vellani V, Zachrisson O, McNaughton PA. Functional bradykinin B1 receptors are expressed in nociceptive neurones and are upregulated by the neurotrophin GDNF. J Physiol 2004;560:391–401.
139. Straub RH, Cutolo M. Involvement of the hypothalamic–pituitary–adrenal/gonadal axis and the peripheral nervous system in rheumatoid arthritis: viewpoint based on a systemic pathogenetic role. Arthritis Rheum 2001;44:493–507.
140. Miller LE, Justen HP, Scholmerich J, Straub RH. The loss of sympathetic nerve fibers in the synovial tissue of patients with rheumatoid arthritis is accompanied by increased norepinephrine release from synovial macrophages. FASEB J 2000;14:2097–2107.
141. Green PG, Janig W, Levine JD. Negative feedback neuroendocrine control of inflammatory response in the rat is dependent on the sympathetic postganglionic neuron. J Neurosci 1997;17:3234–3238.
142. Green PG, Strausbaugh HJ, Levine JD. Annexin I is a local mediator in neural-endocrine feedback control of inflammation. J Neurophysiol 1998;80:3120–3126.
143. Miao FJ, Janig W, Levine JD. Nociceptive neuroendocrine negative feedback control of neurogenic inflammation activated by capsaicin in the rat paw: role of the adrenal medulla. J Physiol 2000;527(Pt 3):601–610.
144. Miao FJ, Janig W, Jasmin L, Levine JD. Spino-bulbo-spinal pathway mediating vagal modulation of nociceptive-neuroendocrine control of inflammation in the rat. J Physiol 2001;532:811–822.
145. Jänig W., Häbler HJ. Specificity in the organization of the autonomic nervous system: a basis for precise neural regulation of homeostatic and protective body functions. Prog Brain Res. 2000;122:351–367.
146. Wang J, Ren Y, Zou X, Fang L, Willis WD, Lin Q. Sympathetic influence on capsaicin-evoked enhancement of dorsal root reflexes in rats. J Neurophysiol 2004;92:2017–2026.
147. Hucho TB, Dina OA, Kuhn J, Levine JD. Estrogen controls PKCepsilon-dependent mechanical hyperalgesia through direct action on nociceptive neurons. Eur J Neurosci 2006;24:527–534.
148. Ma W, Zhang Y, Bantel C, Eisenach JC. Medium and large injured dorsal root ganglion cells increase TRPV-1, accompanied by increased alpha2C-adrenoceptor co-expression and functional inhibition by clonidine. Pain 2005;113:386–394.
149. Barnes PJ. Effect of beta-agonists on inflammatory cells. J Allergy Clin Immunol 1999;104:S10–S17.
150. Miller LE, Grifka J, Scholmerich J, Straub RH. Norepinephrine from synovial tyrosine hydroxylase positive cells is a strong indicator of synovial inflammation in rheumatoid arthritis. J Rheumatol 2002;29:427–435.

151. Pereira da Silva JA, Carmo-Fonseca M. Peptide containing nerves in human synovium: immunohistochemical evidence for decreased innervation in rheumatoid arthritis. J Rheumatol 1990;17:1592–1599.
152. Miao FJ, Janig W, Green PG, Levine JD. Inhibition of bradykinin-induced plasma extravasation produced by noxious cutaneous and visceral stimuli and its modulation by vagal activity. J Neurophysiol 1997;78:1285–1292.
153. Miao FJ, Green PG, Levine JD. Mechanosensitive duodenal afferents contribute to vagal modulation of inflammation in the rat. J Physiol 2004;554:227–235.
154. Ajuebor MN, Swain MG, Perretti M. Chemokines as novel therapeutic targets in inflammatory diseases. Biochem Pharmacol 2002;63:1191–1196.
155. Edwards SW, Hallett MB. Seeing the wood for the trees: the forgotten role of neutrophils in rheumatoid arthritis. Immunol Today 1997;18:320–324.
156. Goulding NJ, Euzger HS, Butt SK, Perretti M. Novel pathways for glucocorticoid effects on neutrophils in chronic inflammation. Inflamm Res 1998;47(Suppl 3):S158–S165.
157. Kasama T, Miwa Y, Isozaki T, Odai T, Adachi M, Kunkel SL. Neutrophil-derived cytokines: potential therapeutic targets in inflammation. Curr Drug Targets Inflamm Allergy 2005;4:273–279.
158. Mohr W, Menninger H. Polymorphonuclear granulocytes at the pannus-cartilage junction in rheumatoid arthritis. Arthritis Rheum 1980;23:1413–1414.
159. Chen M, Lam BK, Kanaoka Y, Nigrovic PA, Audoly LP, Austen KF et al. Neutrophil-derived leukotriene B4 is required for inflammatory arthritis. J Exp Med 2006;203:837–842.
160. Paegelow I, Trzeczak S, Bockmann S, Vietinghoff G. Migratory responses of polymorphonuclear leukocytes to kinin peptides. Pharmacology 2002;66:153–161.
161. Santos DR, Calixto JB, Souza GE. Effect of a kinin B2 receptor antagonist on LPS- and cytokine-induced neutrophil migration in rats. Br J Pharmacol 2003;139:271–278.
162. Lo EJ, Green PG, Miao FJ, Relchling DB, Levine JD. Bradykinin-induced neurogenic migration of neutrophils into the rat knee joint. NeuroReport 1999;10:3821–3824.
163. Sainz IM, Uknis AB, Isordia-Salas I, Dela Cadena RA, Pixley RA, Colman RW. Interactions between bradykinin (BK) and cell adhesion molecule (CAM) expression in peptidoglycan-polysaccharide (PG-PS)-induced arthritis. FASEB J 2004;18:887–889.
164. Cassim B, Naidoo S, Naidoo Y, Williams R, Bhoola KD. Immunolocalisation of the kinin moiety and bradykinin (B2) receptors on synovial fluid neutrophils in rheumatoid arthritis. Immunopharmacology 1996;33:321–324.
165. Gonzales R, Sherbourne CD, Goldyne ME, Levine JD. Noradrenaline-induced prostaglandin production by sympathetic postganglionic neurons is mediated by alpha 2-adrenergic receptors. J Neurochem 1991;57:1145–1150.
166. Ahluwalia A, Perretti M. Involvement of bradykinin B1 receptors in the polymorphonuclear leukocyte accumulation induced by IL-1 beta in vivo in the mouse. J Immunol 1996;156:269–274.
167. Bierhaus A, Humpert PM, Nawroth PP. Linking stress to inflammation. Anesthesiol Clin 2006;24:325–340.
168. Landmann RM, Muller FB, Perini C, Wesp M, Erne P, Buhler FR. Changes of immunoregulatory cells induced by psychological and physical stress: relationship to plasma catecholamines. Clin Exp Immunol 1984;58:127–135.
169. Dhabhar FS. Stress-induced augmentation of immune function–the role of stress hormones, leukocyte trafficking, and cytokines. Brain Behav Immun 2002;16:785–798.

170. O'Leary EC, Marder P, Zuckerman SH. Glucocorticoid effects in an endotoxin-induced rat pulmonary inflammation model: differential effects on neutrophil influx, integrin expression, and inflammatory mediators. Am J Respir Cell Mol Biol 1996;15:97–106.
171. Bilbo SD, Dhabhar FS, Viswanathan K, Saul A, Yellon SM, Nelson RJ. Short day lengths augment stress-induced leukocyte trafficking and stress-induced enhancement of skin immune function. Proc Natl Acad Sci U S A 2002;99:4067–4072.
172. Shephard RJ. Adhesion molecules, catecholamines and leucocyte redistribution during and following exercise. Sports Med 2003;33:261–284.
173. Wrona D. Neural-immune interactions: an integrative view of the bidirectional relationship between the brain and immune systems. J Neuroimmunol 2006;172:38–58.
174. Downing JE, Miyan JA. Neural immunoregulation: emerging roles for nerves in immune homeostasis and disease. Immunol Today 2000;21:281–289.
175. Elenkov IJ, Wilder RL, Chrousos GP, Vizi ES. The sympathetic nerve – an integrative interface between two supersystems: the brain and the immune system. Pharmacol Rev 2000;52:595–638.
176. Straub RH, Westermann J, Scholmerich J, Falk W. Dialogue between the CNS and the immune system in lymphoid organs. Immunol Today 1998;19:409–413.
177. Benschop RJ, Schedlowski M, Wienecke H, Jacobs R, Schmidt RE. Adrenergic control of natural killer cell circulation and adhesion. Brain Behav Immun 1997;11:321–332.
178. Oberbeck R. Catecholamines: physiological immunomodulators during health and illness. Curr Med Chem 2006;13:1979–1989.
179. Barnes PJ. Beta-adrenergic receptors and their regulation. Am J Respir Crit Care Med 1995;152:838–860.
180. Kin NW, Sanders VM. It takes nerve to tell T and B cells what to do. J Leukoc Biol 2006;79:1093–1104.
181. Maestroni GJ. Sympathetic nervous system influence on the innate immune response. Ann N Y Acad Sci 2006;1069:195–207.
182. Wahle M, Neumann RP, Moritz F, Krause A, Buttgereit F, Baerwald CG. Beta2-adrenergic receptors mediate the differential effects of catecholamines on cytokine production of PBMC. J Interferon Cytokine Res 2005;25:384–394.
183. Roupe van der Voort C, Heijnen CJ, Wulffraat N, Kuis W, Kavelaars A. Stress induces increases in IL-6 production by leucocytes of patients with the chronic inflammatory disease juvenile rheumatoid arthritis: a putative role for alpha (1)-adrenergic receptors. J Neuroimmunol 2000;110:223–229.
184. Straub RH, Linde HJ, Mannel DN, Scholmerich J, Falk W. A bacteria-induced switch of sympathetic effector mechanisms augments local inhibition of TNF-alpha and IL-6 secretion in the spleen. FASEB J 2000;14:1380–1388.
185. de Coupade C, Gear RW, Dazin PF, Sroussi HY, Green PG, Levine JD. Beta2-adrenergic receptor regulation of human neutrophil function is sexually dimorphic. Br J Pharmacol 2004;143:1033–1041.
186. de Coupade C, Brown AS, Dazin PF, Levine JD, Green PG. Beta (2)-Adrenergic receptor-dependent sexual dimorphism for murine leukocyte migration. J Neuroimmunol 2007;186:54–62.
187. Tintinger GR, Theron AJ, Anderson R, Ker JA. The anti-inflammatory interactions of epinephrine with human neutrophils in vitro are achieved by cyclic AMP-mediated accelerated resequestration of cytosolic calcium. Biochem Pharmacol 2001;61:1319–1328.
188. Trabold B, Gruber M, Frohlich D. Functional and phenotypic changes in polymorphonuclear neutrophils induced by catecholamines. Scand Cardiovasc J 2007;41:59–64.

189. Malpica MI, Rodriguez AB, Saez MC, Garcia JJ, Barriga C, Ortega E. In-vitro study of the effect of adrenaline on the functional capacity of human neutrophils: role during exercise. J Neuroendocrinol 2002;14:824–828.
190. Dimitrijevic M, Stanojevic S, Micic S, Vujic V, Kovacevic-Jovanovic V, Mitic K et al. Neuropeptide Y (NPY) modulates oxidative burst and nitric oxide production in carrageenan-elicited granulocytes from rat air pouch. Peptides 2006;27:3208–3215.
191. Bedoui S, von Horsten S, Gebhardt T. A role for neuropeptide Y (NPY) in phagocytosis: implications for innate and adaptive immunity. Peptides 2007;28:373–376.
192. Altenburg SP, Martins MA, Silva AR, Cordeiro RS, Castro-Faria-Neto HC. LPS-induced blood neutrophilia is inhibited by alpha 1-adrenoceptor antagonists: a role for catecholamines. J Leukoc Biol 1997;61:689–694.
193. Green PG. Gastrin-releasing peptide, substance P and cytokines in rheumatoid arthritis. Arthritis Res Ther 2005;7:111–113.
194. Grimsholm O, Rantapaa-Dahlqvist S, Forsgren S. Levels of gastrin-releasing peptide and substance P in synovial fluid and serum correlate with levels of cytokines in rheumatoid arthritis. Arthritis Res Ther 2005;7:R416–R426.
195. O'Connor TM, O'Connell J, O'Brien DI, Goode T, Bredin CP, Shanahan F. The role of substance P in inflammatory disease. J Cell Physiol 2004;201:167–180.
196. Zhang C, McDougall JJ. Stimulation of sensory neuropeptide release by nociceptin/orphanin FQ leads to hyperaemia in acutely inflamed rat knees. Br J Pharmacol 2006;148:938–946.
197. McGillis JP, Mitsuhashi M, Payan DG. Immunomodulation by tachykinin neuropeptides. Ann N Y Acad Sci 1990;594:85–94.
198. Payan DG, McGillis JP, Renold FK, Mitsuhashi M, Goetzl EJ. Neuropeptide modulation of leukocyte function. Ann N Y Acad Sci 1987;496:182–191.
199. Shenker NG, Blake DR, McCabe CS, Haigh R, Mapp PI. Symmetry, T cells and neurogenic arthritis. Novartis Found Symp 2004;260:241–252; discussion 252–257, 277–279.
200. Varela N, Chorny A, Gonzalez-Rey E, Delgado M. Tuning inflammation with anti-inflammatory neuropeptides. Expert Opin Biol Ther 2007;7:461–478.
201. Locatelli L, Sacerdote P, Mantegazza P, Panerai AE. Effect of ibuprofen and diclofenac on the chemotaxis induced by substance P and transforming growth factor-beta on human monocytes and polymorphonuclear cells. Int J Immunopharmacol 1993;15:833–838.
202. Roch-Arveiller M, Regoli D, Chanaud B, Lenoir M, Muntaner O, Stralzko S et al. Tachykinins: effects on motility and metabolism of rat polymorphonuclear leucocytes. Pharmacology 1986;33:266–273.
203. Ruff MR, Wahl SM, Pert CB. Substance P receptor-mediated chemotaxis of human monocytes. Peptides 1985;6(Suppl 2):107–111.
204. Saito T, Koshino T. Distribution of neuropeptides in synovium of the knee with osteoarthritis. Clin Orthop Relat Res 2000:172–182.
205. Delgado AV, McManus AT, Chambers JP. Production of tumor necrosis factor-alpha, interleukin 1-beta, interleukin 2, and interleukin 6 by rat leukocyte subpopulations after exposure to substance P. Neuropeptides 2003;37:355–361.
206. Lavagno L, Bordin G, Colangelo D, Viano I, Brunelleschi S. Tachykinin activation of human monocytes from patients with rheumatoid arthritis: in vitro and ex-vivo effects of cyclosporin A. Neuropeptides 2001;35:92–99.

207. Gosain A, Jones SB, Shankar R, Gamelli RL, DiPietro LA. Norepinephrine modulates the inflammatory and proliferative phases of wound healing. J Trauma 2006;60:736–744.
208. Kim LR, Pomeranz B. The sympathomimetic agent, 6-hydroxydopamine, accelerates cutaneous wound healing. Eur J Pharmacol 1999;376:257–264.
209. Delgado AV, McManus AT, Chambers JP. Exogenous administration of Substance P enhances wound healing in a novel skin-injury model. Exp Biol Med (Maywood) 2005;230:271–280.
210. McDougall JJ, Yeung G, Leonard CA, Bray RC. A role for calcitonin gene-related peptide in rabbit knee joint ligament healing. Can J Physiol Pharmacol 2000;78:535–540.
211. Habler HJ, Wasner G, Janig W. Interaction of sympathetic vasoconstriction and antidromic vasodilatation in the control of skin blood flow. Exp Brain Res 1997;113:402–410.
212. Lin Q, Zou X, Ren Y, Wang J, Fang L, Willis WD. Involvement of peripheral neuropeptide Y receptors in sympathetic modulation of acute cutaneous flare induced by intradermal capsaicin. Neuroscience 2004;123:337–347.
213. Lin Q, Zou X, Fang L, Willis WD. Sympathetic modulation of acute cutaneous flare induced by intradermal injection of capsaicin in anesthetized rats. J Neurophysiol 2003;89:853–861.
214. Felten SY, Madden KS, Bellinger DL, Kruszewska B, Moynihan JA, Felten DL. The role of the sympathetic nervous system in the modulation of immune responses. Adv Pharmacol 1998;42:583–587.
215. Tracey KJ. Physiology and immunology of the cholinergic antiinflammatory pathway. J Clin Invest 2007;117:289–296.

Neurogenic Inflammation in Health and Disease
Edited by Gábor Jancsó

Capsaicin-Sensitive Sensory Nerves in Myocardial Ischemia–Reperfusion Injury and Ischemic Stress Adaptation: Role of Nitric Oxide and Calcitonin Gene-Related Peptide

PÉTER FERDINANDY[1] and GÁBOR JANCSÓ[2]

[1]*Cardiovascular Research Group and Pharmahungary™ Group, Department of Biochemistry, University of Szeged, Szeged, Hungary*
[2]*Department of Physiology, University of Szeged, Szeged, Hungary*

ABSTRACT

Ischemic heart disease is a major cause of mortality in civilized societies. Effective therapeutic strategies for protecting the ischemic myocardium are much sought after. Ischemic pre- and postconditioning of myocardium is a well-described adaptive response in which brief exposure to ischemia markedly enhances the ability of the heart to withstand a subsequent ischemic insult. The underlying molecular mechanisms of ischemic stress adaptation have been extensively investigated. Despite the intensive research in the past decade, the exact biochemical mechanism of pre- and postconditioning is still a question of debate. Among a variety of cellular mechanisms, cardiac capsaicin-sensitive sensory nerves are certainly involved in the mechanism of cardiac stress adaptation via their ability to sense chemical signals during ischemia and to release cardioprotective neurotransmitters, such as nitric oxide (NO) and calcitonin gene-related peptide (CGRP).

NO is an important cardioprotective molecule via its vasodilator, antioxidant, antiplatelet, and antineutrophil actions and it is essential for normal heart function. However, NO is detrimental if it combines with superoxide (O_2^-) to form peroxynitrite ($ONOO^-$) which rapidly decomposes to highly reactive oxidant species. There is a critical balance between cellular concentrations of NO, superoxide, and superoxide dismutase, which physiologically favor NO production but in pathological conditions such as ischemia and reperfusion result in $ONOO^-$ formation. However, exposure of the heart to brief episode(s) of ischemia markedly enhances its ability to withstand a subsequent ischemic injury. The activation of this endogenous cardioprotective mechanism known as preconditioning, requires both NO and superoxide synthesis. However, preconditioning in turn attenuates the overproduction of NO, superoxide, and $ONOO^-$ during subsequent ischemia and reperfusion, thereby protecting the heart. However, still very little is known about the role of NO superoxide and $ONOO^-$ in postconditioning.

CGRP is a potent vasodilator, increases heart rate and myocardial contractility, and it has been shown to protect the heart against ischemia and reperfusion. CGRP is an important endogenous cardioprotective substance and it has been shown to contribute to the mechanism of ischemic stress adaptation. However, little is known about the exact role of CGRP in cardiac pathologies.

Here we review the possible roles of capsaicin-sensitive sensory nerves, NO, and CGRP in both ischemia–reperfusion injury and cardiac stress adaptation.

1. MYOCARDIAL ISCHEMIA–REPERFUSION INJURY

Ischemic heart disease, a major cause of mortality in the civilized societies, is characterized by insufficient blood supply to certain regions of the myocardium which leads to tissue necrosis (infarction). Ischemic heart disease develops as a consequence of hypertension, atherosclerosis, dyslipidemia, diabetes, etc. The treatment of this condition has entered a new era where mortality can be approximately halved by procedures which allow the rapid return of blood flow, i.e., reperfusion, to the ischemic zone of the myocardium. Reperfusion, however, may lead to further complications such as diminished cardiac contractile function (stunning) and arrhythmias (see Ref. [1] for review). Therefore, development of cardioprotective agents to improve myocardial function, decrease the incidence of arrhythmias, lessen the necrotic tissue mass, and to delay the onset of necrosis during ischemia–reperfusion are of great clinical importance. Previous attempts to attenuate the consequences of ischemia–reperfusion injury with pharmacological tools have been largely unsuccessful. However, the heart was found to be able to adapt to ischemic stress, and this phenomenon was termed ischemic pre- and post-conditioning [2–4].

2. MYOCARDIAL ADAPTATION TO ISCHEMIA: PRE- AND POSTCONDITIONING

Ischemic preconditioning is a well-described adaptive response in which brief exposure to ischemia markedly enhances the ability of the heart to withstand a subsequent ischemic injury (see Ref. [5] for review). Preconditioning confers a remarkable cardioprotection in a variety of species including humans (see Ref. [5] for review), although the cardioprotective effectiveness of ischemic preconditioning might be attenuated in the heart during aging and some disease states such as hyperlipidemia and diabetes (see Refs [3,6,7] for reviews). Preconditioning can be elicited by different sublethal stress signals, such as brief periods of ischemia, hypoxia, rapid electrical pacing, heat stress, or administration of bacterial endotoxin. The cardioprotective effect of preconditioning shows two distinct phases. The early phase is manifested within minutes after the preconditioning stimulus and has a duration of less than 3 h. The late phase is characterized by a slower onset (20 h) and a duration of up to 72 h. Both phases of preconditioning involve reduction of necrotic tissue mass (infarct size), improvement of cardiac performance, and reduction of the incidence of arrhythmias following ischemia and reperfusion (see Refs [4–6] for reviews).

Moreover, brief cycles of ischemia–reperfusion applied following a longer period of ischemia also confer cardioprotection against the consequences of myocardial ischemia–reperfusion, a phenomenon called ischemic postconditioning [8].

The discovery of this endogenous cardioprotective mechanism encouraged the exploration of the cellular mechanisms responsible for this remarkable adaptation. However, the exact cellular mechanism of pre- and postconditioning is still a question of debate. A variety of substances and ion channels, i.e., adenosine, bradykinin, nitric oxide (NO), superoxide, peroxynitrite ($ONOO^-$) [9], calcitonin gene-related peptide (CGRP), cGMP, protein kinases, norepinephrine, and ATP-sensitive K^+-channels (K_{ATP}) have been implicated in the mechanisms of ischemia–reperfusion injury and in the development of the cardioprotective effect of preconditioning (see Refs [3,4,10,11] for reviews). The inconsistent findings in the literature as to the trigger and the mediator of preconditioning are generally attributed to species differences (see Refs [12–14] for reviews), the different triggers of preconditioning, like rapid pacing and no-flow ischemia [15], and different study end points, i.e., myocardial function, arrhythmias, or infarct size. The mechanisms of early and late pre- and postconditioning seem to be different [3]. Moreover, the possible ischemia sensor, if any exists, and the formation site of the potential mediators of preconditioning are still not known. It is plausible to assume that cardiac sensory nerves are involved in the mechanism of cardiac stress adaptation via their ability to sense chemical signals and to release NO and CGRP.

3. ROLE OF CAPSAICIN-SENSITIVE SENSORY NERVES IN THE MECHANISM OF ISCHEMIA–REPERFUSION INJURY AND IN THE DEVELOPMENT OF ENDOGENOUS CARDIAC STRESS ADAPTATION

In contrast to the adrenergic and cholinergic innervation of the heart, less attention has been paid to the functional significance of the rich sensory innervation of the myocardium and the coronary vascular system. Available experimental evidence shows that most cardiac sensory nerves associated with the vagus nerves and the sympathetic system display polymodal behavior responding to mechanical and chemical stimuli [16,17]. Capsaicin is a highly selective sensory neurotoxin which leads to a selective functional blockade and/or ablation of a morphologically well-defined population of primary sensory neurons [18,19]. Hence, capsaicin has become one of the most important probes for investigations of sensory neural morphology, physiology, pathology, and pharmacology (see Refs [16,20,21] for reviews). A subpopulation of capsaicin-sensitive cardiac C-fiber afferents costore CGRP, substance P (SP), neurokinin A, and NO. Capsaicin-sensitive sensory nerves may exert a strong influence on cardiac function and adaptive responses due to their NO and vasoactive peptide content, such as CGRP and SP [16,22–26]. The roles for CGRP and tachikinins in the local sensory control of cardiac contractility and coronary vascular tone is reviewed in depth in the nonischemic heart by Fraco-Cereceda [16].

Myocardial ischemia, via the resulting decrease in pH, increase in K^+ and lactate concentrations, and release of bradykinin and adenosine, is known to activate the nerve endings of the capsaicin-sensitive subpopulation of the thin afferent fibers (see Refs [16,27] for reviews). These afferent C-fibers show local efferent functions which are attributed to their CGRP, NO, and SP content [19]. NO [28,29] and CGRP [30] were found as important mediators of preconditioning. It is well known that CGRP is located only in afferent axons in the heart [16,20]. The cellular sources of NO in the heart includes cardiac myocytes, vascular and endocardial endothelium [31] as well as specific cardiac neurons. In our previous study, systemic capsaicin-treatment decreased cardiac NO signal approximately as low as the detection limit as assessed by electron spin resonance spectroscopy in left ventricular tissue samples [32]. Moreover, capsaicin-sensitive neurons seem to regulate basal myocardial NO–cGMP signaling [33]. This suggests that a significant portion of the total cardiac NO content may derive from capsaicin-sensitive afferent nerve fibers [32,33]. This is supported by another observation showing that capsaicin-sensitive sensory ganglion cells express NO synthase (NOS) [22]. An earlier study by Pabla and Curtis [34] suggesting a neural origin of cardiac NO release in the rat also strongly supports this assumption. Further studies show strong interaction between sensory neuronal CGRP and NO in the rabbit coronary circulation [35] and also in human internal mammary artery [36]. We have shown earlier that in hearts isolated from capsaicin-treated rats, the protective effect of preceding rapid ventricular pacing against the deterioration of myocardial function and lactate dehydrogenase release due to coronary occlusion was not observed, demonstrating that an intact sensory innervation is a prerequisite to elicit pacing-induced preconditioning in the rat heart [32]. We have also shown that selective sensory denervation by capsaicin aggravates adriamycin-induced cardiomyopathy in rats [37]. Moreover, we have recently shown that capsaicin-sensitive sensory neurons regulate gene expression pattern of rat heart, including the capsaicin receptor gene, as shown by a DNA-microarray study [38].

Based on the aforementioned studies, capsaicin-sensitive nerve endings are obvious candidates to be involved in the mechanism of cardiac stress adaptation as sensors for metabolic (e.g., ischemia) stressors and local effectors via CGRP and NO release.

4. CGRP IN CARDIAC ISCHEMIA AND REPERFUSION

CGRP is a potent vasodilator and positive inotropic and chronotropic peptide that has been shown to have a physiological and/or pathological role in neurogenic inflammation, headaches, thermal injury, circulatory shock, pregnancy and menopause, hypertension and heart failure, and it is known to be cardioprotective. Clinical trials have shown beneficial effects of the vasodilatory action of CGRP in hypertension, angina, heart failure, and Raynaud's disease. However, the clinical potential of CGRP is limited as it has to be given by infusion and is quickly broken down. Oral long-acting CGRP mimetics may have potential in disorders in which CGRP has been shown to be beneficial. CGRP mimetics include capsaicin/vanilloid receptor agonists and gene transfer of an adenoviral vector that encodes prepro-CGRP. CGRP inhibitors have therapeutic potential in conditions in which excessive CGRP-mediated vasodilatation is present. CGRP inhibitors include capsaicin, antagonists at capsaicin/vanilloid receptors, civamide, CGRP receptor antagonists, and 5-HT1D-receptor agonists. CGRP interacts with specific G-protein-coupled receptors; however, still relatively little is known about the different types of CGRP receptors (see Refs [39,40] for reviews). In the heart, significant amounts of specific CGRP binding were identified in atrial and ventricular myocardium, all portions of the conducting system, coronary arteries, the aorta and pulmonary trunk and intracardiac ganglia. The high density of CGRP receptors in the distal conducting system and the presence of CGRP receptors in intracardiac ganglia further suggest that CGRP could have important effects on cardiac conduction velocity and parasympathetic regulation of the heart [41]. Future advances are dependent on a better understanding of the structure and function of CGRP receptor(s) and the concomitant identification of selective and potent agonists and antagonists useful for addressing therapeutic hypotheses [42].

The potential cardioprotective effect of CGRP has been shown in several experimental models of ischemia and reperfusion. CGRP exerts a direct protective effect against severe hypoxia and simulated reperfusion in myocardial cells [43]. In this model, protection of cardiac cells by CGRP was associated with changes in membrane fluidity [44]. CGRP has been also shown to play a role in long-term cardiac preservation with nitroglycerin-containing cardioplegic solutions [45]. In a pig coronary occlusion reperfusion model, an aggravation of myocardial infarction was observed after capsaicin-induced depletion of CGRP [46]. However, exogenously administered CGRP, although caused systemic hypotension and augmented postischemic coronary flow, had no cardioprotective effect in pigs after coronary occlusion and reperfusion, possibly due to the short half-life of the peptide [47]. Nevertheless, the potential cardioprotective effect of CGRP is observed in a rat reperfusion arrhythmia model of short coronary occlusion/reperfusion. In this model, CGRP reduced and delayed the occurrence of ischemic reperfusion-induced arrhythmia such as ventricular tachycardia and fibrillation, especially in the early period, and decreased the mortality [48].

The cardioprotective effect of CGRP against myocardial ischemia is well established in humans [49]. In myocardial ischemia with its clinical manifestations angina pectoris and infarction, C-fiber afferents not only convey the sensation of pain, but there is also a local "efferent" release of CGRP in the heart. After being released, CGRP causes coronary vasodilatation and attenuates the development of myocardial infarction. CGRP may thus represent an endogenous local myocardial protective substance with interesting clinical implications (see Ref. [50] for review). The ischemia-related outflow of CGRP during cardiac surgery indicates that the vasodilating and cardioprotective properties of this peptide may be of importance in myocardial ischemia in humans [51]. Intravenous CGRP has been shown to dilate coronary arteries at the site of atheromatous stenoses and delays the onset of myocardial ischemia during treadmill exercise testing in patients with chronic stable angina [52].

Recent studies with deletion of the mouse α-CGRP gene or TRPV1 gene increase the vulnerability of the heart to ischemia–reperfusion injury [53,54].

5. NITRIC OXIDE IN ISCHEMIA AND REPERFUSION: CARDIOPROTECTIVE OR DETRIMENTAL?

In the normal heart, NO is synthesized by Ca^{2+}-dependent NOSs in cardiac myocytes, vascular and endocardial endothelium (NOS III) as well as in specific cardiac neurons (NOS I) and plays an important role in the regulation of cardiac function by induction of coronary vasodilatation, inhibition of platelet and neutrophil actions, modulation of cardiac contractile function, and inhibiting cardiac oxygen consumption and free radical actions [55,56]. Very little is known concerning the proportion of cardiac NO originating from cardiac sensory fibers. At least in the rat heart, basal NO release is most probably related to the intact function of cardiac sensory nerves [22,32,34,38,57].

NO plays protective roles in the heart by (i) stimulating soluble guanylate cyclase and thus reducing $[Ca^{2+}]_i$ partly through activation of cGMP-dependent protein kinase, (ii) terminating chain propagating lipid radical reactions caused by oxidative stress [58], and (iii) by inhibiting the activation of platelets and neutrophils and their adhesion to the endothelial surface [59,60]. NO (produced either via activation of NOS or through NO donors) protects against the toxic effects of exogenously supplied $ONOO^-$ on the coronary circulation [61], platelets [62], and liposomes [58].

5.1 Toxicity of NO via reaction with superoxide to form $ONOO^-$

NO is necessary for normal cardiac function, but it is potentially toxic in excess concentrations [63–66]. It is now understood that many of the toxic actions of NO are not directly due to NO itself but are mediated via production of $ONOO^-$, the reaction product of NO with superoxide [67–70]. NO and superoxide combine at a reaction rate which is only limited by diffusion to form $ONOO^-$. $ONOO^-$ at pH < 8 is protonated to form the unstable intermediate, peroxynitrous acid, which spontaneously decomposes to yield highly reactive oxidant species. Understanding the balance between local concentrations of NO, superoxide, and superoxide dismutase (SOD) is critical in understanding NO biology and its potential toxicity in the form of $ONOO^-$ [67,68]. One must consider the competition between NO and SOD for superoxide. Under normal physiological conditions in vascular endothelium, [NO] is ~10 nM and [SOD] is ~1 μM. Therefore, the reaction rate of superoxide to form $ONOO^-$ is only a fraction of the dismutation rate of superoxide by SOD, and, as a result, very little $ONOO^-$ is formed. However, at maximal vascular rates of NO production (i.e., which may occur during acute reperfusion of ischemic tissue or during inducible NOS (iNOS) expression), [NO] is 1 μM. Therefore, the formation of $ONOO^-$ will predominate over the dismutation of superoxide. Under a variety of cellular stress conditions where NO production is upregulated in the endothelium and cardiomyocytes, either by acute increase in Ca^{2+}-dependent endothelial NOS (eNOS) activity via changes in shear stress or high $[Ca^{2+}]_i$ which occurs during acute reperfusion and ischemia, or via *de novo* expression of iNOS in cells or tissues as a result of the action of proinflammatory cytokines, one predicts the formation of $ONOO^-$ [70–73].

5.2 Pathophysiology of NO and $ONOO^-$ in myocardial ischemia-reperfusion

Myocardial stunning (acute injury) $ONOO^-$ is produced during the acute reperfusion of ischemic hearts and that drugs which inhibit $ONOO^-$ formation or antagonize its toxicity protect the heart from this injury [74]. Rapid generation of $ONOO^-$ during reperfusion of the ischemic heart has also been detected using luminol chemiluminescence in the perfusate and antinitrotyrosine labeling of myocardial proteins [75]. Low concentrations of the NOS inhibitor L-NMMA: N^G-nitro-mono-methyl-L-arginine, or a cell-permeable SOD mimetic, MnTBAP protected the hearts from ischemia–reperfusion injury. The beneficial effect of L-NMMA fell within a narrow range and was lost at higher concentrations which further reduced coronary flow [74]. Previous studies showed that a NO donor, at subvasodilatory concentration, protected hearts from endogenous $ONOO^-$-mediated injury. This study provided the first mechanistic evidence of how either NO donors or NOS inhibitors reduce ischemia–reperfusion injury [74].

5.2.1 Myocardial infarction (late injury, neutrophil dependent)

What is the importance of the neutrophil-independent free radical release during acute reperfusion, as opposed to the neutrophil-dependent reperfusion events which occur several hours later? It should be noted that the late influx of neutrophils does not substantially contribute to the development of infarction; however, the literature is somewhat controversial in this regard [76–79]. One very important defect in the heart resulting from acute reperfusion following ischemia is endothelial "stunning," which is manifested by reduced endothelium-dependent vasodilator response within the first minute of reperfusion [80]. We speculate that this damage is self-inflicted, resulting from the burst of endogenous $ONOO^-$ from the vascular endothelium immediately at reperfusion. This would cause enhanced susceptibility of the endothelial surface to neutrophil and platelet adhesion, platelet aggregation and neutrophil activation, events which are normally inhibited by the physiological production of endothelium-derived NO [59,60]. Villa et al. [61] showed that bolus injection of $ONOO^-$ into isolated hearts acutely inhibited endothelium-dependent coronary vasodilatation. Myocardial $ONOO^-$ generation has also been shown to occur during the neutrophil-dependent phase of ischemia–reperfusion injury, seen

5 h after reperfusion, using an *in vivo* model of regional ischemia in rats [81]. An understanding of the oxidative damage which occurs during acute reperfusion following ischemia is thus crucial to help devise strategies to reduce a possible subsequent neutrophil-mediated damage in the later stages of reperfusion injury.

5.3 Cellular targets of $ONOO^-$

The possible downstream targets of $ONOO^-$ which mediate its toxicity are several. Its highly reactive decomposition products at physiological or acidic pH can attack protein (oxidation of sulfhydryls, nitration of tyrosine residues), lipids (formation of lipid peroxides), and DNA (strand breakage). This results in the depletion of low-molecular-weight antioxidants such as glutathione and most often the inhibition of several enzyme activities including SOD, aconitase, and other enzymes of the mitochondrial respiratory chain, creatine kinase, Ca^{2+}-ATPase, Na^+–K^+-ATPase, glutathione peroxidase, prostacyclin synthase, antiproteinase, and many others, just to name a few (see Refs [67,68,82,83] for reviews). As a result of DNA strand breakage, activation of the NAD^+ consuming DNA repair enzyme poly-ADP-ribose polymerase contributes further to the depletion of cellular energy stores [82]. An important question is that which of these multiple effects represents an early event in $ONOO^-$-mediated toxicity, at a time point prior to the onset of irreversible energy depletion, cellular damage and death. Interestingly, $ONOO^-$ may also activate some proteins, again due to its ability to react with critical thiol residues. $ONOO^-$ has been shown to activate the zymogen form of matrix metalloproteinases [84–86]. This family of proteolytic enzymes, best known for their ability to degrade and remodel the extracellular matrix, are now recognized to contribute to myocardial ischemia–reperfusion injury [74,87].

5.4 Can $ONOO^-$ protect the heart against myocardial ischemia–reperfusion?

The literature is somewhat controversial as to whether $ONOO^-$ is either cytotoxic or cytoprotective [70,88,89]. It seems that deleterious effects of $ONOO^-$ in the heart are predominantly observed in *in vitro* or *ex vivo* crystalloid buffer-perfused systems; however, $ONOO^-$ is cardioprotective if applied *in vivo*. However, this controversy derives solely from the rapid reactions of exogenously administered $ONOO^-$ with a variety of biomolecules. Simply, exogenous administration of $ONOO^-$ does not accurately reflect pathological conditions in which endogenous generation of $ONOO^-$ is enhanced.

Due to the very short half-life of $ONOO^-$ at physiological pH, it has very little chance to reach its cellular targets when it is applied via the blood, as it rapidly reacts with plasma proteins and thiols such as glutathione and cysteine. Thus, $ONOO^-$ is likely to be detoxified before it has a chance to reach tissues downstream of the injection site, let alone the intracellular compartment [90]. $ONOO^-$ oxidizes thiols [91], e.g., it reacts with glutathione (GSH) to form the nitrosothiol, nitrosoglutathione, a NO donor [61,92,93]. $ONOO^-$ is a vasodilator [61,94] and inhibitor of platelet aggregation [62] by this reaction. Thus nature has a built in a mechanism to return toxic $ONOO^-$ into a NO donor. Indeed, hearts with enhanced endogenous GSH levels are less susceptible to ischemia–reperfusion injury [95] and micromolar concentrations of GSH added to the perfusate protects isolated hearts from stunning injury through the reduced formation of $ONOO^-$ at reperfusion [96]. Exogenously administered $ONOO^-$ was also shown to inhibit leukocyte–endothelial cell interactions and to protect against ischemia–reperfusion injury in rats *in vivo* [97] Intraventricular infusion of $ONOO^-$ reduced myocardial infarct size and preserved coronary endothelium after ischemia and reperfusion in cats [98], which effect was mediated by the intermediate formation of *S*-nitrosothiols [99].

Exogenously applied $ONOO^-$, however, has been shown to be detrimental to cellular functions when it was applied in crystalloid buffer systems, in which the concentrations of extracellular antioxidants and both free and protein-bound thiols are limited. In this case, exogenous $ONOO^-$ and its toxic metabolites have a greater chance to reach their cellular targets and cause injury, this of course being dependent on concentration and the duration of exposure. We have shown that continuous infusion of 40 but not 4 μM $ONOO^-$ into isolated working rat hearts perfused with Krebs–Henseleit buffer impaired cardiac contractile function within 45 min [100]. Authentic $ONOO^-$ inhibited contractile function in cardiac myocytes [101] and in isolated rat papillary muscle [102] Administration of the $ONOO^-$ generator SIN-1: morpholinosydnonimine exerted either cardiodepression in crystalloid buffer-perfused or cardioprotection in blood-perfused rat hearts [103].

One can conclude that exogenous $ONOO^-$ is toxic when applied in crystalloid-perfused hearts; however, it can be protective under experimental conditions in which $ONOO^-$ first reacts with thiol groups, thereby forming NO donors. It is questionable, whether this mechanism of $ONOO^-$ detoxification is always of the capacity to eliminate a sufficient portion of $ONOO^-$ at the site of its endogenous formation, especially under conditions of ischemia–reperfusion. To our knowledge, there is no literature showing any tissue protective effect of endogenously formed $ONOO^-$. In contrast, many studies show that enhanced formation of $ONOO^-$ is the myocardium is cytotoxic to the heart and contributes to: ischemia–reperfusion injury in isolated rat hearts [74,75] and anesthetized rats [81], and in different other pathological models of myocardial dysfunction in several species *in vivo* or *ex vivo* [73,104–108] including myocardial inflammation in humans [109]. Many of these studies showed a correlation between endogenous $ONOO^-$ formation and deterioration of cardiac function.

Taken together, there is a consensus in the literature that endogenously formed $ONOO^-$ contributes to the injury to the heart as a result of ischemia–reperfusion and other cardiac pathologies.

However, recent results also suggest that basal production of $ONOO^-$ in physiological amount in the heart may have important regulatory functions [70]. For example, we have recently shown that cardiac capsaicin-sensitive sensory nerves regulate myocardial relaxation via *S*-nitrosylation of SERCA: sarcoplasmic reticulum Ca^{2+}-ATPase, and that this mechanism involves basal peroxynitrite formation [57].

5.5 NO is cardioprotective in ischemia–reperfusion injury

A number of *ex vivo* studies using isolated crystalloid buffer-perfused hearts have shown that enhancing NO levels, either by applying NO donors, NO-dependent vasodilators, L-arginine supplementation, angiotensin converting enzyme inhibitors, or pretreating the animal with endotoxin analogs which enhance myocardial iNOS activity, functionally protect the heart from acute stunning injury, and/or infarct size development [74,110–113]. Whether this protective effect is due to the antioxidant properties of NO, lowering $[Ca^{2+}]_i$ via guanylate cyclase, or other mechanisms is unclear at the moment.

In vivo studies show that NO donors improve the recovery of mechanical function and/or reduce infarct size following ischemia–reperfusion. Subvasodilatory doses of NO donors [80], L-arginine [114], or agonists of endothelium-derived NO [115] were shown to be protective (see Ref. [116] for review). *In vivo* studies using eNOS knockout mice enhance the notion that the basal release of NO in the heart is an important endogeneous cardiac protectant, particularly in regard to the prevention of neutrophil sticking and platelet activation. Infarct size in eNOS knockout mice was larger than in the wild-type controls, with higher P-selectin expression and significantly more neutrophils in hearts from eNOS knockout mice [117]. Yang et al. [118] showed that the protective effect of an angiotensin converting enzyme inhibitor in ischemia–reperfusion injury in wild-type mice was lost in the corresponding eNOS knockout mouse. In contrast, an *ex vivo* study showed that the functional recovery of hearts from eNOS knockout mouse was improved in comparison to wild-type controls [119], suggesting that NO generated from eNOS contributes to stunning injury seen in early reperfusion.

6. NO, SUPEROXIDE, AND $ONOO^-$ IN EARLY PRECONDITIONING

The role of NO in early preconditioning was suggested first by Vegh et al. [29]. They demonstrated that 10 mg/kg L-NAME, N^G-nitro-L-arginine-methyl esther administered both before and after preconditioning abolished the antiarrhythmic effect of preconditioning in a coronary occlusion model of anesthetized, open chest dogs [29]. Bilinska et al. [120] reported that infusion of the NO donors nitroglycerin (500 μM) or SIN-1 (10 μM) for 5 min before coronary occlusion mimicked the antiarrhythmic effect of ischemic preconditioning in isolated rat hearts. Using the electron spin resonance technique to directly measure cardiac NO content we have shown that a decrease in basal cardiac NO content before preconditioning, due to either *in vivo* pretreatment with 1 mg/kg L-nitroarginine [28], experimental hypercholesterolemia [121], or selective depletion of neurotransmitters including NO from cardiac sensory neurons [32] lead to the loss of pacing-induced preconditioning in isolated working rat hearts. However, preconditioning, in turn, markedly decreased the accumulation of NO in the heart tissue during subsequent ischemia and reperfusion in isolated working rat hearts [72]. In the presence of 4.6 μM L-nitroarginine, a nonvasoactive concentration in this model which reduced basal NO synthesis, preconditioning failed to protect against ischemia–reperfusion and failed to attenuate NO accumulation produced by ischemia–reperfusion. When L-nitroarginine was applied after the preconditioning protocol, the effect of preconditioning on test ischemia–reperfusion and ischemic NO accumulation was not affected [72,121]. These results prove that intact NO biosynthesis is required for the triggering mechanism of preconditioning and further show that the cardioprotection provided by preconditioning involves a mechanism which decreases the accumulation of NO in the myocardium during ischemia and reperfusion. In accordance with this, Woolfson et al. [122] reported that the limitation of infarct size after inhibition of NO synthesis by L-NAME shares a common mechanism with ischemic preconditioning in rabbit hearts. Furthermore, preliminary studies by Wang and Zweier [123] showed that preconditioning decreases NO synthesis during subsequent ischemia and this is associated with the protective effect of preconditioning in isolated rat hearts.

The nature of preconditioning-induced inhibition of NO synthesis is not known. Preconditioning may decrease the rate of enzymatic and/or nonenzymatic [124] NO production during ischemia–reperfusion by altering cellular pH and the availability of cofactors and/or arginine for NO synthesis, or may possibly stimulate the formation of endogenous NOS inhibitors [125].

In contrast to the aforementioned studies, Lu et al. [126] reported that 10 mg/kg L-NMMA or L-NAME did not affect the antiarrhythmic effect of preconditioning in anesthetized rats with coronary occlusion/reperfusion. Weselcouch et al. [127] showed that in rat hearts, 30 μM L-NAME did not interfere with the effect of preconditioning on postischemic myocardial function. In an isolated rabbit heart study, 100 μM L-NAME failed to block the infarct size limiting effect of preconditioning, nor it had any effect on infarct size without preconditioning. However, exogenous NO production by during SNAP: S-nitroso-N-acetylpenicillamine pretreatment was able to induce preconditioning via an antioxidant pathway [128]. In these studies, surprisingly, the different NOS inhibitors neither interfered with preconditioning nor the outcome of ischemia– reperfusion without preceding preconditioning. Since neither NO generation nor NOS activities were determined in these studies and only a single dose of NOS inhibitors were used, it is difficult to interpret these negative results.

The possible role for oxygen free radicals in preconditioning was suggested by Tanaka et al. [129] who showed that the free radical scavengers SOD or mercaptopropionyl-glycine were both able to inhibit the protective effect of preconditioning on infarct size in rabbits. Osada et al. [130] reported that the antiarrhythmic effect of preconditioning was lost when the preconditioning ischemia was applied in the presence of both SOD and catalase in isolated rat hearts. Tritto et al. [131] showed that a 5 min infusion of O_2^- generated by purine–xanthine oxidase prior to ischemia–reperfusion resulted in a reduction of infarct size in rabbit hearts, similar to the effect of preconditioning. Activation of mitochondrial K_{ATP} channels has been shown by several studies to play a role in preconditioning [11,132]; however, it was recently suggested that opening of mitochondrial K_{ATP} channels triggers the preconditioned state by generating free radicals [133]. These studies strongly suggest that the generation of free radicals during preconditioning stimuli is necessary to trigger the protective machinery of preconditioning. Preconditioning, in turn, attenuates the increased free radical synthesis during subsequent ischemia–reperfusion [134]. Others, however, showed that preconditioning was not affected by SOD in rabbit hearts [135,136].

As the majority of studies show that NO and oxygen free radicals are both required to elicit preconditioning, it was plausible to speculate that formation of $ONOO^-$ is an important oxidative stimulus to trigger cellular adaptive mechanisms. Altug et al. [137,138] have found that brief exposure of isolated rat hearts to 1 μM exogenous $ONOO^-$ was capable of mimicking the beneficial effects of ischemic preconditioning, and this was abolished by the administration of the antioxidant *N*-2-mercaptopropionylglycine. Others have shown that administration of $ONOO^-$-generating system triggered early preconditioning in rabbits *in vivo*. As exogenous $ONOO^-$ may not properly reflect the effect of endogenous $ONOO^-$ formation, we have recently measured endogenous $ONOO^-$ formation during preconditioning induced by three brief cycles ischemia and reperfusion and also during subsequent test ischemia–reperfusion in isolated working rat hearts to clarify the possible contribution of endogenous $ONOO^-$ to ischemic preconditioning. When test ischemia and reperfusion was preceded by preconditioning, $ONOO^-$ formation was markedly attenuated upon reperfusion [9]. We have also found that the first brief period of ischemia significantly enhanced endogenous $ONOO^-$ formation upon the first brief period of reperfusion, which was significantly reduced after subsequent cycles of brief ischemia–reperfusion periods [9]. These results show that $ONOO^-$ formation during ischemia–reperfusion might act as a trigger for preconditioning, but preconditioning in turn decreases formation of $ONOO^-$ upon subsequent cycle(s) of ischemia–reperfusion (see Ref. [66] for review).

7. CGRP IN EARLY PRECONDITIONING

CGRP is an endogenous myocardial protective substance which plays an important role in cardiac stress adaptation, i.e., preconditioning (see Ref. [30] for review). We have shown that in capsaicin-pretreated rats, the cardioprotective effect of pacing-induced preconditioning was abolished in rats together with depletion of cardiac CGRP and a decrease in basal NO content [32]. Others have shown that CGRP released by brief periods of ischemia triggers preconditioning in isolated rat hearts, and that exogenous CGRP or low doses of capsaicin are able to induce cardioprotection similar to that of induced by ischemic preconditioning. The mechanism of CGRP-induced preconditioning is not exactly known. CGRP-mediated ischemic preconditioning has been shown to be related to inhibition of cardiac tumor necrosis factor (TNF)-alpha production, but not to activation of the K_{ATP} channel [139]. Whatever the mechanism of ischemic preconditioning is, the aforementioned findings strongly suggest that CGRP is a key endogenous mediator of preconditioning [140].

Drug-induced preconditioning has been also shown to stimulate release of CGRP from the heart. In an isolated rat heart study, the cardioprotection afforded by nitroglycerin was abolished by CGRP-(8-37), a selective CGRP receptor antagonist. Pretreatment with capsaicin, which specifically depletes the transmitter content of sensory nerves, also abolished the protective effects of nitroglycerin and markedly reduced the release of CGRP from the heart during nitroglycerin perfusion. These findings suggest that nitroglycerin-induced preconditioning is related to stimulation of CGRP release in rat hearts [141]. The cardioprotective effect of bradykinin-induced preconditioning was also related to stimulation of CGRP release in the rat [142].

We have previously shown that the cardioprotective effect of ischemic preconditioning is attenuated in some disease states such as hyperlipidemia, diabetes, and aging (see Ref. [6] for review). It has been shown that the decreased protection afforded by ischemic preconditioning in aging and in diabetic hearts is related to reduction of the release and effect of CGRP in the rat heart [143,144]. Moreover, a recent study by Zhong at al. showed that TRPV1 gene knockout impairs preconditioning protection against myocardial injury in isolated perfused hearts in mice.

8. NO AND $ONOO^-$ IN LATE PRECONDITIONING

In a conscious rabbit model of preconditioning with repetitive coronary occlusion/reperfusion, Bolli et al. [145] reported that L-nitroarginine given either during preconditioning or 24 h later abrogated the protective effect of preconditioning and that the selective iNOS inhibitors aminoguanidine or *S*-methylisothiourea abolished preconditioning only when applied 24 h after preconditioning. They also showed that preconditioning induces an increase in iNOS mRNA levels in the ischemic regions of the rabbit heart, and that this induction is triggered by increased generation of NO during the preconditioning stimulus [146]. Targeted disruption of the iNOS gene in mice led to a complete blockade of late preconditioning [147]. Dexamethasone or selective iNOS inhibitors inhibited the late effect of preconditioning on infarct size in anesthetized rabbits [148] and on arrhythmias in anesthetized dogs [149]. These results suggest a dual role of NO in late preconditioning, as intact NO synthesis by eNOS is necessary to trigger late preconditioning and NO derived from iNOS is a mediator of late protection (see Ref. [10] for review). However, the exact role of iNOS in late preconditioning is not known as neither NO level nor NOS activities were determined in the aforementioned studies.

The involvement of oxygen free radicals in late preconditioning was suggested by Sun et al. [150] who showed that a combination of antioxidants (SOD, catalase, mercaptopropionylglycine) infused during the preconditioning stimulus completely abolished the late effect of preconditioning on stunning in conscious pigs. Zhou et al. [151] reported that isolated rat myocytes preconditioned with anoxia or with administration of superoxide both induced late cytoprotection which was characterized by increased Mn SOD activity and decreased superoxide production. Takano et al. [152] demonstrated that i.v. infusion of the NO donors DETA/NO: diethylenetriamine-NONOate or SNAP induces cardioprotection 24 h later in conscious rabbits and this effect was lost when they were infused with mercaptopropionyl-glycine [152]. This suggests that the mechanism whereby NO induces preconditioning involves the generation of oxidant species, possibly $ONOO^-$.

Taken together, these results suggest that both NO and oxygen free radicals are necessary to trigger late preconditioning. Xuan et al. [153] have suggested that eNOS is the source of NO during the trigger phase of delayed preconditioning. Since both free radical scavengers and inhibitors of eNOS abolish delayed preconditioning, it has been proposed that the formation of $ONOO^-$ is an important upstream event upon the triggering mechanism of delayed preconditioning [154]. Preliminary studies by Emani et al. [155] have reported that administration of a $ONOO^-$ generating system to rabbits evoked a delayed protective response 24 h later, although interpretation in this study is difficult because under these conditions $ONOO^-$ could serve as a donor of NO as discussed in detail above. It should be noted that most of the aforementioned conclusions are based on mainly pharmacological studies. To elucidate the exact role of NO, superoxide, and $ONOO^-$ in preconditioning requires further biochemical evidence.

9. CGRP IN LATE PRECONDITIONING

The possible role of CGRP in late preconditioning has not been extensively studied, but available experimental evidence suggests that CGRP release from cardiac sensory nerves may be involved in the cardioprotective effect of late preconditioning.

Monophosphoryl lipid-A (MLA), an analog of bacterial endotoxin which evokes the late phase of preconditioning, caused a significant increase in the expression of CGRP and hemoxigenase-1 (HO-1) and in plasma concentrations of CGRP in rats. The cardioprotection as well as the synthesis and release of CGRP induced by MLA were completely abolished by pretreatment with zinc protoporphrin IX, an inhibitor of HO-1, or by capsaicin. These results suggest that the delayed cardioprotection afforded by MLA is mediated by CGRP via activation of the HO-1 pathway [156].

It has been also shown that MLA-induced delayed preconditioning enhanced preservation with cardioplegia and that the protective effect of MLA was related to stimulation of CGRP release. MLA caused a significant increase in the expression of alpha-CGRP mRNA, but not beta-CGRP mRNA, concomitantly with an increase in plasma concentrations of CGRP. The increased level of CGRP expression happened before stimulation of CGRP release. The effect of MLA was completely abolished by pretreatment with L-NAME, an inhibitor of NOS, or by capsaicin. These results suggest that the delayed cardioprotection afforded by MLA involves the synthesis and release of CGRP via the NO pathway, and that the protection is mainly mediated by alpha-CGRP isoform [157].

Heat stress, similarly to brief ischemia, protects the myocardium against ischemia–reperfusion injury. The early or delayed protection by heat stress has been shown to involve endogenous CGRP synthesis and that can be abolished by pretreatment with capsaicin [158]. Improvement of preservation with cardioplegia induced by heat stress has been also demonstrated to be mediated by CGRP [159].

10. NO, ONOO$^-$, AND CGRP IN POSTCONDITIONING

Very little is known about the role of NO, ONOO, and CGRP in postconditioning, although some preliminary studies has been published recently showing that both CGRP and NO might play a role in postconditioning [3,160–162].

11. CONCLUSIONS/RELEVANCE

Severe myocardial ischemia and reperfusion deteriorates myocardial function and leads to development of infarction. However, brief episodes of ischemia trigger an endogenous cardioprotective mechanism knows as preconditioning. Cardiac capsaicin-sensitive nerves are stimulated during ischemia and reperfusion and they are involved in the development of cardiac stress adaptation known as early and delayed preconditioning via their ability to act as a sensor of ischemia and a local effector due to release of cardioprotective mediators such as NO and CGRP. Development of modulators of cardiac sensory nerve endings is an exciting new strategy to protect the heart from ischemic stress injury and to preserve the endogenous stress adaptation in some disease states.

ACKNOWLEDGMENTS

P.F. acknowledges the support of grants from the Hungarian National Scientific Research Found (OTKA T046417), Hungarian Ministries of Health (ETT 597/2006) and Economy and Transport (GVOP-TST0095/2004), the Wellcome Trust, as well as the National Office for Research and Technology (5let-2006ALAP1-00088/2006, Jedlik-NKFP A1-2006-029, Asboth 2005). G.J. acknowledges the support of grants from the Hungarian National Scientific Research Found (OTKA T046469, 63663) and Hungarian Ministry of Health (ETT 193/ 2006).

REFERENCES

1. Braunwald EK. Myocardial reperfusion: a double-edged sword? J Clin Invest 1985;76:1713–1719.
2. Murry CE, Jennings RB, Reimer KA. Preconditioning with ischemia: a delay of lethal cell injury in ischemic myocardium. Circulation 1986;74:1124–1136.
3. Ferdinandy P, Schulz R, Baxter GF. Interaction of cardiovascular risk factors with myocardial ischemia/reperfusion injury, preconditioning, and postconditioning. Pharmacol Rev 2007;59:418–458.
4. Baxter GF, Ferdinandy P. Delayed preconditioning of myocardium: current perspectives. Basic Res Cardiol 2001;96:329–344.
5. Przyklenk K, Kloner RA. Ischemic preconditioning: exploring the paradox. [Review] [263 refs]. Prog Cardiovasc Dis 1998;40:517–547.
6. Ferdinandy P, Szilvassy Z, Baxter GF. Adaptation to myocardial stress in disease states: is preconditioning a healthy heart phenomenon? Trends Pharmacol Sci 1998;19:223–229.

7. Ferdinandy P. Myocardial ischaemia/reperfusion injury and preconditioning: effects of hypercholesterolaemia/hyperlipidaemia. Br J Pharmacol 2003;138:283–285.
8. Zhao ZQ, Corvera JS, Halkos ME, Kerendi F, Wang NP, Guyton RA et al. Inhibition of myocardial injury by ischemic postconditioning during reperfusion: comparison with ischemic preconditioning. Am J Physiol Heart Circ Physiol 2003;285:H579–H588.
9. Csonka C, Csont T, Onody A, Ferdinandy P. Preconditioning decreases ischemia/reperfusion-induced peroxynitrite formation. Biochem Biophys Res Commun 2001;285:1217–1219.
10. Bolli R, Dawn B, Tang XL, Qiu Y, Ping P, Xuan YT et al. The nitric oxide hypothesis of late preconditioning [review]. Basic Res Cardiol 1998;93:325–338.
11. Schulz R, Cohen MV, Behrends M, Downey JM, Heusch G. Signal transduction of ischemic preconditioning. Cardiovasc Res 2001;52:181–198.
12. Parratt JR. Possibilities for the pharmacological exploitation of ischaemic preconditioning. J Mol Cell Cardiol 1995;27:991–1000.
13. Baxter GF, Marber MS, Yellon DM. Myocardial stress response, cytoprotective proteins and the second window of protection against infarction. In: Wainwright CL, Parratt JR, eds. Myocardial Preconditioning. Austin, Texas: R.G. Landes Company, 1996; pp. 233–250.
14. Walker DM, Yellon DM. Ischaemic preconditioning: from mechanisms to exploitation. Cardiovasc Res 1992;26:734–739.
15. Ferdinandy P, Szilvássy Z, Koltai M, Dux L. Ventricular overdrive pacing-induced preconditioning and no-flow ischemia-induced preconditioning in isolated working rat hearts. J Cardiovasc Pharmacol 1995;25:97–104.
16. Franco-Cereceda A. Calcitonin gene-related peptide and tachykinins in relation to local sensory control of cardiac contractility and coronary vascular tone. Acta Physiol Scand 1988;133:1–63.
17. Hunag MH, Horackova M, Negoescu RM, Wolf S, Armour JA. Polysensory response characteristics of dorsal root ganglion neurones that may serve sensory functions during myocardial ischaemia. Cardiovasc Res 1996;32:503–515.
18. Jancso N. Desensitization with capsaicin as a tool for studying the function of pain receptors. In: Lim RKS, ed. Pharmacology of Pain. Oxford: Pergamon Press, 1968; pp. 33–55.
19. Jancso G, Kiraly E, Jancso-Gabor A. Pharmacologically induced selective degeneration of chemosensitive primary sensory neurones. Nature 1977;270:741–743.
20. Holzer P. Capsaicin: cellular targets, mechanisms of action, and selectivity for thin sensory neurons. Pharmacol Rev 1991;43:143–201.
21. Jancso G. Pathobiological reactions of C-fibre primary sensory neurones to peripheral nerve injury. Exp Physiol 1992;77:405–431.
22. Ren K, Ruda MA. Nitric oxide synthase-containing neurons in sensory ganglia of the rat are susceptible to capsaicin-induced cytotoxicity. Neuroscience 1995;65:505–511.
23. Sosunov AA, Hassall CJS, Loesch A, Turmaine M, Burnstock G. Ultrastructural investigation of nitric oxide synthase-immunoreactive nerves associated with coronary blood vessels of rat and guinea-pig. Cell Tissue Res 1995;280:575–582.
24. Sosunov AA, Hassall CJS, Loesch A, Turmaine M, Burnstock G. Nitric oxide synthase-containing neurones and nerve fibres within cardiac ganglia of rat and guinea-pig: an electron- microscopic immunocytochemical study. Cell Tissue Res 1996;284:19–28.

25. Takaki M, Akashi T, Ishioka K, Kikuta A, Matsubara H, Yasuhara S et al. Effects of capsaicin on mechanoenergetics of excised cross-circulated canine left ventricle and coronary artery. J Mol Cell Cardiol 1994;26:1227–1239.
26. Yaoita H, Sato E, Kawaguchi M, Saito T, Maehara K, Maruyama Y. Nonadrenergic noncholinergic nerves regulate basal coronary flow via release of capsaicin-sensitive neuropeptides in the rat heart. Circ Res 1994;75:780–788.
27. Bevan S, Geppetti P. Protons: small stimulants of capsaicin-sensitive sensory nerves. TINS 1994;17:509–512.
28. Ferdinandy P, Szilvassy Z, Balogh N, Csonka C, Csont T, Koltai M et al. Nitric oxide is involved in active preconditioning in isolated working rat hearts. Ann N Y Acad Sci 1996;793:489–493.
29. Vegh A, Szekeres L, Parratt JR. Preconditioning of the ischaemic myocardium; Involvement of the L-arginine nitric oxide pathway. Br J Pharmacol 1992;107:648–652.
30. Li YJ, Song QJ, Xiao J. Calcitonin gene-related peptide: an endogenous mediator of preconditioning. Acta Pharmacol Sin 2000;21:865–869.
31. Mülsch A, Hecker M, Mordvintcev PI, Vanin AF, Busse R. Enzymic and nonenzymic release of NO accounts for the vasodilator activity of the metabolites of CAS 936, a novel long-acting sydnonimine derivative. Naunyn Schmiedebergs Arch Pharmacol 1993;347:92–100.
32. Ferdinandy P, Csont T, Csonka C, Torok M, Dux M, Nemeth J et al. Capsaicin-sensitive local sensory innervation is involved in pacing-induced preconditioning in rat hearts: role of nitric oxide and CGRP? Naunyn Schmiedebergs Arch Pharmacol 1997;356:356–363.
33. Csont T, Csonka C, Kovacs P, Jancso G, Ferdinandy P. Capsaicin-sensitive sensory neurons regulate myocardial nitric oxide and cGMP signaling. Eur J Pharmacol 2003;476:107–113.
34. Pabla R, Curtis MJ. Endogenous protection against reperfusion-induced ventricular fibrillation: role of neuronal versus non-neuronal sources of nitric oxide and species dependence in the rat versus rabbit isolated heart. J Mol Cell Cardiol 1996;28:2097–2110.
35. Mitchell JA, Williams FM, Williams TJ, Larkin SW. Role of nitric oxide in the dilator actions of capsaicin-sensitive nerves in the rabbit coronary circulation. Neuropeptides 1997;31:333–338.
36. Raddino R, Pelŕ G, Manca C, Barbagallo M, D'Aloia A, Passeri M et al. Mechanism of action of human calcitonin gene-related peptide in rabbit heart and in human mammary arteries. J Cardiovasc Pharmacol 1997;29:463–470.
37. Katona M, Boros K, Santha P, Ferdinandy P, Dux M, Jancso G. Selective sensory denervation by capsaicin aggravates adriamycin-induced cardiomyopathy in rats. Naunyn Schmiedebergs Arch Pharmacol 2004;370:436–443.
38. Zvara A, Bencsik P, Fodor G, Csont T, Hackler L, Jr., Dux M et al. Capsaicin-sensitive sensory neurons regulate myocardial function and gene expression pattern of rat hearts: a DNA microarray study. FASEB J 2006;20:160–162.
39. Doggrell SA. Migraine and beyond: cardiovascular therapeutic potential for CGRP modulators. Expert Opin Investig Drugs 2001;10:1131–1138.
40. Brain SD, Poyner DR, Hill RG. CGRP receptors: a headache to study, but will antagonists prove therapeutic in migraine? Trends Pharmacol Sci 2002;23:51–53.
41. Chang Y, Stover SR, Hoover DB. Regional localization and abundance of calcitonin gene-related peptide receptors in guinea pig heart. J Mol Cell Cardiol 2001;33:745–754.
42. Feuerstein G, Willette R, Aiyar N. Clinical perspectives of calcitonin gene related peptide pharmacology. Can J Physiol Pharmacol 1995;73:1070–1074.

43. Ren YS, Ma TG, Wang HB, Yu SQ. Protective effects of calcitonin gene-related peptide (CGRP) on myocardial cell injury and calcium and magnesium contents following severe hypoxia and simulated reperfusion. Med Sci Res 1993;21:177–178.
44. Ren YS, Ma TG, Wang HB, Yu SQ. Membrane fluidity changes in myocardial cells following severe hypoxia and simulated reperfusion and effects of calcitonin gene-related peptide (CGRP). Med Sci Res 1993;21:627–628.
45. Zhou ZH, Deng HW, Li YJ. Improvement of preservation with cardioplegic solution by nitroglycerin-induced delayed preconditioning is mediated by calcitonin gene-related peptide. Int J Cardiol 2001;81:211–218.
46. Kallner G, Franco-Cereceda A. Aggravation of myocardial infarction in the porcine heart by capsaicin-induced depletion of calcitonin gene-related peptide (CGRP). J Cardiovasc Pharmacol 1998;32:500–504.
47. Kallner G, Gonon A, Franco-Cereceda A. Calcitonin gene-related peptide in myocardial ischaemia and reperfusion in the pig. Cardiovasc Res 1998;38:493–499.
48. Zhang JF, Liu J, Liu XZ, Li MY, Sheng SL, Zhang WJ. The effect of calcitonin gene-related peptide on ischemic reperfusion-induced arrhythmias in rats. Int J Cardiol 1994;46:33–36.
49. Kallner G. Release and effects of calcitonin gene-related peptide in myocardial ischaemia. Scand Cardiovasc J Suppl 1998;49:1–35.
50. Franco-Cereceda A, Liska J. Potential of calcitonin gene-related peptide in coronary heart disease. Pharmacology 2000;60:1–8.
51. Kallner G, Owall A, Franco-Cereceda A. Myocardial outflow of calcitonin gene-related peptide in relation to metabolic stress during coronary artery bypass grafting without cardiopulmonary bypass. J Thorac Cardiovasc Surg 1999;117:447–453.
52. Uren NG, Seydoux C, Davies GJ. Effect of intravenous calcitonin gene related peptide on ischaemia threshold and coronary stenosis severity in humans. Cardiovasc Res 1993;27:1477–1481.
53. Huang R, Karve A, Shah I, Bowers MC, Dipette DJ, Supowit SC et al. Deletion of the mouse (alpha)-calcitonin gene-related peptide gene increases the vulnerability of the heart to ischemia-reperfusion injury. Am J Physiol Heart Circ Physiol 2008;294:H1291–H1297.
54. Wang L, Wang DH. TRPV1 gene knockout impairs postischemic recovery in isolated perfused heart in mice. Circulation 2005;112:3617–3623.
55. Hare JM, Comerford ML. Role of nitric oxide in the regulation of myocardial function. Prog Lipid Res 1995;38:155–166.
56. Xie YW, Wolin MS. Role of nitric oxide and its interaction with superoxide in the suppression of cardiac muscle mitochondrial respiration. Involvement in response to hypoxia/reoxygenation. Circulation 1996;94:2580–2586.
57. Bencsik P, Kupai K, Giricz Z, Gorbe A, Huliak I, Furst S et al. Cardiac capsaicin-sensitive sensory nerves regulate myocardial relaxation via S-nitrosylation of SERCA: role of peroxynitrite. Br J Pharmacol 2008;153:488–496.
58. Rubbo H, Radi R, Trujillo M, Telleri R, Kalyanaraman B, Barnes S et al. Nitric oxide regulation of superoxide and peroxynitrite-dependent lipid peroxidation. Formation of novel nitrogen-containing oxidized lipid derivatives. J Biol Chem 1994;269:26066–26075.
59. Kubes P, Suzuki M, Granger DN. Nitric oxide: an endogenous modulator of leukocyte adhesion. Proc Natl Acad Sci U S A 1991;88:4651–4655.
60. Radomski MW, Palmer RMJ, Moncada S. The anti-aggregating properties of vascular endothelium: interactions between prostacyclin and nitric oxide. Br J Pharmacol 1987;92:639–646.

61. Villa LM, Salas E, Darley-Usmar VM, Radomski MW, Moncada S. Peroxynitrite induces both vasodilatation and impaired vascular relaxation in the isolated perfused rat heart. Proc Natl Acad Sci U S A 1994;91:12383–12387.
62. Moro MA, Darley-Usmar V, Goodwin DA, Read NG, Zamora-Pino R, Feelisch M et al. Paradoxical fate and biological action of peroxynitrite in human platelets. Proc Natl Acad Sci U S A 1994;91:6702–6706.
63. Radak Z, Pucsok J, Csont T, Ferdinandy P. Muscle soreness-induced reduction in force generation is accompanied by increased nitric oxide content and DNA damage in human skeletal muscle. Free Radic Biol Med 1999;26:1059–1063.
64. Varga E, Bodi A, Ferdinandy P, Droy-Lefaix MT, Blasig IE, Tosaki A. The protective effect of EGb 761 in isolated ischemic/reperfused rat hearts: a link between cardiac function and nitric oxide production. J Cardiovasc Pharmacol 1999;34:711–717.
65. Ferdinandy P, Szilvássy Z, Csont T, Csonka C, Nagy E, Koltai M et al. Nitroglycerin-induced direct protection of the ischaemic myocardium in isolated working hearts of rats with vascular tolerance to nitroglycerin. Br J Pharmacol 1995;115:1129–1131.
66. Ferdinandy P, Schulz R. Nitric oxide, superoxide, and peroxynitrite in myocardial ischaemia-reperfusion injury and preconditioning. Br J Pharmacol 2003;138:532–543.
67. Beckman JS, Koppenol WH. Nitric oxide, superoxide, and peroxynitrite: the good, the bad and ugly. Am J Physiol 1996;271:C1424–C1437.
68. Rubbo H, Darley-Usmar V, Freeman BA. Nitric oxide regulation of tissue free radical injury. [Review]. Chem Res Toxicol 1996;9:809–820.
69. Ferdinandy P, Das DK, Tosaki A. Pacing-induced ventricular fibrillation leading to oxygen free radical formation in aerobically perfused rat heart. J Mol Cell Cardiol 1993;25:683–692.
70. Ferdinandy P. Peroxynitrite: just an oxidative/nitrosative stressor or a physiological regulator as well? Br J Pharmacol 2006;148:1–3.
71. Depre C, Hue L. Cyclic GMP in the perfused rat heart. Effect of ischaemia, anoxia and nitric oxide synthase inhibitor. FEBS Lett 1994;345:241–245.
72. Csonka C, Szilvassy Z, Pali T, Blasig IE, Tosaki A, Schulz R et al. Classic preconditioning decreases the harmful accumulation of nitric oxide during ischemia and reperfusion in rat hearts. Circulation 1999;100:2260–2266.
73. Ferdinandy P, Danial H, Ambrus I, Rothery RA, Schulz R. Peroxynitrite is a major contributor to cytokine-induced myocardial contractile failure. Circ Res 2000;87:241–247.
74. Yasmin W, Strynadka KD, Schulz R. Generation of peroxynitrite contributes to ischemia-reperfusion injury in isolated rat hearts. Cardiovasc Res 1997;33:422–432.
75. Wang PH, Zweier JL. Measurement of nitric oxide and peroxynitrite generation in the postischemic heart – evidence for peroxynitrite-mediated reperfusion injury. J Biol Chem 1996;271:29223–29230.
76. Williams FM, Kus M, Tanda K, Williams TJ. Effect of duration of ischaemia on reduction of myocardial infarct size by inhibition of neutrophil accumulation using an anti-CD18 monoclonal antibody. Br J Pharmacol 1994;111:1123–1128.
77. Baxter GF. The neutrophil as a mediator of myocardial ischemia-reperfusion injury: time to move on. Basic Res Cardiol 2002;97:268–375.
78. Gao F, Yue TL, Shi DW, Christopher TA, Lopez BL, Ohlstein EH et al. p38 MAPK inhibition reduces myocardial reperfusion injury via inhibition of endothelial adhesion molecule expression and blockade of PMN accumulation. Cardiovasc Res 2002;53:414–422.

79. Jordan JE, Zhao ZQ, Vinten-Johansen J. The role of neutrophils in myocardial ischemia-reperfusion injury. Cardiovasc Res 1999;43:860–878.
80. Siegfried MR, Erhardt J, Rider T, Xin-Liang Ma, Lefer AM. Cardioprotection and attenuation of endothelial dysfunction by organic nitric oxide donors in myocardial ischemia-reperfusion. J Pharmacol Exp Ther 1992;260:668–675.
81. Liu PT, Hock CE, Nagele R, Wong PYK. Formation of nitric oxide, superoxide, and peroxynitrite in myocardial ischemia-reperfusion injury in rats. Am J Physiol Heart Circ Physiol 1997;272:H2327–H2336.
82. Szabó C. The pathophysiological role of peroxynitrite in shock, inflammation, and ischemia-reperfusion injury. Shock 1996;6:79–88.
83. Ronson RS, Nakamura M, Vinten-Johansen J. The cardiovascular effects and implications of peroxynitrite. Cardiovasc Res 1999;44:47–59.
84. Okamoto T, Akaike T, Sawa T, Miyamoto Y, van der Vliet A, Maeda H. Activation of matrix metalloproteinases by peroxynitrite-induced protein S-glutathiolation via disulfide S-oxide formation. J Biol Chem 2001;276:29596–29602.
85. Okamoto T, Akaike T, Nagano T et al. Activation of human neutrophil procollagenase by nitrogen dioxide and peroxynitrite: a novel mechanism for procollagenase activation involving nitric oxide. Arch Biochem Biophys 1997;342:261–274.
86. Dorman G, Kocsis-Szommer K, Spadoni C, Ferdinandy P. MMP inhibitors in cardiac diseases: an update. Recent Patents Cardiovasc Drug Discov 2007;2:186–194.
87. Giricz Z, Lalu MM, Csonka C, Bencsik P, Schulz R, Ferdinandy P. Hyperlipidemia attenuates the infarct size-limiting effect of ischemic preconditioning: role of matrix metalloproteinase-2 inhibition. J Pharmacol Exp Ther 2006;316:154–161.
88. Vinten-Johansen J. Physiological effects of peroxynitrite. Potential products of the environment. Circ Res 2000;87:170–172.
89. Ferdinandy P, Schulz R. Peroxynitrite: toxic or protective in the heart? Circ Res 2001;88:e12–e13.
90. Ishida H, Genka C, Nakazawa H. Application of authentic peroxynitrite to biological materials. Meth Enzymol 1999;301:402–409.
91. Cheung PY, Danial H, Jong J, Schulz R. Thiols protect the inhibition of myocardial aconitase by peroxynitrite. Arch Biochem Biophys 1998;350:104–108.
92. Prendergast BD, Sagach VF, Shah AM. Basal release of nitric oxide augments the Frank-Starling response in the isolated heart. Circulation 1997;96:1320–1329.
93. Mayer B, Schrammel A, Klatt P, Koesling D, Schmidt K. Peroxynitrite-induced accumulation of cyclic GMP in endothelial cells and stimulation of purified soluble guanylyl cyclase. Dependence on glutathione and possible role of S-nitrosation. J Biol Chem 1995;270:17355–17360.
94. Wu M, Pritchard KAJ, Kaminski PM, Fayngersh RP, Hintze TH, Wolin MS. Involvement of nitric oxide and nitrosothiols in relaxation of pulmonary arteries to peroxynitrite. Am J Physiol 1994;266:H2108–H2113.
95. Kirshenbaum LA, Singal PK. Increase in endogenous antioxidant enzymes protects hearts against reperfusion injury. Am J Physiol Heart Circ Physiol 1993;265:H484–H493.
96. Cheung PY, Wang W, Schulz R. Glutathione protects against myocardial ischemia-reperfusion injury by detoxifying peroxynitrite. J Mol Cell Cardiol 2000;32:1669–1678.
97. Lefer DJ, Scalia R, Campbell B, Nossuli TO, Hayward R, Salamon M et al. Peroxynitrite inhibits leukocyte-endothelial cell interactions and protects against ischemia-reperfusion injury in rats. J Clin Invest 1997;99:684–691.

98. Nossuli TO, Hayward R, Scalia R, Lefer AM. Peroxynitrite reduces myocardial infarct size and preserves coronary endothelium after ischemia and reperfusion in cats. Circulation 1997;96:2317–2324.
99. Nossuli TO, Hayward R, Jensen D, Scalia R, Lefer AM. Mechanisms of cardioprotection by peroxynitrite in myocardial ischemia and reperfusion injury. Am J Physiol 1998;275:H509–H519.
100. Schulz R, Dodge KL, Lopaschuk GD, Clanachan AS. Peroxynitrite impairs cardiac contractile function by decreasing cardiac efficiency. Am J Physiol Heart Circ Physiol 1997;272:H1212–H1219.
101. Ishida H, Ichimori K, Hirota Y, Fukahori M, Nakazawa H. Peroxynitrite-induced cardiac myocyte injury. Free Radic Biol Med 1996;20:343–350.
102. Digerness SB, Harris KD, Kirklin JW, Urthaler F, Viera L, Beckman JS et al. Peroxynitrite irreversibly decreases diastolic and systolic function in cardiac muscle. Free Radic Biol Med 1999;27:1386–1392.
103. Ma XL, Gao F, Lopez BL, Christopher TA, Vinten-Johansen J. Peroxynitrite, a two-edged sword in post-ischemic myocardial injury – dichotomy of action in crystalloid- versus blood-perfused hearts. J Pharmacol Exp Ther 2000;292:912–920.
104. Ferdinandy P, Panas D, Schulz R. Peroxynitrite contributes to spontaneous loss of cardiac efficiency in isolated working rat hearts. Am J Physiol 1999;276:H1861–H1867.
105. Oyama J, Shimokawa H, Momii H, Cheng X, Fukuyama N, Arai Y et al. Role of nitric oxide and peroxynitrite in the cytokine-induced sustained myocardial dysfunction in dogs in vivo. J Clin Invest 1998;101:2207–2214.
106. Weinstein DM, Mihm MJ, Bauer JA. Cardiac peroxynitrite formation and left ventricular dysfunction following doxorubicin treatment in mice. J Pharmacol Exp Ther 2000;294:396–401.
107. Ishiyama S, Hiroe M, Nishikawa T, Abe S, Shimojo T, Ito H et al. Nitric oxide contributes to the progression of myocardial damage in experimental autoimmune myocarditis in rats. Circulation 1997;95:489–496.
108. Sakurai M, Fukuyama N, Iguchi A et al. Quantitative analysis of cardiac 3-L-nitrotyrosine during acute allograft rejection in an experimental heart transplantation. Transplantation 1999;68:1818–1822.
109. Kooy NW, Lewis SJ, Royall JA, Ye YZ, Kelly DR, Beckman JS. Extensive tyrosine nitration in human myocardial inflammation: evidence for the presence of peroxynitrite. Crit Care Med 1997;25:812–819.
110. Masini E, Bianchi S, Mugnai L, Gambassi F, Lupini M, Pistelli A et al. The effect of nitric oxide generators on ischemia reperfusion injury and histamine release in isolated perfused guinea-pig heart. Agents Actions 1991;33:53–56.
111. Schoelkens BA, Linz W. Bradykinin-mediated metabolic effects in isolated perfused rat hearts. Agents Actions Suppl 1992;38:36–42.
112. Massoudy P, Becker BF, Gerlach E. Nitric oxide accounts for postischemic cardioprotection resulting from angiotensin-converting enzyme inhibition: indirect evidence for a radical scavenger effect in isolated guinea pig heart. J Cardiovasc Pharmacol 1995;25:440–447.
113. Xi L, Jarrett NC, Hess ML, Kukreja RC. Essential role of inducible nitric oxide synthase in monophosphoryl lipid A-induced late cardioprotection: evidence from pharmacological inhibition and gene knockout mice. Circulation 1999;99:2157–2163.
114. Weyrich AS, Ma X, Lefer AM. The role of L-arginine in ameliorating reperfusion injury after myocardial ischemia in the cat. Circulation 1992;86:279–288.

115. Richard V, Blanc T, Kaeffer N, Tron C, Thuillez C. Myocardial and coronary endothelial protective effects of acetylcholine after myocardial ischaemia and reperfusion in rats: role of nitric oxide. Br J Pharmacol 1995;115:1532–1538.
116. Grisham MB, Granger DN, Lefer DJ. Modulation of leukocyte-endothelial interactions by reactive metabolites of oxygen and nitrogen: relevance to ischemic heart disease [review]. Free Radic Biol Med 1998;25:404–433.
117. Jones SP, Girod WG, Palazzo AJ, Granger DN, Grisham MB, Jourd'Heuil D et al. Myocardial ischemia-reperfusion injury is exacerbated in absence of endothelial cell nitric oxide synthase. Am J Physiol 1999;276:H1567–H1573.
118. Yang XP, Liu YH, Shesely EG, Bulagannawar M, Liu F, Carretero OA. Endothelial nitric oxide gene knockout mice: cardiac phenotypes and the effect of angiotensin-converting enzyme inhibitor on myocardial ischemia/reperfusion injury. Hypertension 1999;34:24–30.
119. Flogel U, Decking UK, Godecke A, Schrader J. Contribution of NO to ischemia-reperfusion injury in the saline-perfused heart: a study in endothelial NO synthase knockout mice. J Mol Cell Cardiol 1999;31:827–836.
120. Bilinska M, Maczewski M, Beresewicz A. Donors of nitric oxide mimic effects of ischaemic preconditioning on reperfusion induced arrhythmias in isolated rat heart. Mol Cell Biochem 1996;160–161:265–271.
121. Ferdinandy P, Szilvassy Z, Horvath LI, Csont T, Csonka C, Nagy E et al. Loss of pacing-induced preconditioning in rat hearts: role of nitric oxide and cholesterol-enriched diet. J Mol Cell Cardiol 1997;29:3321–3333.
122. Woolfson RG, Patel VC, Neild GH, Yellon DM. Inhibition of nitric oxide synthesis reduces infarct size by an adenosine-dependent mechanism. Circulation 1995;91: 1545–1551.
123. Wang P, Zweier JL. Ischemic preconditioning decreases nitric oxide (NO) formation and NO mediated injury in the postischemic heart. Circulation 1997;96(Suppl 1):72.
124. Zweier JL, Wang P, Samuilov A, Kuppusamy P. Enzyme-independent formation of nitric oxide in biological tissues. Nat Med 1995;1:804–809.
125. Vallance P, Leone A, Calver A, Collier J, Moncada S. Accumulation of an endogenous inhibitor of nitric oxide synthesis in chronic renal failure. Lancet 1992;339:572–575.
126. Lu HR, Remeysen P, De Clerck F. Does the antiarrhythmic effect of ischemic preconditioning in rats involve the L-arginine nitric oxide pathway. J Cardiovasc Pharmacol 1995;25:524–530.
127. Weselcouch EO, Baird AJ, Sleph P, Grover GJ. Inhibition of nitric oxide synthesis does not affect ischemic preconditioning in isolated perfused rat hearts. Am J Physiol Heart Circ Physiol 1995;268:H242–H249.
128. Nakano A, Liu GS, Heusch G, Downey JM, Cohen MV. Exogenous nitric oxide can trigger a preconditioned state through a free radical mechanism, but endogenous nitric oxide is not a trigger of classical ischemic preconditioning. J Mol Cell Cardiol 2000;32:1159–1167.
129. Tanaka M, Fujiwara H, Yamasaki K, Sasayama S. Superoxide dismutase and N-2-mercaptopropionyl glycine attenuate infarct size limitation effect of ischaemic preconditioning in the rabbit. Cardiovasc Res 1994;28:980–986.
130. Osada M, Sato T, Komori S, Tamura K. Protective effect of preconditioning on reperfusion induced ventricular arrhythmias of isolated rat hearts. Cardiovasc Res 1991;25:441–444.
131. Tritto I, D'Andrea D, Eramo N, Scognamiglio A, De Simone C, Violante A et al. Oxygen radicals can induce preconditioning in rabbit hearts. Circ Res 1997;80:743–748.
132. Liu Y, Sato T, O'Rourke B, Marban E. Mitochondrial ATP-dependent potassium channels: novel effectors of cardioprotection? Circulation 1998;97:2463–2469.

133. Pain T, Yang XM, Critz SD, Yue Y, Nakano A, Liu GS et al. Opening of mitochondrial K(ATP) channels triggers the preconditioned state by generating free radicals. Circ Res 2000;87:460–466.
134. Tosaki A, Cordis GA, Szerdahelyi P, Engelman RM, Das DK. Effects of preconditioning on reperfusion arrhythmias, myocardial functions, formation of free radicals, and ion shifts in isolated ischemic/reperfused rat hearts. J Cardiovasc Pharmacol 1994;23:365–373.
135. Iwamoto T, Miura T, Adachi T, Noto T, Ogawa T, Tsuchida et al. Myocardial infarct size-limiting effect of ischemic preconditioning was not attenuated by oxygen free-radical scavengers in the rabbit. Circulation 1991;83:1015–1022.
136. Omar BA, Hanson AK, Bose SK, McCord JM. Ischemic preconditioning is not mediated by free radicals in the isolated rabbit heart. Free Radic Biol Med 1991;11:517–520.
137. Altug S, Demiryurek AT, Ak D, Tungel M, Kanzik I. Contribution of peroxynitrite to the beneficial effects of preconditioning on ischaemia-induced arrhythmias in rat isolated hearts. Eur J Pharmacol 2001;415:239–246.
138. Altug S, Demiryurek AT, Kane KA, Kanzik I. Evidence for the involvement of peroxynitrite in ischaemic preconditioning in rat isolated hearts. Br J Pharmacol 2000;130:125–131.
139. Peng J, Xiao J, Ye F, Deng HW, Li YJ. Inhibition of cardiac tumor necrosis factor-alpha production by calcitonin gene-related peptide-mediated ischemic preconditioning in isolated rat hearts. Eur J Pharmacol 2000;407:303–308.
140. Li YJ, Xiao ZS, Peng CF, Deng HW. Calcitonin gene-related peptide-induced preconditioning protects against ischemia-reperfusion injury in isolated rat hearts. Eur J Pharmacol 1996;311:163–167.
141. Hu CP, Li YJ, Deng HW. The cardioprotective effects of nitroglycerin-induced preconditioning are mediated by calcitonin gene-related peptide. Eur J Pharmacol 1999;369:189–194.
142. Song QJ, Li YJ, Deng HW. Cardioprotective effect of bradykinin-induced preconditioning mediated by calcitonin gene-related peptide in isolated rat heart. Zhongguo Yao Li Xue Bao 1999;20:162–166.
143. Lu R, Hu CP, Deng HW, Li YJ. Calcitonin gene-related peptide-mediated ischemic preconditioning in the rat heart: influence of age. Regul Pept 2001;99:183–189.
144. Lu R, Hu CP, Peng J, Deng HW, Li YJ. Role of calcitonin gene-related peptide in ischaemic preconditioning in diabetic rat hearts. Clin Exp Pharmacol Physiol 2001;28:392–396.
145. Bolli R, Manchikalapudi S, Tang XL, Takano H, Qiu Y, Guo Y et al. The protective effect of late preconditioning against myocardial stunning in conscious rabbits is mediated by nitric oxide synthase. Evidence that nitric oxide acts both as a trigger and as a mediator of the late phase of ischemic preconditioning. Circ Res 1997;81:1094–1107.
146. Jones WK, Flaherty MP, Tang XL, Takano H, Qiu Y, Banerjee S et al. Ischemic preconditioning increases iNOS transcript levels in conscious rabbits via a nitric oxide-dependent mechanism. J Mol Cell Cardiol 1999;31:1469–1481.
147. Guo Y, Jones WK, Xuan YT, Tang XL, Bao W, Wu WJ et al. The late phase of ischemic preconditioning is abrogated by targeted disruption of the inducible NO synthase gene [see comments]. Proc Natl Acad Sci U S A 1999;96:11507–11512.
148. Imagawa J, Yellon DM, Baxter GF. Pharmacological evidence that inducible nitric oxide synthase is a mediator of delayed preconditioning. Br J Pharmacol 1999;126:701–708.
149. Kis A, Vegh A, Papp JG, Parratt JR. Repeated cardiac pacing extends the time during which canine hearts are protected against ischaemia-induced arrhythmias: role of nitric oxide. J Mol Cell Cardiol 1999;31:1229–1241.

150. Sun JZ, Tang XL, Park SW, Qiu YM, Turrens JF, Bolli R. Evidence for an essential role of reactive oxygen species in the genesis of late preconditioning against myocardial stunning in conscious pigs. J Clin Invest 1996;97:562–576.
151. Zhou XB, Zhai X, Ashraf M. Direct evidence that initial oxidative stress triggered by preconditioning contributes to second window of protection by endogenous antioxidant enzyme in myocytes. Circulation 1996;93:1177–1184.
152. Takano H, Tang XL, Qiu Y, Guo Y, French BA, Bolli R. Nitric oxide donors induce late preconditioning against myocardial stunning and infarction in conscious rabbits via an antioxidant-sensitive mechanism. Circ Res 1998;83:73–84.
153. Xuan YT, Tang XL, Qiu Y, Banerjee S, Takano H, Han H et al. Biphasic response of cardiac NO synthase isoforms to ischemic preconditioning in conscious rabbits. Am J Physiol Heart Circ Physiol 2000;279:H2360–H2371.
154. Bolli R. The late phase of preconditioning. Circ Res 2000;87:972–983.
155. Emani VR, Ockaili RA, Brown MH, Krottapalli K, Hess ML, Kukreja RC. Peroxynitrite directly triggers early and delayed preconditioning *in vivo* via opening of mitochondrial KATP channels in the rabbit heart. Circulation 1999;100:1–259. Abstract.
156. Peng J, Lu R, Ye F, Deng HW, Li YJ. The heme oxygenase-1 pathway is involved in calcitonin gene-related peptide-mediated delayed cardioprotection induced by monophosphoryl lipid A in rats. Regul Pept 2002;103:1–7.
157. Peng J, Lu R, Deng HW, Li YJ. Involvement of alpha-calcitonin gene-related peptide in monophosphoryl lipid A-induced delayed preconditioning in rat hearts. Eur J Pharmacol 2002;436:89–96.
158. Song QJ, Li YJ, Deng HW. Early and delayed cardioprotection by heat stress is mediated by calcitonin gene-related peptide. Naunyn Schmiedebergs Arch Pharmacol 1999;359:477–483.
159. Song QJ, Li YJ, Deng HW. Improvement of preservation with cardioplegia induced by heat stress is mediated by calcitonin gene-related peptide. Regul Pept 1999;79:141–145.
160. Li D, Li NS, Chen QQ, Guo R, Xu PS, Deng HW et al. Calcitonin gene-related peptide-mediated cardioprotection of postconditioning in isolated rat hearts. Regul Pept 2008;147:4–8.
161. Wang HC, Zhang HF, Guo WY, Su H, Zhang KR, Li QX et al. Hypoxic postconditioning enhances the survival and inhibits apoptosis of cardiomyocytes following reoxygenation: role of peroxynitrite formation. Apoptosis 2006;11:1453–1460.
162. Burley DS, Ferdinandy P, Baxter GF. Cyclic GMP and protein kinase-G in myocardial ischaemia-reperfusion: opportunities and obstacles for survival signaling. Br J Pharmacol 2007;152:855–869.

Neurogenic Inflammation in Health and Disease
Edited by Gábor Jancsó

Tachykinins and Neurogenic Inflammation at Visceral Level

RICCARDO PATACCHINI[1] and CARLO A. MAGGI[2]

[1] *Department of Pharmacology, Chiesi Pharmaceuticals SpA, Parma, Italy*
[2] *Pharmacology Department, Menarini Ricerche, Florence, Italy*

ABSTRACT

Tachykinin-containing projections of capsaicin-sensitive primary afferent neurons widely distribute to visceral organs of the respiratory, gastrointestinal, and genitourinary systems. At this level, tachykinins, along with calcitonin gene-related peptide and other neuropeptides, can be released by adequate stimuli having pathological rather than physiological relevance. Such stimuli may represent chemical irritants to the respiratory tree or alimentary tract, or noxious agents secreted into the urinary system. Tachykinins produced in the visceral organs are part of the responses obtained at somatic level by antidromic activation of capsaicin-sensitive nerve terminals and collectively known as "neurogenic inflammation": that is, increase in vascular permeability and plasma protein leakage followed by edema. In addition, tachykinins produced at visceral level elicit smooth muscle contraction, neuronal stimulation, mucus secretion, and recruitment/activation of immune cells. The physiological significance of these effects is that to afford protection toward chemical irritants or noxious stimuli by facilitating their removal from the body. Nonetheless, a large body of preclinical evidence exists indicating that tachykinins participate to the genesis and/or maintenance of symptoms accompanying various human diseases, such as asthma/bronchial hyperreactivity, cystitis of various etiologies, inflammatory bowel diseases, and irritable bowel syndrome. Tachykinin receptor antagonists, which are proved to prevent/reduce tachykinin-mediated effects in experimental animal models of these pathologies, are expected to afford therapeutically relevant effects in humans.

1. INTRODUCTION

The term "neurogenic inflammation" is generally used to describe the inflammatory response evoked by certain neurotransmitters released from the peripheral terminals of a subpopulation of capsaicin-sensitive primary afferent neurons. Among the neurotransmitters which are stored and released from the nerve terminals of these sensory neurons, the tachykinins (substance P (SP) and neurokinin A (NKA)), and calcitonin gene-related peptide (CGRP) produce the majority of the neurogenic inflammatory effects. Such inflammation occurs in peripheral tissues, as documented by a large amount of studies appeared during the last 20 years. Here we review the available evidence on tissue distribution of tachykinin-containing neurons, the nature of the stimuli required for evoking the release of neuropeptides from capsaicin-sensitive nerve fibers, and the responses produced by tachykinins in organs belonging to the respiratory, gastrointentinal, and genitourinary tracts, with indication of the tachykinin receptors involved. The identification of the receptors mediating the effects of tachykinins has been made possible by the introduction of receptor-selective antagonists since the last 10 years [1–5] and by the use of tachykinin (NK_1) receptor knockout animals. A list of most important studies with these latter animals is summarized in Table 1. An attempt has also been made to distinguish between a

Table 1 List of studies investigating the role of tachykinins and tachykinin NK_1 receptors (NK_1R) in visceral neurogenic inflammation by the use of genetically modified mice

Knockout	Animal model/results	Reference
NK_1 R	*Chemically induced cystitis.* Cyclophosphamide fails to induce plasma protein extravasation in the urinary bladder of NK_1 receptor (–/–) mice. NK_1 receptor-deprived mice also lack to exhibit typical behavioral responses associated with painful cystitis.	[192]
NK_1 R	*Antigen-induced cystitis.* NK_1 receptor (–/–) mice are protected from plasma extravasation, edema, and migration of neutrophils elicited by intravesical instillation of dinitrophenol 4 (DNP_4)-ovalbumin. NK_1 receptor-deprived mice still exhibit antigen-induced mast cells degranulation.	[193]
Preprotachykinin A gene (PPT-A)	*Chemically induced cystitis.* Irritation of lower urinary tract (by intravesical acetic acid) in SP/NKA null mice induces a lower number (compared to normal animals) of *fos*-positive neurons in spinal cord areas in which bladder afferents terminate. Mutant mice also exhibit higher percentage of urinary retention and overflow incontinence during cystometograms.	[251]
NK_1 R	*Chemically induced colitis.* Intracolon instillation of acetic acid or capsaicin fails to induce tissue edema, acute (cardiovascular) reflex responses, and behavioral spontaneous reactions to acuté visceral stimulus in NK_1 receptor (–/–) mice.	[192]
NK_1 R	*Clostridium difficile-induced enteritis.* NK_1 receptor (–/–) mice are protected from secretory and inflammatory reactions and epithelial damage caused by intraluminal *C. difficile* toxin A administration.	[235]
Neutral endopeptidase (NEP)	*Chemically induced cholitis.* NEP (–/–) mice undergo fourfold higher basal protein plasma extravasation in the colon, and a more severe inflammatory reaction to DNBS-induced colitis. The exacerbated inflammation in knockout mice was prevented by administration of exogenous NEP or NK_1 receptor antagonist.	[252]
Neutral endopeptidase (NEP)	*C. difficile-induced enteritis.* NEP (–/–) mice undergo a twofold higher inflammatory responses caused by intraluminal *C. difficile* toxin A administration. The exacerbated inflammation in knockout mice was prevented by administration of exogenous NEP.	[253]
NK_1 R	*Caerulein-induced pancreatitis.* NK_1 receptor (–/–) mice undergo a reduced hyperlipasemia, neutrophil sequestration in the pancreas, and acinar cell necrosis induced by caerulein, in comparison with wild-type mice.	[237]
NK_1 R	*Diet-induced hemorragic pancreatitis.* Genetic deletion of the NK_1 receptor in (–/–) mice significantly improved survival (100% vs. 8%) and reduced pancreatic myeloperoxidase (MPO) and acinar cell necrosis, in comparison with wild-type mice.	[238]
NK_1 R	*Caerulein-induced pancreatitis.* NK_1 receptor (–/–) mice undergo a reduced pancreatic plasma protein extravasation caused by caerulein, while they are totally resistant to SP-induced plasma protein extravasation. Both caerulein-induced hyperamylasemia and pancreatic MPO are also reduced in knockout mice.	[239]

physiological (protective or trophic) role of the tachykinin-mediated inflammatory process from more pathological conditions, although sharp boundaries cannot be easily drawn at the present state of knowledge. Finally, the ability of tachykinin receptor antagonists to afford anti-inflammatory effects and their possible use in human pathological conditions are discussed.

2. SOURCE OF TACHYKININS IN PERIPHERAL TISSUES

Tachykinins are a family of neuropeptides, which share the common C-terminal sequence Phe–Xaa–Gly–Leu–Met–NH_2. This sequence is crucial for their interaction with specific receptors and for producing the most of their biological effects. Nevertheless, other domains in tachykinin peptidic sequence are responsible for the production of certain effects; e.g., mast cell degranulation is an effect encoded by the N-terminal sequence of SP [6]. SP, the first peptide of tachykinin family, was discovered by Von Euler and Gaddum as early as in 1931, whereas two other mammalian tachykinins, NKA and neurokinin B (NKB) were disclosed more than 50 years later [7]. SP, NKA, and NKB are the mammalian tachykinins that have actually reached an established status of neurotransmitters, whereas neuropeptide K and neuropeptide γ (two natural N-terminally extended forms of NKA) although being capable to produce full biological responses (e.g., Refs [8,9]) have not been proved to play a neurotransmitter role yet. Two genes have been identified in mammals encoding peptides of the tachykinin family: the preprotachykinin I gene (PPT-I or PPT-A) which encodes both SP and NKA, and the preprotachykinin II gene (PPT-II or PPT-B) which encodes NKB. The PPT-I gene has been detected in both central and peripheral nervous system, in enteric neurons of the gut, and in various cells of the immune system. In contrast, the PPT-II gene is expressed almost exclusively in the central nervous system [7,10–14]. The lack of PPT-II gene expression in peripheral tissues explains why SP and NKA are the only two tachykinins detected in appreciable amounts at this level. Recently, the molecular cloning of a further preprotachykinin gene (termed PPT-C) has been described from cDNA of murine hematopoietic cells [15]. The novel tachykinin encoded by PPT-C gene (termed hemokinin I) was suggested to act as an autocrine factor for the growth of hematopoietic cells. Zhang and coworkers [15] suggested also that hemokinin I would stimulate a specific receptor, distinct from any other known tachykinin receptors. However, further *in vitro* and *in vivo* experimentation has established that hemokinin I behaves as a tachykinin NK_1 receptor-preferring agonist [16]. In normal, noninflamed tissues, tachykinins originate from neuronal cells, being present in (i) peripheral endings of capsaicin-sensitive primary afferent neurons [14] and (ii) enteric neurons of both submucousal and myenteric plexuses innervating all layers of the gut [17]. In addition, in inflammatory processes certain activated immune cells synthesize and possibly release tachykinins, thus representing a nonneuronal source of releasable tachykinins in inflamed tissues [18–20].

3. TACHYKININ RECEPTORS

Following the discovery of NKA and NKB, evidence was provided for the existence of three distinct receptors (termed NK_1, NK_2, and NK_3) mediating the biological actions encoded by the common C-terminal sequence of tachykinins. SP, NKA, and NKB were initially considered to be the "preferred" ligands for the tachykinin NK_1, NK_2, and NK_3 receptors, respectively [6,14,21–23]. However, this concept has been challenged by radioligand binding and functional experiments that have shown both NKA and NKB to be as potent as SP at stimulating the tachykinin NK_1 receptor [24–26]. Moreover, all three mammalian tachykinins (SP, NKA, and NKB) possess the ability to act as full agonists at each one of the three receptors, albeit at different concentrations. The cloning of further tachykinin receptor showing a high degree of structural homology and a similar binding profile to the previously identified NK_3 receptor protein, has been reported [27,28]. However, a recent attempt to isolate this receptor (referred to as "NK_4" or "NK_{3B}" in literature) or identify the corresponding gene by the use of various molecular biological techniques in humans and other species, was unsuccessful [29]. The tachykinin NK_1, NK_2, and NK_3 receptors belong to the superfamily of rhodopsin-like, G-protein-coupled receptors with seven transmembrane spanning segments. Stimulation of phosphoinositol (Pi) breakdown is a common effector system coupled to each one of the three tachykinin receptors, although other intracellular pathways are triggered upon occupation of tachykinin receptors as well, e.g., leading to cAMP formation and increased arachidonic acid metabolism [30–34]. Marked species-related differences in pharmacology exist for all three tachykinin receptors, revealed by the use of selective tachykinin receptor antagonists [14,35].

4. TACHYKININ-MEDIATED NEUROGENIC INFLAMMATION IN THE AIRWAYS

4.1 Source of tachykinins in the airways

The presence of a sensory innervation of the airways, from which tachykinins and other neuropeptides are released from peripheral nerve terminals, has been originally described in the guinea pig [36–41]. Extrinsic nerve fibers containing tachykinins have also been detected in human airways [38,40,42], although their density is much lower compared to that found in other animal species [38,43]. SP and NKA have been found in bronchoalveolar lavage fluid of both asthmatic and nonasthmatic subjects [44,45]. Application of irritant stimuli thought to activate capsaicin-sensitive primary afferent neurons provokes production of secretions in which tachykinin-like immunoreactivity has been detected. It is worth noting that the amount of tachykinins present in secretions produced by inflamed airways has been found higher than in normal airways. For example, SP levels in bronchial and nasal lavage fluids of patients suffereing for allergic asthma challanged with allergens were higher than those measured in normal subjects [44,46]. Likewise, SP content measured in sputum induced by inhalation of hypertonic saline was higher in asthmatic patients and in patients suffering for chronic bronchitis than in healthy subjects [47]. A possible upregulation of tachykinin-containing nerves in asthmatics is suggested by certain studies (e.g., Ref. [19]), although other studies have failed to confirm this finding [48–50]. Lilly et al. [49], failed to detect any increase of SP-like immunoreactivity (SP-LI) in tracheal lung extracts from asthmatic subjects, and suggested that this could be due to augmented release of SP followed by degradation. An alternative explanation of the discrepancy between the reported increase of tachykinin-LI detected in secretions collected from patients affected by inflammatory airway diseases on one hand, and the apparent lack of overexpression of tachykinin-containing nerve profiles in airways of these latter subjects on the other, is that large amounts of tachykinins are released by cells belonging to the immune system, that are attracted in respiratory tissues by the inflammatory process [51]. This latter nonneuronal source of tachykinins is now considered to play a significant role in tachykinin-mediated inflammatory diseases of the airways (and possibly of other organs). In this regard, there is an accumulating evidence that immune cells such as eosinophils, lymphocytes, macrophages, and monocytes are actually able to synthesize and possibly release SP and other tachykinins in peripheral tissues [20,52]. For example, Tramontana and coworkers [53] have recently shown that the late phase of airway hyperreactivity in guinea pigs induced by i.v. Sephadex beads is accompanied by increased inflammatory cells (eosinophils and macrophages) and tachykinins (SP and NKA) in bronchoalveolar lavage fluid. As the late Sephadex-induced bronchial hyperreactivity was prevented by a tachykinin NK_2 receptor-selective antagonist (nepadutant [53]) but not by capsaicin pretreatment of the animals. It was concluded that tachykinins, released from inflammatory cells recruited/activated by Sephadex challenge, were responsible for the observed bronchial hypersensitivity. Furthermore, the observations that (i) autacoids such as histamine, serotonin, nerve growth factor, and cytokines (generated by mast cells, macrophages, eosinophils, and lymphocytes) are capable of stimulate capsaicin-sensitive sensory nerves terminals, which in turn release tachykinins (along with other neuropeptides) [20,54,55], and (ii) that tachykinins in turn possess the ability to recruit and stimulate the function of attracted/resident immune cells in inflamed tissues [14], add further complexity to the existing relationships among immune cells, tachykinins, and capsaicin-sensitive primary afferent neurons during the inflammatory process.

4.2 Stimuli capable of releasing tachykinins in the airways

A number of stimuli thought to provoke the release of tachykinins in the airways of several animal species have been described (Table 2). The biological responses produced by these stimuli are dependent on the presence of intact capsaicin-sensitive sensory innervation, and are mimicked by exogenous administration of tachykinins and are prevented by tachykinin receptor antagonists. Since most stimuli capable of releasing tachykinins from sensory nerves are of

Table 2 Examples of adequate stimuli producing tachykinin release from nerve endings of capsaicin-sensitive primary afferent neurons at visceral level

Respiratory system
Exposure to *allergens* (e.g., ovalbumin) of sensitized animals.
Inhalation of chemical–physical *irritants* such as capsaicin, ozone, acrolein, cigarette smoke, industrial pollutants (e.g., toluene diisocyanate), gastric juice, hydrocloric and other acids, hypertonic saline, and cold air.
Stimulation afforded by proinflammatory *autacoids* such as histamine, serotonin, bradykinin, cytokins, and nerve growth factor.
Genitourinary system
Exposure to *allergens* (e.g., ovalbumin) of sensitized animals.
Administration (e.g., intravesical instillation) of chemical *irritants* such as capsaicin and capsaicin-like substances, xylene, cyclophosphamide, acrolein, industrial pollutants (e.g., toluene diisocyanate), acidic media, and hypertonic media.
Stimulation afforded by *products of bacterial metabolism* (e.g., FMLP).
Stimulation afforded by *constituents of urine* such as potassium and bradykinin.
Application of *mechanical stimuli* (e.g., bladder distension, ureteral catheterization).
Gastrointestinal system
Exposure to *allergens* (e.g., ovalbumin) of sensitized animals.
Administration (e.g., intracolon instillation) of chemical *irritants* such as capsaicin, castor oil, dinitro- or trinitro-benzensulfonic acid, dextran sulfate, formalin, acetic acid, and other acidic media.
Bacterial infection (e.g., *Trichinella spiralis*, *Nyppostrongylus brasiliensis*, *Salmonella*)
Administration of *bacterial toxins* (e.g., *Clostridium difficile* toxin A)
Application of physical (e.g., γ-irradiation) or *mechanical* stimuli (intestinal wall distension); caveat: depending on the intensity of the stimulus either intrinsic neurons or extrinsic nerve fibers can be activated.

pathological rather than physiological relevance, the idea that tachykinins may participate in the genesis of asthma/bronchial hyperreactivity and rhinitis was advanced [56]. Several chemical/physical irritants have been shown to provoke asthmatic-like reactions in laboratory animals, such as capsaicin [36,57], toluene diisocyanate [58], ciragette smoke [59], gastric juice [60], hydochloric acid [60,61], ozone [62–64], and cold air [65]. Exposure of sensitized animals to antigens, like ovalbumin, represents a well-experimented challenge, which produces tachykinin-mediated bronchoconstriction, plasma protein extravasation, and bronchial hyperresponsiveness [66–68].

4.3 Biological effects of tachykinins in the airways

4.3.1 Airway smooth muscle contraction

One of the most impressive effects produced by tachykinins in the airways of certain mammalian species, humans included, is tracheal and bronchial smooth muscle contraction (Table 3). In guinea pig isolated airways, atropine-resistant contractions can be evoked by electrical nerve stimulation, which is a noncholinergic response and is prevented by capsaicin pretreatment [36,41]. It was also demonstrated that caspsaicin itself produces atropine-resistant contractions insensitive to tetrodotoxin [69]. These facts served as the basis for proving the existence of sensory nerves in the airways that work in an "efferent" mode. Subsequent *in vitro* and *in vivo* studies have demonstrated that both SP and NKA are released from capsaicin-sensitive primary afferents to produce tracheo-bronchial contraction and that tachykinin NK_2 and, at a lesser extent, NK_1 receptors mediate this effect in the guinea pig [14,70,71]. In rabbit isolated airways, capsaicin is able to produce a contractile response that is somewhat smaller than that obtained by activation of cholinergic nerves [72,73]. In this species, SP produces tracheo-bronchial contractions that are reduced by atropine [74], suggesting an additional facilitatory role of tachykinins on acetylcholine release from cholinergic nerve terminals. Subsequently, Belvisi et al. [75] have shown that the facilitatory role of tachykinins on cholinergic neurotransmission in the rabbit is mediated by both NK_1 and NK_2 receptors, whereas no facilitatory role could be demonstrated in human isolated bronchus [75]. The failure of tachykinin agonists to enhance cholinergic neurotransmission in the human bronchus, as reported by Belvisi and coworkers [75], is at variance with the results obtained by Black et al. [76] who observed a potentiation of electrically induced cholinergic contractions. However, this latter study was performed in the presence of K^+ channel blockade [76]. In human isolated bronchus, tachykinins behave as powerful spasmogens [77,78]. Early investigations performed with natural tachykinins (SP, NKA, and NKB) and certain receptor-selective agonists suggested that tachykinin-induced contractions of human isolated bronchial smooth muscle are exclusively mediated by receptors of the NK_2 type [79,80]. This conclusion was reinforced by the use of tachykinin receptor-selective antagonists [81,82]. However, more recent studies performed in human isolated small-size bronchi (about 1 mm diameter [83] and medium-size bronchi 92–95 mm diameter [84]) have proved a role also for the NK_1 receptor, by showing that tachykinin NK_1 receptor activation leads to prostanoid-dependent and prostanoid-independent contractions, in small-size vs. medium-size bronchi, respectively. Inhalation of SP or NKA has been reported by several groups to produce bronchoconstriction in asthmatic patients, NKA being more potent than SP [85–87], whereas inhalation of these tachykinins failed to change specific airway conductance in healthy nonsmoking volunteers [85]. Since the bronchoconstriction produced by inhaled tachykinins in asthmatics can be prevented by either nedocromil sodium or sodium

Table 3 Main pathophysiological effects produced by tachykinins at visceral level and tachykinin receptors involved

Effect	Receptor
Respiratory system	
Tracheo-bronchial smooth muscle contraction	NK_2/NK_1
Plasma protein extravasation	NK_1 (NK_2/NK_3)
Mucus secretion (from both upper and lower airways)	NK_1
Recruitment/stimulation of resident inflammatory cells	NK_1/NK_2 (NK_3)
Neuronal stimulation	NK_3
Genitorurinary system	
Smooth muscle contraction of urinary bladder, urethra (corpus spongiosum), corpus cavernosum, ureter, renal pelvis, vas deferens, myometrium	NK_2/NK_1
Smooth muscle relaxation of corpus spongiosum and cavernosum	NK_1
Plasma protein extravasation	NK_1
Recruitment/stimulation of resident inflammatory cells	NK_1/NK_2
Activation of chemically induced detrusor muscle hyperreflexia	NK_2
Gastrointestinal system	
Gastrointestinal smooth muscle contraction of both circular and longitudinal layers. Contribution to peristalsis (under physiological conditions) or to exaggerated intestinal motility (associated with inflammatory or infectious diseases: e.g., diarrhea)	$NK_1/NK_2/NK_3$
Gastrointestinal smooth muscle relaxation (afforded by inhibitory transmitters released from enteric neurons). Contribution to postsurgical intestinal atony	NK_1/NK_3
Neuro-neuronal communications in enteric plexuses, leading to release of excitatory (Ach, tachykinins) or inhibitory (NO, VIP) transmitters from enteric neurons	$NK_3/NK_1/NK_2$
Water/ions secretion from intestinal epithelium	$NK_1/NK_2/NK_3$
Genesis and/or maintainance of symptoms associated with acute or chronic inflammatory diseases (Crohn's disease, ulcerative colitis, etc.) and pancreatitis.	NK_1
Pain arising from gastrointestinal system (visceral nociception)	NK_2 (NK_3)

Note: See the text for references.

cromoglycate [88,89] or partially prevented by zafirlukast [90], the involvement of other inflammatory mediators in the above responses has been postulated (see Ref. [90] for review).

4.3.2 Plasma protein extravasation

Increase in vascular permeability and plasma leakage are most peculiar signs of neurogenic inflammation occurring in the skin in response to antidromic activations of capsaicin-sensitive sensory nerves, as reported by the pioneering studies of Jancsò and coworkers (see Ref. [14] for review). In rat and guinea pig airways, administration of capsaicin and other irritants or exogenous tachykinins produce plasma protein extravasation, a response that is prevented by capsaicin pretreatment of the animals [14,91]. Neurogenic plasma leakage induced by capsaicin or SP has recently been demonstrated also in mouse airways [92] but only in the presence of inhibitors of peptidases. In contrast, direct evidence for plasma protein extravasation elicited by capsaicin in human airways has not been reported yet [93]. The use of tachykinin receptor-selective antagonists has allowed to identify the NK_1 receptor as the main mediator of plasma protein extravasation provoked in rat and guinea pig airways by tachykinin receptor-selective agonists, capsaicin, and other irritant agents [14,51,91,94,95]. In addition, tachykinin NK_2 receptors have been reported to mediate plasma extravasation in the lower respiratory tract (i.e., secondary bronchi and intraparenchymal airways) of the guinea pig [96,97]. Daoui et al. [98–100] have shown that the NK_3 receptor-selective antagonist SR 142801 prevents the potentiation of histamine-induced microvascular leakage produced by aerosol pretreatment with SP, citric acid, or NK_3 receptor-selective agonists in the guinea pig, thus suggesting a role also for the tachykinin NK_3 receptor in mediating neurogenic inflammatory responses in the airways.

4.3.3 Mucus secretion

Another component of the neurogenic inflammatory response observed in both upper and lower airways following sensory nerves stimulation is production of secretions. Evidence has been provided that also this effect is mediated by endogenous tachykinins released from sensory nerves upon application of irritant stimuli. For example, capsaicin inhaled by animals [101] or human volunteers (both healthy or suffering from vascular/allergic rhinitis) causes abundant nasal secretion accompanied by sneezing, pain [102,103], and neurogenic plasma protein extravasation [104]. This set of responses enlights a defensive role played by capsaicin-sensitive sensory nerves in the nasal mucosa against environmental inhaled irritant agents. The involvement of tachykinins in capsaicin-induced nasal responses is suggested by the observation that inhaled tachykinins [101,102] mimic the effects of capsaicin, and can be blocked by a tachykinin receptor antagonist [101]. In the lower respiratory tract, tachykinins are able to elicit mucus secretion from both submucous glands [105,106] and from goblet cells [107], via activation of the tachykinin NK_1 receptors [106,108].

4.3.4 Further effects of tachykinins in the airways

SP possesses the ability to degranulate mast cells, leading to release of reactive agents such as histamine, serotonin, prostaglandins, and others: this effect would be encoded by the N-terminal sequence of SP [109,110]. Furthermore, tachykinins possess the ability to attract and/or stimulate various resident inflammatory cells in the airways; this ability is nowadays regarded as an important step in development and maintenance of neurogenic inflammation [14,20,52,55]. In particular, tachykinins have been shown to enhance lymphocyte proliferation [111], stimulate alveolar macrophages [112] and monocytes [113], and recruit neutrophils into the airways [114]. Tachykinin NK_1 and NK_2 receptors have been indicated as mediators of both recruitment and activation of leukocytes in the airways [115]. This concept has recently been reinforced by experiments performed with tachykinin NK_1 (SR 140333) and NK_2 (SR 48968) receptor-selective antagonists given by aerosol to ovalbumin-sensitized conscious unrestrained guinea pigs, before ovalbumin challenge of the animals [116,117]. These authors reported that SR 140333 reduced allergen-induced infiltration of eosinophils, neutrophils, and lymphocytes (collected in bronchoalveolar lavage fluid) [116] while SR 48968 inhibited the allergen-induced infiltration of neutrophils and lymphocytes in guinea pig airways [117]. Additional evidence has recently been provided for a possible role played by the NK_3 receptor in inflammatory cell recruitment by the use of SR 142801 (NK_3 receptor-selective antagonist), which significantly reduced both neutrophils and eosinophils accumulation in ovalbumin-challenged mice [118].

4.4 Clinical trials with tachykinin antagonists in respiratory diseases

The available clinical studies reporting on effectiveness of tachykinin receptor antagonists in human pathological conditions of the airways, consist in few investigations in which these substances were used to prevent asthmatic reactions provoked by various irritating stimuli (Table 4). In the first of these reports, Ichinose and coworkers [119] claimed the effectiveness of the dual (NK_1/NK_2) tachykinin receptor antagonist FK 224, in preventing both cough and bronchoconstriction elicited by inhaled bradykinin in asthmatic patients. However, further investigations with FK 224 failed to prove this compound capable to block NKA-induced bronchoconstriction in asthmatics [120] or to ameliorate symptoms/lung function in patients suffering of mild to moderate asthma [121]. The use of a tachykinin antagonist selective for the NK_1 receptor, the nonpeptide compound CP 99994, did not afford better results, since CP 99994 was proved unable to prevent bronchoconstriction and cough induced by hypertonic saline in mild asthmatic patients [122]. Likewise, the tachykinin NK_1 receptor-selective antagonist FK 888 failed to afford beneficial effects on exercise-induced bronchoconstriction in asthmatics, but significantly shortened the recovery times [123], suggesting a possible role of tachykinin NK_1 receptors in the late bronchoconstrictor response to exercise. On the other hand, SR 48968 (saredutant), a tachykinin NK_2 receptor-selective antagonist [124], has been the first tachykinin antagonist compound found active (by oral route) in reducing bronchoconstriction provoked by inhaled NKA, in mild asthmatic patients [125]. The capability of tachykinin NK_2 receptor-selective antagonists to prevent NKA-induced bronchoconstriction in asthmatics has been further demonstrated by the use of nepadutant [126], in a double-blind, placebo-controlled cross-over trial [127]. Nevertheless, a recent study [128] has reported that saredutant (100 mg once a day, *per os*) was unable to reduce airways hyperresponsiveness caused by adenosine 5′-monophosphate in allergic asthmatic patients. Encouraging results have recently been obtained by the use of a dual tachykinin NK_1/NK_2 receptor antagonist (DNK 333A, 100 mg *per os*) [129], in preventing bronchoconstriction evoked by inhaled NKA in mild asthmatic patients. Further trials are awaited to prove the efficacy of tachykinin receptor blockers to counteract bronchoconstriction and inflammation occurring in asthma, bronchial hyperactivity, and other inflammatory diseases of the airways.

Table 4 Human visceral diseases in which tachykinins are thought to play a role during the genesis and/or in the maintenance

Respiratory system
Asthma
Bronchial hyperreactivity
Rhinitis
Genitourinary system
Cystitis of different etiologies
Gastrointestinal system
Inflammatory bowel diseases
Pancreatitis
Irritable bowel syndrome

Note: See the text for references.

5. TACHYKININ-MEDIATED NEUROGENIC INFLAMMATION IN THE GENITOURINARY TRACT

5.1 Source of tachykinins in the genitourinary tract

All the organs of upper and lower urinary tract and the genital organs of various species, including humans, are densely innervated by peripheral projections of capsaicin-sensitive primary afferent neurons, that represent the main if not the sole source of tachykinins in noninflamed tissues of this area [14,130]. In the urinary bladder of rats and guinea pigs, perivascular nerve fibers containing SP-LI and NKA-LI (along with CGRP-LI) enter the base and distribute throughout the detrusor muscle up to the dome: at this level, the density of these fibers is significantly less than that observed in the neck [131–134]. Systemic administration of capsaicin to rats and guinea pigs leads to quite complete loss of fibers containing SP- and CGRP-LI in both the bladder and urethra, thus proving their sensory origin [132,133,135]. The human urinary bladder bears tachykinin-containing nerve fibers which follow a similar distribution to that found in rodents, although their density and levels of extractable tachykinin-LI in human tissues is lower than that present in small rodents [14,136,137]. Interestingly, the number of fibers containing tachykinins, as well as tissue content of tachykinins, were increased by chronic inflammation of the urinary bladder or bladder dysfunction (outflow obstruction) in rats [138–140]. Tachykinin-containing nerve fibers were also increased in the bladder of patients affected by idiopathic micturition disorders such as interstitial cystitis [141,142]. Interstitial cystitis is further characterized by an increase in the expression of tachykinin (NK_1/NK_2) receptors [143,144]. Tachykinin-containing nerve fibers have been described in other human organs belonging to the genitourinary tract such as the urethra [145,146], renal pelvis, and ureter [147–149], uterus and Fallopian tubes [150,151], and erectile tissues of the penis [145,146,152]. The density is normally lower than that found in corresponding tissues of other species.

5.2 Stimuli capable of releasing tachykinins in the genitourinary tract

As in the airways, several stimuli of different nature (chemical, physical, or mechanical) of pathological rather than physiological relevance, have been shown to produce neurogenic inflammatory responses in the genitourinary tract of several animal species (Table 2). These inflammatory responses are dependent on the presence of intact capsaicin-sensitive sensory innervation, and are mediated by tachykinin receptors as shown by the use of either tachykinin receptor agonists, which mimic these effects, or by tachykinin receptor antagonists which can prevent them. In addition, the development of tachykinin NK_1 receptor knockout mice has shown that these animals are insensitive/less sensitive to these stimuli (Table 1). Among the stimuli described to induce tachykinin-mediated inflammatory responses, there are chemical irritants such as capsaicin and capsaicin congeners [153–155], acrolein [156], xylene [157], and cyclophosphamide [158]. A local release of sensory neuropeptides, producing inflammatory effects in the mammalian urinary bladder and urethra, has been reported in response to low pH media [159], hypertonic media [160], formyl-methionyl-leucyl-phenylalanine (FMLP) (a product of bacterial metabolism [161]), and certain constituents of urine such as potassium [162] and bradykinin [163]. Examples of mechanical stimuli capable of producing responses through activation of capsaicin-sensitive sensory nerves are bladder distension (muscle stretch) [164] and ureteral catheterization [165] (see Refs [14,167] for reviews).

5.3 Biological effects of tachykinins in the genitourinary tract

5.3.1 Smooth muscle contraction

Smooth muscle contraction is one of the most investigated effects produced by tachykinins throughout the genitourinary tract (Table 3). The urinary bladder detrusor muscle and urethra of all mammalian species thus far examined respond with a contraction to exogenously applied tachykinins, which stimulate NK_1 or NK_2 receptors or both, depending on the species considered (see Ref. [14] for review). In humans, tachykinin-induced urinary bladder contraction is exclusively mediated by the NK_2 receptor [167], as in the human urethra [168,169] and human ureter [170]. In the urinary bladder of small rodents, capsaicin elicits contractile responses through release of endogenous tachykinins from sensory nerves [154,171], while it does not produce any motor effect in bladders from other species, humans included [172]. In the rat bladder, capsaicin-sensitive sensory nerves can also be activated by application of electrical stimuli of sufficient long duration [173]. In the spontaneously active guinea pig renal pelvis, both inotropic and chronotropic effects can be obtained by administration of exogenous tachykinins, as well as by administration of capsaicin or application of electrical nerve stimulation: these latter treatments being reported to induce release of endogenous tachykinins [174]. Exogenously applied tachykinins also elicit rhythmic phasic contractions in the quiescent ureter of several species, including humans [170]. However, unlike the renal pelvis, administration of capsaicin or application of electrical field stimulation to the ureter do not mimic the spasmogenic effect of tachykinins. Rather, in the presence of induced ureteral motility, either treatments produce inhibitory effects [14,175]. This is thought to be due to the higher sensitivity of the ureter (compared to the renal pelvis) to the inhibitory action of CGRP, that is normally coreleased with tachykinins in these organs [14,175]. On this basis, Santicioli and Maggi [175] have suggested that the pathophysiological significance of sensory nerve activation in the pyeloureteral system is to facilitate the removal of chemical irritants, or noxious stimuli, from the urinary tract. For example, bacterial chemotactic peptides generated during infections and other harmful constituents or urine may surve as irritants (whose diffusion to the urothelium may be caused by tissue lesions produced by stones). This goal would be made possible by the predominant ability of the released tachykinins to increase motor activity in the renal pelvis and by the concomitant ability of CGRP to inhibit latent pacemakers in the ureter; two effects which ensure unidiretional peristalsis of urine from kidney to the bladder [175].

Tachykinins have been shown to produce motor effects in smooth muscles of various genital organs. For example, in the human corpus spongiosum urethrae and corpus cavernosum, tachykinins produce contraction via NK_2 receptors but, in the same tissues, NK_1 receptor-mediated relaxations of precontracted tissues can be evoked by tachykinins, through formation of endogenous NO of nonneurogenic origin (possibly released from endothelial cells) [176]. In the vas deferens of various species, tachykinins exert a facilitatory effect on twitch contractions elicited by electrical stimulation [177–179]. Nevertheless, capsaicin produces inhibition of twitch contractions of the electrically driven vas deferens; an effect due to a higher sensitivity of the organ to CGRP over the coreleased tachykinins (see Ref. [14] for discussion). Tachykinins produce excitatory motor responses in the isolated myometrium from nonpregnant rats and mice (see Ref. [180] for review). Recently it has been shown that tachykinins elicit NK_2 receptor-mediated contractions of human myometrium obtained from pregnant women near term [181]; this finding suggests a possible role of tachykinins in modulating uterine contractility in women at term [180,181].

Exogenously administered tachykinins, besides stimulating ureteral peristalsis, are able to activate the micturition reflex in anesthetized animals (see Refs [166,175] for reviews). Thus, it could be hypothesized that endogenous tachykinins, released into the bladder wall, take part into the voiding mechanism of the urinary bladder. However, Lecci and coworkers [164,182] provided evidence against a peripheral role of tachykinins in initiating volume-evoked micturition reflex in rats, by showing that peripherally acting tachykinin NK_1 and/or NK_2 receptor selective antagonists are unable to modify the urodynamic parameters of distension-induced micturition, under physiological conditions. Rather, endogenous tachykinins acting via NK_2 receptors have been shown to contribute to detrusor hyperreflexia generated by noxious stimuli, as proved by the ability of tachykinin NK_2 (but not NK_1) receptor-selective antagonists to correct altered urodynamic parameters (micturition frequency and bladder tone) following intravesical administration of irritant agents in rats (such as capsaicin, prostaglandin E_2, xylene, and *Escherichia coli* lipopolysaccharide) [183–187]. It may be speculated that endogenous tachykinins mediate chemically induced hyperreflexia of the detrusor muscle by directly stimulating afferent nerve fibers in the mucosal layers of the bladder, on which specific (NK_2) receptors are expressed. Alternatively, tachykinins might act indirectly by stimulating generation of other local transmitters such as prostanoids (see Ref. [188] for discussion).

5.3.2 *Plasma protein extravasation*

As observed in the airways, plasma protein extravasation can be produced in an acute inflammatory response by exogenously applied tachykinins in organs of the urinary tract, or by agents that stimulate the release of tachykinins from sensory nerves, such as capsaicin [94], xylene [189], cyclophosphamide [156], or antigens in sensitized animals [190]. It has been reported that the intensity of tachykinin-induced plasma protein extravasation decreases from the ureters and urethra to the bladder: at this level it is greater in the base than in the dome (see Ref. [14] for review). As in the airways, tachykinin-mediated plasma protein extravasation in the urinary tract requires activation of tachykinin receptors of the NK_1 type. The involvement of tachykinin NK_1 receptors in mediating this effect has been proved since the introduction of the first receptor-selective agonists and antagonists [94]. Recently, the involvement of the NK_1 receptor has been confirmed by the failure of cyclophosphamide either to produce plasma protein extravasation in the urinary bladder or NK_1 receptor knockout mice, or to induce pain-related behavioral responses in these genetically modified animals [191] (Table 1). Likewise, Saban et al. [192] reported that NK_1 receptor knockout mice are protected from plasma extravasation, bladder edema, and migration of neutrophils in a model of antigen-induced cystitis (Table 1). However, mice lacking the NK_1 receptor exhibited a similar bladder mast cell degranulation in response to antigen challenge, as did wild-type mice [192]. This latter finding is in keeping with previous observations showing that SP-induced plasma protein extravasation (in rat bladder) can be reduced by indomethacin or compound 48/80 [157]. Thus, the evidence so far collected indicates that party of SP-induced plasma protein extravasation is an effect independent from NK_1 receptor activation, possibly due to mast cell degranulation and release of prostanoids.

5.4 Pathophysiological significance of tachykinin-induced neurogenic inflammation in the genitourinarty tract

As outlined in the above sections, the physiological significance of the acute effects produced by tachykinins, released in the genitourinary tract along with CGRP upon stimulation of capsaicin-primary afferent nerve terminals, seems to be that to afford protection by facilitating the removal of the noxae (Table 4). On the other hand, the bulk of data collected in *in vitro* and in *in vivo* studies on animal models of urinary diseases, suggests that tachykinins could be involved in producing symptoms accompanying cystitis of different etiologies. Nonetheless, the therapeutical potential of tachykinin receptor antagonists in human diseases of the urinary tract has not been investigated yet, despite these latter compounds have been proved effective in the preclinical trials.

6. TACHYKININ-MEDIATED NEUROGENIC INFLAMMATION IN THE GASTROINTESTINAL TRACT

6.1 Source of tachykinins in the gastrointestinal tract

Alike respiratory and genitourinary systems, the quantitatively most abundant source of tachykinins in the gastrointestinal tract is represented by intrinsic (enteric) neurons, localized in the myenteric and submucosal plexuses and projecting to all tissue layers of the gastrointestinal channel, including circular and longitudinal smooth muscle layers. Spinal and vagal primary afferent neurons and immune cells contribute the rest of tachykinin content in gastrointestinal organs. Unlike tachykinin-containing peripheral fibers of primary afferent neurons, tachykinin-containing enteric neurons are insensitive to the excitatory and neurotoxin actions of capsaicin [14,17,193]. In keeping with the reported distribution of tachykinins in the gut, the levels of SP-LI in tissue extracts from the gastrointestinal tract do not change significantly in capsaicin-pretreated animals, compared to controls [194]. Likewise, tachykinin-immunoreactive neural network innervating the common bile duct of the guinea pig is not modified by capsaicin pretreatment of the animals [195].

6.2 Stimuli capable of releasing tachykinins in the gastrointestinal tract

Tachykinins stored in primary afferent nerves can be released by stimuli, mainly or irritative nature, similar to those provoking their release in the airways and genitourinary tracts (Table 2). The most specific substance capable of stimulating the primary afferent neurons, capsaicin, has been reported to induce inflammatory responses (plasma extravasation) in the gastrointestinal tract, with the small intestine being more sensitive than stomach or colon [196]. The administration of antigens (e.g., ovalbumin), bacterial toxins (e.g., *Clostridium difficile* toxin A), acids, and other irritating substances (e.g., castor oil), provokes the release of tachykinins and other neuropeptides from extrinsic primary afferent neurons in experimental animal models, which contribute to the establishment of pathological-like conditions, as documented by a large number of studies (see Ref. [197] for review). On the other hand, tachykinins stored in enteric neurons can be released by mechanical (i.e., intestinal wall distension) or chemical (synaptic input) stimuli having a physiological relevance (see Refs [14,17] for reviews).

6.3 Biological effects of tachykinins in the gastrointestinal tract having physiological/trophic relevance

Alike the organs belonging to respiratory or genitourinary tract, all gastrointestinal smooth muscles are contracted by tachykinins (Table 3). Most of tachykinin-induced contractile effects originate from direct stimulation of tachykinin NK_1 and/or NK_2 receptors present on smooth muscle cells/cells of Cajal of the circular and longitudinal muscle layers [14,17]. In addition, stimulation of tachykinin receptors (mainly of the NK_3 type) present on gastrointestinal neurons may induce indirect contractile responses mediated by either acetylcholine [198] and/or tachykinins themselves [199,200]. Most of spasmogenic effects produced by tachykinins in human gastrointestinal organs such as the ileum [200], colon [201], and esophagus [202] are mediated by NK_2 receptors, while NK_1 receptors have been reported to contribute tachykinin-induced contraction in the circular muscle of the human ileum [200]. Several *in vitro* studies were aimed at investigating the endogenous mediators of both excitatory and inhibitory neurotransmission in the human intestine [203–205], biliary tract [206], and esophagus [207]. Tachykinins have been identified as major noncholinergic excitatory neurotransmitters at gastrointestinal level. This raised the question whether tachykinins contribute to (gastro) intestinal peristalsis under physiological conditions. In this regard, *in vivo* studies in which intestinal contractions were evoked by mechanical stimuli (e.g., intestinal wall distension) have shown that endogenous tachykinins are actually involved in these effects (see Ref. [208] for review). However, it should be noted that the contribution of endogenous tachykinins to intestinal peristalsis is hardly appreciable under normal condition, unless muscarinic receptors are blocked. Giuliani et al. [209,210] have shown that in atropine-pretreated animals, distension-evoked intestinal reflex contractions can be elicited and that tachykinin NK_1 and NK_2 receptor-selective antagonists inhibit these atropine-resistant responses, thus proving the involvement of endogenous tachykinins. Likewise, Lecci et al. [211], studying colonic propulsive activity in anesthetized atropine-pretreated guinea pigs, and Holzer et al. [212], studying *in vitro* peristalsis of isolated segments of guinea pig small intestine, arrived at similar conclusions. The former studies also proved that tachykinins probably do not play a major role in intestinal motor activity under normal conditions, due to the modest/null ability of tachykinin receptor antagonists to modify normal peristalsis. This concept has been recently reinforced by studies performed in unanesthetized dogs [213] and in healthy volunteers [214], in which the tachykinin NK_2 receptor-selective antagonist nepadutant [126] did not affect gastrointestinal normal motor activity *per se*. Nevertheless, Tonini and coworkers [215] have shown that simultaneous administration of three tachykinin antagonists, each one selective for the NK_1, NK_2, or NK_3 receptor type, produced a 50% inhibition of velocity of propulsion of a balloon inserted in isolated segments of guinea pig distal colon. This study indicates that for a conclusive statement on the contribution of tachykinins to normal peristalsis, further studies in which all tachykinin receptors are simultaneously blocked are required.

Another effect of tachykinins of possible physiological/trophic significance is regulation (enhancement) of ion and fluid secretions from both small and large intestines. At this level, tachykinins would act as endogenous modulators by stimulating secretomotor neurons, which in turn release acetylcholine and other noncholinergic transmitters, which are the final mediators of ion and fluid secretion (see Refs [14,216] for reviews). All three tachykinin receptor types seem to be involved in mediating the facilitatory effects of tachykinins on intestinal secretions [217]. In this regard, it should be mentioned that in the rat colon mucosa the tachykinin NK_2 receptors mediating increases in short-circuit current (I_{SC}) responses have been recently found functionally distinguishable from those mediating contractions in the muscularis mucosae [218]. Thus, tachykinin-containing primary afferent neurons could play a homeostatic role in the intestine as well as in the stomach. At this level, the neuropeptide CGRP (coreleased from sensory nerve terminals with tachykinins) has been reported to play a trophic role by increasing blood flow in the gastric mucosa to facilitate removal of the noxae (see Ref. [14] for discussion).

6.4 Biological effects of tachykinins in the gastrointestinal tract having pathological relevance

6.4.1 Abnormal motor activity

While the role of tachykinins in contributing to physiological peristalsis is still matter of debate, less doubts exist on their contribution to exaggerated intestinal motility, associated with various inflammatory and infectious intestinal diseases [197] (Table 3). For example, Croci et al. [219] have clearly shown that tachykinin NK_2 (and partially NK_1) receptor antagonists may prevent increased fecal excretion and colonic giant contraction in castor oil-induced diarrhea in rats, without producing constipation. Carini and coworkers [220] have recently contributed to the identification of the site of action of tachykinins producing exaggerated intestinal motility, by showing that tachykinins, released by chemical irritation in the rat colonic mucosa, increase cholinergic excitatory neurotransmission by stimulating NK_2 receptors possibly located on cholinergic motoneurons. Besides producing excitatory motor responses of pathophysiological significance, tachykinins have also been reported to produce inhibitory motor effects following stimulation of tachykinin NK_1, NK_2, or NK_3 receptors [208]. Stimulation of tachykinin (NK_2) receptors present on enteric inhibitory neurons, which in turn release nitric oxide, has been proposed as possible mechanisms of tachykinin-induced inhibitory motor effects at the intestinal level [216,221,222]. In this regard, the observed reduction of the duration of experimental postsurgical ileus in capsaicin-treated rats [223] has provided evidence for a role of capsaicin-sensitive primary afferents in postsurgical intestinal atony. Tolouse et al. [224] have recently shown that a tachykinin NK_2 receptor-selective antagonist (nepadutant [126]) significantly shortens surgery-induced inhibition of intestinal motility (ileus) in the rat, thus proving the involvement of tachykinins (via NK_2 receptors) in this model and suggesting a possible use of tachykinin NK_2 receptor antagonists in postoperative ileus.

6.4.2 Inflammatory gastrointestinal diseases

Tachykinins are thought to be most important mediators involved in the genesis and/or maintenance of several inflammatory gastrointestinal diseases (Table 4). This concept arises from the dramatic changes of both tachykinin content and tachykinin receptor expression measured in inflamed tissues obtained from either animals, or from patients suffering from inflammatory bowel diseases. Several animal models of ileocolitis are presently available, in which inflammatory reactions are produced by administration of chemical irritants (e.g., trinitrobenzensulfonic acid (TNBS), dextran sulfate, formalin), γ-irradiations, parasitic – and bacterial infections (e.g., *Trichinella spiralis*, *Salmonella*, *Nyppostrongylus brasiliensis*) or toxins (e.g., *C. difficile* toxin A and *T. spiralis* toxin A). While conflicting results have been reported on the variations of tachykinin content measured in tissue specimens obtained from patients affected by Crohn's disease or ulcerative colitis [197], a more precise picture arises from the animal studies. Thus, in the rat model of TNBS-induced colitis tachykinin levels (immunoreactivity) decreased in the early phase of the injury (possibly due to their massive release caused by the noxae), which was followed by return to basal values and/or increase of tachykinin mRNAs transcription and content [225–227]. In the same animal model, a significant decrease of tachykinin NK_1 and NK_2 receptor mRNAs has been reported, as an early consequence of the inflammatory reaction [227,228]. The involvement of endogenous tachykinins in mediating hypersecretory and inflammatory reactions of the gut induced by these agents has been further supported by demonstration that tachykinin NK_1 receptor antagonists exert protective effects in several models of intestinal injury/inflammation [188,216,229]. As an example, Stucchi and coworkers [230] have recently reported that chronic administration of the tachykinin NK_1 receptor antagonist CP 96345 significantly attenuated colonic inflammation and oxidative stress in an animal model (dextran sulfate-induced colitis in the rat), which resembles chronic ulcerative colitis in humans. In rats treated with intraluminal *C. difficile* toxin A tachykinins, released from capsaicin-sensitive primary afferents, greatly stimulate NK_1 receptors present on enteric neurons, and subsequently massive internalization (endocytosis) takes place. This latter phenomenon is absent in rats that are functionally deprived of sensory fibers (by neonatal capsaicin administration), or in rats pretreated with a tachykinin NK_1 receptor antagonist [231]. However, in chronic inflammatory bowel diseases, tachykinin NK_1 receptors have been found markedly upregulated, compared to their density measured in the intestine of normal subjects. Upregulation involves either sites where NK_1 receptors are normally present (e.g., vasculature, epithelial cells), or even sites where they are normally absent (ectopic expression: e.g., in intestinal lymphoid aggregates) [197,232]. Recently, Renzi and coworkers [233] have shown that, in addition to NK_1, also NK_2 receptor expression is markedly increased on inflammatory cells (eosinophils) of the lamina propria of patients with Crohn's disease or ulcerative colitis, suggesting their possible involvement in the above pathologies. The role of the tachykinin NK_1 receptor in mediating inflammatory reactions of the gut has recently been supported by the use of tachykinin NK_1 receptor-knockout mice, which showed reduced intestinal symptoms and tissue damage produced by *C. difficile* toxin A compared to control animals [234] (Table 1).

Mounting evidence has been accumulating in recent years showing that tachykinins and tachykinin NK_1 receptors take part in the genesis/maintenance of pancreatitis. Nerve fibers containing SP-LI have been found in the pancreas of several species including humans [14]. Furthermore, increased expression of these fibers (putatively of sensory origin) in the pancreas of patients suffering from chronic pancreatitis had been reported [235]. More recently, Bhatia et al. [236] have shown that both SP levels, and pancreatic acinar cell expression of NK_1 receptors, were increased in caerulein-induced pancreatitis in mice. Direct evidence for the involvement of tachykinins in pancreatitis has recently been provided by the use of mice genetically deprived of the NK_1 receptor (Table 1). Thus, NK_1 receptor (–/–) animals underwent a significantly lower degree of tissue damage associated with either caerulein-induced pancreatitis [236] or with diet-induced hemorrhagic pancreatitis [237], than normal animals. Genetic deletion of the NK1 receptor also improved survival (100% vs. 8% at 120 h) of (–/–) mice in a lethal model of hemorrhagic pancreatitis [237]. Grady and coworkers [238] provided further evidence by clearly demonstrating that SP or caerulein are able to produce pancreatic plasma extravasation in rats and mice, which resulted in pancreatic edema. This observation was considered to have clinical significance in the development of acute pancreatitis [238]. These effects were blocked by a tachykinin NK_1 receptor antagonist in rats and completely absent in NK_1 receptor knockout mice [238].

6.4.3 Visceral nociception

Much experimental evidence has been accumulating in the past few years, obtained by the use of tachykinin receptor-selective antagonists in various animal models of visceral nociception, showing that tachykinins mediate pain arising from the gastrointestinal system. In particular, tachykinin NK_2 receptor antagonists have been found effective in reducing reflex responses (e.g., abdominal contractions) caused by painful stimuli like rectal wall distension

[239] or intraperitoneal administration of acetic acid [240]. Moreover, tachykinin NK_2 receptor antagonists (SR 48968 or nepadutant) have been proved effective in animal models of visceral hyperalgesia induced by inflammation (e.g., caused by nematode infection [241]) or stress [242]. In particular, the results obtained by Tolouse et al. [242], showing the ability of nepadutant to inhibit rectal hypersensitive responses in rats pretreated with TNBS or previously subjected to restraint, suggest that the tachykinin NK_2 receptor is a main target mediating visceral allodynia/hyperalgesia. The role of tachykinin NK_2 receptors if further supported by the studies of Kiss et al. [243], who showed that the increased expression of either *c-fos* or *c-jun* proto-oncogene markers in spinal cord and dorsal root ganglia neurons of rats pretreated with TNBS is prevented by nepadutant. Laird and coworkers [244] also found that nepadutant was capable of preventing the hypersensitivity of single spinal cord neurons responding to colorectal distension or pelvic nerve stimulation in rats pretreated with intracolonic acetic acid. In addition, Laird and coworkers [191] have shown that intracolonic instillation of acetic acid or capsaicin fails to produce either acute (cardiovascular) responses or primary hyperalgesia. Also the NK_3 receptor (located at peripheral or/and central level) could play a role in visceral hyperalgesia, due to the reported effectiveness of either intraperitoneal [245] or intrathecal [246] administration of SR 142801 (a tachykinin NK_3 receptor-selective antagonist) in reducing reflex abdominal contractions elicited by various nociceptive stimuli.

6.5 Clinical perspectives for tachykinin receptor antagonists in gastrointestinal diseases

Although the precise site at which endogenous tachykinins act to mediate visceral pain is still matter of debate [247], tachykinin receptor antagonists have been proposed for the treatment of functional gastrointestinal disorders characterized by visceral pain such as irritable bowel syndrome [248,249] (Table 4). In this regard, it should be mentioned that a NK_2 receptor antagonist (nepadutant) has been proved effective in preventing the increase of gastrointestinal motility provoked by systemic administration of NKA, in healthy volunteers [214]. Interestingly, the action of NKA on gastrointestinal motility was pain, accompanied by a number of adverse effects, part of which resembling symptoms characterizing the irritable bowel syndrome, like borborygmi, abdominal pain, headache, nausea, vomiting, and flush (of the face or whole body). All NKA-induced adverse effects were absent in the group of subjects receiving nepadutant, except for the flush, thought to be due to cutaneous NK_1 receptor-mediated vasodilatation [214]. Inflammatory bowel diseases, in which tachykinins via NK_1 receptors have been clearly shown to play a key role in preclinical studies, are further possible targets for tachykinin NK_1 receptor-blocking drugs. Nevertheless, no clinical trials have been reported yet on patients affected by gastrointestinal diseases.

7. CONCLUSIONS

In conclusion, numerous studies have been performed to investigate the pathophysiological role of tachykinins at the visceral level. The available evidence indicates that these neuropeptides are released, along with other neurotransmitters, from peripheral endings of capsaicin-sensitive primary afferent neurons by irritant/noxious stimuli that penetrate the body through the respiratory tree, the alimentary channel, or the urinary tract. The inflammatory reaction triggered by the released tachykinins is aimed at the neutralization/removal of the noxae as documented by the reported ability of tachykinins to increase ureteral and gastrointestinal peristalsis, increase bronchial and intestinal secretions, induce plasma protein extravasation, and attract/activate cells belonging to the immune system. These latter cells, in turn, have been reported to synthesize and possibly release tachykinins in inflamed tissues. In addition, tachykinins released from intrinsic capsaicin-insensitive neurons of the gastrointestinal tract have been proposed to play a physiological/trophic role at this level, although their relevance seems to be lower as compared to acetylcholine and other neurotransmitters. On the other hand, a large body of preclinical evidence indicates that tachykinins participate in the genesis and/or maintenance of symptoms accompanying various human diseases, such as asthma/bronchial hyperreactivity, rhinitis, cystitis of various etiologies, inflammatory bowel diseases, and irritable bowel syndrome. Although very few clinical trials have been reported in literature on the efficacy of tachykinin receptor antagonists in peripheral pathological conditions, their ability to prevent/reduce tachykinin-mediated effects in experimental animal models of these diseases justify the expectation that tachykinin receptor-based drugs will be therapeutically useful in patients.

REFERENCES

1. Quartara L, Rovero P, Maggi CA. Peptide-based tachykinin NK_2 receptor antagonists. Med Res Rev 1995;15:139–155.
2. Quartara L, Maggi CA. The tachykinin NK-1 receptor. Part I: ligands and mechanisms of cellular activation. Neuropeptides 1997;31:537–553.
3. Quartara L, Maggi CA. The tachykinin NK-1 receptor. Part II: distribution and pathophysiological roles. Neuropeptides 1998;32:1–49.
4. Lowe JA III. Nonpeptide tachykinin antagonists: medicinal chemistry and molecular biology. Med Res Rev 1996;16:527–545.
5. Leroy V, Mauser P, Gao Z, Peet NP. Neurokinin receptor antagonists. Expert Opin Investig Drugs 2000;9:735–746.
6. Maggi CA, Patacchini R, Rovero P, Giachetti A. Tachykinin receptors and receptor antagonists. J Auton Pharmacol 1993;13:23–93.
7. Maggio JE. Tachykinins. Annu Rev Neurosci 1998;11:13–28.
8. Takeda Y, Krause JE. Neuropepide K potently stimulates salivary gland secretion and potentiates substance P-induced salivation. Proc Natl Acad Sci U S A 1989;86:392–396.
9. Takeda Y, Krause JE. Γ-preprotachykinin-(72-94)-peptide amide potentiates substance P-induced salivation. Eur J Pharmacol 1989;161:267–271.
10. Nawa H, Hirose T, Takashima H, Inayama S, Nakanishi S. Neucleotide sequences of cloned cDNAs for two types of bovine substance P precursor. Nature 1983;306:32–36.
11. Nawa H, Kotani H, Nakanishi S. Tissue specific generation of two preprotachykinin mRNAs from one gene by alternative RNA splicing. Nature 1983;312:729–734.
12. Krause JE, Chirgwin JM, Carter MS, Xu ZS, Hershey AD. Three rat preprotachykinin mRNAs encode the neuropeptides substance P and neurokinin A. Proc Natl Acad Sci U S A 1987;84:811–885.
13. Nakanishi S. Substance P precursor and kininogen: their structures, gene organization and regulation. Physiol Rev 1987;67:1117–1172.
14. Maggi CA. Tachykinins and calcitonin gene-related peptide (CGRP) as co-transmitters released from peripheral endings of sensory nerves. Prog Neurobiol 1995;45:1–98.

15. Zhang Y, Lu L, Furlonger C, Wu GE, Paige CJ. Hemokinin is a hematopoietic-specific tachykinin that regulates B lymphopoiesis. Nat Immunol 2000;1:392–397.
16. Bellucci F, Carini F, Catalani C, Cucchi P, Lecci A, Meini S et al. Pharmacological profile of the novel mammalian tachykinin, hemokinin I. Br J Pharmacol 2002;135:266–274.
17. Holzer P, Holzer-Petsche U. Tachykinins in the gut. Part I. Expression, release and motor function. Pharmacol Ther 1997;73:173–217.
18. De Giorgio R, Tazzari PL, Barbara G, Stanghellini V, Corinaldesi R. Detection of substance P immunoreactivity in human peripheral lymphocytes. J Neuroimmunol 1998;82:175–181.
19. Lai JP, Douglas SD, Ho WZ. Human lymphocytes express SP and its receptor. J Neuroimmunol 1998;86:80–86.
20. Maggi CA. The effects of tachykinins on inflammatory and immune cells. Regul Pept 1997;70:75–90.
21. Regoli D, Drapeau G, Dion S, D'Orleans-Juste P. Receptors for substance P and related neurokinins. Pharmacology 1989;38:1–15.
22. Guard S, Watson SP. Tachykinin receptor types: classification and membrane signalling mechanisms. Neurochem Int 1991;18:149–165.
23. Mussap CJ, Geraghty DP, Burcher E. Tachykinin receptors: a radioligand binding perspective. J Neurochem 1993;6:1987–2009.
24. Hastrup H, Schwartz TX. Septide and neurokinin A are high affinity ligands on the NK_1 receptor: evidence from homologous versus heterologous binding analysis. FEBS Lett 1996;399:264–266.
25. Maggi CA, Schwartz TW. The dual nature of the tachykinin NK_1 receptor. Trends Pharmacol Sci 1997;18:351–355.
26. Bremer AA, Tansky MF, Wu M, Boyd ND, Leeman SE. Direct evidence for the interaction of neurokinin A with the tachykinin NK1 receptor in tissue. Eur J Pharmacol 2001;423:143–147.
27. Donaldson LF, Haskell CA, Hanley MR. Functional characterization by heterologous expression of a novel cloned tachykinin peptide receptor. Biochem J 1996;320:1–5.
28. Krause JE, Staveteig PT, Mentzer JN, Schmidt SK, Tucker JB, Brodbeck RM et al. Functional expression of a novel human neurokinin-3 receptor homolog that binds (3H)senktide and (125-I-MePhe-7) neurokinin B, and is responsive to tachykinin peptide agonists. Proc Natl Acad Sci U S A 1997;94:310–315.
29. Sarau HM, Mooney JL, Schmidt DB, Foley JJ, Buckley PT, Giardina GAM et al. Evidence that the proposed novel human "neurokinin-4" receptor is pharmacologically similar to the human neurokinin-3 receptor but is not of human origin. Mol Pharmacol 2000;58:552–559.
30. Henderson AK, Lai J, Buck SHJ, Fujiwara Y, Singh G, Yamamura MS et al. A cloned NK2 receptor mediates phosphatidylinositol hydrolysis in a transfected murine fibroblast. Life Sci 1990;47:PL7–PL12.
31. Eistetter HR, Church DJ, Mills A, Godfrey PP, Capponi AM, Brewster R et al. Recombinant bovine NK_2 receptor stably expressed in CHO cells couples to multiple signal transduction pathways. Cell Regul 1991;2:767–779.
32. Eistetter HR, Mills A, Arkinstall SJ. Signal transduction mechanisms of recombinant bovine NK2 receptor stably expressed in baby hamster kidney cells. J Cell Biochem 1993;52:84–91.
33. Arkinstall S, Emergy I, Church D, Chollet A, Kawashima E. Ca influx and protein kinase Ca activation mediate arachidonic acid mobilization by the human NK2 receptor expressed in CHO cells. FEBS Lett 1994;338:75–80.

34. Catalioto R-M, Cucchi P, Renzetti AR, Criscuoli M, Maggi CA. Independent coupling of the human tachykinin NK_2 receptor to phospholipases C and A_2 in transfected Chinese hamster ovary cells. Naunyn Schmied Arch Pharmacol 1998;358:395–403.
35. Patacchini R, Maggi CA. Tachykinin receptors and receptor subtypes. Arch Int Pharm Ther 1995;329:161–184.
36. Lundbert JM, Saria A. Bronchial smooth muscle contraction induced by stimulation of capsaicin-sensitive sensory neurons. Acta Physiol Scand 1982;116:473–476.
37. Lundberg JM, Brodin E, Saria A. Effects and distribution of vagal capsaicin-sensitive SP neurons with special reference to the trachea and lungs. Acta Physiol Scand 1983;119:243–252.
38. Lundberg JM, Hokfelt T, Martling CR, Saria A, Cuello C. Substance P immunoreactive sensory nerves in the lower respiratory tract of various mammals including man. Cell Tissue Res 1984;235:251–261.
39. Lundblad L, Lundberg JM, Brodin E, Anggard A. Origin and distribution of capsaicin-sensitive SP-immunoreactive nerves in the nasal mucosa. Acta Otolaryngol (Stockh) 1983;96:485–493.
40. Martling CR, Saria A, Fischer JA, Hokfelt T, Lundberg JM. CGRP and the lung: neuronal coexistence with SP, release by capsaicin and vasodilatory effects. Regul Pept 1988;20:125–139.
41. Szolcsanyi J, Barthò L. Capsaicin-sensitive non-cholinergic excitatory innervation of the guinea-pig tracheobronchial smooth muscle. Neurosci Lett 1982;34:247–250.
42. Luts A, Uddman R, Alm P, Basterra J, Sundler P. Peptide-containing nerve fibres in human airways: distribution and coexistence pattern. Int Arch Immunol 1993;101:52–60.
43. Bowden J, Gibbins IL. Relative density of substance P-immunoreactive nerve fibres in the tracheal epithelium of range of species. FASEB J 1992;6:1276A.
44. Nieber K, Baumgarten CR, Rathsack R, Furkert J, Oehme P, Kunkel G. Substance P and β-endorphine-like immunoreactivity in lavage fluids of subjects with and without allergic asthma. J Allergy Clin Immunol 1992;90:646–652.
45. Heaney LG, Cross LJM, McGarvey LPA, Buchanan KD, Ennis M, Shaw C. Neurokinin A is the predominant tachykinin in human bronchoalveolar lavage fluid in normal and asthmatic subjects. Thorax 1998;53:357–362.
46. Nieber K, Baumgarten C, Witzel A, Rathsack R, Oehme P, Brunee T et al. The possible role of SP in the allergic reaction based on two different provocation models. Int Arch Allergy Appl Immunol 1991;94:334–338.
47. Tomaki M, Ichinose M, Miura M, Hirayama Y, Yamauchi H, Nakajima N et al. Elevated substance P content in induced sputum from patients with asthma and patients with chronic bronchitis. Am J Respir Crit Care Med 1995;151:613–617.
48. Howarth PH, Djukanovic R, Wilson JW, Holgate ST, Springall DR, Polak JM. Mucosal nerves in endobronchial biopsies in asthma and non-asthma. Int Arch Allergy Clin Immunol 1991;94:330–333.
49. Lilly CM, Bai TR, Shore SA, Hall AE, Drazen JM. Neuropeptide content of lungs from asthmatic and nonasthmatic patients. Am J Respir Crit Care Med 1995;151:548–553.
50. Chanez P, Springall DR, Vignola A, Moradoghi-Hattvani A, Polak J, Godard P et al. Bronchial mucosal immunoreactivity of sensory neuropeptides in severe airway diseases. Am J Respir Crit Care Med 1998;158:985–990.
51. Joos GF, DeSwert KO, Pauwels RA. Airway inflammation and tachykinins: prospects for the development of tachykinin receptor antagonists. Eur J Pharmacol 2001;429: 230–250.

52. Joos GF, Pauwels RA. Role of tachykinins in asthma. Allergy 2000;55:321–337.
53. Tramontana M, Santicioli P, Giuliani S, Catalioto R-M, Lecci A, Carini F et al. Role of tachykinins in sephadex-induced airway hyperreactivity and inflammation in guinea-pigs. Eur J Pharmacol 2002;439:149–158.
54. Maggi CA. The pharmacology of the efferent function of sensory nerves. J Autonom Parmacol 1991;11:173–208.
55. Kraneveld AD, James DE, De Vries A, Nijkamp FP. Excitatory non-adrenergic, non-cholinergic neuropeptides: key players in asthma. Eur J Pharmacol 2000;405:113–129.
56. Maggi CA, Meli A. The sensory-efferent function of capsaicin-sensitive sensory neurons. Gen Pharmacol 1988;19:1–43.
57. Hsiue TR, Garland A, Ray DW, Hershenson MB, Leff AR, Solway J. Endogenous sensory neuropeptide release enhances nonspecific airway responsiveness in guinea-pigs. Am Rev Respir Dis 1992;146:148–153.
58. Thompson JE, Scypinski LA, Gordon T, Sheppard D. Tachykinins medicate the acute increase in airway responsiveness caused by toluene diisocyanate in guinea-pigs. Am Rev Respir Dis 1987;136:43–49.
59. Daffonchio L, Hernandez A, Gallico L, Omini C. Airway hyperreactivity induced by active cigarette smoke exposure in guinea-pigs: possible role of sensory neuropeptides. Pulmonary Pharmacol 1990;3:161–166.
60. Martling CR, Lundberg JM. Capsaicin-sensitive afferents contribute to acute airway edema following tracheal instillation of hydrochloric or gastric juice in the rat. Anesthesiology 1988;68:350–356.
61. Hamamoto J, Kohrogi H, Kawano O, Iwagoe H, Fujii K, Hirata N et al. Esophageal stimulation by hydrochloric acid causes neurogenic inflammation in the airways in guinea pigs. J Appl Physiol 1997;82:738–745.
62. Hazbun ME, Hamilton R, Holian A, Eschenbacher WL. Ozone-induced increases in substance P and 8-eip-protraglandin F2α in the airways of human subjects. Am J Respir Cell Mol Biol 1993;9:568–572.
63. Graham RM, Friedman M, Hoyle GW. Sensory nerves promote ozone-induced lung inflammation in mice. Am J Respir Crit Care Med 2001;164:307–313.
64. Fu L, Tkanko T, Idena H, Nishiyama H, Suzuki S, Okubo T et al. Tachykinins via Tachykinin NK_2 receptor activation mediate ozone-induced increase in the permeability of the tracheal mucosa in guinea-pigs. Br J Pharmacol 2002;135:1331–1335.
65. Yoshihara S, Geppetti P, Hara M, Linden A, Ricciardolo FL, Chan B et al. Cold-air induced bronchoconstriction is mediated by tachykinin and kinin release in guinea-pigs. Eur J Pharmacol 1996;296:291–296.
66. Saria A, Lundbert JM, Skofitsch G, Lembeck F. Vascular protein leakage in various tissues induced by SP, capsaicin, bradykinin, serotonin, histamine and by antigen challenge. Naunyn Schmiedebergs Arch Pharmacol 1983;324:212–218.
67. Lundberg JM, Alving K, Karlsson JA, Matran R, Nilsson G. Sensory neuropeptide involvement in animal models of airway irritation and of allergen-evoked asthma. Am Rev Respir Dis 1991;143:1429–1431.
68. Maghni K, Taha R, Afif W, Hamid Q, Martin JG. Dichotomy between neurokinin receptor actions in modulating allergic airway responses in an animal model of helper T cell type 2 cytokine-associated inflammation. Am J Respir Crit Care Med 2000;162:1068–1074.
69. Szolcsanyi J. Tetrodotoxin-resistant noncholinergic neurogenic contraction evoked by capsaicinoids and piperine on the guinea-pig trachea. Neurosci Lett 1983;42:83–88.

70. Maggi CA, Patacchini R, Quartara L, Rovero P, Santicioli P. Tachykinin receptors in the guinea-pig isolated bronchi. Eur J Pharmacol 1991;197:167–174.
71. Maggi CA, Patacchini R, Rovero P, Santicioli P. Tachykinin receptors and noncholinergic bronchoconstriction in the guinea-pig isolated bronchi. Am Rev Respir Dis 1991;144:363–367.
72. Spina D, McKenniff MU, Coyle AJ, Seeds EAM, Perretti F, Tramontana M et al. Effect of capsaicin on PAF-induced bronchial hyperresponsiveness and pulmonary cell accumulation in the rabbit. Br J Pharmacol 1991;103:1268–1275.
73. Riccio MM, Manzini S, Page CP. The effect of neonatal capsaicin on the development of bronchial hyperresponsiveness in allergic rabbits. Eur J Pharmacol 1993;232:89–97.
74. Tanaka DT, Grundstein NM. Effect of substance P on neurally-mediated contraction of rabbit airway smooth muscle. J Appl Physiol 1986;60:455–463.
75. Belvisi MG, Patacchini R, Barnes PJ, Maggi CA. Facilitatory effects of selective agonists for tachykinin receptors on cholinergic neurotransmission: evidence for species differences. Br J Pharmacol 1994;111:102–110.
76. Black JL, Johnson PRA, Alouan L, Armour CL. Neurokinin A with potassium channel blockade potentiates contraction to electrical stimulation in human bronchus. Eur J Pharmacol 1990;180:311–317.
77. Lundberg JM, Martling CR, Saria A. Susbtance P and capsaicin induced contraction of human bronchi. Acta Physiol Scand 1983;119:49–53.
78. Naline E, Devilier P, Drapeau G, Toty L, Bakdach H, Regoli D et al. Characterization of neurokinin effects and receptor selectivity in human isolated bronchi. Am Rev Respir Dis 1989;140:679–686.
79. Advenier C, Naline E, Drapeau G, Regoli D. Relative potencies of neurokinins in guinea-pig trachea and human bronchus. Eur J Pharmacol 1987;139:133–137.
80. Dion S, Rouissi N, Natel F, Drapeau G, Regoli D, Naline E et al. Receptors for neurokinins in human bronchus and urinary bladder are of the NK-2 type. Eur J Pharmacol 1990;178:215–219.
81. Ellis JL, Undem BJ, Kays JS, Ghanekar SV, Barthlow HG, Buckner CK. Pharmacological examination of receptors mediating contractile responses to tachykinins in airways isolated from human, guinea pig and hamster. J Pharmacol Exp Ther 1993;267:95–101.
82. Sheldrick RL, Rabe KF, Fischer A, Magnussen H, Coleman RA. Further evidence that tachykinin-induced contraction of human isolated bronchus is mediated only by NK_2-receptors. Neuropeptides 1995;29:281–292.
83. Naline E, Molimard M, Regoli D, Emonds-Alt X, Bellamy JF, Advenier C. Evidence for functional tachykinin NK_1 receptors on human isolated small bronchi. Am J Physiol 1996;271:L763–L767.
84. Amedesi S, Moreau J, Tognetto M, Springer J, Trevisani M, Naline E et al. NK_1 Receptor stimulation causes contraction and inositol phosphate increase in medium-size human isolated bronchi. Am J Respir Crit Care Med 2001;163:1206–1211.
85. Joos G, Pauwels R, Van Der Straeten M. Effect of inhaled substance P and neurokinin A on the airways of normal and asthmatic subjects. Thorax 1987;42:779–783.
86. Joos GF, Germonpré PR, Kips JC, Peleman RA, Pauwels RA. Sensory neuropeptides and the human lower airways: present state and future directions. Eur Respir J 1994;7:1161–1171.
87. Advenier C, Joos G, Molimard M, Lagente V, Pauwels R. Role of tachykinins as contractile agonists of human airways in asthma. Clin Exp Allergy 1999;29:579–584.

88. Crimi N, Palermo F, Olivieri R, Palermo B, Vancheri C, Polosa R et al. Effect of nedocromil on bronchospasm induced by inhalation of substance P in asthmatic subjects. Clin Allergy 1988;18:375–382.
89. Joos GF, Pauwels RA, Van Der Straeten ME. The effect of nedocromil sodium on the bronchoconstrictor effect of neurokinin A in subjects with asthma. J Allergy Clin Immunol 1989;83:663–668.
90. Joos G, Van de Velde V, Schelfhout V, Pauwels R. The leukotriene receptor antagonist zafirlukast inhibits neurokinin A-induced bronchoconstriction in patients with asthma. Am J Respir Crit Care Med 2001;163:A418.
91. Abelli L, Maggi CA, Rovero P, Del Bianco E, Regoli D, Drapeau G et al. Effect of synthetic tachykinin analogues on airway microvascular leakage in rats and guinea-pigs: evidence for the involvement on NK-1 receptors. J Auton Pharmacol 1991;11:267–275.
92. Baluk P, Thurston G, Murphy TJ, Bunnett NW, McDonald DM. Neurogenic plasma leakage in mouse airways. Br J Pharmacol 1999;126:522–528.
93. Greiff L, Svensson C, Andersson M, Persson CGA. Effects of topical capsaicin in seasonal rhinitis. Thorax 1995;50:225–229.
94. Eglezos A, Giuliani S, Viti G, Maggi CA. Direct evidence that capsaicin-induced plasma protein extravasation is mediated through tachykinin NK-1 receptors. Eur J Pharmacol 1991;209:277–279.
95. Bowden JJ, Garland AM, Baluk P, Lefevre P, Grady EF, Vigna SR et al. Direct observation of substance P-induced internalization of neurokinin 1 (NK1) receptors at sites of inflammation. Proc Natl Acad Sci U S A 1994;91:8964–8968.
96. Tousignant C, Chan CC, Guevremont D, Brideau C, Hale JJ, MacCoss M, Rodger IW. NK-2 receptor mediate plasma extravasation in guinea-pig lower airways. Br J Pharmacol 1993;108:383–396.
97. Savoie C, Chan CC, Tousignant C, Rodger IW. Involvement of NK1 and NK2 receptors in pulmonary responses elicited by NANC vagal stimulation in anesthetized guinea-pigs. Am Rev Respir Dis 1993;147:A708.
98. Daoui S, Cui YY, Lagente V, Emonds-Alt X, Advenier C. A tachykinin NK_3 receptor antagonist SR 142801 (Osanetant) prevents substance P-induced bronchial hyperreactivity in guinea-pigs. Pulm Pharmacol Ther 1997;10:261–270.
99. Daoui S, Cognon C, Naline E, Emonds-Alt X, Advenier C. Involvement of tachykinin NK_3 receptors in citric acid-induced cough and bronchial responses in guinea-pigs. Am J Respir Crit Care Med 1998;158:42–48.
100. Daoui S, Ahnaou A, Naline E, Emonds-Alt X, Lagente V, Advenier C. Tachykinin NK_3 receptor agonists induced microvascular leakage hypersensitivity in the guinea-pig airways. Eur J Pharmacol 2001;433:199–207.
101. Petersson G, Malm L, Ekman R, Hakanson R. Capsaicin evokes secretion of nasal fluid and depletes SP and CGRP from the nasal mucosa in the rat. Br J Pharmacol 1989;98:930–936.
102. Geppetti P, Fusco B, Marabini S, Maggi CA, Fanciullacci M, Sicuteri F. Secretion pain and sneezing induced by the application of capsaicin to the nasal mucosa in man. Br J Pharmacol 1988;93:509–514.
103. Stjarne P, Lundblad L, Lundbert JM, Angaard A. Capsaicin and nicotine-sensitive afferent neurons and nasal secretion in healthy human volunteers and in patients with vasomotor rhinitis. Br J Pharmacol 1989;96:693–701.

104. Sanico AM, Atsuta S, Proud D, Togias A. Plasma Extravasation through neuronal stimulation in human nasal mucosa in the setting of allergic rhinitis. J Appl Physiol 1998;84:537–543.
105. Coles SJ, Neill KH, Reid LM. Potent stimulation of glycoprotein secretion in canine trachea by SP. J Appl Physiol 1984;57:1323–1327.
106. Meini S, Mak J, Rohde JAL, Rogers DF. Tachykinin control of ferret airways: mucus secretion, bronchoconstriction and receptor mapping. Neuropeptides 1993;24: 81–89.
107. Kuo HP, Rohde JA, Tokuyama K, Barnes PJ, Rogers DF. Capsaicin and sensory neuropeptide stimulation of goblet cell secretion in guinea-pig trachea. J Physiol (Lond) 1990;431:629–641.
108. Geppetti P, Bertrand C, Bacci E, Snider RM, Maggi CA, Nadel JA. The NK-1 receptor antagonist CP 96,345 blocks SP evoked mucus secretion from the ferret trachea in vitro. Neuropeptides 1992;22:25.
109. Fewtrell CMS, Foreman JC, Jordan CC, Oehma P, Rener H, Stewart JM. The effect of substance P on histamine and 5-hydroxytryptamine release in the rat. J Physiol 1982;330:393–411.
110. Mousli M, Bueb JL, Bronner C, Rouot B, Landry Y. G-protein activation: a receptor-independent mode of action for cationic amphiphilic neuropeptides and venom peptides. Trends Pharmacol Sci 1992;11:358–362.
111. Payan DG, Brewster Dr, Goetzl EJ. Specific stimulation of human T lymphocytes by SP. J Immunol 1983;131:1613–1615.
112. Brunelleschi S, Vanni L, Ledda F, Giotti A, Maggi CA, Fantozzi R. Tachykinins activate guinea-pig alveolar macrophages: involvement of NK_2 and NK_1 receptors. Br J Pharmacol 1990;100:417–420.
113. Wagner F, Fink R, Hart R, Dancygier H. Substance P enhances interferon-g production by human peripheral blood mononuclear cells. Regul Pept 1987;19:355–364.
114. Hoshino H, Lotvall J, Skoogh BE, Linden A. Neutrophil recruitment by interleukin-17 into rat airways in vivo. Am J Respir Crit Care Med 1999;159:1423–1428.
115. Advenier C, Lagente V, Biochot E. The role of tachykinin receptor antagonists in the prevention of bronchial hyperresponsiveness, airway inflammation and cough. Eur Respir J 1997;10:1892–1906.
116. Schuiling M, Zuidhof AB, Meurs H, Zaagsma J. Role of tachykinin NK_2 receptor activation in the allergen-induced late asthmatic reaction, airway hyperreactivity and airway inflammatory cell influx in conscious unrestrained guinea-pigs. Br J Pharmacol 1999;127:1030–1038.
117. Schuiling M, Zuidhof AB, Zaagsma J, Meurs H. Involvement of tachykinin NK_1 receptor in the development of allergen-induced airway hyperreactivity and airway inflammation in conscious, unrestrained guinea-pigs. Am J. Respir Crit Care Med 1999;159:423–430.
118. Nenan S, Germain N, Legente V, Emonds-Alt X, Advenier C, Boichot E. Inhibition of inflammatory cell recruitment by the tachykinin NK_3-receptor antagonist, SR 142801, in a murine model of asthma. Eur J Pharmacol 2001;421:201–205.
119. Ichinose M, Nakajima N, Takahashi T, Tamauchi H, Inoue H, Takishima T. Protection against bradykinin-induced bronchoconstriction in asthmatic patients by neurokinin receptor antagonists. Lancet 1992;340:1248–1251.
120. Joos G, Van Schoor J, Kips JC, Pauwels RA. The effect of inhaled FK 224, a tachykinin NK-1 and NK-2 receptor antagonist on neurokinin A-induced bronchoconstriction in asthmatics. Am J Respir Crit Care Med 1996;153:1781–1784.

121. Lunde H, Hedner J, Svedmyr N. Lack of efficacy of 4 weeks treatment with the neurokinin receptor antagonist FK 224 in mild to moderate asthma. Eur Respir J 1994;7:15S.
122. Fahy JV, Wong HH, Geppetti P, Reis JM, Harris SC, Maclean DB et al. Effect of an NK1 receptor antagonist (CP-99,994) on hypertonic saline-induced bronchoconstriction and cough in male asthmatic subjects. Am J Respir Crit Care Med 1995;152:879–884.
123. Ichinose M, Miura M, Yamauchi H, Kageyama N, Tomaki M, Oyake T et al. A neurokinin 1-receptor antagonist improves exercise-induced airway narrowing in asthmatic patients. Am J Respir Crit Care Med 1996;153:936–941.
124. Edmond-Alt X, Vilain P, Goulaoic P, Van Broeck D, Advenier C, Naline E et al. A potent and selective non-peptide antagonist of the neurokinin A (NK2) receptor. Life Sci 1992;50:101–106.
125. Van Schoor V, Joos GF, Chasson BL, Brouard RJ, Pauwels RA. The effect of the NK2 tachykinin receptor antagonist SR 48968 (saredutant) on neurokinin A induced bronchoconstriction in asthmatics. Eur Respir J 1998;12:17–23.
126. Catalioto R-M, Criscuoli M, Cucchi P, Giachetti A, Giannotti D, Giuliani S et al. MEN 11420 (Nepadutant), a novel glycosylated bicyclic peptide tachykinin NK2 receptor antagonist. Br J Pharmacol 1998;123:1–91.
127. Joos G, Schelfhout V, Van De Velde V, Maggi CA, Pauwels RA. The effect of the tachykinin NK_2 receptor antagonist MEN 11420 (nepadutant) on neurokinin-A induced bronchoconstriction in patients with asthma. "Tachykinins 2000," La Grande Motte, France, October 17–20, 2000.
128. Kraan J, Vink-Klooster H, Postma DS. The NK-2 receptor antagonist SR 48968C does not improve adenosine hyperresponsiveness and airway obstruction in allergic asthma. Clin Exp Allergy 2001;31:274–278.
129. Joos G, Vincken W, Louis R, Wang J, Shaw M, Pauwels R. The effect on the dual NK1/NK2 tachykinin receptor antagonist DNK 333A on neurokinin A-induced bronchoconstriction in patients with asthma. Am J Respir Crit Care Med 2001;163:A628.
130. Maggi CA. The dual, sensory and 'efferent' function of the capsaicin-sensitive primary sensory neurons in the urinary bladder and urethra. In: Maggi CA, ed. Nervous Control of the Urogenital System, vol. 3. Chur, Switzerland: Harwood Academic Publisher, 1993; pp. 338–422.
131. Alm P, Alumets J, Brodin E, Hakanson R, Nilsson G, Sjoberg NO et al. Peptidergic (SP) nerves in the genitourinary tract. Neuroscience 1978;3:419–425.
132. Sharkey KA, Williams RG, Schultzberg M, Dockray GJ. Sensory SP innervation of the urinary bladder: possible site of action of capsaicin in causing urine retention. Neuroscience 1983;10:861–868.
133. Yokokawa K, Sakanaka M, Shiosaka S, Tohyama M, Shiotani Y, Sonoda T. Three dimensional distribution of SP-like immunoreactivity in the urinary bladder of the rat. J Neural Transmission 1985;63:209–222.
134. Tateishi K, Kishimoto S, Kobayashi H, Kobuke K, Matsuoka Y. Distribution and localization of neurokinin A-like immunoreactivity and neurokinin B-like immunoreactivity in rat peripheral tissue. Regul Pept 1990;30:193–200.
135. Su HC, Sharton J, Polak JM, Mulderry PK, Ghatei MA, Gibson SJ et al. CGRP immunoreactivity in afferent neurons supplying the urinary tract: combined retrograde tracing and immunohistochemistry. Neuroscience 1986;18:727–747.
136. Edyvane KA, Marshall VR. Neuropeptides in the human urinary tract. Neurourol Urodyn 1990;9:346–347.

137. Wakabayashi Y, Tomoyoshi T, Fujimiya M, Arai R, Maeda T. Substance P-containing axon terminals in the mucosa of the human urinary bladder: pre-embedding immunohistochemistry using cryostat sections for electron microscopy. Histochemistry 1993;100:401–407.
138. Andersson PO, Andersson K-E, Fahrenkrug J, Mattiasson A, Sjorgen C, Uvelius B. Contents and effects of substance P and vasoactive intestinal polypeptide in the bladder of rats with and without intravesical outflow obstruction. J Urol 1988;140:168–172.
139. Crowe R, Vale J, Troot K, Soedonio P, Robson T, Burnstock J. Radiation-induced changes in neuropeptides in the rat urinary bladder. J Urol 1996;156:2062–2066.
140. Callsen-Cencic P, Mense S. Expression of neuropeptides and nitric oxide synthase in neurons innervating the inflamed rat urinary bladder. J Auton Nerv Syst 1997;65:33–44.
141. Pang X, Marchand J, Sant GR, Kream RM, Theoharides TC. Increased number of substance P positive nerve fibres in interstitial cystitis. Br J Urol 1995;75:744–750.
142. Smet PJ, Moore KH, Jonavicious J. Distribution and co-localization of calcitonin gene-related peptide, tachykinins and vasoactive intestinal peptide in normal and idiopathic unstable human urinary bladder. Lab Invest 1997;77:37–49.
143. Marchand JE, Sant GR, Kream RM. Increased expression of substance P receptor-encoding mRNA in bladder biopsies from patients with interstitial cystitis. Br J Urol 1998;81:224–228.
144. Buffington CAT, Wolfe SA Jr. High affinity binding sites for [3H] substance P in urinary bladders of cats with interstitial cystitis. J Urol 1998;160:605–611.
145. Andersson K-E, Hedlund H, Mattiasson A, Sjorgen C, Sundler F. Relaxation of isolated human corpus spongiosum induced by vasoactive intestinal polypeptide, substance P, carbachol and electrical field stimulation. World J Urol 1983;1:203–208.
146. Hauser-Kronberg C, Hacker GW, Graf A-H, Mack D, Sundler F, Dietze O et al. Neuropeptides in the human penis: an immunohistochemical study. J Androl 1994;15:510–520.
147. Hua X-Y, Theodorsson-Norheim E, Lundbert JM, Kinn A-C, Hökfelt T, Cuello AC. Co-localization of tachykinins and calcitonin gene-related peptide in capsaicin-sensitive afferents in relation to motility effects on the human ureter in vitro. Neuroscience 1987;23:693–703.
148. Edyvane KA, Trussell DC, Jonavicius J, Henwood A, Marshall VR. Presence and regional variation in peptide-containing nerves in the human ureter. J Auton Nerv Syst 1992;39:127–137.
149. Edyvane KA, Smet PJ, Trussell DC, Jonavicius J, Marshall VR. Patterns of neuronal colocalisation of tyrosine hydroxylase, neuropeptide Y, vasoactive intestinal polypeptide, calcitonin gene-related peptide and substance P in human ureter. J Auton Nerv Syst 1994;48:241–255.
150. Reinecke M, Gauwerky JF, Schneider K. Peptidergic (NPY, NT, VIP, SP, CGRP) innervation of the functional systems of the uterus and fallopian tube in the human. Arch Gynecol Obstet 1989;245:399–401.
151. Heinrich D, Reinecke M, Forssmann WG. Peptidergic innervation of the human and guinea pig uterus. Arch Gynecol 1986;237:213–219.
152. Gu J, Polak M, Probert L, Islam N, Marangos PJ, Mina S et al. Peptidergic innervation of the human male genital tract. J Urol 1983;130:386–391.
153. Maggi CA, Giuliani S, Santicioli P, Abelli L, Geppetti P, Somma V et al. Species-related variations in the effects of capsaicin on urinary bladder functions: relation to bladder content of SP-LI. Naunyn Schmiedebergs Arch Pharmacol 1987;336:546–555.

154. Maggi CA, Patacchini R, Tramontana M, Amann R, Giuliani S, Santicioli P. Similarities and differences in the action of resininferatoxin and capsaicin on central and peripheral endings of primary sensory neurons. Neuroscience 1990;37:531–539.
155. Patacchini R, Maggi CA, Meli A. Capsaicin-like activity of some natural pungent substances on peripheral endings of visceral primary afferents. Naunyn Schmiedebergs Arch Pharmacol 1990;352:72–77.
156. Ahluwalia A, Maggi CA, Santicioli P, Lecci A, Giuliani S. Characterization of the capsaicin-sensitive component of cyclophosphamide induced inflammation in the rat urinary bladder. Br J Pharmacol 1994;111:1017–1022.
157. Costa M, Furness JB, Llewellyn-Smith IJ. Histochemistry of the enteric nervous system. In: Johnson LR, ed. Physiology of the Gastrointestinal Tract. New York: Raven Press, 1987; pp. 1–40.
158. Maggi CA, Lecci A, Santicioli P, Del Bianco E, Giuliani S. Cyclophosphamide cystitis in rats: involvement of capsaicin-sensitive primary afferents. J Auton Nerv Syst 1992;38:201–208.
159. Geppetti P, Tramontana M, Patacchini R, Del Bianco E, Santicioli P, Maggi CA. Neurochemical evidence for the activation of the efferent function of capsaicin-sensitive nerves by lowering of the pH in the guinea-pig urinary bladder. Neurosci Lett 1990;114:101–106.
160. Maggi CA, Abelli L, Giuliani S, Somma V, Furio M, Patacchini R et al. Motor and inflammatory effect of hyperosmolar solutions on the rat urinary bladder in relation to capsaicin-sensitive sensory nerves. Gen Pharmacol 1990;21:97–103.
161. Giuliani S, Santicioli P, Tramontana M, Geppetti P, Maggi CA. Peptide N-formyl-methionyl-leucyl-phenylalanine (FLMP) activates capsaicin-sensitive primary afferent nerves in guinea-pig atria and urinary bladder. Br J Pharmacol 1991;102:730–734.
162. Maggi CA, Santicioli P, Geppetti P, Parlani M, Astolfi M, Del Bianco E et al. The effect of calcium free medium and nifedipine on the release of substance P-like immunoreactivity and contractions induced by capsaicin in the isolated guinea-pig and rat bladder. Gen Pharmacol 1989;20:445–456.
163. Maggi CA, Patacchini R, Santicioli P, Geppetti P, Cecconi R, Giuliani S et al. Multiple mechanisms in the motor responses of the guinea-pig isolated urinary bladder to bradykinin. Br J Pharmacol 1989;98:619–629.
164. Lecci A, Giuliani S, Tramontana M, Santicioli P, Criscuoli M, Dion S et al. Bladder distension and activation of the efferent function of sensory fibres: similarities with the effect of capsaicin. Br J Pharmacol 1998;124:259–266.
165. Abelli L, Conte B, Somma V, Parlani M, Geppetti P, Maggi CA. Mechanical irritation induces neurogenic inflammation in the rat urethra. J Urol 1991;146:1624–1626.
166. Leci A, Maggi CA. Tachykinins as modulators of the micturition reflex in the central and peripheral nervous system. Regul Pept 2001;101:1–18.
167. Maggi CA, Santicioli P, Patacchini R, Cellerini M, Turini D, Barbanti G et al. Contractile response of the human isolated urinary bladder to neurokinins: involvement of NK_2 receptors. Eur J Pharmacol 1988;145:335–340.
168. Palea S, Corsi M, Artibani W, Ostardo E, Pietra C. Pharmacological characterization of tachykinin NK_2 receptors on isolated human urinary bladder prostatic urethra and prostate. J Pharmacol Exp Ther 1996;277:700–705.
169. Parlani M, Conte B, Majmone S, Maggi CA, Rovero P, Giachetti A. The contractile effect of tachykinins on human prostatic urethra: involvement of NK_2 receptors. J Urol 1990;144:1543–1545.

170. Patacchini R, Santicioli P, Zagorodnyuk V, Lazzeri M, Turini D, Maggi CA. Excitatory motor and electrical effects produced by tachykinins in the human and guinea-pig isolated ureter and guinea-pig renal pelvis. Br J Pharmacol 1998;125:987–996.
171. Maggi CA, Patacchini R, Santicioli P, Giuliani S. Tachykinin antagonists and capsaicin-induced contraction of the rat isolated urinary bladder: evidence for tachykinin-mediated cotransmission. Br J Pharmacol 1991;102:1535–1541.
172. Maggi CA, Patacchini R, Santicioli P, Turini D, Barbanti G, Beneforti P et al. Further studies on the motor response of the human isolated urinary bladder to tachykinins, capsaicin and electrical field stimulation. Gen Pharmacol 1989;20:663–669.
173. Meini S, Maggi CA. Evidence for capsaicin-sensitive, tachykinin mediated, component in the NANC contraction of the rat urinary bladder to nerve stimulation. Br J Pharmacol 1994;112:1123–1131.
174. Maggi CA, Patacchini R, Eglezos A, Quartara L, Giuliani S, Siachetti A. Tachykinin receptors in the guinea-pig renal pelvis: activation by exogenous and endogenous tachykinins. Br J Pharmacol 1992;107:27–33.
175. Santicioli P, Maggi CA. Myogenic and neurogenic factors in the control of pyeloureteral motility and ureteral peristalsis. Pharmacol Rev 1998;50:683–721.
176. Patacchini R, Barbagli G, Palminteri E, Lazzeri M, Turini D, Maggi CA. Tachykinin NK-1 and NK-2 receptors mediate inhibitory vs. excitatory motor responses in human isolated corpus cavernosum and spongiosum. Br J Pharmacol 2002;135:1351–1354.
177. Tousignant C, Dion S, Drapeau G, Regoli D. Characterization of pre- and postjunctional receptors for neurokinins and kinins in the rat vas deferens. Neuropeptides 1987;9:333–343.
178. Patacchini R, Maggi CA, Rovero P, Regoli D, Drapeau G, Meli A. Effect of thiorphan on tachykinin-induced potentiation of nerve-mediated contractions of the rat isolated vas deferens. J Pharmacol Exp Ther 1989;250:678–681.
179. Hall JM, Morton IKM. Novel selective agonists and antagonists confirm neurokinin NK-1 receptors in guinea-pig vas deferens. Br J Pharmacol 1991;102:511–517.
180. Patak EN, Pennefather JN, Story ME. Effects of tachykinins on uterine smooth muscle. Clin Exp Pharmacol Physiol 2000;27:922–927.
181. Patek EN, Ziccone S, Story ME, Fleming AJ, Lilley A, Pennefather JN. Activation of neurokinin NK_2 receptors by tachykinin peptides causes contraction of uterus in pregnant women near term. Mol Hum Reprod 2000;6:549–554.
182. Lecci A, Giuliani S, Patacchini R, Maggi CA. Evidence against a peripheral role of tachykinins in the initiation of micturition reflex in rats. J Pharmacol Exp Ther 1993;264:1327–1332.
183. Pietra C, Bettellini R, Hagan RM, Ward P, EcElroy A, Trist D. Effect of selective antagonists in tachykinin NK_1 and NK_2 receptors on xylene-induced cystitis in rats. Neuropeptides 1992;22:52.
184. Ishizuka O, Igawa Y, Mattiasson A, Andrsson K-E. Capsaicin-induced bladder hyperreactivity in normal conscious rats. J Urol 1994;152:525–530.
185. Ishizuka O, Mattiasson A, Andrsson K-E. Prostaglandin E2-induced bladder hyperreactivity in normal conscious rats: involvement of tachykinins. J Urol 1995;153:2034–2038.
186. Lecci A, Giuliani S, Tramontana M, Criscuoli M, Maggi CA. MEN 11420, a peptide tachykinin NK_2 receptor antagonist, reduces motor responses induced by intravesical administration of capsaicin in vivo. Naunyn Schmiedebergs Arch Pharmacol 1997;356:182–188.
187. Lecci A, Tramontana M, Giuliani S, Criscuoli M, Maggi CA. Effect of tachykinin NK_2 receptor blockade on detrusor hyperreflexia induced by bacterial toxin in rats. J Urol 1998;160:206–209.

188. Lecci A, Giuliani S, Tramontana M, Carini F, Maggi CA. Peripheral actions of tachykinins. Neuropeptides 2000;34:303–313.
189. Giuliani S, Santicioli P, Lippe IT, Lecci A, Maggi CA. Effect of bradykinin and tachykinin receptor antagonist on xylene-induced cystitis in rats. J Urol 1993;150:1014–1017.
190. Ahluwalia A, Giuliani S, Scotland R, Maggi CA. Ovalbumin-induced neurogenic inflammation in the bladder of sensitized rats. Br J Pharmacol 1998;124:190–196.
191. Laird JM, Olivar T, Roza C, De Filipe C, Hunt SP, Cervero F. Deficits in visceral pain and hyperalgesia of mice with a disruption of the tachykinin NK_1 receptor gene. Neuroscience 2000;98:345–352.
192. Saban R, Saban MR, Nguyen NB, Lu B, Gerard C, Gerard NP et al. Neurokinin-1 (NK-1) receptor is required in antigen-induced cystitis. Am J Pathol 2000;156:775–780.
193. Lecci A, Maggi CA. In: Geppetti P, Holzer P, eds. Neurogenic Inflammation. Boca Raton: CRC Press, 1996; pp. 201–209.
194. Holzer P, Gamse R, Lembeck F. Distribution of SP in the rat of effect of capsaicin pretreatment. Eur J Pharmacol 1980;61:303–307.
195. Patacchini R, Barthò L, DeGiorgia R, Lénard L, Stanghellini V, Barbara G et al. Involvement of endogenous tachykinins and CGRP in the motor responses produced by capsaicin in the guinea-pig common bile duct. Naunyn Schmiedebergs Arch Pharmacol 1999;360:344–353.
196. Sann H, Dux M, Schemann M, Jancsò G. Neurogenic inflammation in the gastrointestinal tract of the rat. Neurosci Lett 1996;219:147–150.
197. Holzer P. Tachykinins as targets of gastroenterological pharmacotherapy. Drugs News Perspect 1998;11:394–401.
198. Guard S, Watson SP. Evidence for neurokinin-3 receptor-mediated tachykinin release in the guinea-pig ileum. Eur J Pharmacol 1987;144:409–412.
199. Patacchini R, Holzer P, Maggi CA. Tachykinin autoreceptors in the gut. Trends Pharmacol Sci 2000;21:166.
200. Maggi CA, Patacchini R, Santicioli P, Giuliani S, Turini D, Barbanti G et al. Human isolated ileum: motor responses of the circular muscle to electrical field stimulation and exogenous neuropeptides. Naunyn Schmiedebergs Arch Pharmacol 1990;341:256–261.
201. Giuliani S, Barbanti G, Turini D, Quartara L, Rovero P, Digchetti A et al. NK_2 tachykinin receptors and contraction of the circular muscle of the human colon: characterization of the NK_2 receptor subtype. Eur J Pharmacol 1991;203:365–370.
202. Huber O, Bertrand C, Bunnett NW, Pellegrini CA, Nadel JA, Nakazato P et al. Tachykinins contract the circular muscle of the human esophageal body in vitro via NK_2 receptors. Gastroenterology 1993;105:981–987.
203. Maggi CA, Giuliani S, Patacchini R, Santicioli P, Theodorsson E, Barbanti G et al. Tachykinin antagonists inhibit nerve-mediated contractions in the circular muscle of the human ileum. Gastroenterology 1992;102:88–96.
204. Zagorodnyuk V, Santicioli P, Turini D, Maggi CA. Tachykinin NK_1 and NK_2 receptors mediate non-adrenergic non-cholinergic excitatory neuromuscular transmission in the human ileum. Neuropeptides 1997;31:265–271.
205. Patacchini R, Giuliani S, Turini A, Navarra G, Maggi CA. Effect of nepadutant at tachykinin NK_2 receptors in human intestine and urinary bladder. Eur J Pharmacol 2000;398:389–397.
206. Patacchini R, DeGiorgia R, Barthò L, Barbara G, Corinaldesi R, Maggi CA. Evidence that tachykinins are the main NANC excitatory neurotransmitters in the guinea-pig common bile duct. Br J Pharmacol 1998;124:1703–1711.

207. Krysiak PS, Preiksaitis HG. Tachykinins contribute to nerve-mediated contractions in the human esophagus. Gastroenterology 2001;120:39–48.
208. Maggi CA, Catalioto R-M, Criscuoli M, Cucchi P, Giuliani S, Lecci A et al. Tachykinin receptors and intestinal motility. Can J Physiol Pharmacol 1997;75:696–703.
209. Giuliani S, Lecci A, Giachetti A, Maggi CA. Tachykinins and reflexy-evoked atropine-resistant motility in the guinea-pig colon in vivo. J Pharmacol Exp Ther 1993;265: 1224–1231.
210. Giuliani S, Tramontana M, Lecci A, Maggi CA. Tachykinin receptors mediate atropine-resistant rat duodenal reflex contractions in vivo. Naunyn Schmiedebergs Arch Pharmacol 1996;354:327–335.
211. Lecci A, Giuliani S, Tramontana M, De Giorgio R, Maggi CA. The role of tachykinin NK_1 and NK_2 receptors in atropine-resistant colonic propulsion in anaesthetized guinea-pigs. Br J Pharmacol 1998;124:27–34.
212. Holzer P, Lippe IT, Heinemann A, Barthò L. Tachykinin NK_1 and NK_2 receptor-mediated control of peristaltic propulsion in the guinea-pig small intestine, in vitro. Neuropharmacol 1998;37:131–137.
213. Giuliani S, Guelfi M, Tolouse M, Bueno L, Lecci A, Tramontana M et al. Effect of tachykinin NK_2 receptor antagonist, nepadutant, on cardiovascular and gastrointestinal function in rats and dogs. Eur J Pharmacol 2001;415:61–71.
214. Lördal M, Navalesi G, Theodorsson E, Maggi CA, Hellström PM. A novel tachykinin NK_2 receptor antagonist prevents motility-stimulating effects of neurokinin A in small intestine. Br J Pharmacol 2001;134:215–223.
215. Tonini M, Spelta V, De Ponti F, De Giorgio R, D'Agostino G, Stanghellini V et al. Tachykinin-dependent and independent components of peristalsis in the guinea-pig isolated distal colon. Gastroenterology 2001;210:938–945.
216. Holzer P, Holzer-Petsche U. Tachykinins in the gut. Part II. Roles in neural excitation, secretion and inflammation. Pharmacol Ther 1997;73:219–263.
217. Cox HM, Tough IR, Grayson K, Yarrow S. Pharmacological characterization of neurokinin receptors mediating anion secretion in rat descending colon mucosa. Naunyn Schmiedebergs Arch Pharmacol 1993;348:172–177.
218. Patacchini R, Cox HM, Stahl S, Tough IR, Maggi CA. Tachykinin NK_2 receptor mediates contraction and ion transport in rat colon by different mechanisms. Eur J Pharmacol 2001;415:227–283.
219. Croci T, Landi M, Emonds-Alt X, Le Fur G, Maffrand JP, Manara L. Role of tachykinins in castor oil diarrhoea in rats. Br J Pharmacol 1997;121:375–380.
220. Carini F, Lecci A, Tramontana M, Giuliani S, Maggi CA. Tachykinin NK2 receptors and enhancement of cholinergic transmission in the inflamed rat colon: an in vivo motility study. Br J Pharmacol 2001;133:1107–1113.
221. Zagorodnyuk V, Maggi CA. Neuronaltachykinin NK_2 receptors mediate release of non-adrenergic noncholinergic inhibitory transmitters in the circular muscle of guinea-pig colon. Neuroscience 1995;69:643–650.
222. Lecci A, De Giorgio R, Barthò L, Sternini C, Tramontana M, Corinaldesi R et al. Tachykinin NK_1 receptor-mediated inhibitory responses in the guinea-pig small intestine. Neuropeptides 1999;33:91–97.
223. Riviere PJM, Pascaud X, Chevalier E, Junien JL. Fedotozine reversal of peritoneal-irritation-induced ileus in rats: possible peripheral action of sensory afferents. J Pharmacol Exp Ther 1994;270:846–850.

224. Tolouse M, Fioramonti J, Maggi CA, Bueno L. Role of NK_2 receptors in gastric barosensitivity and in experimental ileus in rats. Neurogastroenterol Motil 2001;13:45–53.
225. Renzi D, Tramontana M, Panerai C, Surrenti C, Evangelista S. Decrease of calcitonin-gene related peptide, but not vasoactive intestinal polypeptide and substance P in the TNB-induced experimental colitis in rats. Neuropeptides 1992;22:56–57.
226. Renzi D, Calabrò A, Panerai C, Tramontana M, Evangelista S, Milani S et al. Preprotachykinin mRNA expression in the colonic tissue during experimental colitis in rats. Digestion 1994;55:36.
227. Evangelista S, Maggi M, Renzetti AR. Downregulation of substance P receptors during colitis induced by trinitrobenzene sulfonic acid in rats. Neuropeptides 1996;30: 425–428.
228. Renzi D, Calabrò A, Panerai C, Tramontana M, Evangelista S, Surrenti C. NK_1 and NK_2 receptor gene expression during TNB-induced colitis in rats. Gut 1996;39:A137.
229. Evangelista S. Involvement of tachykinins in intestinal inflammation. Curr Pharm Des 2001;7:19–30.
230. Stucchi AF, Shofer S, Leeman S, Materne O, Beer E, McClung J et al. NK-1 antagonist reduces colonic inflammation and oxidative stress in dextran sulfate-induced colitis in rats. Am J Physiol Gastrointest Liver Physiol 2000;279:1298–1306.
231. Mentyh CR, Pappas TN, Lapp JA, Washington MK, Neville LM, Ghilardi JR et al. Substance P activation of enteric neurons in response to intraluminal Clostridium difficile toxin A in the rat ileum. Gastroenterology 1996;111: 1272–1280.
232. Mantyh PW, Catton MD, Boehmer CG, Welson ML, Passaro EP, Maggio JE et al. Receptors for sensory neuropeptides in human inflammatory diseases: implications for the effector role of sensory neurons. Peptides 1989;10:627–645.
233. Renzi D, Pellegrini B, Tonelli F, Surrenti C, Calabrò A. Substance P (NK1) and neurokinin A (NK2) receptor gene and protein expression in healthy and inflammed human intestine. Am J Pathol 2000;157:1511–1522.
234. Castagliuolo I, Riegler M, Pasha A, Nikulasson S, Lu B, Gerard C et al. Neurokinin-1 (NK-1) receptor is required in Clostridium difficile-induced enteritis. J Clin Invest. 1998;101:1547–1550.
235. Weihe E, Buchler M, Muller S, Friess H, Zentel HJ, Yanaihara N. Peptidergic innervation in chronic pancreatitis. In: Beger E, ed. Chronic Pancreatitis. Berlin: Springer Verlag, 1990; pp. 83–105.
236. Bhatia M, Saluka AK, Hofbauer B, Frossard JL, Lee HS, Castagliuolo I et al. Role of substance P and the neurokinin 1 receptor in acute pancreatitis and pancreatitis-associated lung injury. Proc Natl Acad Sci U S A 1998;95:4760–4765.
237. Maa J, Grady EF, Yoshimi SKI, Drasin TE, Kjm EH, Hutter MM et al. Substance P is a determinant of lethality in diet-induced hemorrhagic pancreatitis in mice. Surgery 2000;128:232–239.
238. Grady EF, Yoshimi SK, Maa J, Valeroso D, Vartanian RK, Rahim S et al. Substance P mediates inflammatory oedema in acute pancreatitis via activation of the neurokinin-1 receptor in rats and mice. Br J Pharmacol 2000;130:505–512.
239. Julia V, Morteau O, Bueno L. Involvement of neurokinin 1 and 2 receptors in viscerosensitive response to rectal distension in rats. Gastroenterology 1994;107:94–102.
240. Julia V, Bueno L. Tachykininergic mediation of viscerosensitive responses to acute inflammation in rats: role of CGRP. Am J Physiol 1997;272:G141–G146.
241. McLean PG, Picard C, Garcia-Villar R, Morè J, Fioramonti J, Bueno L. Effects of nematode infection on sensitivity to intestinal distension: role of tachykinin NK_2 receptors. Eur J Pharmacol 1997;337:279–282.

242. Tolouse M, Coelho AM, Fioramonti J, Lecci A, Maggi CA, Bueno L. Role of tachykinin NK_2 receptors in normal and altered rectal sensitivity in rats. Br J Pharmacol 2000;129: 193–199.
243. Kiss S, Lecci A, De Groat WC, Maggi CA, Birder LA. The effect of the NK2 receptor antagonist, MEN 11420, on proto-oncogene expression following experimental colitis. Soc Neurosci Abs 1999;25:411.
244. Laird MJ, Olivar Lopez-Garcia JA, Maggi CA, Cervero F. Responses of rat spinal neurons to distension of inflamed colon: role of tachykinin NK_2 receptors. Neuropharmacol 2001;40:696–701.
245. Julia V, Su X, Gueno L, Gebhart GF. Role of neurokinin 3 receptors on responses to colorectal distension in the rat: electrophysiological and behavioral studies. Gastroenterology 1999;116:1124–1131.
246. Kamp EH, Back DR, Gebhart GF. Combinations of neurokinin receptor antagonists reduce visceral hyperalgesia. J Pharmacol Exp Ther 2001;299:105–113.
247. Holzer P. Gastrointestinal afferents as targets of novel drugs for the treatment of functional bowel disorders and visceral pain. Eur J Pharmacol 2001;429:177–193.
248. Camileri M. Management of the irritable bowel syndrome. Gastroenterology 2001;120: 652–668.
249. Patacchini R, Maggi CA. Peripheral tachykinin receptors as targets for new drugs. Eur J Pharmacol 2001;429:13–21.
250. Kiss S, Yoshiyama M, Cao YQ, Basbaum AI, de Groat WC, Lecci A et al. Impaired response to chemical irritation of the urinary tract in mice with disruption of the preprotachykinin gene. Neurosci Lett 2001;313:57–60.
251. Sturaile S, Barbara G, Qiu B, Figini M, Geppetti P, Gerard N et al. Neutral endopeptidase (EC 3.4.24.11) terminates colitis by degrading substance P. Proc Natl Acad Sci U S A 1999;96:11653–11658.
252. Kirkwood KS, Bunnett NW, Maa J, Castagliolo I, Liu B, Gerard N et al. Deletion of neutral endopeptidase exacerbates intestinal inflammation induced by Clostridium difficile toxin A. Am J Physiol Gastrointest Liver Physiol 2001;281:G544–551.
253. Ollerenshaw SL, Jarvis D, Sullivan CE, Woolcock AJ. Substance P immunoreactive nerves in airways from asthmatics and nonasthmatics. Eur Respir J 1991;4:673–682.
254. Maggi CA, Abelli L, Giuliani S, Santicioli P, Geppetti P, Somma V et al. The contribution of sensory nerves to xylene-induced cystitis in rats. Neuroscience 1988;26:709–723.
255. Giuliani S, Maggi CA, Rovero P, Meli A. Neurokinins induce a relaxation of the rat duodenum 'in vivo' by activating postganglionic sympathetic elements in prevertebral ganglia: involvement of a NK_2 type of neurokinin receptor. J Pharmacol Exp Ther 1988;246:322–327.

Neurogenic Inflammation in Health and Disease
Edited by Gábor Jancsó

The Role of the Vagus Nerve in Afferent Signaling and Homeostasis During Visceral Inflammation

PETER HOLZER

Research Unit of Translational Neurogastroenterology, Institute of Experimental and Clinical Pharmacology, Medical University of Graz, Graz, Austria

ABSTRACT

Unlike other visceral organs, the gastrointestinal tract has several communication channels transmitting to the spinal cord and brain, which include vagal afferent neurons, spinal afferent neurons, intestinofugal enteric neurons, and endocrine mechanisms. The vagus nerves have long been thought to play only a role in the efferent reflex regulation of digestion. It has turned out, however, that the majority of fibers in the vagus nerves is afferent in nature and subserves an unprecedented variety of physiological and pathophysiological roles. Thus, vagal afferents have a function in chemonociception and interoception (the sense of the physiological state of the body), modify affective-emotional processes in the brain, can give rise to nausea and vomiting, and carry messages relevant to appetite and nutrition. In the context of this book, it is particularly relevant that vagal afferents serve as a sensory interface between the peripheral immune system and the central as well as autonomic nervous system. Vagal afferents can respond to proinflammatory cytokines such as interleukin-1beta and contribute to the behavioral sickness response to infection. In addition, they elicit a vagovagal anti-inflammatory reflex that results in the stimulation of cholinergic vagal efferents. Upon peripheral release, acetylcholine activates alpha7 subunit-containing nicotinic receptors on immune cells, which inhibits further release of proinflammatory cytokines and suppresses inflammation. These substantial advances in the understanding of vagal nerve function offer novel therapeutic opportunities for the management of nociception and inflammation.

1. THE EMERGENCE OF VAGAL NOCICEPTORS IN THE GUT

While somatic tissues and abdominal organs such as the urogenital system have only a single afferent interface with the central nervous system, the gastrointestinal tract has several afferent communication channels with the spinal cord and brain. These multiple afferent connections are constituted by vagal afferent neurons, spinal afferent neurons, intestinofugal enteric neurons, and endocrine mechanisms [1]. These multiple information channels reflect the functional importance of the gut–brain communication which is vital to the homeostasis of the body. Spinal afferent neurons originating from the dorsal root ganglia have been extensively studied in their properties to respond to innocuous and noxious stimuli. By virtue of their sensory

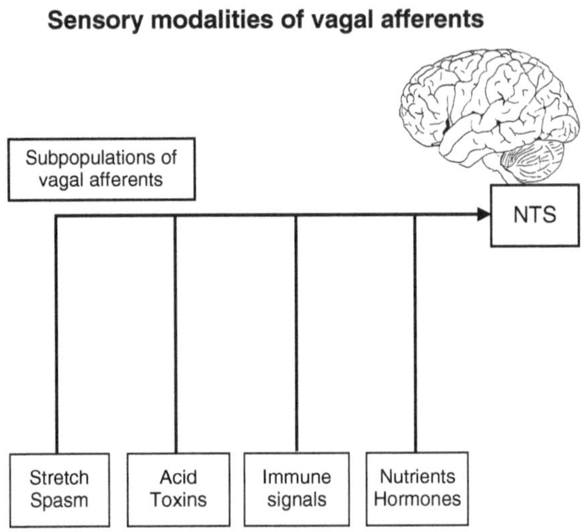

Figure 1. Sensory modalities of vagal afferent neurons. The diagram shows multiple populations of abdominal vagal afferents that respond to different stimuli and project to the nucleus tractus solitarii (NTS) in a viscerotopic manner.

modalities they are not only prominently involved in visceral nociception but also in mediating neurogenic inflammatory responses [1]. In contrast, afferent neurons in the vagus nerves have long been thought to be exclusively engaged in the reflex regulation of digestive processes in the upper gastrointestinal tract, without any role in gastrointestinal pain and other pathological processes [2,3].

This concept is no longer tenable in view of the emerging evidence that vagal neurons contribute to chemonociception (Fig. 1) and that they are sensitized during inflammatory processes [4–8]. Vagal afferent neurons contribute to interoception (the sense of the physiological state of the body) and appear to be of particular relevance to the affective-emotional component of visceral pain [9–12]. In addition, vagal afferents convey messages that give rise to nausea and vomiting (Fig. 1), which reflects a particular type of nociceptive reflex that protects the body from harmful ingesta [13]. Similarly, vagal afferents are able to respond to immune challenge in the gastrointestinal tract (Fig. 1) and in this way contribute to the sickness behavior that ensues following infection and systemic inflammation [14,15]. It is particularly intriguing that, following activation by immune stimuli, vagal afferents seem to initiate an anti-inflammatory reflex that feeds back on the pathological processes in the periphery and may be taken use of in anti-inflammatory treatment [16]. Apart from their control of digestive functions, vagal afferents also carry messages that are relevant to appetite and nutrient/energy homeostasis (Fig. 1).

The objective of this article is to give a brief overview of the pleiotropic roles of vagal afferent neurons in body homeostasis, with particular emphasis on their role in nociception and inflammation.

2. INNERVATION OF THE GUT BY VAGAL AFFERENT NEURONS

The vagal afferent neurons innervating abdominal organs originate from the nodose and jugular ganglion complex and reach the organs in the abdominal cavity via the common hepatic branch, two gastric branches and two celiac branches of the vagus nerve [17,18]. Their central fibers project to the lower brainstem where they terminate in the nucleus tractus solitarii (NTS) and area postrema in a viscerotopic manner [17–22]. Most of vagal afferent neurons are associated with unmyelinated fibers conducting in the C-fiber range, although some thinly myelinated fibers conducting in the Adelta-fiber range have also been found. Electrophysiological characterization has revealed that they show relative specificity with respect to the adequate mechanical stimuli (distension, contraction, shear stress) and chemical stimuli (acidosis, osmolarity, nutrients) that they are able to respond to [2,3,22]. This specificity is related to the anatomical arrangement of their endings in the wall of the alimentary canal and to their transduction mechanisms [22].

The major targets of vagal afferent neurons in the wall of the gastrointestinal tract are the myenteric plexus, muscle, and mucosa where they monitor changes in the chemical and physical environment. Intramural endings of vagal afferents form intraganglionic laminar endings in the myenteric plexus that have functional characteristics of tension receptors and intramuscular arrays that have properties of stretch receptors [17,22,23]. The mucosal endings of vagal afferents project into the lamina propria but do not penetrate the basal lamina. Given this situation, the responses of vagal afferents to chemical stimuli in the lumen of the alimentary canal are typically transduced by endocrine cells of the mucosa. For instance, mechanical, osmotic, nutrient, and toxic stimuli can activate enterochromaffin cells to release 5-hydroxytryptamine (5-HT) which via 5-HT_3 receptors excites vagal afferent neurons. Protein and long-chain lipids activate other endocrine cells that release cholecystokinin (CCK) which stimulates CCK_1 receptors on vagal afferents [1,22].

3. ROLE OF VAGAL AFFERENTS IN GASTRIC CHEMO- AND MECHANONOCICEPTION

3.1 Clinical evidence

It has long been a general contention that visceral pain arising from pelvic and abdominal organs is exclusively mediated by spinal afferents, without any contribution from vagal afferent neurons [2,3]. There are anecdotal reports that acute ulcer perforation or distension can cause pain in patients with complete lesions of the cervical spinal cord, but it has never been systematically investigated whether gastritis, peptic ulcer disease, or gastric distension gives rise to pain in patients with cervical cord lesions [22]. DeVault et al. [24] report that patients with a C6 or C7 spinal cord injury are still able to experience pain caused by balloon distension above the lower esophageal sphincter. Furthermore, patients who underwent proximal gastric vagotomy as treatment for peptic ulcer disease have been found to report fewer dyspeptic complaints lasting for more than 1 week than patients who were treated with histamine H_2 blockers [25]. As Lindsetmo et al. [25] point out, there was actually no significant difference between vagotomized patients and community controls in reporting abdominal or dyspeptic complaints. An antinociceptive effect of vagotomy is in keeping with reports that strong electrical stimulation of the vagus nerve for the treatment of medically intractable epilepsy can give rise to the sensation of pain [26].

3.2 Experimental evidence

The clinical findings affirm that the vagus nerve plays a role in nociception in the upper gastrointestinal tract, a contention that finds unequivocal support in recent experimental studies [7,8].

The major lines of evidence for vagal nerve-mediated nociception are fourfold. (i) Vagotomy abolishes the pseudoaffective pain-related reaction that is elicited by acid challenge of the rat stomach and recorded via an increase of the myoelectrical activity in the acromiotrapezius muscle of the neck [8]. (ii) Similarly, vagotomy prevents gastric acid challenge from inducing the transcription factor c-fos in neurons of the NTS and area postrema which under normal conditions receive massive input from the stomach [10,27]. (iii) Inflammation induces signs of sensitization in the vagal afferent system, given that experimentally induced gastritis and gastric ulceration or cytokine exposure increase the pseudoaffective pain-related reaction and the medullary c-fos response to gastric acid challenge [8,28,29]. (iv) Bilateral transection of the cervical vagus nerve diminishes pseudoaffective pain-related reflexes due to distension or acid stimulation of the upper cervical esophagus of the rat [7].

The role of vagal afferent neurons in nociception in the upper gastrointestinal tract is consistent with their sensitivities to chemical and mechanical stimuli in the noxious range [3,4]. Electrophysiologically, two functional classes of vagal afferents have been distinguished: in-series tension receptors in the outer layers of the gastrointestinal wall and mucosal afferents [30]. The in-series tension receptors are excited by distension or contraction of the gut wall, while mucosal afferents respond to chemical and mechanical stimuli applied to the mucosa [31–34]. Unlike gastrointestinal mechanoreceptors of spinal afferents which may have low or high mechanical thresholds, vagal mechanoreceptors have in general low mechanical thresholds and encode a wide range of stimulus strengths corresponding to physiological and noxious modalities. Contrary to a previously held concept, vagal afferents become sensitized in response to inflammation and mucosal ulceration [31,32,34].

3.3 The vagus nerve is a sensory nerve

The emerging implications of vagal sensory neurons in nociception and related functions are consistent with the important role of the vagus nerve in the communication between the gut and the brain (interoception) and in the maintenance of homeostasis (Fig. 2). The overwhelmingly afferent function is consistent with the fact that the vast majority (80–90%) of the axons in the vagus nerve are afferent nerve fibers [1,13,22], making the vagus nerve the largest visceral sensory nerve in the body [11]. Vagal afferents are thought to tonically deliver information from the alimentary canal to the brain, and this massive sensory input is thought to be relevant not only to the autonomic regulation of digestion and energy balance [35] but also to the interpretation of external sensory inputs, attitude, and behavior [11]. In this aspect, the afferent part of the vagus nerve makes a significant contribution to interoception [12], a function that sometimes is also referred to as the “sixth sense” [11].

3.4 Vagal afferents have a different role in gastric, chemonociception, and mechanonociception

In view of the long-standing contention that vagal afferent neurons are not relevant to visceral pain [2,3], the question arises as to whether the sensory nerve fibers in the vagus nerve contribute to the sensory-discriminative aspect of abdominal nociception or whether they are primarily of relevance to the emotional-affective qualities of visceral pain. As alluded to above, there is increasing evidence that vagal afferents play an important role in acid surveillance of the gastric mucosa, an issue that has been examined by visualization of gastric afferent input to the brainstem by functional c-fos neuroanatomy. Exposure of the rat gastric mucosa to

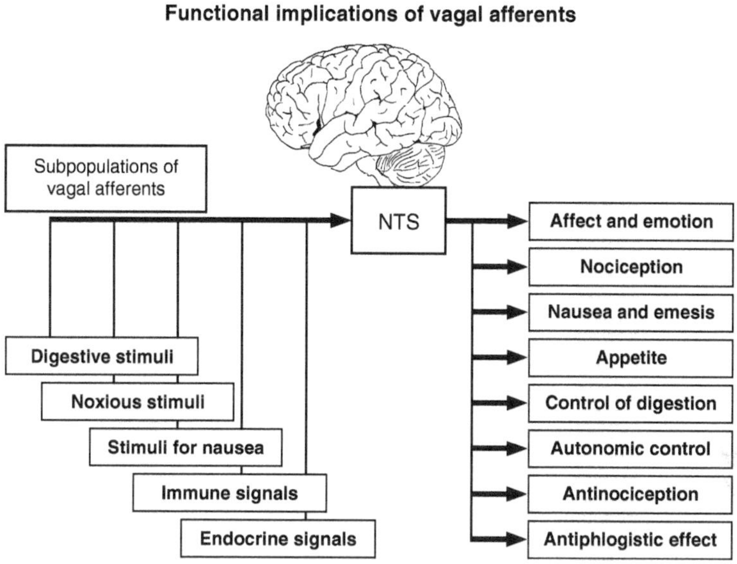

Figure 2. Functional implications of abdominal vagal afferents. The diagram shows multiple populations of abdominal afferents that have different sensory modalities and depicts various effector functions that depend on vagal afferent input from the peritoneal/abdominal cavity. NTS, nucleus tractus solitarii.

minimally injurious concentrations of hydrochloric acid leads to rapid expression of c-fos in the NTS and area postrema, but not spinal cord [27]. Like the pseudoaffective pain-related reaction to gastric acid challenge [8], the medullary c-fos response is largely mediated by vagal afferent neurons because it is suppressed by chronic bilateral subdiaphragmatic vagotomy [10,27]. It follows that the major pathway that signals a gastric acid insult to the brainstem is vagal afferent in nature although it needs to be considered that the NTS receives visceral input not only via vagal afferent neurons [19,20] but also via a spinosolitary pathway [36,37].

While the pseudoaffective pain-related reaction to gastric acid challenge is abolished by vagotomy but left unaltered by splanchnic nerve transection, the analogous response to gastric distension at noxious intraluminal pressures is prevented by prior splanchnic nerve resection but remains unchanged following vagotomy [8]. This observation indicates that chemonociceptive and mechanonociceptive reflexes arising from the rat stomach are mediated by two fundamentally different afferent pathways. Vagal afferents are essential for the reflex response to gastric acid challenge, whereas spinal afferents are responsible for the distension-induced reflex. It is important to note that the transcription of c-fos in the brainstem does not fully match the nociceptive reflex responses. Whereas acid challenge induces c-fos in the brainstem but not spinal cord [27], noxious distension causes c-fos expression in both spinal cord and brainstem [9]. It follows that vagal afferents are also activated by noxious distension of the stomach although the vagal system does not seem to contribute to the pseudoaffective reaction in the neck muscle.

Unlike intragastric acid, intragastric administration of capsaicin activates both vagal and spinal afferent neurons as visualized by expression of c-fos messenger ribonucleic acid in both brainstem and spinal cord [38]. Relative to vehicle, intragastric hydrochloric acid (0.5 M) and capsaicin (3.2 mM) were found to increase c-fos transcription in the NTS by factors of 7.0

and 2.1, respectively. Capsaicin also caused a 5.2-fold rise of c-fos transcription in lamina I of the caudal thoracic spinal cord, whereas intragastric acid failed to induce a c-fos response in the spinal cord [38]. It would thus appear that gastric mucosal exposure to acid and capsaicin is differentially transmitted to the brainstem and spinal cord. Since only acid blocks gastric emptying, it is hypothesized that the acid stimulus is exclusively signaled by gastric vagal afferents, whereas the capsaicin stimulus is processed both by gastric vagal and intestinal spinal afferents [38].

3.5 Processing of vagal afferent input in the brain

The vagal afferent input from the acid-threatened stomach is processed in the medullary brainstem such that the information is passed on to the lateral parabrachial nucleus, the thalamic and hypothalamic paraventricular nuclei, the supraoptic nucleus, the central amygdala, and the mediolateral habenula [10]. There is, however, no activation of the insular cortex, the major cerebral representation area of afferent input from the stomach. Thus, vagal afferent signaling of gastric acid challenge does not seem to give rise to perception of pain but leads to activation of subcortical brain nuclei that are involved in emotional, behavioral, autonomic, and neuroendocrine reactions to noxious stimuli [10]. However, the possible implications of vagal afferents in the sensory-discriminative and affective-emotional aspects of visceral nociception cannot be differentiated before real-time functional imaging studies reveal the processing of gastric noxae in the cerebral cortex.

4. SENSITIZATION OF VAGAL AFFERENTS BY GASTRITIS

It has been a long-standing argument that vagal afferents do not contribute to visceral nociception because they do not sensitize in response to inflammation [2,3]. Recent studies, however, have shown that the vagal afferent system becomes hypersensitive to acid and distension after experimental induction of inflammation or ulceration in the rat stomach [8,31,32,34,39,40]. In particular, iodoacetamide-induced gastritis as well as acetic acid-induced ulceration in the rat stomach increase the pseudoaffective pain-related reflex response to gastric acid challenge in the neck muscle [8]. As this response is abolished by vagotomy [8], it needs to be inferred that there are dynamic changes in the vagal afferent system that account for the hyperresponsiveness to acid. This conclusion is borne out by the findings that gastric ulceration alters the kinetics of tetrodotoxin-resistant sodium currents as well as acid-induced currents in nodose ganglion neurons [34,40]. Similarly, acute exposure of the esophagus to hydrochloric acid plus pepsin sensitizes brainstem neurons to esophageal distension [33].

Sensitization of vagal afferent pathways has also been observed by studying gastric acid-evoked expression of medullary c-Fos immunoreactivity in two models of experimentally induced gastritis [29]. In this study, mild gastritis was induced by addition of dextran sulfate sodium (molecular weight 8000; 5%) or iodoacetamide (0.1%) to the drinking water and verified by an increase in gastric myeloperoxidase activity and histological injury of the mucosal surface. Pretreatment with either agent for 1 week did not alter the c-Fos response to intragastric saline but significantly enhanced the response to acid by 54% and 74%, respectively. It has been argued that the gastritis-evoked increase in the gastric acid-evoked c-Fos expression in the NTS is related to disruption of the gastric mucosal barrier, mucosal inflammation, mucosal acid influx, and enhanced activation of vagal afferents [29]. There is evidence, however, that inflammation can enhance the responsiveness of vagal afferent neurons to gastric acid independently of injury to the gastric mucosa. Thus, systemic administration of the proinflammatory cytokines interleukin-1beta and tumor necrosis factor-alpha leads to a prolonged increase in the gastric acid-evoked expression of c-Fos in the brainstem [41].

The acid sensitivity of vagal afferent neurons in the stomach is regulated not only by pathological circumstances but also by the physiological conditions in the gastric mucosa. For instance, the ability of exogenous acid, administered into the stomach, to induce c-Fos expression in the rat NTS is attenuated by inhibition of endogenous acid secretion with cimetidine or omeprazole [42]. *Vice versa*, stimulation of endogenous gastric acid secretion by pentagastrin is *per se* able to induce c-Fos expression in the brainstem and to enhance the c-Fos response to gastric acid challenge [42]. These findings suggest that endogenous gastric acid enhances the sensitivity of vagal afferents to an exogenous acid stimulus, much as acid is able to sensitize mechanosensitive afferent neurons to distension [43].

5. MOLECULAR ACID SENSORS OF VAGAL AFFERENT NEURONS

Vagal afferents in the stomach have long been known to discharge action potentials when their peripheral terminals are exposed to acid [34,44–46], but their molecular sensors of H^+ ions have only recently come to light. Afferent neurons express many acid sensors, among which the cation channel transient receptor potential vanilloid 1 (TRPV1) and the acid-sensing ion channels (ASICs) take a prominent place [47–50]. The ASICs belong to the voltage-insensitive, amiloride-sensitive epithelial Na^+ channel/degenerin family of cation channels [48,49]. The H^+-gated members of this family are encoded by three different genes, ASIC1, ASIC2, and ASIC3, with ASIC1 and ASIC2 each having alternative splice variants termed ASIC1a, ASIC1b, ASIC2a, and ASIC2b. Functional channels are made up of different ASIC subunits which form homo- or heteromultimeric channels that differ in their pH sensitivity and pharmacological properties [48,49]. While ASIC1 and ASIC2 are widely distributed in the nervous system, ASIC3 (previously termed DRASIC for dorsal root ASIC) is restricted to primary afferent neurons. Most ASIC subunits have been localized to both nodose ganglion and dorsal root ganglion neurons of the rat and mouse although to different degrees [47,50–52]. Retrograde tracing has revealed that 75% of the nodose ganglion neurons and 82% of the dorsal root ganglion neurons projecting to the rat stomach express ASIC3 [51].

Unlike ASIC3 which responds to mild acidosis, TRPV1 and the related cation channel TRPV4 are activated only if the extracellular pH is reduced to values below 6, in which case a sustained channel current is generated [47,50,53,54]. Apart from H^+, noxious heat, vanilloids such as capsaicin and resiniferatoxin, and some arachidonic acid-derived lipid mediators can also gate TRPV1. Importantly, mild acidosis (pH 7–6) is able to sensitize TRPV1 to other stimuli such as capsaicin and heat and to lower its temperature threshold such that the channel becomes active at normal body temperature. TRPV1-positive nerve fibers occur in mucosa, musculature, and enteric nerve plexuses of the rat, guinea pig, and mouse gut. Since enteric neurons usually do not stain for TRPV1, it follows that the TRPV1-positive nerve fibers in the GI tract represent processes of spinal afferents and, in the stomach, of some vagal afferents [47,50,51]. Using retrograde tracing and immunohistochemistry, Schicho et al. [51] have analyzed the origin of TRPV1-like immunoreactivity in the rat stomach. Most stomach-innervating neurons in the nodose ganglia were immunoreactive for TRPV1 (80%), this result being similar in the dorsal root ganglia (71%). Combined vagotomy and sympathetic ganglionectomy abolished expression of TRPV1, which indicates that TRPV1 in the rat gastrointestinal tract derives from an extrinsic source.

In an attempt to determine the role of ASIC2 and ASIC3 in vagal chemonociception, Wultsch et al. [55] examined whether knockout of the ASIC2 or ASIC3 gene modifies afferent signaling of a gastric acid insult in the normal and inflamed stomach. Exposure of the murine gastric mucosa to hydrochloric acid (0.25 M) caused a threefold increase in the number of c-Fos-positive neurons in the NTS. This afferent input to the NTS remained unchanged by ASIC3 knockout, whereas ASIC2 knockout augmented the c-Fos response to gastric challenge by 33% [55]. This observation is reminiscent of the increase in the mechanosensitivity of gastro-esophageal mucosal nerve endings seen in ASIC2 knockout mice, whereas the mechanosensitivity of other populations of afferent nerve fibers in the gut is increased or decreased to various degrees [52,56]. It follows that ASIC2 dampens afferent signaling of a gastric acid insult and that hence knockout of ASIC2 results in disinhibition of afferent input to the NTS [55].

Unlike ASIC2 gene deletion, knockout of ASIC3 did not alter the gastric acid-evoked input to the NTS. It was concluded, therefore, that ASIC3 does not contribute to the acid sensitivity of afferent neurons in the normal stomach [55]. Although the expression of ASIC1 and ASIC2 is not altered in ASIC3 knockout mice [57], it need be considered that developmental changes in other acid sensors may have masked the functional implications of ASIC2 and ASIC3 in germline knockout mice. This argument is consistent with the expression of multiple acid sensors on afferent neurons, including TRPV1 [34,47,50,51,58].

Although ASIC3 knockout mice did not exhibit any deficit in the afferent signaling of an acid insult to the normal stomach, they failed to become hyperresponsive to gastric acid following induction of mild gastritis with iodoacetamide (0.1% in the drinking water for 7 days). In wild-type animals, gastric acid enhanced the number of NTS neurons responding to gastric acid challenge by 41% [55], which is in keeping with the ability of iodoacetamide-induced gastritis to augment pseudoaffective pain responses to gastric distension or an acid insult [8,39]. The gastric acid hyperresponsiveness, as assessed by c-Fos expression in the NTS, was absent in ASIC3 knockout mice but fully preserved in ASIC2 knockout mice. These data indicate that ASIC3 plays a major role in the acid hyperresponsiveness associated with experimental gastritis [55]. Since ASIC3 is primarily expressed by primary afferent neurons [47–52], it appears likely that the ASIC3-mediated hyperresponsiveness to gastric acid takes place at the level of these neurons, either by a rise of channel expression or through an increase in channel sensitivity to acid [55]. Because ASIC3 also contributes to mechanical hyperresponsiveness of the inflamed colon [58], it is tempting to propose that ASIC3 is an important target for the therapeutic management of hyperalgesia associated with gastrointestinal inflammation.

6. NEUROTRANSMITTERS OF GASTROINTESTINAL VAGAL AFFERENT NEURONS IN THE BRAINSTEM

Despite the emerging role of nodose ganglion neurons in abdominal nociception and other homeostatic reactions, the chemical coding of vagal afferent neurons innervating the alimentary canal has not been fully characterized [17,18,59,60]. This gap of knowledge hampers the design of rational approaches to the pharmacological control of the transmission between vagal afferents and their projection neurons in the NTS. Glutamate is among the major transmitters of nodose ganglion neurons, and neuropeptides such as calcitonin gene-related peptide and the tachykinins substance P and neurokinin A occur in subpopulations of nodose ganglion neurons [17,18,59,60]. In an attempt to examine the participation of glutamate and tachykinins in the afferent signaling of a gastric acid insult to the NTS, the N-methyl-D-aspartate (NMDA) glutamate receptor antagonist MK-801 as well as the NK_1 tachykinin receptor antagonist GR-205,171 and the NK_2 tachykinin receptor antagonist SR-144,190 were employed [61]. When given alone, each of these receptor antagonists failed to significantly reduce the medullary c-fos transcription in response to intragastric acid. In contrast, the triple combination of MK-801, GR-205,171, and SR-144,190 inhibited the c-fos response by 45% [61]. These data demonstrate that glutamate acting via NMDA receptors and tachykinins acting via NK_1 and NK_2 receptors cooperate in the vagal afferent input from the acid-threatened stomach to the NTS.

The contribution of NMDA receptors to the transmission process was increased by pre-exposure of the stomach to an injurious concentration of acid 48 h before the actual acid challenge. Under these conditions, MK-801 attained the ability to significantly attenuate the gastric acid-evoked induction of c-fos in the NTS [61]. This increase in NMDA receptor function in the NTS is likely to have a bearing on the hyperresponsiveness of the vagal afferent–NTS system under conditions of gastritis and gastric hyperalgesia.

7. VAGAL AFFERENTS AS INTERFACE BETWEEN PERIPHERAL IMMUNE SYSTEM AND BRAIN

The ability of proinflammatory cytokines such as interleukin-1beta to sensitize vagal afferent pathways to gastric acid [41] is consistent with the emerging role that nodose ganglion neurons play in the communication between the peripheral immune system and the brain [1,6,14,15,22] (Figs 1 and 2). The function of vagal afferents as interface between the peripheral immune system and the brain is supported by a particular proximity of vagal afferent nerve fibers to immunologically relevant structures in the abdominal cavity [1,6]. A series of studies has revealed that, after intraperitoneal administration, bacterial lipopolysaccharide (endotoxin) is primarily transported to the liver where it induces the release of interleukin-1beta from Kupffer cells (macrophage-like cells to screen blood and lymph) [1,6,14]. The cytokine, in turn, is thought to excite afferent nerve fibers in the hepatic branch of the vagus nerve or to enhance their afferent signaling [62,63]. In addition, the abdominal vagus is associated with paraganglia and connective tissue containing macrophages and dendritic cells that respond to intraperitoneal administration of endotoxin with synthesis of interleukin-1beta [64,65]. The abdominal paraganglia of the vagus nerve contain glomus-like cells which have interleukin-1 receptors [66], are innervated by vagal afferents [67], and appear to act as chemosensory accessory cells [14]. Furthermore, vagal afferents innervate abdominal lymph nodes which represent another interface with the visceral immune system [14].

Electrophysiological recordings have demonstrated that peripheral administration of interleukin-1beta is indeed able to increase action potential firing in vagal afferents [68,69] and to induce c-Fos expression in the NTS [41,70]. The expression of interleukin-1 receptors of type I by nodose ganglion cells makes it likely that the cytokine is capable of exciting vagal afferents by a direct action on the axons, although prostaglandins acting via EP_3 receptors and CCK acting via CCK_1 receptors have also been implicated [69,71]. Moreover, interleukin-1beta has been shown to increase the sensitivity of gastric vagal afferents to fire in response to PGE_2 and CCK [69,72].

Proinflammatory cytokines induced by infection and inflammation are known to trigger a complex behavioral response, encompassed in the terms "sickness behavior" or "illness response," which comprises fever, anorexia, somnolence, decrease in locomotor activity, decrease in social exploration, and hyperalgesia [14,15]. A number of studies has revealed that vagal afferents, being responsive to peripheral endotoxin and interleukin-1beta, participate in the behavioral illness response to infection and inflammation [14,15]. Accordingly, certain features of the sickness behavior are attenuated by subdiaphragmatic vagotomy although circulating proinflammatory cytokines can access the brain via circumventricular organs that are devoid of a blood–brain barrier [14,15].

The role of vagal afferents as an interface between the peripheral immune system and the brain represents another aspect of interoception [12]. This view is consistent with the role of nodose ganglion neurons being the first link to carry information from the viscera to autonomic and limbic circuitries in the brain [11,12,22]. As a consequence, afferent input conveyed by vagal afferents can have a far-reaching impact on emotional-affective processes and autonomic regulation of bodily processes although the clinical significance of these implications awaits to be elucidated. It appears reasonable to assume, however, that the adverse symptoms of both acute infection and inflammation of the gastrointestinal tract as well as those associated with postinfectious and postinflammatory dyspepsia and irritable bowel syndrome involve vagal afferent neurons excited and sensitized by proinflammatory cytokines.

8. ANTINOCICEPTIVE ROLE OF VAGAL AFFERENT NEURONS

Apart from signaling adverse events from the gut to the brainstem, subpopulations of vagal afferents also seem to play an antinociceptive role (Fig. 2). This implication was first discovered when stimulation of the cervical vagus nerve was found to reduce nociceptive impulse transmission in the spinal dorsal horn and attenuate nociceptive behavior [73,74]. In the rat, this antinociceptive effect depends on vagal afferent input to the NTS and subsequent activation of noradrenergic neurons in the locus ceruleus–subceruleus–parabrachial complex as well as serotonergic neurons in the nucleus magnus of the rostral ventromedial medulla [75]. These nuclei issue descending projections to the dorsal horn of the spinal cord where they inhibit nociceptive transmission through activation of adrenoceptors, 5-HT receptors, and opioid receptors [73,75]. The vagally driven antinociceptive system appears to operate not only following stimulation of nodose ganglion neurons but may also be maintained by spontaneous activity in vagal afferents [76].

Further analysis has shown that chemical, electrical, or physiological activation of cardio-pulmonary (cervical, thoracic, or cardiac), diaphragmatic, or subdiaphragmatic vagal afferents can result in either facilitation or inhibition of nociception in a species-dependent manner [75]. In monkeys, for instance, stimulation of cervical – but not subdiaphragmatic – vagal afferents has been found to suppress impulse activity to spinothalamic relay neurons [75]. It has been argued that differences in the antinociceptive impact of vagal afferents may reflect the modality of noxious input (e.g., visceral versus cutaneous), the type of neuronal activity recorded (e.g., resting versus noxious stimulus-evoked), the location of recording in the spinal cord (e.g., thoracic versus lumbosacral), and/or the parameters of stimulation [75].

Apart from activating descending pathways that inhibit nociceptive transmission in the spinal cord, vagal afferents can induce antinociception also via endocrine mechanisms [76]. This antinociceptive pathway was revealed by studying the effect of subdiaphragmatic vagotomy to

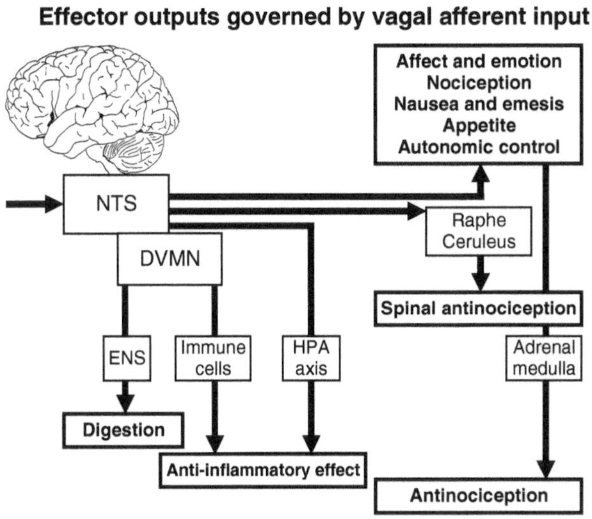

Figure 3. Effector output governed by vagal afferent input. The diagram portrays various outputs that leave the nucleus tractus solitarii (NTS) and their functional implications. Ceruleus, locus ceruleus; DVMN, dorsal vagal motor nucleus; ENS, enteric nervous system; HPA axis, hypothalamic–pituitary–adrenal cortex axis.

decrease the mechanical pain threshold in the rat paw at baseline and after intradermal injection of bradykinin [22,76]. Most of the increase in mechanical responsiveness caused by vagotomy is thought to be generated by an endocrine signal released from the adrenal medulla (Fig. 3), because denervation or removal of the adrenal medulla prevents or reverses the pronociceptive effect of vagotomy [76].

Vagal afferent nerve activity has not only a bearing on the control of nociception but also seems to regulate inflammatory processes in remote tissues. Jänig, Levine, and their colleagues [22,76,77] have found that stimulation of cutaneous and spinal visceral nociceptive afferents reduces bradykinin-induced plasma extravasation in the knee joint of the rat. This anti-inflammatory effect is mediated by the hypothalamic–pituitary–adrenal axis (Fig. 3) and potentiated by subdiaphragmatic interruption of the celiac – but not gastric and hepatic – branches of the abdominal vagus [77]. The effect of vagotomy has been taken to propose that activity in vagal afferents tonically inhibits ascending impulse transmission in the neuraxis projecting to the hypothalamus. By removing such inhibition, vagotomy leads to enforced depression of bradykinin-evoked plasma protein extravasation [77]. Based on the antinociceptive and anti-inflammatory implications, vagal afferents are attributed a widespread role in body protection [22,76]. In this concept, vagal afferent input from the viscera to the brain is thought to regulate the sensitivity of nociceptors throughout the body via neuroendocrine mechanisms [76].

9. THE VAGAL ANTI-INFLAMMATORY REFLEX

The hypothesis that the vagus nerve plays a role in body protection [22] has been further borne out by the discovery that vagal efferent neurons mediate an anti-inflammatory effect [16,78]. This role of parasympathetic efferent neurons complements the ability of vagal afferents to respond to proinflammatory cytokines generated in the course of infection and inflammation, to trigger a variety of illness responses and to instigate a sickness-like behavior [14,15]. Activation of vagal efferent neurons triggers what has been termed the cholinergic anti-inflammatory reflex [78]. By releasing acetylcholine acting via distinct acetylcholine receptors on abdominal immune cells (Fig. 3), vagal efferents suppress proinflammatory cytokine release and inhibit inflammation [16,78]. This cholinergic anti-inflammatory action was first envisaged when acetylcholine was observed to concentration-dependently inhibit the release of the proinflammatory cytokines tumor necrosis factor, interleukin-1, interleukin-6, and interleukin-8 from endotoxin-activated primary human macrophages, whereas the production of the anti-inflammatory cytokine interleukin-10 was left unaltered [16,78].

In an attempt to put these findings into pathophysiological perspective, it was found that electrical stimulation of the efferent vagus nerve decreases serum and hepatic levels of tumor necrosis factor and attenuates the development of severe hypotension and systemic shock due to hypovolemia, endotoxin exposure, or ischemia–reperfusion injury [16,78]. Tissue macrophages and other cytokine-synthesizing cells in the organs of the reticuloendothelial system are among the most plausible cellular targets of the immunomodulatory effect of the efferent vagus. The effect of acetylcholine to inhibit cytokine release from immune cells is mediated primarily by nicotinic, but not muscarinic, acetylcholine receptors. Macrophages express alpha7 subunit-containing nicotinic receptors, and there is convincing evidence that this subunit plays an essential role in the effect of acetylcholine released from vagal efferents to suppress cytokine release [16,78]. Thus, antisense oligonucleotides to the alpha7 subunit abolish the effect of nicotine to suppress the release of tumor necrosis factor and other cytokines from stimulated human macrophages, and knockout of the alpha7 subunit prevents nicotine, acetylcholine, and efferent vagal nerve stimulation from blocking tumor necrosis factor release by peritoneal macrophages of the mouse [16,78,79].

The specific involvement of alpha7 subunit-containing nicotinic receptors makes it conceivable that an agonist selectively targeting this receptor may represent a new type of anti-inflammatory agent that is devoid of an action on autonomic ganglia in which transmission is mediated primarily by alpha3/beta4 subunit-containing nicotinic receptors [78]. This concept has received significant support by the ability of GTS-21, a selective alpha7 nicotinic acetylcholine receptor agonist, to inhibit the release of proinflammatory cytokines *in vitro* and *in vivo* and to improve survival in a murine endotoxemia and sepsis model [80].

Tracey [16] has envisaged that the anti-inflammatory action of the efferent vagus is part of a vagal anti-inflammatory reflex [16]. Although the functional characteristics of this reflex have not yet been worked out, a hypothetical model of the anti-inflammatory reflex has been elaborated [78]. In this model, peripheral immune and inflammatory signals generate an input to the brain both via vagal afferents and circumventricular organs that are devoid of a blood–brain barrier. These signals are processed by the NTS and central autonomic circuitries to provide an integrated output via cholinergic vagal efferents [78]. Acetylcholine, released from cholinergic axon terminals in the periphery, activates alpha7 subunit-containing nicotinic receptors on immune cells, which results in inhibition of proinflammatory cytokine release and restoration of immune homeostasis [78]. The exact mechanism by which acetylcholine approaches nicotinic receptors on immune cells has not yet been fully explored, because the lifetime of acetylcholine diffusing through the parenchyma of organs of the reticuloendothelial system to nearby immune cells is extremely limited by the catabolic activity of acetylcholinesterase [78].

There is increasing evidence that the vagovagal anti-inflammatory reflex has a particular role in controlling inflammation within the peritoneal/abdominal cavity. For instance, septic peritonitis induced in mice by intraperitoneal injection of live *Escherichia coli* is attenuated by nicotine and exacerbated by vagotomy [81]. Likewise, murine pancreatitis induced by cerulein is ameliorated by the alpha7 nicotinic acetylcholine receptor agonist GTS-21, whereas vagotomy or treatment with the nicotinic receptor antagonist, mecamylamine, exacerbates pancreatitis as reflected by histology, edema, and interleukin-6 levels [82]. Dextrane sulfate sodium-induced colitis is associated with increased tissue levels of several proinflammatory cytokines in the intestinal wall. These and other indices of inflammation have been found to be exaggerated by vagotomy and hexamethonium and attenuated by nicotine [83]. The counterinflammatory action of the vagus nerve in colitis involves a macrophage-dependent mechanism, because vagotomy fails to exacerbate colitis in mice that are deficient of macrophage colony-stimulating factor [83]. The extent of inflammation following abdominal surgery is a factor that has an important bearing on the severity and duration of postoperative ileus. Stimulation of nicotinic acetylcholine receptors has been reported to inhibit macrophage activation, ameliorate surgery-induced inflammation, and reduce postoperative ileus through downstream activation of the Jak2–STAT3 signaling pathway [84].

10. CONCLUSIONS

The vagus nerve has long been considered to be an exclusively efferent communication channel that is involved in the physiological regulation of digestion. More recent research has revealed, however, that the vagus nerve is primarily an afferent nerve and subserves an unprecedented variety of physiological and pathophysiological functions [85,86] (Figs 2 and 3). Abdominal vagal afferents carry a wealth of information to the brainstem, including signals relevant to nutrient homeostasis, feeding and drinking behavior [87,88], noxious challenge of the tissue [7,8,55] as well as infection and inflammation [14,15]. Through this pleiotropic sensory role, vagal afferents participate in a wide range of physiological reactions aimed at maintenance of homeostasis (Fig. 2). It seems very likely that vagal afferents comprise a multitude of neuronal subpopulations that have different sensory modalities (Fig. 1) and functional implications. One of the challenges ahead is to dissect these distinct subpopulations of afferent neurons, to expand the understanding of vagus nerve physiology and to take use of this knowledge in the design and development of novel therapeutic strategies. Advances in vagal nerve pharmacology offer therapeutic opportunities to address the control of food intake and the treatment of functional gastrointestinal disorders, nausea and emesis, gastroesophageal reflux disease, abdominal pain and hyperalgesia, disturbances of affective-emotional behavior, gastrointestinal infection, and gastrointestinal inflammation (Fig. 3).

ACKNOWLEDGMENTS

Work in the author's laboratory was supported by the Jubilee Funds of the Austrian National Bank (grant 9858), the Zukunftsfonds Steiermark (grant 262), and the Austrian Scientific Research Funds (FWF grant L25-B05).

REFERENCES

1. Holzer P. Sensory neurone responses to mucosal noxae in the upper gut: relevance to mucosal integrity and gastrointestinal pain. Neurogastroenterol Motil 2002;14:1–17.
2. Cervero F. Sensory innervation of the viscera: peripheral basis of visceral pain. Physiol Rev 1994;74:95–138.
3. Grundy D. Neuroanatomy of visceral nociception: vagal and splanchnic afferent. Gut 2002;51(Suppl. 1):i2–i5.
4. Gebhart GF. Pathobiology of visceral pain: molecular mechanisms and therapeutic implications. IV. Visceral afferent contributions to the pathobiology of visceral pain. Am J Physiol 2000;278:G834–G838.
5. Holzer P, Michl T, Jocic M, Heinemann A, Schuligoi R. Essential role of vagal afferents in the central signalling of a gastric mucosal acid insult. In: Krammer HJ, Singer MV, eds. Neurogastroenterology. From the Basics to the Clinics. Dordrecht: Kluwer, 2000; pp. 697–707.
6. Holzer P, Danzer M, Schicho R, Lippe IT. Regulation of vagal afferent sensitivity to gastric acid. In: Holtmann G, Talley NJ, eds. Gastrointestinal Inflammation and Disturbed Gut Function: The Challenge of New Concepts. Dordrecht: Kluwer, 2003; pp. 126–135.
7. Hummel T, Barz S, Holscher T, Neuhuber WL. Differences in responses to nociceptive stimulation of the oral and aboral oesophagus. J Clin Neurosci 2003;10:223–225.
8. Lamb K, Kang YM, Gebhart GF, Bielefeldt K. Gastric inflammation triggers hypersensitivity to acid in awake rats. Gastroenterology 2003;125:1410–1418.
9. Traub RJ, Sengupta JN, Gebhart GF. Differential c-fos expression in the nucleus of the solitary tract and spinal cord following noxious gastric distention in the rat. Neuroscience 1996;74:873–884.
10. Michl T, Jocic M, Heinemann A, Schuligoi R, Holzer P. Vagal afferent signaling of a gastric mucosal acid insult to medullary, pontine, thalamic, hypothalamic and limbic, but not cortical, nuclei of the rat brain. Pain 2001;92:19–27.
11. Zagon A. Does the vagus nerve mediate the sixth sense? Trends Neurosci 2001;24:671–673.
12. Craig AD. How do you feel? Interoception: the sense of the physiological condition of the body. Nat Rev Neurosci 2002;3:655–666.
13. Andrews PLR. 5-HT3 receptor antagonists and antiemesis. In: King FD, Jones BJ, Sanger GJ, eds. 5-Hydroxytryptamine-3 Receptor Antagonists. Boca Raton: CRC Press, 1994; pp. 255–317.
14. Goehler LE, Gaykema RPA, Hansen MK, Anderson K, Maier SF, Watkins LR. Vagal immune-to-brain communication: a visceral chemosensory pathway. Auton Neurosci Basic Clin 2000;85:49–59.
15. Konsman JP, Parnet P, Dantzer R. Cytokine-induced sickness behaviour: mechanisms and implications. Trends Neurosci 2002;25:154–159.
16. Tracey KJ. The inflammatory reflex. Nature 2002;420:853–859.
17. Berthoud HR, Neuhuber WL. Functional and chemical anatomy of the afferent vagal system. Auton Neurosci 2000;85:1–17.
18. Neuhuber WL, Raab M, Berthoud HR, Wörl J. Innervation of the mammalian esophagus. Adv Anat Embryol Cell Biol 2006;185:1–73.
19. Norgren R, Smith GP. Central distribution of subdiaphragmatic vagal branches in the rat. J Comp Neurol 1988;273:207–223.

20. Altschuler SM, Bao XM, Bieger D, Hopkins DA, Miselis RR. Viscerotopic representation of the upper alimentary tract in the rat: sensory ganglia and nuclei of the solitary and spinal trigeminal tracts. J Comp Neurol 1989;283:248–268.
21. Grundy D, Scratcherd T. Sensory afferents from the gastrointestinal tract. In: Schultz SG, Wood JD, Rauner BB, eds. Handbook of Physiology, Section 6: The Gastrointestinal System, Volume1: Motility and Circulation, Part 1. Bethesda: American Physiological Society, 1989; pp. 593–620.
22. Jänig W. Visceral afferent neurons and autonomic regulations. In: Jänig W, ed. The Integrative Action of the Autonomic Nervous System. Neurobiology of Homeostasis. Cambridge: Cambridge University Press, 2006; pp. 35–83.
23. Zagorodnyuk VP, Chen BN, Costa M, Brookes SJ. Mechanotransduction by intraganglionic laminar endings of vagal tension receptors in the guinea-pig oesophagus. J Physiol 2003;553:575–587.
24. DeVault KR, Beacham S, Castell DO, Streletz LJ, Ditunno JF. Esophageal sensation in spinal cord-injured patients: balloon distension and cerebral evoked potential recording. Am J Physiol 1996;271:G937–G941.
25. Lindsetmo RO, Johnsen R, Revhaug A. Abdominal and dyspeptic symptoms in patients with peptic ulcer treated medically or surgically. Br J Surg 1998;85:845–849.
26. Liporace J, Hucko D, Morrow R, Barolat G, Nei M, Schnur J et al. Vagal nerve stimulation: adjustments to reduce painful side effects. Neurology 2001;57:885–886.
27. Schuligoi R, Jocic M, Heinemann A, Schöninkle E, Pabst MA, Holzer P. Gastric acid-evoked c-fos messenger RNA expression in rat brainstem is signaled by capsaicin-resistant vagal afferents. Gastroenterology 1998;115:649–660.
28. Holzer P, Danzer M, Schicho R, Samberger C, Painsipp E, Lippe IT. Vagal afferent input from the acid-challenged rat stomach to the brainstem: enhancement by interleukin-1beta. Neuroscience 2004;129:439–445.
29. Holzer P, Wultsch T, Edelsbrunner M, Mitrovic M, Shahbazian A, Painsipp E et al. Increase in gastric acid-induced afferent input to the brainstem in mice with gastritis. Neuroscience 2007;145:1108–1119.
30. Blackshaw LA, Brookes SJ, Grundy D, Schemann M. Sensory transmission in the gastrointestinal tract. Neurogastroenterol Motil 2007;19(Suppl. 1):1–19.
31. Kang YM, Bielefeldt K, Gebhart GF. Sensitization of mechanosensitive gastric vagal afferent fibers in the rat by thermal and chemical stimuli and gastric ulcers. J Neurophysiol 2004;91:1981–1989.
32. Dang K, Bielfeldt K, Lamb K, Gebhart GF. Gastric ulcers evoke hyperexcitability and enhance P2X receptor function in rat gastric sensory neurons. J Neurophysiol 2005;93:3112–3119.
33. Medda BK, Sengupta JN, Lang IM, Shaker R. Response properties of the brainstem neurons of the cat following intra-esophageal acid-pepsin infusion. Neuroscience 2005;135:1285–1294.
34. Sugiura T, Dang K, Lamb K, Bielefeldt K, Gebhart GF. Acid-sensing properties in rat gastric sensory neurons from normal and ulcerated stomach. J Neurosci 2005;25:2617–2627.
35. Grundy D, Schemann M. Enteric nervous system. Curr Opin Gastroenterol 2007;23:121–126.
36. Menetrey D, Basbaum AI. Spinal and trigeminal projections to the nucleus of the solitary tract: a possible substrate for somatovisceral and viscerovisceral reflex activation. J Comp Neurol 1987;255:439–450.

37. Gamboa-Esteves FO, Kaye JC, McWilliam PN, Lima D, Batten TF. Immunohistochemical profiles of spinal lamina I neurones retrogradely labelled from the nucleus tractus solitarii in rat suggest excitatory projections. Neuroscience 2001;104:523–538.
38. Holzer P, Painsipp E, Schuligoi R. Differential effects of intragastric acid and capsaicin on gastric emptying and afferent input to the rat spinal cord and brainstem. BMC Neurosci 2005;6:60.
39. Ozaki N, Bielefeldt K, Sengupta JN, Gebhart GF. Models of gastric hyperalgesia in the rat. Am J Physiol 2002;283:G666–G676.
40. Bielefeldt K, Ozaki N, Gebhart GF. Experimental ulcers alter voltage-sensitive sodium currents in rat gastric sensory neurons. Gastroenterology 2002;122:394–405.
41. Holzer P, Danzer M, Schicho R, Samberger C, Painsipp E, Lippe IT. Vagal afferent input from the acid-challenged rat stomach to the brainstem: enhancement by interleukin-1beta. Neuroscience 2004;129:439–445.
42. Danzer M, Jocic M, Samberger C, Painsipp E, Bock E, Pabst MA et al. Stomach-brain communication by vagal afferents in response to luminal acid backdiffusion, gastrin, and gastric acid secretion. Am J Physiol 2004;286:G403–G411.
43. Coffin B, Chollet R, Flourié B, Lémann M, Franchisseur C, Rambaud JC et al. Intraluminal modulation of gastric sensitivity to distension: effects of hydrochloric acid and meal. Am J Physiol 2001;280:G904–G909.
44. Iggo A. Gastric mucosal chemoreceptors with vagal afferent fibres in the cat. Q J Exp Physiol Cogn Med Sci 1957;42:398–409.
45. Clarke GD, Davison JS. Mucosal receptors in the gastric antrum and small intestine of the rat with afferent fibres in the cervical vagus. J Physiol (Lond) 1978;284:55–67.
46. Hillsley K, Grundy D. Sensitivity to 5-hydroxytryptamine in different afferent subpopulations within mesenteric nerves supplying the rat jejunum. J Physiol (Lond) 1998;509:717–727.
47. Holzer P. Acid-sensitive ion channels in gastrointestinal function. Curr Opin Pharmacol 2003;3:618–625.
48. Kress M, Waldmann R. Acid sensing ionic channels. Curr Top Membr 2006;57:241–276.
49. Wemmie JA, Price MP, Welsh MJ. Acid-sensing ion channels: advances, questions and therapeutic opportunities. Trends Neurosci 2006;29:578–586.
50. Holzer P. Taste receptors in the gastrointestinal tract. V. Acid sensing in the gastrointestinal tract. Am J Physiol 2007;292:G699–G705.
51. Schicho R, Florian W, Liebmann I, Holzer P, Lippe IT. Increased expression of TRPV1 receptor in dorsal root ganglia by acid insult of the rat gastric mucosa. Eur J Neurosci 2004;19:1811–1818.
52. Page AJ, Brierley SM, Martin CM, Price MP, Symonds E, Butler R et al. Different contributions of ASIC channels 1a, 2, and 3 in gastrointestinal mechanosensory function. Gut 2005;54:1408–1415.
53. Caterina MJ, Julius D. The vanilloid receptor: a molecular gateway to the pain pathway. Annu Rev Neurosci 2001;24:487–517.
54. Clapham DE. TRP channels as cellular sensors. Nature 2003;426:517–524.
55. Wultsch T, Painsipp E, Shahbazian A, Mitrovic M, Edelsbrunner M, Waldmann R et al. Deletion of the acid-sensing ion channel ASIC3 prevents gastritis-induced acid hyperresponsiveness of the stomach-brainstem axis. Pain 2008;134:245–253.
56. Roza C, Puel JL, Kress M, Baron A, Diochot S, Lazdunski M et al. Knockout of the ASIC2 channel in mice does not impair cutaneous mechanosensation, visceral mechanonociception and hearing. J Physiol (Lond) 2004;558:659–669.

57. Price MP, McIlwrath SL, Xie J, Cheng C, Qiao J, Tarr DE et al. The DRASIC cation channel contributes to the detection of cutaneous touch and acid stimuli in mice. Neuron 2001;32:1071–1083.
58. Jones RC, Xu L, Gebhart GF. The mechanosensitivity of mouse colon afferent fibers and their sensitization by inflammatory mediators require transient receptor potential vanilloid 1 and acid-sensing ion channel 3. J Neurosci 2005;25: 10981–10989.
59. Zhuo H, Ichikawa H, Helke CJ. Neurochemistry of the nodose ganglion. Prog Neurobiol 1997;52:79–107.
60. Andrews PL, Sanger GJ. Abdominal vagal afferent neurones: an important target for the treatment of gastrointestinal dysfunction. Curr Opin Pharmacol 2002;2:650–656.
61. Jocic M, Schuligoi R, Schöninkle E, Pabst MA, Holzer P. Cooperation of NMDA and tachykinin NK1 and NK2 receptors in the medullary transmission of vagal afferent input from the acid-threatened rat stomach. Pain 2001;89:147–157.
62. Watkins LR, Wiertelak EP, Goehler LE, Smith KP, Martin D, Maier SF. Characterization of cytokine-induced hyperalgesia. Brain Res 1994;654:15–26.
63. Simons CT, Kulchitsky VA, Sugimoto N, Homer LD, Szekely M, Romanovsky AA. Signaling the brain in systemic inflammation: which vagal branch is involved in fever genesis? Am J Physiol 1998;275:R63–R68.
64. Berthoud HR, Kressel M, Raybould HE, Neuhuber WL. Vagal sensors in the rat duodenal mucosa: distribution and structure as revealed by in vivo DiI-tracing. Anat Embryol 1995;191:203–212.
65. Goehler LE, Gaykema RPA, Nguyen KT, Lee JE, Tilders FJH, Maier SF et al. Interleukin-1β in immune cells of the abdominal vagus nerve: a link between the immune and nervous systems? J Neurosci 1999;19:2799–2806.
66. Goehler LE, Relton JK, Dripps D, Kiechle R, Tartaglia N, Maier SF et al. Vagal paraganglia bind biotinylated interleukin-1 receptor antagonist: a possible mechanism for immune-to-brain communication. Brain Res Bull 1997;43:357–364.
67. Berthoud HR, Kressel M, Neuhuber WL. Vagal afferent innervation of the rat abdominal paraganglia as revealed by anterograde DiI-tracing and confocal microscopy. Acta Anat 1995;152:127–132.
68. Niijima A. The afferent discharges from sensors for interleukin 1β in the hepatoportal system in the anesthetized rat. J Auton Nerv Syst 1996;61:287–291.
69. Kurosawa M, Uvnäs-Moberg K, Miyasaka K, Lundeberg T. Interleukin-1 increases activity of the gastric vagal afferent nerve partly via stimulation of type A CCK receptor in anesthetized rats. J Auton Nerv Syst 1997;62:72–78.
70. Ericsson A, Kovács KJ, Sawchenko PE. A functional anatomical analysis of central pathways subserving the effects of interleukin-1 on stress-related neuroendocrine neurons. J Neurosci 1994;14:897–913.
71. Ek M, Kurosawa M, Lundeberg T, Ericsson A. Activation of vagal afferents after intravenous injection of interleukin-1β: role of endogenous prostaglandins. J Neurosci 1998;18:9471–9479.
72. Kreis ME, Haupt W, Grundy D. Interleukin-1β (IL-1β) selectively augments mesenteric afferent sensitivity to prostaglandin E_2. Gastroenterology 2000;118:A431.
73. Ren K, Randich A, Gebhart GF. Electrical stimulation of cervical vagal afferents. I. Central relays for modulation of spinal nociceptive transmission. J Neurophysiol 1990;64:1098–1114.

74. Randich A, Ren K, Gebhart GF. Electrical stimulation of cervical vagal afferents. II. Central relays for behavioral antinociception and arterial blood pressure decreases. J Neurophysiol 1990;64:1115–1124.
75. Randich A, Gebhart GF. Vagal afferent modulation of nociception. Brain Res Rev 1992;17:77–99.
76. Jänig W, Khasar SG, Levine JD, Miao FJ. The role of vagal visceral afferents in the control of nociception. Prog Brain Res 2000;122:273–287.
77. Miao FJ, Jänig W, Levine JD. Vagal branches involved in inhibition of bradykinin-induced synovial plasma extravasation by intrathecal nicotine and noxious stimulation in the rat. J Physiol 1997;498:473–481.
78. Pavlov VA, Tracey KJ. The cholinergic anti-inflammatory pathway. Brain Behav Immun 2005;19:493–499.
79. Wang H, Yu M, Ochani M, Amella CA, Tanovic M, Susarla S et al. Nicotinic acetylcholine receptor alpha7 subunit is an essential regulator of inflammation. Nature 2003;421:384–388.
80. Pavlov VA, Ochani M, Yang LH, Gallowitsch-Puerta M, Ochani K, Lin X et al. Selective alpha7-nicotinic acetylcholine receptor agonist GTS-21 improves survival in murine endotoxemia and severe sepsis. Crit Care Med 2007;35:1139–1144.
81. van Westerloo DJ, Giebelen IA, Florquin S, Daalhuisen J, Bruno MJ, de Vos AF et al. The cholinergic anti-inflammatory pathway regulates the host response during septic peritonitis. J Infect Dis 2005;191:2138–2148.
82. van Westerloo DJ, Giebelen IA, Florquin S, Bruno MJ, Larosa GJ, Ulloa L et al. The vagus nerve and nicotinic receptors modulate experimental pancreatitis severity in mice. Gastroenterology 2006;130:1822–1830.
83. Ghia JE, Blennerhassett P, Kumar-Ondiveeran H, Verdu EF, Collins SM. The vagus nerve: a tonic inhibitory influence associated with inflammatory bowel disease in a murine model. Gastroenterology 2006;131:1122–1130.
84. de Jonge WJ, van der Zanden EP, The FO, Bijlsma MF, van Westerloo DJ, Bennink RJ et al. Stimulation of the vagus nerve attenuates macrophage activation by activating the Jak2-STAT3 signaling pathway. Nat Immunol 2005;6:844–851.
85. Grundy D. Signalling the state of the digestive tract. Auton Neurosci 2006;125:76–80.
86. Fioramonti J, Gebhart GF. In vivo and transgenic animal models used to study visceral hypersensitivity. Neurogastroenterol Motil 2007;19(Suppl. 1):20–28.
87. Raybould HE. Visceral perception: sensory transduction in visceral afferents and nutrients. Gut 2002;51(Suppl. 1):i11–i14.
88. Li Y, Owyang C. Musings on the wanderer: what's new in our understanding of vago-vagal reflexes? V. Remodeling of vagus and enteric neural circuitry after vagal injury. Am J Physiol 2003;285:G461–G46.

Subject Index

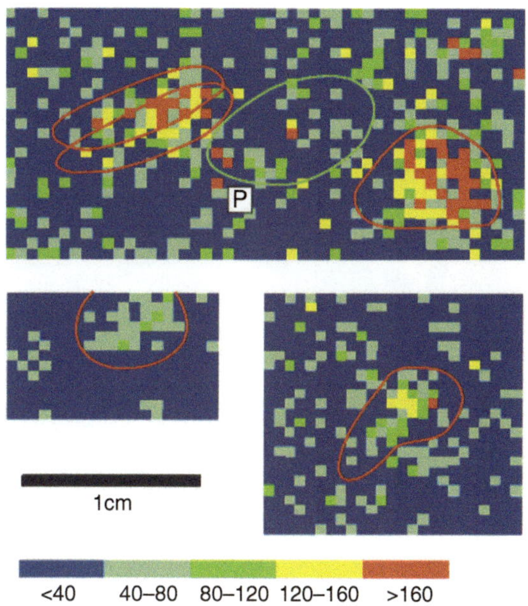

Plate 1. Blood flow images during stimulation of fine nerve filaments containing C-heat nociceptor axons in the pig. Upper field: filament contained three identified heat nociceptors – afferent receptive fields outlined in red and one polymodal nociceptor (P) receptive field outlined in green. Note that only the heat nociceptors gave vasodilator responses. Lower two fields: filaments with one heat nociceptor axon. Note the good relation between afferent receptive field and area of vasodilatation. Images have been scaled so that color levels represent fixed percentages above the average baseline level and have been oriented so that proximal is to the right and anterior at the top. Scans took 1–3 min. Stimulation was 1–2 Hz for the duration of the scans (from Ref. [16]). (See Figure 1 of Chapter 2, p. 41).

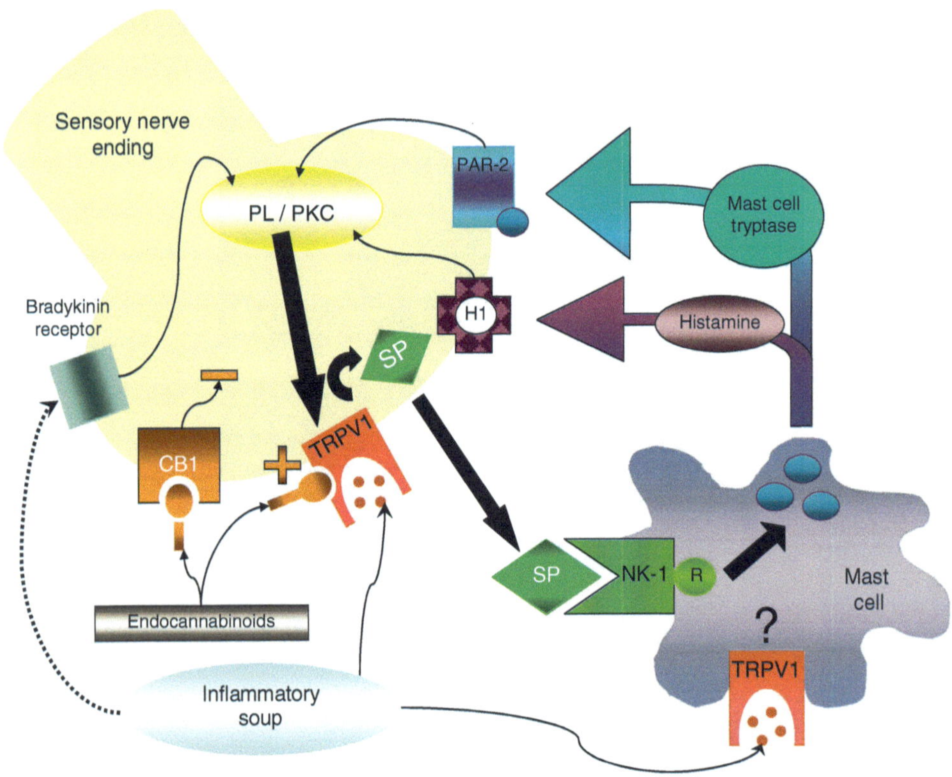

Plate 2. Hypothetical interaction of TRPV1-expressing sensory afferents with mast cells. During inflammation, numerous mediators are generated, including bradykinin, prostaglandin, and protons. This "inflammatory soup" can directly activate TRPV1 (e.g., low pH) or indirectly sensitize TRPV1 (e.g., bradykinin through its receptor, bradykinin R). Activation of TRPV1 in sensory afferents results in the release of neuropeptides, including substance P (SP). SP via NK1 can activate mast cells resulting in the release of mast cell tryptase – an activator of PAR2. Sensitization of TRPV1 via PAR2 and/or histamine receptor (H1) would complete a positive regulatory loop between mast cells and sensory afferents that drives neurogenic inflammation. Endocannabinoids, such as anandamide, may activate sensory neurons through direct interaction with TRPV1(+) or decrease sensory neuron activity via cannabinoid receptor, CB1(–). PL, phospholipase; PKC, protein kinase C. (See Figure 1 of Chapter 5, p. 108).

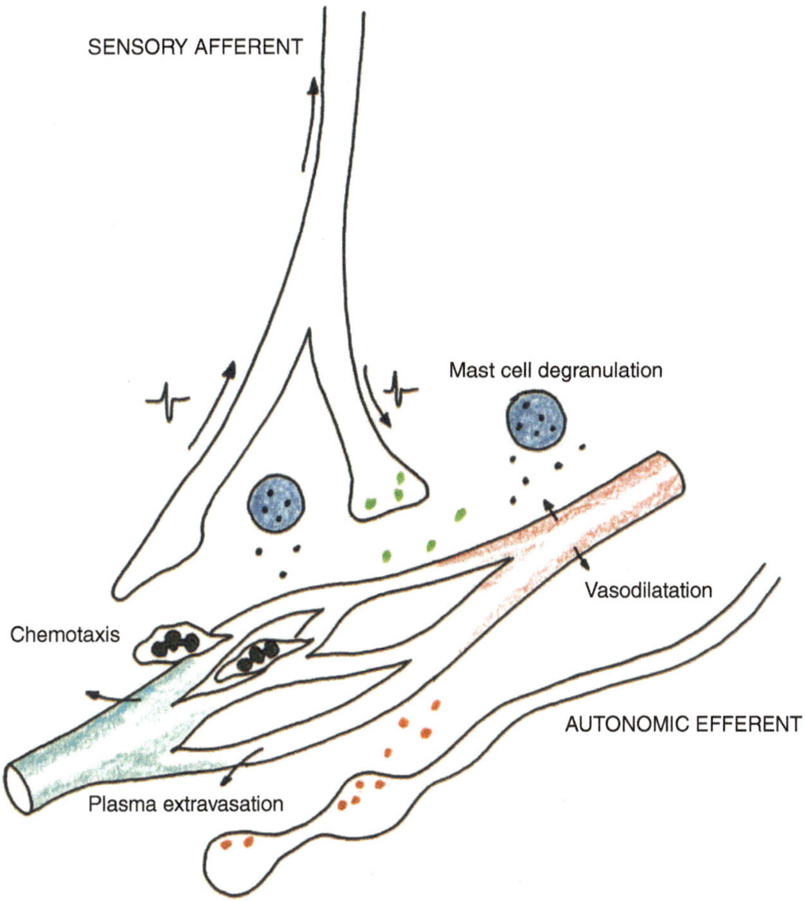

Plate 3. The axon reflex as first proposed by Lewis. Stimulation of primary afferent terminals results in orthodromic action potentials conducted to the CNS. Invasion of primary afferent terminal branches by antidromic action potentials (arrows) results in the release of proinflammatory neuropeptides such as substance P and calcitonin gene-related peptide (CGRP). These act on arterioles to cause vasodilatation (CGRP), on capillaries and postcapillary venules to cause plasma extravasation, on mast cell resulting in degranulation, and attract leukocytes out of the bloodstream (chemotaxis). Autonomic fibers may also release neuropeptide Y and vasoactive intestinal polypeptide, which also cause vasodilatation. (See Figure 1 of Chapter 10, p. 212).

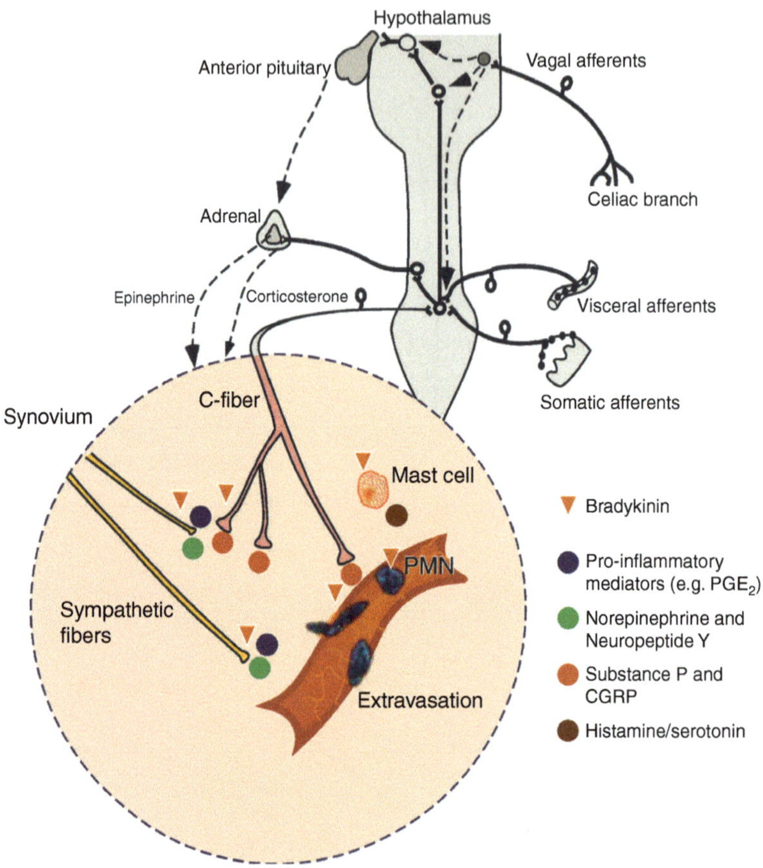

Plate 4. Production of synovitis by bradykinin. Bradykinin acts on B_2 or B_1 bradykinin receptors on the sympathetic postganglionic neuron to release pro-inflammatory mediators. These inflammatory mediators (e.g., prostaglandin E_2) are released independent of electrical activity of the sympathetic neuron, and act directly on the postcapillary venule and/or on other cell types associated with the sympathetic neuron. Bradykinin can also act directly on its receptors on C-fiber neurons to release proinflammatory mediators (e.g., substance P). Sympathetic C-fiber peptidergic coupling is bidirectional, with norepinephrine/neuropeptide Y acting on C-fibers and substance P/CGRP acting on sympathetic neuron terminals. Two neuroendocrine pathways, HPA and sympathoadrenal, that are activated by ongoing noxious input from inflammatory sites, mediate a negative feedback control of inflammation by releasing mediators (e.g., glucocorticoids and epinephrine); these circuits are tonically inhibited by activity in duodenal vagal afferents, exerted at the level of the spinal cord [144]. Immune cells, which play an essential role in the inflammatory response, are attracted to the synovium and activated by mediators released from sympathetic and C-fiber afferent neurons as well as by bradykinin and other tissue-generated inflammatory mediators. Other mechanisms, including the proinflammatory and anti-inflammatory cytokine network, sympathetic innervation of lymphoid organs [214], and the anti-inflammatory cholinergic efferent vagal pathway [215], contribute to the pathophysiology of synovial inflammation, but for the sake of clarity are not included in this schematic figure. (See Figure 4 of Chapter 11, p. 253).

FSC
www.fsc.org
MIX
Papier | Fördert
gute Waldnutzung
FSC® C083411

Kolibri 360 GmbH
Ferdinand-Jühlke-Straße 7
99095 Erfurt, Deutschland
produktsicherheit@kolibri360.de